Sphingolipids as Signaling and Regulatory Molecules

ADVANCES IN EXPERIMENTAL MEDICINE AND BIOLOGY

Recent Volumes in this Series

Volume 680
ADVANCES IN COMPUTATIONAL BIOLOGY
Edited by Hamid R. Arabnia

Volume 681
MELANOCORTINS: MULTIPLE ACTIONS AND THERAPEUTIC POTENTIAL
Edited by Anna Catania

Volume 682
MUSCLE BIOPHYSICS: FROM MOLECULES TO CELLS
Edited by Dilson E. Rassier

Volume 683
INSECT NICOTINIC ACETYLCHOLINE RECEPTORS
Edited by Steeve Hervé Thany

Volume 684
MEMORY T CELLS
Edited by Maurizio Zanetti and Stephen P. Schoenberger

Volume 685
DISEASES OF DNA REPAIR
Edited by Shamim I. Ahmad

Volume 686
RARE DISEASES EPIDEMIOLOGY
Edited by Manuel Posada de la Paz and Stephen C. Groft

Volume 687
BCL-2 PROTEIN FAMILY: ESSENTIAL REGULATORS OF CELL DEATH
Edited by Claudio Hetz

Volume 688
SPHINGOLIPIDS AS SIGNALING AND REGULATORY MOLECULES
Edited by Charles Chalfant and Maurizio Del Poeta

Sphingolipids as Signaling and Regulatory Molecules

Edited by

Charles Chalfant, PhD
Department of Biochemistry, Virginia Commonwealth University Richmond, Virginia, USA

Maurizio Del Poeta, MD
Department of Biochemistry, Medical University of South Carolina Charleston, South Carolina, USA

Springer Science+Business Media, LLC

Landes Bioscience

Springer Science+Business Media, LLC
Landes Bioscience

Printed in the USA.

Springer Science+Business Media, LLC, 233 Spring Street, New York, New York 10013, USA
http://www.springer.com

Please address all inquiries to the publishers:
Landes Bioscience, 1002 West Avenue, Austin, Texas 78701, USA
Phone: 512/ 637 6050; FAX: 512/ 637 6079
http://www.landesbioscience.com

The chapters in this book are available in the Madame Curie Bioscience Database.
http://www.landesbioscience.com/curie

Sphingolipids as Signaling and Regulatory Molecules, edited by Charles Chalfant and Maurizio Del Poeta. Landes Bioscience / Springer Science+Business Media, LLC dual imprint / Springer series: Advances in Experimental Medicine and Biology.

ISBN: 978-1-4419-6740-4

While the authors, editors and publisher believe that drug selection and dosage and the specifications and usage of equipment and devices, as set forth in this book, are in accord with current recommendations and practice at the time of publication, they make no warranty, expressed or implied, with respect to material described in this book. In view of the ongoing research, equipment development, changes in governmental regulations and the rapid accumulation of information relating to the biomedical sciences, the reader is urged to carefully review and evaluate the information provided herein.

Library of Congress Cataloging-in-Publication Data

Sphingolipids as signaling and regulatory molecules / edited by Charles Chalfant, Maurizio Del Poeta.
p. ; cm. -- (Advances in experimental medicine and biology ; v. 688)
Includes bibliographical references and index.
ISBN 978-1-4419-6740-4
1. Sphingolipids--Physiological effect. 2. Cellular signal transduction. I. Chalfant, Charles, 1970- II. Del Poeta, Maurizio. III. Series: Advances in experimental medicine and biology ; v. 688. 0065-2598.
[DNLM: 1. Sphingolipids. W1 AD559 v.688 2010 / QU 85 S7537 2010]
QP752.S6S685 2010
571.7'4--dc22

2010019450.

DEDICATION

To my wife Laura and two children, Isabella and Alec, whose support is instrumental in my research career. —Charles Chalfant

To my parents, Foglia Celsa and Pierino Del Poeta, for their understanding, my wife Chiara for her support, and to Prof. Giorgio Scalise who helped me to find my path. —Maurizio Del Poeta

PREFACE

The realization that sphingolipids are important biological mediators that regulate a broad range of cellular processes is widely regarded as one of the major advances in modern biology in the past 20 years. In addition to their role in energy source and in membrane structure, sphingolipids function also as signaling molecules.

This book attempts to analyze the latest discoveries in sphingolipid biology and how the alteration of their metabolism leads to altered signaling events and to the development of pathobiological disorders, such as cancer, cardiovascular diseases, asthma, diabetes, inflammation and infectious diseases. The volume also provides additional chapters covering studies of sphingolipids in different system models, such as *Saccharomyces cerevisiae*, and plants, and also includes a chapter dedicated on how bioinformatics can help to decipher specific function(s) of the network of the sphingolipid pathway(s). The study of sphingolipid functions in these models has dramatically helped the study of the sphingolipid signaling in mammalian cells.

The book includes a series of mouse animal models currently available for studying the role of sphingolipids in embryonic development, aging and a series of pathological processes. Finally, an Appendix is dedicated to current tools and techniques in sphingolipid research that could be particularly useful for current and new students and postdoctoral fellows who want to start a research career in sphingolipids signaling.

Each chapter of this work is self-contained to enable the reader to follow any individual chapter in isolation. Alternatively, the entire book could be read as one continuous text.

We hope this book will represent a comprehensive review of the current knowledge of sphingolipid signaling and we believe, research scientists can use it as a reference for their studies. We have worked particularly to include areas of current excitement. Any book is ultimately judged by how well it stands the test of time, but another measure of success is when a book stimulates and advances a field so well that it is ultimately outdated. We invite you to peruse and study the current state of knowledge here and hope that it will spur you to further advance the field.

Charles Chalfant, PhD
Maurizio Del Poeta, MD

ABOUT THE EDITORS...

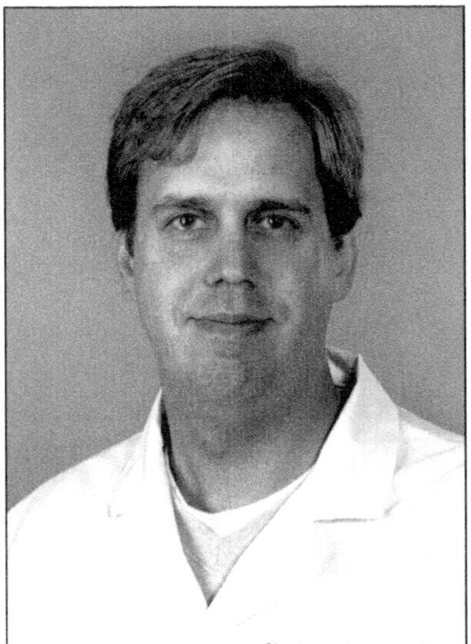

Charles E. Chalfant, PH.D.

CHARLES CHALFANT is a tenured Associate Professor of Biochemistry and Molecular Biology at Virginia Commonwealth University-School of Medicine. He is also a Research Career Scientist in the Veterans Administration based at the Hunter Holmes McGuire Veterans Administration Medical Center in Richmond, Virginia. His main research interests lie in lipid signaling, specifically the role of intracellular lipid second messengers (e.g., ceramide-1-phosphate) in regulating eicosanoid synthesis. He also studies the role of oncogenic signaling pathways in regulating the alternative splicing of tumor suppressor genes such as caspase 9. Dr. Chalfant serves on the Editorial Board of *The Journal of Lipid Research* as well as on several National Panels of Scientific Review including formal membership on the Cancer Molecular Pathobiology Study Section for the National Institutes of Health and the Oncology A Study Section for the Veterans Administration. He received his PhD degree from the University of South Florida-College of Medicine and did his postdoctoral studies at Duke University and the Medical University of South Carolina.

ABOUT THE EDITORS...

MAURIZIO DEL POETA, MD is a tenured Associate Professor of Biochemistry and Molecular Biology, Microbiology and Immunology, Craniofacial Biology, and for the Division of Infectious Diseases at the Medical University of South Carolina, Charleston, South Carolina, USA. He is also the Director of the Graduate Program in Biochemistry and Molecular Biology at MUSC. His main research interests include the study of the role of sphingolipids in fungal pathogenesis and the development of new diagnostic and therapeutic strategies against fungal infections. He received many awards for his research discoveries and teaching. He is also a member of many national and international societies such as the American Society for Clinical Investigation (ASCI), the American Society for Biochemistry and Molecular Biology, and the American Society for Microbiology. Dr. Del Poeta is a Burroughs Wellcome New Investigator in the Pathogenesis of Infectious Diseases.

PARTICIPANTS

Fernando Alvarez-Vasquez
Deptartments of Biostatistics,
Bioinformatics and Epidemiology
and
Deptartments of Biochemistry
and Molecular Biology
Medical University of South Carolina
Charleston, South Carolina
USA

Lide Arana
Departments of Biochemistry
and Molecular Biology
Faculty of Science and Technology
University of the Basque Country
Bilbao
Spain

James D. Bangs
Department of Medical Microbiology
and Immunology
University of Wisconsin School
of Medicine and Public Health
Madison, Wisconsin
USA

Stephen M. Beverley
Department of Molecular Microbiology
Washington University School
of Medicine
St. Louis, Missouri
USA

Alicja Bielawska
Departments of Biochemistry
and Molecular Biology
Medical University of South Carolina
Charleston, South Carolina
USA

Jacek Bielawski
Departments of Biochemistry
and Molecular Biology
Medical University of South Carolina
Charleston, South Carolina
USA

Myles C. Cabot
John Wayne Cancer Institute
Santa Monica, California
USA

Charles Chalfant
Department of Biochemistry
Virginia Commonwealth University
Richmond, Virginia
USA

Maurizio Del Poeta
Department of Biochemistry
Medical University of South Carolina
Charleston, South Carolina
USA

Robert C. Dickson
The Lucille P. Markey Cancer Center
Departments of Molecular and Cellular Biochemistry
University of Kentucky College of Medicine
Lexington, Kentucky
USA

Todd E. Fox
Department of Pharmacology
Penn State College of Medicine
Hershey, Pennsylvania
USA

Anthony H. Futerman
Department of Biological Chemistry
Weizmann Institute of Science
Rehovot
Israel

Patricia Gangoiti
Departments of Biochemistry and Molecular Biology
Faculty of Science and Technology
University of the Basque Country
Bilbao
Spain

Christopher R. Gault
Departments of Biochemistry and Molecular Biology
Medical University of South Carolina
Charleston, South Carolina
USA

Paola Giussani
Departments of Medical Chemistry, Biochemistry and Biotechnology
The Medical Faculty
University of Milan, LITA
Segrate, Milan
Italy

Antonio Gómez-Muñoz
Departments of Biochemistry and Molecular Biology
Faculty of Science and Technology
University of the Basque Country
Bilbao
Spain

María H. Granado
Departments of Biochemistry and Molecular Biology
Faculty of Science and Technology
University of the Basque Country
Bilbao
Spain

Nitai C. Hait
Departments of Biochemistry and Molecular Biology
Virginia Commonwealth University School of Medicine
Richmond, Virginia
USA

Yusuf A. Hannun
Departments of Biochemistry and Molecular Biology
Medical University of South Carolina
Charleston, South Carolina
USA

Vania Hinkovska-Galcheva
Department of Pediatrics
Division of Critical Care Medicine
University of Michigan Medical School
Ann Arbor, Michigan
USA

Joost C.M. Holthuis
Membrane Enzymology
Bijvoet Center and Institute of Biomembranes
Utrecht University
Utrecht
The Netherlands

Toshihiko Kawamori
Department of Pathology & Laboratory Medicine
Medical University of South Carolina
Charleston, South Carolina
USA

Mark Kester
Department of Pharmacology
Penn State College of Medicine
Hershey, Pennsylvania
USA

Kazuyuki Kitatani
Department of Clinical Laboratory/
Hematology
Tottori University Hospital
Yonago, Tottori
Japan
and
Departments of Biochemistry
and Molecular Biology
Medical University of South Carolina
Charleston, South Carolina
USA

Chiara Luberto
Departments of Biochemistry
and Molecular Biology
Medical University of South Carolina
Charleston, South Carolina
USA

Michael Maceyka
Departments of Biochemistry
and Molecular Biology
Virginia Commonwealth University
School of Medicine
Richmond, Virginia
USA

Maria C. Messner
John Wayne Cancer Institute
Santa Monica, California
USA

Sheldon Milstien
National Institute of Mental Health
Bethesda, Maryland
USA

Mariana N. Nikolova-Karakashian
Department of Physiology
University of Kentucky College
of Medicine
Lexington, Kentucky
USA

Lina M. Obeid
Departments of Biochemistry, Molecular
Biology and Medicine
Medical University of South Carolina
Charleston, South Carolina
USA

Babak Oskouian
Children's Hospital Oakland Research
Institute
Oakland, California
USA

Alberto Ouro
Departments of Biochemistry
and Molecular Biology
Faculty of Science and Technology
University of the Basque Country
Bilbao
Spain

Jason S. Pierce
Departments of Biochemistry
and Molecular Biology
Medical University of South Carolina
Charleston, South Carolina
USA

Alessandro Prinetti
Center of Excellence on
Neurodegenerative Diseases
Departments of Medical Chemistry,
Biochemistry and Biotechnology
University of Milan
Segrate
Italy

Barbara Rembiesa
Departments of Biochemistry
and Molecular Biology
Medical University of South Carolina
Charleston, South Carolina
USA

Ryan Rhome
Departments of Biochemistry
and Molecular Biology
Medical University of South Carolina
Charleston, South Carolina
USA

Laura Riboni
Departments of Medical Chemistry,
Biochemistry and Biotechnology
The Medical Faculty
University of Milan, LITA
Segrate, Milan
Italy

Krassimira A. Rozenova
Department of Physiology
University of Kentucky College of Medicine
Lexington, Kentucky
USA

Julie D. Saba
Children's Hospital Oakland Research Institute
Oakland, California
USA

James A. Shayman
Department of Internal Medicine
Nephrology Division
University of Michigan Medical School
Ann Arbor, Michigan
USA

Justin Snider
Departments of Biochemistry and Molecular Biology
Medical University of South Carolina
Charleston, South Carolina
USA

Sandro Sonnino
Center of Excellence on Neurodegenerative Diseases
Departments of Medical Chemistry, Biochemistry and Biotechnology
University of Milan
Segrate
Italy

Sarah Spiegel
Departments of Biochemistry and Molecular Biology
Virginia Commonwealth University School of Medicine
Richmond, Virginia
USA

Johnny Stiban
Department of Biological Chemistry
Weizmann Institute of Science
Rehovot
Israel

Graham M. Strub
Departments of Biochemistry and Molecular Biology
Virginia Commonwealth University School of Medicine
Richmond, Virginia
USA

Zdzislaw M. Szulc
Departments of Biochemistry and Molecular Biology
Medical University of South Carolina
Charleston, South Carolina
USA

Phillipp Ternes
Albrecht-von-Haller-Institute for Plant Sciences
Department for Plant Biochemistry
University of Goettingen
Goettingen
Germany

Rotem Tidhar
Department of Biological Chemistry
Weizmann Institute of Science
Rehovot
Israel

Paola Viani
Departments of Medical Chemistry, Biochemistry and Biotechnology
The Medical Faculty
University of Milan, LITA
Segrate, Milan
Italy

Eberhard O. Voit
Integrative BioSystems Institute and Wallace H. Coulter Department of Biomedical Engineering
Georgia Institute of Technology
Atlanta, Georgia
USA

Dirk Warnecke
Biocenter Klein Flottbek
University of Hamburg
Hamburg
Germany

Simone Zäuner
Departments of Biochemistry
and Molecular Biology
Michigan State University
East Lansing, Michigan
USA

Kai Zhang
Department of Biological Sciences
Texas Tech University
Lubbock, Texas
USA

CONTENTS

14. THERAPEUTIC STRATEGIES FOR DIABETES AND COMPLICATIONS: A ROLE FOR SPHINGOLIPIDS? 206

Todd E. Fox and Mark Kester

15. ROLES FOR SPHINGOLIPIDS IN *SACCHAROMYCES CEREVISIAE* 217

Robert C. Dickson

16. SPHINGOLIPID SIGNALING IN FUNGAL PATHOGENS 232

Ryan Rhome and Maurizio Del Poeta

17. SPHINGOLIPIDS IN PARASITIC PROTOZOA 238

Kai Zhang, James D. Bangs and Stephen M. Beverley

ACKNOWLEDGEMENTS

We are extremely grateful to the authors and co-authors of the chapters included in this book.

Charles Chalfant, PhD
Maurizio Del Poeta, MD

Chapter 1

An Overview of Sphingolipid Metabolism:

From Synthesis to Breakdown

Christopher R. Gault, Lina M. Obeid and Yusuf A. Hannun*

Abstract

Sphingolipids constitute a class of lipids defined by their eighteen carbon amino-alcohol backbones which are synthesized in the ER from nonsphingolipid precursors. Modification of this basic structure is what gives rise to the vast family of sphingolipids that play significant roles in membrane biology and provide many bioactive metabolites that regulate cell function. Despite the diversity of structure and function of sphingolipids, their creation and destruction are governed by common synthetic and catabolic pathways. In this regard, sphingolipid metabolism can be imagined as an array of interconnected networks that diverge from a single common entry point and converge into a single common breakdown pathway.

In their simplest forms, sphingosine, phytosphingosine and dihydrosphingosine serve as the backbones upon which further complexity is achieved. For example, phosphorylation of the C1 hydroxyl group yields the final breakdown products and/or the important signaling molecules sphingosine-1-phosphate, phytosphingosine-1-phosphate and dihydrosphingosine-1-phosphate, respectively. On the other hand, acylation of sphingosine, phytosphingosine, or dihydrosphingosine with one of several possible acyl CoA molecules through the action of distinct ceramide synthases produces the molecules defined as ceramide, phytoceramide, or dihydroceramide. Ceramide, due to the differing acyl CoAs that can be used to produce it, is technically a class of molecules rather than a single molecule and therefore may have different biological functions depending on the acyl chain it is composed of.

At the apex of complexity is the group of lipids known as glycosphingolipids (GSL) which contain dozens of different sphingolipid species differing by both the order and type of sugar residues attached to their headgroups. Since these molecules are produced from ceramide precursors, they too may have differences in their acyl chain composition, revealing an additional layer of variation. The glycosphingolipids are divided broadly into two categories: glucosphingolipids and galactosphingolipids. The glucosphingolipids depend initially on the enzyme glucosylceramide synthase (GCS) which attaches glucose as the first residue to the C1 hydroxyl position. Galactosphingolipids, on the other hand, are generated from galactosylceramide synthase (GalCerS), an evolutionarily dissimilar enzyme from GCS. Glycosphingolipids are further divided based upon further modification by various glycosyltransferases which increases the potential variation in lipid species by several fold. Far more abundant are the sphingomyelin species which are produced in parallel with glycosphingolipids, however they are defined by a phosphocholine headgroup rather than the addition of sugar residues. Although sphingomyelin

*Corresponding Author: Yusuf A. Hannun—Departments of Biochemistry and Molecular Biology, Medical University of South Carolina, Charleston, South Carolina 29425, USA. Email: hannun@musc.edu

Sphingolipids as Signaling and Regulatory Molecules, edited by Charles Chalfant and Maurizio Del Poeta.

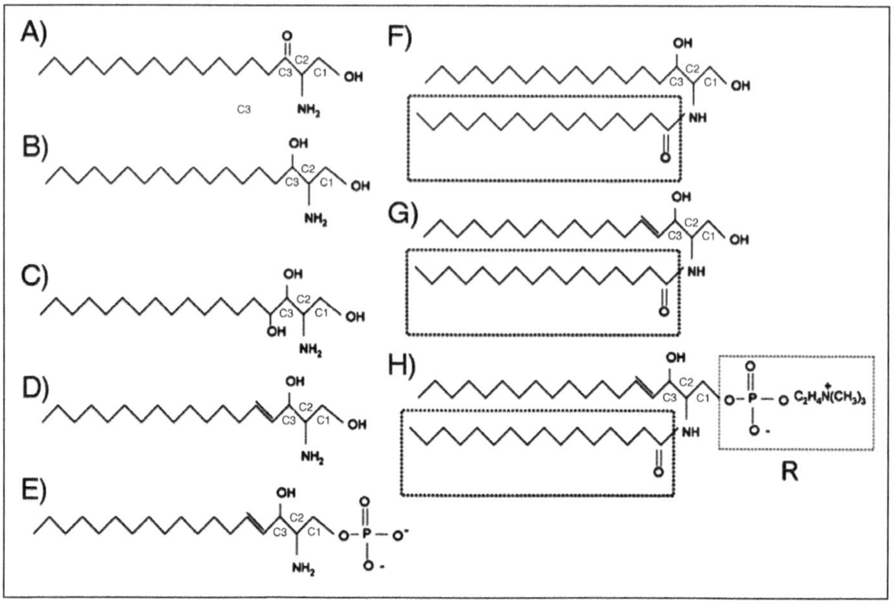

Figure 1. Simple and complex sphingolipid structures. Structures shown: (A) 3-Ketodihydrosphingosine, (B) Dihydrosphingosine, (C) Phytosphingosine, (D) Sphingosine, (E) Sphingosine-1-Phosphate, (F) Dihydroceramide: Boxed region shows variable acyl chain, (G) Ceramide, (H) Complex Sphingolipids: Sphingomyelin shown with phosphocholine R group. Substitute R for glucose=Glucosylceramide, Substitute R for galactose=Galactosylceramide.

species all share a common headgroup, they too are produced from a variety of ceramide species and therefore can have differing acyl chains attached to their C-2 amino groups. Whether or not the differing acyl chain lengths in SMs dictate unique functions or important biophysical distinctions has not yet been established. Understanding the function of all the existing glycosphingolipids and sphingomyelin species will be a major undertaking in the future since the tools to study and measure these species are only beginning to be developed (see Fig 1 for an illustrated depiction of the various sphingolipid structures).

The simple sphingolipids serve both as the precursors and the breakdown products of the more complex ones. Importantly, in recent decades, these simple sphingolipids have gained attention for having significant signaling and regulatory roles within cells. In addition, many tools have emerged to measure the levels of simple sphingolipids and therefore have become the focus of even more intense study in recent years. With this thought in mind, this chapter will pay tribute to the complex sphingolipids, but focus on the regulation of simple sphingolipid metabolism.

Sphingolipid Properties in Membranes

An important feature of lipid biology is that many of these molecules are restricted to biological membranes and therefore lipids are governed by a set of rules based on their biophysical properties. For example, compartmentalization of a lipid can mean access to both sides of a membrane, access to only a single leaflet of a bilayer and therefore only a single compartment, or a molecule could be sufficiently amphipathic that it could diffuse from a membrane and freely traverse the cytosol, the lumen of an organelle, or enter the extracellular space. This becomes important to keep in mind when understanding how compartmentalized enzymes only have effects on a specific pool of lipid metabolites. It is also important to understand the enzymes that regulate the lipids since cytosolic proteins are generally restricted to a single compartment and transmembrane domain containing enzymes generally have their catalytic sites facing only one of the leaflets in which a bilayer divides.

In the case of sphingolipids, sphingosine and dihydrosphingosine are sufficiently amphipathic to diffuse between membranes and to flip between membrane leaflets; however, they also are likely to accumulate in acidic pH organelles due to ionization of their free amino group. All of the known enzymes to act upon sphingosine have their catalytic sites facing the cytosolic compartment suggesting that only cytosolic sphingosine is in a modifiable form. Ceramide, on the other hand is restricted to membranes, but has a relatively rapid flip rate. Ceramide is therefore likely to be restricted to the organelle in which it was created, but may have access to enzymes or binding proteins on either side of the bilayer in which it was produced. This is important because these enzymes are distributed in discrete compartments and therefore ceramide will most likely be modified by whichever enzyme is within the same compartment in which the ceramide was generated or transported to. Sphingomyelins and glycosphingolipids are the most spatially restricted sphingolipids of all since their bulky head-groups make flipping between membrane leaflets extremely unlikely without the aid of specific flippases. One such flippase is thought to be present in the Golgi apparatus to aid glucosylceramide in gaining access to lumenal glycosyltransferases. In the absence of a flippase, sphingomyelin and glycosphingolipids are restricted to whichever leaflet they are generated in and since they are generated in the Golgi lumen, they are mainly present in the lumenal Golgi leaflet or on the outer leaflet of the plasma membrane after vesicular transport to that location. Finally, the ultimate catabolic products of all sphingolipids are sphingosine-1-phosphate and dihydrosphingosine-1-phosphate which are soluble in a hydrophilic environment, but are unable to traverse membranes without the aid of lipid transporters. Therefore, S1P and DHS1P are also restricted to the hydrophilic compartments in which they are generated, but can be exported to the extracellular space with the aid of specific transporters or be dephosphorylated into a more hydrophobic compound.

Sphingolipid structures differ significantly from kingdom to kingdom and sphingolipid diversity within the animal kingdom itself has been recognized. Despite this diversity being extremely fascinating for its implications in evolutionary biology, this chapter will only discuss mammalian enzymes involved in sphingolipid metabolism. Although many of the critical enzymes in mammalian systems could not have been identified without the aid of yeast and other model organisms, for the sake of brevity, unless otherwise stated, the enzymes discussed in this chapter are to be assumed to be the mammalian form. In addition, the reader may find it useful to utilize figure 2 as an illustrated reference of sphingolipid metabolism as they progress through this chapter. By discussing the synthesis and the catabolism of mammalian sphingolipids we hope to bring to light an understanding of sphingolipids, how sphingolipids are created and destroyed and to define more clearly the process by which sphingolipids become distributed to their respective membranes within mammalian cells.

De Novo Synthesis in the ER

De novo sphingolipid synthesis begins at the cytosolic leaflet of the ER where a set of four enzyme groups coordinately generate ceramides of different acyl chain lengths from nonsphingolipid precursors. Through the coordinated action of Serine palmitoyltransferase, 3-Ketodihydrosphingosine Reductase and (dihydro)Ceramide Synthases, the ER is able to convert cytosolic serine and palmitoyl CoA molecules into a single membrane bound lipid, dihydroceramide. After its generation, dihydroceramide is acted on by a desaturase which introduces a double bond. This coordinated anabolic pathway generates the precursors to complex sphingolipids that can serve such diverse functions as providing electrical insulation to axons, act as an important hydrophobic barrier within the epidermis essential for decreasing water loss and regulate red blood cell surface charge to prevent agglutination, to name just a few important functions.

In this phase of synthesis, complex sphingolipids begin to be diversified through differential addition of fatty acyl chains at the C2-amino group of the dihydrosphingosine/sphingosine backbone through the action of ceramide synthases. Variations in ceramide acyl chain length as well as the use of alpha hydroxylated fatty acids could potentially alter membrane bilayer dynamics, or have differential signaling properties by recruiting different binding partners. The effects of different acyl chain lengths on ceramide or complex sphingolipid biology are not yet understood.

Serine Palmitoyltransferase and 3-Ketodihydrosphingosine Reductase

The initial reaction in sphingolipid synthesis requires the enzyme serine palmitoyltransferase (SPT). This reaction occurs through cytosolic serine and palmitoyl CoA condensation to produce 3-ketodihydrosphingosine.[1] SPT is encoded by the genes *SPTLC1*, *SPTLC2* and the recently identified *SPTLC3*.[2] Each SPT subunit contains several putative transmembrane domains and displays Type I topology with its N-terminus directed into the ER lumen, C-terminus facing the cytosol and its catalytic site facing the cytosol.[3] SPT1 and SPT2 form a heterodimer in the ER membrane which is likely the active form of the enzyme.[4] Although SPT3 shares 40% homology to SPT2 and likely can substitute for the SPT2 subunit in the SPT complex, it has not yet been shown to dimerize with SPT1 and still needs further investigation.[2] In yeast, a third subunit of SPT, *TSC3* plays a major role in regulation of SPT activity by forming a heterotrimer with the SPT1 and SPT2 homologues, however, no mammalian homologue to *TSC3* has yet been identified.[5] SPT is a member of the α-oxoamine synthase family, a group of enzymes that catalyze the condensation of amino acids with carboxylic acid CoA thioesters.[6] Like other members of this family, SPT requires the cofactor pyridoxal 5′-phosphate (PLP) for catalysis.[7] SPT2 is the only subunit which binds PLP, however both subunits are required for catalytic activity.[4] Hanada et al have proposed that SPT2 requires SPT1 for stabilization and that up regulation of SPT1 expression subsequently leads to an increase in SPT2 expression through protein stabilization. Therefore, a secondary function of SPT1 other than its contribution to catalytic activity may be simply to stabilize SPT2 in the ER.[6] Whether or not regulation of SPT1 is the primary means by which SPT activity is regulated is yet to be determined.

Some insight into the functional significance of SPT comes from human patients with Hereditary Sensory and Autonomic Neuropathy Type I (HSNI), a disease which, in some families, has been mapped to a mutation in the *SPTLC1* gene on chromosome 9.[8] HSNI is an autosomal dominant disease that is characterized by progressive degeneration of motor neurons and dorsal ganglia with symptoms initiating after the first or second decade.[9,10] Measurements of total SPT activity in patient lymphoblasts showed less than 50% SPT activity within these cells.[11] It is significant that patients still have a significant amount of SPT activity since in mice, complete lack of either SPT1 or SPT2 was shown to be embryonic lethal.[4,12] The mutations responsible for this disease have been identified as point mutations in Cys133 or Val144 in the SPT1 protein which act in a dominant negative fashion on SPT activity.[8,11] These mutations have been predicted to be localized, based on tertiary modeling of other known α-oxoamine synthases, near the catalytic interface of SPT1 and SPT2. HSNI mutant forms of SPT1 are able to form heterodimers with SPT2, but lack catalytic activity.[13] Interestingly, mutations in the *SPTLC2* gene were not identified in any of the tested families affected by HSNI.[14] The recently identified *SPTLC3* gene product has homology to *SPTLC2* suggesting that SPT3 may be able to functionally substitute for SPT2 if expressed.[2] It is possible that a functional SPT3 subunit could mask a partial defect in the SPT2 subunit if they are functionally redundant. Although speculative, this could be one explanation for why *SPTLC2* mutants are not associated with HSNI. Future studies into the interplay between the three SPT subunits will provide further insights into how SPT function is determined by its components.

The second step in the synthesis of all sphingolipids is performed by the enzyme 3-Ketodihydrosphingosine Reductase (KDHR). 3-Ketodihydrosphingosine (KDHSph), the direct product of SPT is reduced at its ketone group to a hydroxyl group by KDHR in a NADPH dependent manner. Only recently were the human and murine genes cloned for this enzyme based on a homology screen for the yeast gene *TSC10*. *TSC10* deficient yeast were identified based on their build up of KDHSph and their inability to grow on media deficient in KDHSph.[15] In humans the KDHR gene was identified through a homology screen as *FVT-1*, a gene which was originally identified and named for its juxtaposition to the Ig-κ gene in a human follicular lymphoma.[16] Whether or not the FVT-1 translocation was merely coincidental or that it conferred any advantage to the follicular lymphoma is unclear. KDHR is predicted to have three transmembrane domains and display Type I topology. Like SPT, KDHR has its catalytic site on the cytosolic leaflet of the ER where it is likely to encounter newly generated KDHSpH.[16] KDHSph is a minor lipid within cells due to the rapid conversion of KDHSph into dihydrosphingosine by the action of KDHR. Although poorly studied, KDHR is a critical step in the synthesis of sphingolipids. Its importance to mammalian physiology has been highlighted by a breed of cattle, recently identified with a missense mutation in *FVT-1*, that become afflicted with bovine spinal muscular atrophy and die shortly after birth.[17]

Dihydroceramide Synthases/Ceramide Synthases and Dihydroceramide Desaturase

Dihydrosphingosine (DHSph) is further acylated by the action of six distinct (Dihydro) ceramide synthases. In mammals, six distinct (dihydro)ceramide synthases abbreviated as CerS1-6 have been identified and are encoded by six distinct genes.[18,19] No other step in sphingolipid metabolism has as many genes devoted to it as dihydroceramide synthesis, suggesting that the different CerS have distinct functions. There is a significant amount of evidence that each CerS has a distinct, but overlapping acyl CoA preference and that each CerS can produce different dihydroceramide/ceramide species profiles. For example, CerS1 has been shown to prefer stearoyl CoA as a substrate and mainly produces C18-ceramide species.[20] On the other hand, CerS2 utilizes C20-C26 acyl CoA species and is one of the major CerS responsible for very long chain ceramide species.[21] Cers5 and CerS6 both prefer palmitoyl CoA as substrates and generate predominantly C16-ceramide species.[22,23] Finally, CerS3, which is predominantly expressed in the testis and weakly in the epidermis, prefers middle and long chain acyl CoAs and is thought to be a contributor to large structural sphingolipid molecules that maintain the water barrier in the epidermis.[24,25] It is still an outstanding question whether or not the bioactive properties associated with ceramide are sensitive to differences in acyl chain length and hence different CerS can influence the cellular fate of cells by modulating bioactive molecules. Another possibility is that the six different CerS exist because differences in the biophysical properties of various ceramide species provide advantages for specific tissue functions (e.g., epidermal barrier maintenance or myelination). It has already been established that the CerS have a significant variation in their tissue expression and that this correlates with differences in their sphingolipid acyl chain compositions.[21]

All CerS studied to date have been localized to the ER with their catalytic sites facing the cytosol. In this manner, CerS are in a position to acylate newly generated DHSph molecules at their C2-amino groups in the presence of available fatty acyl CoAs.

Very little is known about the regulation of CerS activity in cells although there are clear differences in CerS expression between different tissues. It is unclear if CerS are predominantly regulated at the transcriptional level or if significant posttranslational regulation also occurs. Recently, a S1P binding site was identified on CerS2, which in vitro inhibited CerS2 activity. This suggests that sphingolipid breakdown could negatively regulate a specific subset of CerS activity within cells, in this case very long chain ceramide synthesis. Although de novo synthesis of dihydroceramide has repeatedly been shown to occur in response to various stress stimuli, the mechanisms by which this occurs remain opaque. Those that have studied it have suggested that the regulation is posttranslational (e.g., not inhibited by cycloheximide).[26]

Although the family name of sphingolipids was named after the molecule sphingosine, this molecule is not actually generated during de novo synthesis. Only through the desaturation of dihydroceramide is the molecule sphingosine eventually generated.

Dihydroceramide Δ4-desaturase (DES) is the member of the desaturase family responsible for converting the dihydrosphingosine backbone within ceramide into a sphingosine backbone.[27] DES utilizes molecular oxygen to first introduce a hydroxyl group into the C4 position of the dihydrosphingosine backbone and following a dehydration reaction, with the aid of NADPH, produces a double bond in the C4-C5 position of dihydroceramide.[28-30] Dihydroceramide with a double bond introduced at this position is referred to as ceramide. Dihydroceramide Δ4-desaturase

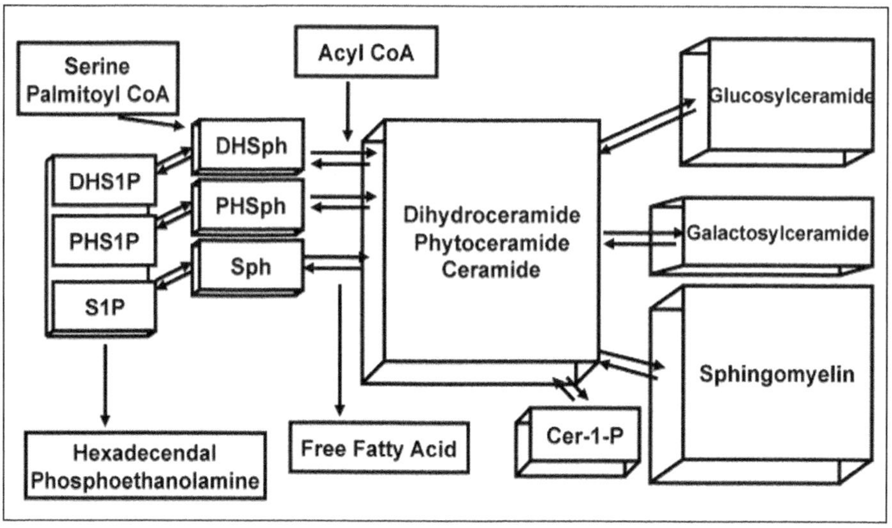

Figure 2. The sphingolipid metabolic network.

(DES1), encoded by the *DES1* gene contains multiple transmembrane domains and was recently shown to require myristoylation on its N-terminus for full activity.[31,32] Like the previous three enzymes in the sphingolipid biosynthetic pathway, DES1 is embedded in the ER membrane where it has access to newly synthesized dihydroceramide species.[27] It is interesting to note that an intermediate reaction product in the conversion of dihydroceramide to ceramide is 4-hydroxyceramide which is also known as phytoceramide. Phytoceramide is the predominant ceramide species in plants and yeast. Although the enzyme DES1 only converts dihydroceramide species into fully desaturated ceramide, a second family member, dihydroceramide C4 hydroxylase/Δ4-desaturase (DES2), is capable of creating either phytoceramide or ceramide from dihydroceramide precursors.[33] Therefore, it is not surprising that DES2 is highly expressed in the intestines, kidneys and skin where phytoceramides are present in high abundance.[34,33] The differences in biophysical properties between dihydroceramide, ceramide and phytoceramide are not entirely clear, however, one may speculate that the addition of a hydroxyl group into the sphingosine backbone may increase lipid packing in the membrane by increasing the amount of hydrogen bonding at the interfacial region of the membrane. It has been shown repeatedly that ceramide has distinct signaling properties from dihydroceramide and phytoceramide, suggesting that, if nothing else, cells have evolved to recognize ceramide as a more significant determinant to initiate a cellular response to in most cells.[35] The recent report of the phenotype of the *Des1*$^{-/-}$ mice suggests that the inability to form ceramide leads to serious consequences for mammalian physiology. *Des1*$^{-/-}$ mice have highly elevated dihydroceramide, low levels of ceramide, multi-organ dysfunction and failure to thrive.[36]

Ceramide Transport from the ER to the Golgi

Ceramide is a membrane bound molecule that has very low solubility in an aqueous environment and therefore, a cell must find a way to transport it from one membrane to another. The cell employs two major mechanisms to mobilize ceramide; either through vesicular transport or through the protein ceramide transfer protein (CERT). CERT is a cytosolic protein that transfers ceramide from the ER, where it is generated, to the Golgi apparatus where it can be modified into sphingomyelins and possibly glycosphingolipids. The CERT protein is composed of at least four functional domains that determine its function. The N-terminus of CERT contains a PH domain which is able to recognize PI4P on acceptor Golgi membranes and therefore allows for directed transport to the Golgi. A FFAT domain in the middle of the protein serves an analogous function to the PH domain but for donor membrane recognition. The FFAT domain is thought to allow its binding to ER resident VAP proteins and therefore CERT can only accept ceramides from the ER, something that may have implications for cellular signaling.[37] The C-terminus of CERT contains a START domain which provides a hydrophobic pocket responsible for the direct binding of ceramide and allows for its delivery to the Golgi through an aqueous environment. In vitro studies with CERT have shown that phosphorylation of CERT at multiple serine residues, by an unidentified kinase, result in an autoinhibitory binding event that occurs between both the START domain and the PH domain.[38] The in vivo significance of CERT phosphorylation is unclear and remains to be seen. On the other hand, a globular domain between the PH and START domain has been shown to be responsible for homotrimer formation during UV stress in keratinocytes, however, this was shown to be phosphorylation independent. It is unclear if oligomerization, or potentially phosphorylation, is a general mechanism by which cellular stresses can inactivate CERT.[39]

CERT was originally identified as the responsible mutant in a CHO cell line, LY-A, that was resistant to hemolysis by the sphingomyelin-dependent celomate toxin lysenin.[40,41] CERT displays a preference for ceramide species with acyl chains less than C22. Although CERT still transfers C22 and C24:1 ceramide, it does so with 40% the efficiency of shorter chain species.[38,42] In addition, CERT showed minimal to no transfer of C24 ceramide. CERT is also able to recognize dihydroceramide and phytoceramide although less effectively than ceramide.[38] Ceramide which is transported to the Golgi by CERT is preferentially incorporated into SM over glycosphingolipids.[41] Since CERT has preference for specific chain lengths, this may have implications for which forms of ceramide are preferentially utilized for SM synthesis and which ceramide species are preferred for glycosphingolipid utilization. If this is true, then one could speculate that relative SM and glycosphingolipid synthesis could be regulated by shifting CerS expression from predominantly long chain specific CerS to very long chain specific CerS and vice versa.

An alternative pathway exists for the transport of ceramide species to the Golgi which is coatomer protein dependent and is based on vesicular transport.[43] Less is known about how this pathway is regulated, however, this is thought to be the major pathway responsible for delivering ceramide to the cis-Golgi for glycosphingolipid synthesis. Clearly, our knowledge of how ceramide species can be transported from the ER to the Golgi for regulated glycosphingolipid synthesis is incomplete.

Synthesis of Complex Sphingolipids

Complex sphingolipids are divided into three major groups based on the primary residue attached to their C1-hydroxy headgroup. This classification also captures the three biosynthetic pathways, spatially separated within the ER and the Golgi complex, that generate an immense diversity of glycosphingolipids and sphingomyelins. The three major enzymes that regulate complex sphingolipid biosynthesis are ceramide galactosyltransferase, glucosylceramide synthase and sphingomyelin synthase.

Ceramide Galactosyltransferase and Galactosphingolipids

Ceramide galactosyltransferase (CGT) utilizes UDP-galactose and ceramide to create galactosylceramide. CGT is an ER transmembrane protein that has its catalytic site facing the lumen of the ER. It is structurally related to UDP-glucuronosyltransferases, an enzyme critical to Type II biotransformation of xenobiotics and porphyrin metabolism.[44] CGT has a limited tissue distribution with expression being detected primarily in schwann cells, oligodendrocytes, kidneys, testis and intestines. In the central nervous system, the product galactosylceramide (and its subsequent metabolite, sulfatide) is highly enriched in myelin. CGT knockout mice display a tremor phenotype, severe motor weakness due to loss of nerve conduction, male infertility and premature death.[45,46] Interestingly, the neuronal phenotype in mice lacking CGT can be rescued by expression of an oligodendrocyte specific CGT gene suggesting that galactosylceramide is extremely important for oligodendrocyte function.[47] Galactosylceramide is a precursor for sulfatides and many of the myelination defects may be due to a lack of sulfatide production. Evidence for this comes from mice deficient in the enzyme galactosylceramide sulfotransferase, the enzyme responsible for sulfatide production from galactosylceramide, which have major defects in myelination, although their pathology is less severe than an outright CGT knockout mouse.[48]

Glucosylceramide Synthase and Derivatives of Glucosylceramide

Glucosylceramide is synthesized in the cis-Golgi from ceramide and UDP-glucose by the enzyme glucosylceramide synthase (GCS).[49] GCS is a transmembrane protein present on the cis-Golgi and it has its catalytic site facing the cytosol where newly produced glucosylceramide can be recognized by the lipid transport protein FAPP2.[50,51] Some reports suggest that FAPP2 transports glucosylceramide back to the ER where it is translocated from the inner leaflet to the outer leaflet, however, this point remains to be resolved.

Unlike galactosylceramide, glucosylceramide (GC) is an absolutely essential sphingolipid for the development of mammals.[52] Mice lacking GCS do not survive to term. The loss of GCS results in embryonic lethality at embryonic day 6.5-7.5 when gastrulation is occurring.[52] This specific defect can be rescued by the addition of exogenous GC to the embryos. Glucosylceramide is the precursor for the majority of all glycosphingolipids that can be produced by a mammal and these glycosphingolipids are likely to play an essential role in cell-cell recognition during embryonic and postnatal development.[52]

Tissue specific knockouts of GCS within the nervous system and the skin have been created. Absence of GCS in the epidermis leads to defects in lamellar body formation which are major contributors to the hydrophobic barrier of the skin. These lamellar body defects lead to rapid water loss due to excessive evaporation and eventual lethality several days after birth.[53] Absence of GCS specifically in neuronal tissue, on the other hand, leads to premature death 11-24 days after birth, suggesting that glycosphingolipids are necessary for neuronal function and proper brain maturation. Unusually, no histological defects could be identified in the brains of the neuron specific GCS knockout mice through light microscopy examination or electron microscopic evaluation of synapses despite their obvious phenotypic differences. However, in vitro studies of primary hippocampal neurons from GCS deficient mice showed defects in neurite outgrowth in culture.[54,55]

Sphingomyelin Synthesis

The most abundant complex sphingolipids in mammalian cells are the sphingomyelin species. Evidence for the essential role that sphingomyelin has in eukaryotic cell viability is displayed by the inability of mammalian or yeast cells to survive in culture when they are unable to produce sphingomyelin either through CERT mutation or defects in de novo sphingolipid synthesis. It is interesting to note that this absolute requirement for a sphingolipid is not true for glucosylceramide or galactosylceramide which, although critical for mammalian development and tissue specific functions are not required for the viability of cells in culture. The precise single function that sphingomyelin fulfills which is absolutely necessary for cell survival is not clear due to sphingomyelin's many known functions in membrane biology.

Sphingomyelin is produced by the action of sphingomyelin synthases. There are at least two members of the sphingomyelin synthase family in most mammalian species[56] and possibly a third family member known as SMSr.[57] The sphingomyelin synthases (SMS) are evolutionarily similar to the lipid phosphate phosphatase family which have six transmembrane domains and have their catalytic domains facing the luminal or exoplasmic leaflet of the membrane.[57] Like their cousins, the SMS family constituents also have six transmembrane domains and are oriented with their catalytic sites facing the Golgi lumen or extracellular space. Sphingomyelin synthases 1 and 2 are both present in the trans-Golgi, however, SMS2 is also localized to the plasma membrane. Therefore, SMS2 may also have a unique function in maintaining plasma membrane sphingomyelin content directly at the plasma membrane. Generation of sphingomyelin occurs through the transfer of a phosphocholine headgroup from phosphatidylcholine to ceramide yielding the products diacylglycerol (DAG) and sphingomyelin (SM). Since both ceramide and DAG have been identified as bioactive lipids with opposing effects on cellular proliferation and survival, SMS has also been proposed to play an essential role in regulating cellular fate. A recent study has implicated SMS in generating DAG in the Golgi with effects on PKC.[58] Because SMS activity directly regulates the level of sphingomyelin, ceramide, DAG and PC simultaneously, direct effects of SMS products on biological processes have been difficult to elucidate to date.

Ceramide Kinase and Ceramide-1-Phosphate

Although ceramide is primarily converted into more complex sphingolipids in the Golgi, ceramide can also be phosphorylated to produce ceramide-1-phosphate (C1P). C1P is produced in the trans-Golgi and potentially the plasma membrane, by ceramide kinase (CERK). CERK, a member of the DAG kinase family, was originally identified based on its homology to sphingosine kinase. Unlike the sphingosine kinases, CERK only utilizes ceramide as a substrate and has no activity for sphingosine or DAG (Sugiura 2002). CERK activity is enhanced in the presence of calcium or magnesium and contains a putative calmodulin-like domain. In addition, ceramide kinase has specificity for sphingosine containing ceramides since it has very low activity against dihydroceramide and phytoceramide species.[59, 60] Among the ceramide species it recognizes, CERK prefers ceramide species with acyl chain lengths greater than 12 carbons long, however, no preference was observed for the degree of saturation.[60] The measurement of C1P levels in A549 lung adenocarcinoma cells revealed an enrichment of C1P species containing acyl chain lengths of C16, C18 and C20 relative to their respective ceramide species.[61] The enrichment for particular ceramide species for C1P was suggested to be due to specific delivery of ceramides to the trans-Golgi by CERT, a lipid transport protein which is biased for ceramide species with acyl chain lengths less than 22 carbons.[62] Knockdown of CERT using RNA interference led to a decrease in C1P levels in A549 cells. A separate study using a pharmacological approach for CERT inhibition, found that inhibiting CERT had no effect on C1P production, but still inhibited sphingomyelin production.[61,63] It is unclear if these differences were due to intrinsic differences in the cell types studied (human lung versus mouse macrophage) or due to nonspecific effects of either approach. Future studies are needed to resolve this discrepancy.

CERK displays significant homology to other DAG kinases, however it also contains a N-terminal myristoylation site and a pleckstrin homology (PH) domain.[64] The PH domain targets CERK to PIP_2 containing membranes, but is also necessary for enzymatic activity. Several studies have shown that CERK is localized to the trans-Golgi in a PH dependent manner. C1P generated in the Golgi can act as a docking site for cytosolic PLA_2 and enhances arachidonic acid release. In addition, CERK translocates to the plasma membrane in response to osmotic swelling, an insult that enhances PIP2 on the plasma membrane. Translocation of CERK to the plasma membrane was shown to also be dependent on its PH domain.

The generation of a *Cerk*$^{-/-}$ mouse has provided some clues to CERK function in vivo. *Cerk*$^{-/-}$ mice are fully viable and show no gross phenotypic changes suggesting that CERK is not essential for development. Lipid analysis of serum from *Cerk*$^{-/-}$ revealed greatly elevated ceramides, but decreased dihydroceramides suggesting that CERK contributes significantly to ceramide metabolism in the serum.[59] Upon closer examination, *Cerk*$^{-/-}$ mice were found to have a normal level of C1P in their brains, despite a lack of CERK activity, suggesting that C1P can be produced through a mechanism independent of CERK.[65] Behavioral testing of the mice did, however, show abnormal emotional behavior based on an increase in ambulation and defecation frequencies in an open field test.[65] More recently, the *Cerk*$^{-/-}$ mice were found to have significant neutropenia under basal conditions. When these mice were challenged with *S. pneumoniae* they succumbed to lethal pneumonia earlier than wildtype mice and had a higher bacterial burden in their lungs.[66]

A CERK homologue formerly named retinitis pigmentosa 26, RP26, but recently renamed CERKL, or CERK-like, has been identified. Since this gene has been implicated in a human form of retinitis pigmentosa it was initially suspected that C1P might play an essential role in retinal biology. Initial expression studies of CERKL expressed in cells failed to detect any CERK activity drawing into question if CERKL was an actual ceramide kinase.[67,68] After generation of a CERKL knockout mouse, it was determined that *Cerkl*$^{-/-}$ mice had no alterations in retinal C1P, ceramide, or CERK activity.[59] *CERK*$^{-/-}$ mice, on the other hand, had 80% less C1P levels and elevated ceramides suggesting that CERK is the major enzyme responsible for C1P in the retina.[59] Taken together, it is unlikely that CERKL is a true ceramide kinase, but may have a separate function which is essential for retinal cells. Also, it is interesting to note that this study found that the retina of *Cerk*$^{-/-}$ mice had greatly reduced C1P levels, whereas previous studies have shown that whole brain C1P in these mice was unchanged.[59,65] Future studies will help address the contribution of CERK to C1P levels and potentially identify other enzymes involved in C1P production.

C1P levels are regulated both by its synthesis through CERK, but also by its dephosphorylation back into ceramide. It is unclear how ceramide-1-phosphate is dephosphorylated, however, several groups have reported C1P phosphatase activity in plasma membrane fractions in both the liver and brain.[69-71] C1P has been shown to traffic through the secretory pathway to reach the plasma membrane where it could potentially be dephosphorylated by C1P phosphatases.[63] Also, C1P appears to be a substrate for nonspecific lipid phosphatases of the LPP family.

Catabolizing Complex Sphingolipids and Sphingomyelins into Ceramide

There is a significant tradeoff between the ability of an organism to produce a novel advantageous lipid and the potential that the organism is incapable of catabolizing the same lipid, resulting in accumulation of a lipid product in cells or tissues. Lipids pose an additional problem in that they can not be excreted as readily as other more hydrophilic molecules and therefore tend to accumulate within cells when they can not be destroyed. As such, it is not surprising that for every enzyme capable of generating a specific sphingolipid, there exists an 'opposing' enzyme capable of breaking down the generated product. Indeed, the mutation of specific catabolic enzymes is the general principle behind lipid storage diseases and defects in sphingolipid catabolizing enzymes are responsible for a significant number of these diseases. Due to the limited scope of this chapter, the reader is referred to several recent reviews on lipid storage diseases for more expansive information on the variety of lipid storage diseases due to glycosphingolipid enzyme mutations.[72,73] Suffice it to say that many lysosomal hydrolases are necessary for the coordinated breakdown of complex glycosphingolipids and the absence of any single one of these results in the accumulation of its respective substrate.

Sphingomyelin is the most abundant complex sphingolipid in human cells. Therefore, coordinated breakdown of sphingomyelin is an essential part of membrane homeostasis. Breakdown of sphingomyelin occurs through the hydrolysis of the phosphocholine headgroups by the sphingomyelinase family. The direct result of sphingomyelin hydrolysis is the production of ceramide and free phosphocholine. The mammalian sphingomyelinases fall into three major categories based upon their pH optimum: acid sphingomyelinase, alkaline sphingomyelinase and the neutral sphingomyelinases. Although all three forms of sphingomyelinases catalyze a similar reaction, these three groups of enzymes are evolutionarily unrelated and have different subcellular distributions. Alkaline sphingomyelinase, which is exclusively expressed in the intestine and liver, plays a role in the digestion of dietary sphingomyelin and will not be discussed further in this section.[74] Acid sphingomyelinase and neutral sphingomyelinase are ubiquitously expressed and serve as the major regulators of SM catabolism in most tissues and will be discussed in more detail.

Acid sphingomyelinase (ASMase) was the first sphingomyelinase to be characterized in mammalian cells. ASMase is predominantly a lysosomal protein which metabolizes sphingomyelin present on endosomal membranes. Lysosomal ASMase becomes N-glycosylated on at least six residues within the ER which stabilizes the enzyme structure and provides protection from proteolysis within the lysosomes. ASMase also becomes mannose-6 phosphorylated within the Golgi which directs it into the lysosomal compartment where it is most active.[75] The ability of ASMase to reach the lysosomal compartment is essential for its ability to catabolize sphingomyelin.

ASMase is also secreted into the extracellular space (where it is often referred to as secretory SMase) where it has access to sphingomyelin-containing lipoproteins which are abundant in the plasma.[76] In addition, secretory ASMase can metabolize outer leaflet SM on the plasma membrane. Secretory ASMase, unlike the lysosomal form, requires zinc for sphingomyelinase activity. It is unclear how the secretion of ASMase is regulated, but it appears to be through the constitutive secretory pathway. Moreover, the function of secretory ASMase is still unclear, but is thought to play a role in reducing plasma SM content and may play a role in cellular stress responses by generating plasma membrane localized ceramide with a specific signaling role.

Insights into the significance of the acid sphingomyelinase protein, encoded by the gene *SMPD1*, come from a human lysosomal storage disorder Niemann Pick Disease Types A (NPD A) and B (NPD B).[77] Complete absence of a functional ASMase gene product results in NPD A which is characterized by a progressive neurodegenerative disease with psychomotor retardation, retinal cherry red spots, hepatosplenomegaly, lung disease and premature death. Patients afflicted with NPD A usually do not live past the age of three. A mouse model of NPD A, *Smpd1*$^{-/-}$, exhibits growth defects similar to the human disease and also dies prematurely around 4 months of age.[78] NPD B, on the other hand, is a less severe form of the disease that lacks neuronopathic symptoms, however, hepatosplenomegaly and lung disease still occur. Fortunately, individuals afflicted with NPD B are able to survive into adulthood. The NPD B phenotype has been mimicked in mice by fusing a *Smpd1* transgene to the *LAMP1* gene.[79] In these mice there is complete absence of secretory SMase, but low level lysosomal ASMase activity. Future studies will dissect which symptoms of NPD are likely due to a lack of secretory ASMase function and which symptoms are associated with lysosomal ASMase function, thereby elucidating further some of the specific functions of each form of the enzyme.

Within the past decade, three different mammalian neutral sphingomyelinase (NSMase) genes have been identified, *SMPD2, SMPD3* and *SMPD4*, although, the first NSMase gene discovered, *SMPD2*, is not likely to function as a SMase, but rather as a lyso-PAF phospholipase C.[80] To further muddy the point, mice lacking the *Smpd2* gene have a decreased tissue SMase activity, but have no alteration in tissue SM levels.[81]

The best characterized NSMase to date is NSmase 2. NSMase 2 contains two highly hydrophobic domains, that may function as membrane anchors but not full transmembrane domains.[82] Unusually, N-SMase2 is thought to have its catalytic site facing the cytosolic leaflet of either the Golgi or plasma membrane. This orientation for a SMase is unusual since SM is thought to be relatively excluded from the cytosolic leaflet.[82] Interestingly, NSMase2 localizes to the Golgi under subconfluent conditions, however upon reaching confluence, NSMase2 translocates to the plasma membrane.[83] It was later shown that plasma membrane association is highly dependent upon its palmitoylation on multiple cysteine residues.[82]

NSMase2 overexpression, in the absence of a specific stimulus causes degradation of sphingomyelin into ceramide with a preference for C24:0 and C24:1 species.[83] Generation of C24 and C24:1 ceramide during confluence dependent growth arrest was dependent on NSMase 2 suggesting that generation of specific ceramide species could have specific effects on growth arrest.[83] These results also suggest that NSMase2 either has specificity for very long chain sphingomyelin species or that very long chain sphingomyelins are enriched in the inner leaflet of the plasma membrane. Since most sphingomyelin species are thought to be localized to the outer leaflet of the plasma membrane or the luminal side of the Golgi, it is a mystery how NSMase2 with an inner leaflet catalytic site could access its substrate. The condundrum of how NSMase2 reaches its substrate undoubtedly puts into question some existing paradigms about sphingomyelin localization in membrane leaflets.

The generation of a *Smpd3*$^{-/-}$ mice has provided some insights into the physiological role of NSMase 2.[84] *Smpd3*$^{-/-}$ mice display severe growth retardation, organ hypoplasia, delayed puberty and skeletal defects. Upon further examination, a combined pituitary hormone deficiency was discovered in these mice and a significant decrease in serum IGF-1, TSH and GnRH was found. Due to the complex interactions between the pituitary hormones and target tissues it is difficult to dissect which symptoms are due to hormone deficiency and which defects are due to primary organ defects. Some insight into this was provided by the generation of a chondrocyte specific SMPD3 expression mouse. When the chondrocyte specific SMPD3 mouse was crossed with the *Smpd3*$^{-/-}$ mouse, there was an absence of skeletal defects, suggesting that chondrocyte specific NSMase2 activity is essential for skeletal development.[85] Despite the dramatic pathology observed with the *Smpd3*$^{-/-}$ mouse, it is still difficult to understand how the NSMase2 knockout phenotype correlates with an intracellular role of NSMase 2 in sphingomyelin generation. Further characterizing the cellular dysfunction in the *Smpd3*$^{-/-}$ mouse in parallel with further characterizing N-SMase 2 in vitro will shed light on a murky area of sphingolipid biology and provide new insights into cell biology as a whole.

Finally, a third NSMase, NSMase3 has recently been identified which is encoded by the *SMPD4* gene.[86] Interestingly, NSMase3 is localized to the ER and possibly the Golgi. It was found to contain at least one transmembrane domain and potentially more along with an ER retention signal.[87] NSMase3 is predominantly expressed in skeletal and cardiac muscle with minor expression in many other tissues. It will be interesting to see how an ER localized SMase can affect sphingolipid metabolism since SM is not thought to be present in the ER compartment.

The Catabolism of Ceramides and the Final Common Breakdown Pathway

Just as a few sphingolipid precursors are generated to produce hundreds of different sphingolipids, all sphingolipids are eventually catabolized to ceramide, sphingosine and finally, sphingosine-1-phosphate. The deacylation of ceramide species is achieved through the family of enzymes known as ceramidases. These ceramidases have organelle specific expression and may have specificity for different forms of ceramide to bias a cell towards the generation of complex sphingolipids with specific sphingoid bases. Organelle specific expression also allows for the possibility to serve as negative regulators of organelle specific ceramide signaling. After ceramide is deacylated into sphingosine, the conversion of sphingosine to sphingosine-1-phosphate is achieved through one of two sphingosine kinases localized in the cytosol or peripherally associated with specific membrane compartments. In the final step of sphingolipid breakdown, sphingosine-1-phosphate is degraded by the enzyme sphingosine-1-phosphate lyase in the ER to produce hexadecenal and phosphoethanolamine.

Acid, Neutral and Alkaline Ceramidases

The ceramidases, like many other sphingolipid enzymes, have been classified biochemically according to their pH optima. Acid ceramidase (AC), as its name suggests, is a lysosomal enzyme which deacylates ceramide species produced from the degradation of plasma membrane sphingolipids. AC, encoded by the *ASAH1* gene, is a member of the N-terminal nucleophile (Ntn) hydrolase superfamily. Members of the Ntn family are characterized by their ability to undergo autoproteolytic cleavage through cysteine dependent proteolysis. Interestingly, this processing was shown to be accelerated at pH 4.5 when compared to neutral pH. This pH dependent maturation is likely a mechanism that has evolved to prevent premature activation of AC prior to it reaching the lysosomal compartment. AC activity was shown to be higher against C12 and C14 ceramide species compared with C6 or C18 ceramide species.[88] Although it has been reported that AC has the greatest activity against medium and long-chain ceramide species, it is difficult to imagine how very long-chain ceramides would be processed in a compartment where AC is the only known ceramidase.[88] Therefore, it is likely that very long-chain ceramide containing sphingolipids are either excluded from the endolysosomal compartment or that AC is able to degrade these very long chain ceramides albeit at a slower rate or with the aid of an accessory protein.

The significance of AC in mammalian systems is reinforced by its role in the human lipid storage disease Farber lipogranulomatosis. Farber disease is an autosomal recessive disorder due to a dysfunctional AC gene product. Farber's disease is characterized by early onset arthritis, swollen lymph nodes, psychomotor difficulties and vocal cord pathology. Complete absence of AC expression in a mouse model results in embryonic lethality at a very early stage in development.[89] Mice heterozygous for AC develop a progressive lipid storage disease due to ceramide accumulation in the liver, skin, lungs and bones. The striking phenotype of the AC heterozygous mice raises questions about its implications for human disease. One could imagine that a heterozygous defect in AC within a human population may lead to a progressive lipid storage disorder that may only express itself with advanced age. Answers to questions like these will most likely be revealed when human genome sequencing becomes more widespread.

Neutral ceramidase, the most active ceramidase at neutral pH is encoded by the *ASAH2* gene. Neutral ceramidase (NC) is synthesized through the secretory pathway as a Type II integral membrane protein. NC can be cleaved at its N-terminus to produce a soluble protein that peripherally associates with the outer leaflet of the plasma membrane. NC contains mucin box domains which are highly O-glycosylated and are necessary for plasma membrane association.[90] NC associated with the plasma membrane is an important regulator of sphingosine and sphingosine-1-phosphate (S1P) production and for S1P release.[91] In addition, NC is highly expressed in the intestinal epithelium and contributes to the digestion of dietary sphingolipids.[92,93] *Asah2*$^{-/-}$ mice showed an inability to metabolize dietary ceramides. Although, sphingosine and ceramide levels were normal in the brain, liver and kidney of *Asah2*$^{-/-}$ mice, the intestines had a significantly increased C16:0 ceramide content but a reduced sphingosine content. Due to an absence of visible pathology in the *Asah2*$^{-/-}$ mice, it is unclear what role NC plays in nonintestinal tissues, if any.[92] Future studies with the *Asah2*$^{-/-}$ mice may reveal subtle or previously uncharacterized, but important functions for this gene under specific stresses.

The alkaline ceramidases (ACERs) contain three separate family members: alkaline ceramidases 1, 2 and 3 are encoded by the *ASAH3*,[94] *ASAH3L*[95] and *PHCA*[96] genes respectively, which share significant homology to one another. These ACERs differ in subcellular localizations and substrate specificity although they have alkaline pH optima for their in vitro activity and are activated by calcium ion in vitro.[97] ACER1 is mainly expressed in the epidermis, whereas ACER2 and ACER3 are expressed in various tissues. ACER1 has multiple putative transmembrane domains and is localized to the ER.[94,98] Interestingly, ACER1 has marked substrate specificity for C24 and C24:1 ceramides, but has no activity against dihydroceramides or phytoceramides.[94,98] This highly restricted substrate specificity may prevent phytoceramide species from being degraded because the skin is one of the limited tissues that is enriched in these types of ceramides.

ACER2, also referred to as Golgi alkaline ceramidase, is highly expressed in the placenta, but its low expression can be detected in most tissues. Like its homologue ACER1, ACER2 has several putative transmembrane domains, however, ACER2 is localized to the Golgi complex. ACER2 has a less restricted substrate specificity since it also metabolizes other long-chain ceramides C_{16}, C_{18} and C_{20} ceramide species and long-chain dihydroceramides and phytoceramides with an unsaturated acyl chain, in addition to C24 and C24:1 species (Cungui Mao personal communication). ACER2 requires calcium ions but not other cations for its activity.

ACER3 was previously called phytoceramidase because it was found to have higher in vitro activity towards the artificial fluorescent phytoceramide D-*ribo*-C_{12}-NBD-phytoceramide, than towards NBD-ceramide or NBD-dihydroceramide.[96] Mao et al recently found that ACER3 only catalyzes the hydrolysis of natural phytoceramide, dihydroceramide and ceramides carrying an unsaturated fatty acid (≤ C20) (Cungui Mao personal communication). ACER3 appears to be the only ceramidase identified in mammals which has a preference for phytoceramide species. Its tissue expression is widespread, but, like ACER2, is highly expressed in the placenta. ACER3 is localized to both the ER and Golgi complex, with a C-terminal ER retention sequence.[96] Interestingly, its activity is inhibited by sphingosine, but not by dihydrosphingosine or phytosphingosine. The significance of this mammalian ceramidase remains to be seen, but its unusual specificity for phytoceramide raises questions about the unique functions of phytoceramide containing sphingolipids.

Although investigation has begun to classify and characterize the different biochemical properties of the ceramidases, much still remains to be done to define the specific functions of each alkaline ceramidase. The generation of the three alkaline ceramidase knockout mice will undoubtedly shed further light on this subject.

Sphingosine-1-Phosphate and Sphingosine Kinases 1 and 2

The two sphingosine kinases (SK), SK1 and SK2 are members of the DAG kinase family.[99] Both enzymes utilize ATP to phosphorylate the C-1 hydroxy group of free sphingosine, dihydrosphingosine, or, in the case of SK2, also phytosphingosine. Both sphingosine kinases are cytosolic enzymes that peripherally associate with membranes. Regulation of sphingosine kinase localization within the cell is thought to be the primary mode by which these enzymes acutely affect sphingolipid metabolism since only modest changes in their activity can be detected after stimulation by various agonists.[100,101] In addition, SK enzymes are regulated transcriptionally by a variety of stimuli,[102-104] Although SK1 and SK2 both catalyze the same reaction, they have slight differences in their substrate specificities and have distinct, but overlapping, subcellular localizations which determine their effects on specific sphingolipid compartments within cells.

Sphingosine kinase 1 (SK1) is a cytosolic enzyme, but it can associate with the plasma membrane, move into the nucleus and even be secreted from cells.[100,105-108] The first major insight into the regulation of SK1 activity first came when it was shown that SK1 translocates from the cytoplasm to the plasma membrane in response to phorbol ester treatment.[100] Shortly after this, translocation to the plasma membrane was shown to be phosphorylation dependent and that ERK2 was the kinase responsible for this phosphorylation.[101] Later work showed that phosphorylation of SK1 lead to its enhanced affinity to anionic phospholipids such as PS, PI and PA which are in high abundance on the inner leaflet of the plasma membrane.[109] The metabolic significance of SK1 translocation to the plasma membrane was shown to be enhancement of S1P production and extracellular release of S1P.[100,110] This agonist induced translocation of SK1 through phosphorylation by ERK2, has become the general paradigm for S1P signaling and receptor activation in response to various agonists. It is worth noting that many different growth factors converge on SK1 as an intracellular target for downstream signaling. These include, but are not limited to, PDGF, VEGF, NGF, EGF, Insulin and IGF-1. Activation of SK1 was shown to be necessary for the proliferative effects of many different growth factors.[111-116] Whether or not all of these growth factors act through the same intracellular signaling pathway to activate SK1 remains to be tested. The reader is referred to several good reviews on SK1 for more detailed information on its signaling role in proliferation and survival.[35,117,118]

Although phosphorylation dependent translocation of SK1 to the plasma membrane has been the best defined mechanism by which SK1 activity is regulated, SK1 also traffics through the nucleus. SK1 was shown to have two putative nuclear export sequences (NES) which are necessary for SK1 to leave the nucleus.[107] It is unclear how SK1 introduction into the nucleus is regulated, but deletion of the two nuclear export sequences present on SK1, lead to nuclear accumulation of SK1. The effect nuclear trafficking of SK1 has on sphingolipid metabolism is unclear, but attachment of a nuclear localization sequence on SK1 inhibited cellular proliferation through an unknown mechanism.[119]

In addition, SK1 has been shown to be secreted from endothelial cells, however, the physiological significance of this is not clear. An interesting discovery will be the route by which an exclusively cytosolic protein gets secreted from endothelial cells. Analysis of SK in endothelial cells showed that secretion was constitutive and not regulated by agonist stimulation.[106] It has been suggested that extracellular SK can generate S1P in the extracellular space following coordinated hydrolysis of plasma SM by secretory SMase and NCDase.[120] Some evidence for SK secretion occurring in vivo comes from Venkataraman et al. A significant amount of soluble SK activity was found in the serum of wildtype, but not *SphK1*$^{-/-}$ mice suggesting that SK1 is indeed released into the blood of mice.[121] Once again, the physiological significance of SK1 release remains to be determined.

In addition to its effects on promoting proliferation, SK1 has plays an important role in survival and resistance to chemotherapy.[122] Overexpression of SK1 correlates with a resistance to chemotherapeutic agents.[123] In addition, overexpressing SK1 in sensitive cells can induce resistance to chemotherapy.[105] Importantly, it has been shown that SK1 degradation is a downstream event in the DNA damage response.[124] Apoptosis inducing agents, such as TNF-α or various DNA damaging agents lead to degradation of SK1 through a Cathepsin B dependent mechanism.[125] In vitro characterization of SK1 degradation has shown that cathepsin B degrades SK1 through a stepwise cleavage, first at Histidine 122 followed by cleavage at Arginine 199.[126] How DNA damaging agents or death inducing ligands, such as TNF-α, induce degradation of cytosolic SK1 through lysosomal localized cathepsin B is not yet clear. Future studies will better define how SK1 contributes to survival and how its degradation is involved in apoptosis.

Sphingosine kinase 2, the lesser studied isoform of the two, is predominantly localized in the nucleus or perinuclear region of a cell. SK2 has broader substrate specificity than SK1 and is able to phosphorylate phytosphingosine as well as sphingosine and dihydrosphingosine. Early reports on SK2 function suggest that it enhances apoptotic responses and can induce apoptosis through overexpression.[127] This is in contrast to SK1, which enhances survival of cells when overexpressed and allows for resistance to apoptosis inducing agents.[122] Due to its perinuclear localization, SK2 is thought to produce phosphate bases distal from the plasma membrane, where ABC transporters are less likely to export them into the extracellular space.[128] Another explanation for their disparate functions comes from a recent study suggesting that SK2 uniquely couples to sphingosine phosphate phosphatase 1 (SPP1) which generates free sphingosine for ceramide synthesis.[129] Interestingly, predominantly C16, C18 and C20 ceramide species were produced through this pathway, suggesting that the recycling pathway through SK2 and SPP1 may couple to specific ceramide synthases. Importantly, SK1 was found not to have the same function in enhancing sphingosine recycling into ceramide. Therefore, SK2 may have a distinct function from SK1 by having an enhanced ability to recycle sphingoid bases for ceramide synthesis. The structural basis for this difference has not yet been characterized, but will help elucidate how two enzymes with the same catalytic activity can have dramatic differences on cell biology.[127]

Insight into the physiological role of S1P was hoped to be gained through generation of sphingosine kinase 1 knockout mice, however, in the absence of stresses they displayed no obvious phenotypic abnormalities.[130] In contrast, *SphK1*$^{-/-}$ mice develop less intestinal adenomas when crossed with the *Apc*$^{min/+}$ mice, a model of intestinal adenocarcinoma.[131] *SphK1*$^{-/-}$ mice also develop less adenocarcinomas when treated with DSS/AOM.[132] Moreover, *SphK1*$^{-/-}$ mice are less susceptible to DSS induced colitis and have a marked decrease in intestinal inflammation in that model.[133]

Sphingosine kinase 2 knockout mice have also been generated and these mice too do not display any obvious abnormalities.[134] Crossing of *SphK1*$^{-/-}$ and *SphK2*$^{-/-}$ mice revealed that complete absence of SK activity and therefore S1P, results in embryonic lethality due to improper neural and vascular development.[134] These defects are likely to be due to a lack of S1P and not due to sphingosine accumulation since sphingosine levels were actually below wildtype levels in double knockout embryos. In addition, the vascular defects observed in the double knockout mice resembled previously observed defects in S1P1 receptor knockout mice, suggesting that S1P is absolutely necessary for vascular development.[134,135] Finally, female *SphK1*$^{-/-}$, *SphK2*$^{+/-}$ mice are infertile due to defects in decidualization.[136] Together these studies suggest that SK1 and SK2 have redundant functions during development since neither SK knockout displays any developmental defects alone, but display severe defects when both enzymes are absent. On the other hand, each SK isozyme may have unique functions in mature tissues since *SphK1*$^{-/-}$ mice have specific defects in inflammatory responses.[133]

Lipid Phosphate Phosphatases, S1P Phosphatases and the Salvage Pathway

Sphingosine-1-phosphate can be dephosphorylated at the cell surface by a family of broad specificity lipid phosphate phosphatases, LPP1-3. LPPs have six transmembrane domains and have their catalytic sites facing the extracellular space. The LPP family is important for sphingolipid metabolism because they are thought to be the primary mechanism by which extracellular S1P signaling is attenuated.[137,138] Overexpression of specific LPPs reduces S1P-dependent signaling events.[139,140] LPP effects on S1P metabolism is complicated by the fact that LPPs also affect phosphatidic acid (PA) levels which have been shown to regulate SK1 localization in cells.[138,141] In addition, S1P may require LPP dependent dephosphorylation prior to its uptake by cells, although a separate mechanism for uptake requiring the CFTR protein has also been proposed.[142] For additional information on LPPs, the reader is referred to some recent reviews.[138,143]

In addition to LPPs, cytosolic S1P can be dephosphorylated at the ER by S1P specific phosphatases, SPP1 and SPP2.[144-146] SPP1 and perhaps SPP2, plays a role in regulating the reintroduction of sphingoid bases into ceramide species at the ER.[129,147] Distantly related to LPPs, SPP1 and 2 each contain eight putative transmembrane domains.[145,148,149] Interestingly, overexpression of SPP1 results in an increase in ceramide accumulation suggesting that dephosphorylation of S1P is a rate limiting step in the salvage pathway.[129,145,150] This increase in ceramide can be exacerbated further by the addition of extracellular S1P. Therefore, regulation of SPP1 levels can change the metabolic fate of S1P to be predominantly recycled into ceramide. This has been shown to have dramatic effects on the biological response of cells to S1P.[150] Recently, SPP1 mediated conversion to ceramide was shown to be enhanced by overexpression of SK2, suggesting that SK2 and SPP1 work coordinately to increase sphingosine salvage.[129] In addition, higher expression of SPP1 reduced extracellular release of S1P, suggesting that SPP1 can negatively regulate extracellular signaling of S1P.[149]

A second sphingosine phosphate phosphatase, SPP2, was identified based on homology to SPP1. Both SPP isoforms are expressed ubiquitously with high expression in the kidney. On the other hand, SPP1 is highly expressed in the placenta whereas SPP2 is highly expressed in the heart.[146] More recently, SPP2 was shown to be upregulated during inflammatory responses, however, it is unclear what effect this has on sphingolipid metabolism.[151] It remains to be seen if SPP2 plays an important role in regulating the sphingolipid recycling pathway like SPP1.

S1P Lyase in the Removal of Sphingoid Bases

Sphingosine-1-phosphate lyase (SPL) is the enzyme responsible for the conversion of phosphorylated sphingoid bases to hexadecenal and phosphoethanolamine, thereby serving as the final step in sphingolipid degradation. SPL is a single pass transmembrane protein, displaying Type I topology and is exclusively localized to the ER.[152] The catalytic site of SPL faces the cytosolic surface of the ER where it has access to cytosolically produced S1P. Since SPL is exclusively localized to the ER, all sphingoid phosphate bases must reach the ER for their final degradation by SPL. SPL has broad substrate specificity since it can utilize sphingosine-1-phosphate, dihydrosphingosine-1-phosphate and phytosphingoinse-1-phosphate as substrates and therefore is able to catabolize all sphingoid bases found in mammals.[153] Like SPT, SPL is dependent on pyridoxal 5′-phosphate (PLP) as a cofactor for enzymatic activity, making both the initial and the final steps in the sphingolipid pathway PLP dependent. The relationship between symptoms associated with vitamin B6 deficiency, the dietary precursor to PLP and alterations in sphingolipid metabolism are unclear, but will be important to assess since inhibition of SPT or S1P lyase could result in demyelination and altered immune function respectively.[8,154]

SPL has a wide tissue distribution with its highest expression in the thymus and intestines and its lowest expression in the brain and skeletal muscle.[152] In addition, no SPL activity can be detected in platelets or red blood cells since these cells lack ER membranes.[152] Immunohistochemical staining for SPL within the intestine revealed high expression on differentiated enterocytes, but low expression in the intestinal cypt cells.[155] In addition, intestinal adenomas from $Apc^{Min/+}$ mice showed a marked decrease in SPL expression compared with adjacent normal mucosa.[131] Together these data suggest that SPL upregulation is part of the normal differentiation of enterocytes. SPL expression in the thymus was also assessed and was localized almost exclusively to thymic epithelium with low expression in lymphocytes. Thymic expression of SPL seems to be essential for proper lymphocyte trafficking between the blood, primary lymphoid tissue and secondary lymphoid tissue due its role in establishing S1P gradients between the different trafficking compartments.[154]

Several studies have identified S1P lyase as an essential gene for development in *Drosophila melanogaster*, *Caenorhibitis elegans* and *Dictyostelium discoideum*.[155-157] Recently, the phenotype of the S1P lyase knockout mouse, $Sgpl1^{-/-}$, was published.[158] These SPL defective mice display an inability to gain weight and premature death by eight weeks of age. These mice also displayed an array of defects which closely resembled those seen with the PDGF receptor, *Pdgfra*$^{-/-}$ and *Pdgfrb*$^{-/-}$, mutant mice. Some of these defects include vascular hemorrhaging, defective glomeruli formation and skeletal defects. It is unclear if *Sgpl1*$^{-/-}$ mice had defects in any other systems since the study was interested only in PDGF dependent phenotypes.[158] Regardless, this study provides clear evidence that SPL plays an essential role in the development of many tissues in mammals. Future studies will help identify the role of SPL in adult tissues and assess its potential as a therapeutic target for the treatment of disorders ranging from immunological disorders to neoplasias.

Conclusion

Sphingolipids are a diverse group of lipids which serve a variety of functions in both mammalian development and physiology. Only a few of these functions have been highlighted in the chapter, but the scope of sphingolipid biology is vast. New functions for sphingolipids in mammalian physiology continue to be discovered each year and will likely increase as our understanding of sphingolipid biology in whole organisms improves.

It is clear that sphingolipids play an essential role in mammalian systems, therefore, it is essential to understand how sphingolipids are synthesized and degraded to maintain their functional levels at both an organismal and cellular level. Thanks to the hard work of many pioneers in the sphingolipid field, the biochemical characterization of many sphingolipid enzymes was laid out before any genes were identified. Aided by the advancement of recombinant DNA technology and the human genome project, significant progress has been made to clone and characterize the enzymes responsible for sphingolipid metabolism. At least one gene and in most cases two or more, has been cloned for each of the known enzymatic steps required from the condensation of serine and palmitoyl CoA by SPT to the production of hexadecenal and phosphoethanolamine by S1P lyase. Although it is easy to assume that we have a near complete list of the enzymes involved in sphingolipid metabolism, new isoforms of these enzymes are still being discovered and validated. While having a list of enzymes is comforting, our understanding of their regulation is still lacking. We still have a lot of work to understand how the different sphingolipid enzymes work in concert to determine the sphingolipid composition of the plasma membrane and subcellular organelles. In addition, the breakthrough discoveries of the sphingolipid transfer proteins CERT and FAPP2 have provided a new paradigm for nonvesicular sphingolipid trafficking. These proteins are likely to only be the first members of a class of sphingolipid transfer proteins which regulate the trafficking of sphingolipids to specific compartments within cells.

Although this chapter aimed to provide an abbreviated overview of our current understanding of sphingolipid metabolism in mammalian systems, it also highlights some areas of lipid biology that are poorly understood. It is a healthy scientific practice to regularly reflect on the progress that has been achieved within a field, both, to bring to light significant faults or assumptions we have about it, but also for the sense of awe one gets about our collective accomplishment.

Acknowledgements

This work was supported in part by NIH grants GM-43825 and CA-87584 to YAH and AG-016583, GM-062887 and a MERIT Award by the Office of Research and Development, Dept. of Veterans Affairs to LMO. Special thanks to Thomas D. Mullen and Cungui Mao for their helpful comments.

References

1. Mandon EC, Ehses I, Rother J et al. Subcellular localization and membrane topology of serine palmitoyltransferase, 3-dehydrosphinganine reductase and sphinganine N-acyltransferase in mouse liver. J Biol Chem 1992; 267(16):11144-8.
2. Hornemann T, Richard S, Rütti MF et al. Cloning and initial characterization of a new subunit for mammalian serine-palmitoyltransferase. J Biol Chem 2006; 281(49):37275-81.
3. Yasuda S, Nishijima M, Hanada K. Localization, topology and function of the LCB1 subunit of serine palmitoyltransferase in mammalian cells. J Biol Chem 2003; 278(6):4176-83.
4. Hojjati MR, Li Z, Jiang XC. Serine palmitoyl-CoA transferase (SPT) deficiency and sphingolipid levels in mice. Biochim Biophys Acta 2005; 1737(1):44-51.

5. Gable K et al. Tsc3p is an 80-amino acid protein associated with serine palmitoyltransferase and required for optimal enzyme activity. J Biol Chem 2000; 275(11):7597-603.
6. Hanada K. Serine palmitoyltransferase, a key enzyme of sphingolipid metabolism. Biochim Biophys Acta 2003; 1632(1-3):16-30.
7. Merrill AH Jr. Characterization of serine palmitoyltransferase activity in Chinese hamster ovary cells. Biochim Biophys Acta 1983; 754(3):284-91.
8. Dawkins JL et al. Mutations in SPTLC1, encoding serine palmitoyltransferase, long chain base subunit-1, cause hereditary sensory neuropathy type I. Nat Genet 2001; 27(3):309-12.
9. Spring PJ, Kok C, Nicholson GA et al. Autosomal dominant hereditary sensory neuropathy with chronic cough and gastro-oesophageal reflux: clinical features in two families linked to chromosome 3p22-p24. Brain 2005; 128(Pt 12):2797-810.
10. Lindahl AJ, Lhatoo SD, Campbell MJ et al. Late-onset hereditary sensory neuropathy type I due to SPTLC1 mutation: autopsy findings. Clin Neurol Neurosurg 2006; 108(8):780-3.
11. Bejaoui K et al. SPTLC1 is mutated in hereditary sensory neuropathy, type 1. Nat Genet 2001; 27(3):261-2.
12. McCampbell A et al. Mutant SPTLC1 dominantly inhibits serine palmitoyltransferase activity in vivo and confers an age-dependent neuropathy. Hum Mol Genet 2005; 14(22):3507-21.
13. Bejaoui K, Wu C, Scheffler MD et al. Hereditary sensory neuropathy type 1 mutations confer dominant negative effects on serine palmitoyltransferase, critical for sphingolipid synthesis. J Clin Invest 2002; 110(9):1301-8.
14. Dawkins JL, Brahmbhatt S, Auer-Grumbach M et al. Exclusion of serine palmitoyltransferase long chain base subunit 2 (SPTLC2) as a common cause for hereditary sensory neuropathy. Neuromuscul Disord 2002; 12(7-8):656-8.
15. Beeler T, Bacikova D, Gable K et al. The Saccharomyces cerevisiae TSC10/YBR265w gene encoding 3-ketosphinganine reductase is identified in a screen for temperature-sensitive suppressors of the Ca^{2+}-sensitive csg2Delta mutant. J Biol Chem 1998; 273(46):30688-94.
16. Kihara A, Igarashi Y. FVT-1 is a mammalian 3-ketodihydrosphingosine reductase with an active site that faces the cytosolic side of the endoplasmic reticulum membrane. J Biol Chem 2004; 279(47):49243-50.
17. Krebs S, Medugorac I, Röther S et al. A missense mutation in the 3-ketodihydrosphingosine reductase FVT1 as candidate causal mutation for bovine spinal muscular atrophy. Proc Natl Acad Sci USA 2007; 104(16):6746-51.
18. Pewzner-Jung Y, Ben-Dor S, Futerman AH. When do Lasses (longevity assurance genes) become CerS (ceramide synthases)? Insights into the regulation of ceramide synthesis. J Biol Chem 2006; 281(35):25001-5.
19. Lahiri S, Lee H, Mesicek J et al. Kinetic characterization of mammalian ceramide synthases: determination of K(m) values towards sphinganine. FEBS Lett 2007; 581(27):5289-94.
20. Venkataraman K, Riebeling C, Bodennec J et al. Upstream of growth and differentiation factor 1 (uog1), a mammalian homolog of the yeast longevity assurance gene 1 (LAG1), regulates N-stearoyl-sphinganine (C18-(dihydro)ceramide) synthesis in a fumonisin B1-independent manner in mammalian cells. J Biol Chem 2002; 277(38):35642-9.
21. Laviad EL, Albee L, Pankova-Kholmyansky I et al. Characterization of ceramide synthase 2: tissue distribution, substrate specificity and inhibition by sphingosine-1-phosphate. J Biol Chem 2008; 283(9):5677-84.
22. Riebeling C, Allegood JC, Wang E et al. Two mammalian longevity assurance gene (LAG1) family members, trh1 and trh4, regulate dihydroceramide synthesis using different fatty acyl-CoA donors. J Biol Chem 2003; 278(44):43452-9.
23. Mizutani Y, Kihara A, Igarashi Y. Mammalian Lass6 and its related family members regulate synthesis of specific ceramides. Biochem J 2005; 390(Pt 1):263-71.
24. Mizutani Y, Kihara A, Igarashi Y. LASS3 (longevity assurance homologue 3) is a mainly testis-specific (dihydro)ceramide synthase with relatively broad substrate specificity. Biochem J 2006; 398(3):531-8.
25. Mizutani Y et al. 2-Hydroxy-ceramide synthesis by ceramide synthase family: enzymatic basis for the preference of FA chain length. J Lipid Res 2008; 49(11):2356-64.
26. Bose R, Verheij M, Haimovitz-Friedman A et al. Ceramide synthase mediates daunorubicin-induced apoptosis: an alternative mechanism for generating death signals. Cell 1995; 82(3):405-14.
27. Cadena DL, Kurten RC, Gill GN. The product of the MLD gene is a member of the membrane fatty acid desaturase family: overexpression of MLD inhibits EGF receptor biosynthesis. Biochemistry 1997; 36(23):6960-7.
28. Geeraert L, Mannaerts GP, van Veldhoven PP. Conversion of dihydroceramide into ceramide: involvement of a desaturase. Biochem J 1997; 327(Pt 1):125-32.

29. Michel C, van Echten-Deckert G, Rother J et al. Characterization of ceramide synthesis. A dihydroceramide desaturase introduces the 4,5-trans-double bond of sphingosine at the level of dihydroceramide. J Biol Chem 1997; 272(36):22432-7.
30. Savile CK, Fabrias G, Buist PH. Dihydroceramide delta(4) desaturase initiates substrate oxidation at C-4. J Am Chem Soc 2001; 123(19):4382-5.
31. Ternes P, Franke S, Zähringer U et al. Identification and characterization of a sphingolipid delta 4-desaturase family. J Biol Chem 2002; 277(28):25512-8.
32. Beauchamp E, Goenaga D, Le Bloc'h J et al. Myristic acid increases the activity of dihydroceramide Delta4-desaturase 1 through its N-terminal myristoylation. Biochimie 2007; 89(12):1553-61.
33. Mizutani Y, Kihara A, Igarashi Y. Identification of the human sphingolipid C4-hydroxylase, hDES2 and its up-regulation during keratinocyte differentiation. FEBS Lett 2004; 563(1-3):93-7.
34. Omae F, Miyazaki M, Enomoto A et al. DES2 protein is responsible for phytoceramide biosynthesis in the mouse small intestine. Biochem J 2004; 379(Pt 3):687-95.
35. Hannun YA, Obeid LM. Principles of bioactive lipid signalling: lessons from sphingolipids. Nat Rev Mol Cell Biol 2008; 9(2):139-50.
36. Holland WL, Brozinick JT, Wang LP et al. Inhibition of ceramide synthesis ameliorates glucocorticoid-, saturated-fat- and obesity-induced insulin resistance. Cell Metab 2007; 5(3):167-79.
37. Hanada K, Kumagai K, Tomishige N et al. CERT and intracellular trafficking of ceramide. Biochim Biophys Acta 2007; 1771(6):644-53.
38. Kumagai K, Kawano M, Shinkai-Ouchi F et al. Interorganelle trafficking of ceramide is regulated by phosphorylation-dependent cooperativity between the PH and START domains of CERT. J Biol Chem 2007; 282(24):17758-66.
39. Charruyer A, Bell SM, Kawano M et al. Decreased ceramide transport protein (CERT) function alters sphingomyelin production following UVB irradiation. J Biol Chem 2008; 283(24):16682-92.
40. Hanada K, Hara T, Fukasawa M et al. Mammalian cell mutants resistant to a sphingomyelin-directed cytolysin. Genetic and biochemical evidence for complex formation of the LCB1 protein with the LCB2 protein for serine palmitoyltransferase. J Biol Chem 1998; 273(50):33787-94.
41. Hanada K, Kumagai K, Yasuda S et al. Molecular machinery for nonvesicular trafficking of ceramide. Nature 2003; 426(6968):803-9.
42. Kudo N, Kumagai K, Tomishige N et al. Structural basis for specific lipid recognition by CERT responsible for nonvesicular trafficking of ceramide. Proc Natl Acad Sci USA 2008; 105(2):488-93.
43. Watson P, Stephens DJ. ER-to-Golgi transport: form and formation of vesicular and tubular carriers. Biochim Biophys Acta 2005; 1744(3):304-15.
44. Stahl N, Jurevics H, Morell P et al. Isolation, characterization and expression of cDNA clones that encode rat UDP-galactose: ceramide galactosyltransferase. J Neurosci Res 1994; 38(2):234-42.
45. Coetzee T, Li X, Fujita N et al. Molecular cloning, chromosomal mapping and characterization of the mouse UDP-galactose:ceramide galactosyltransferase gene. Genomics 1996; 35(1):215-22.
46. Fujimoto H, Tadano-Aritomi K, Tokumasu A et al. Requirement of seminolipid in spermatogenesis revealed by UDP-galactose: Ceramide galactosyltransferase-deficient mice. J Biol Chem 2000; 275(30):22623-6.
47. Zoller I et al. Oligodendrocyte-specific ceramide galactosyltransferase (CGT) expression phenotypically rescues CGT-deficient mice and demonstrates that CGT activity does not limit brain galactosylceramide level. Glia 2005; 52(3):190-8.
48. Marcus J, Honigbaum S, Shroff S et al. Sulfatide is essential for the maintenance of CNS myelin and axon structure. Glia 2006; 53(4):372-81.
49. Ichikawa S, Sakiyama H, Suzuki G et al. Expression cloning of a cDNA for human ceramide glucosyltransferase that catalyzes the first glycosylation step of glycosphingolipid synthesis. Proc Natl Acad Sci USA 1996; 93(10):4638-43.
50. Jeckel D, Karrenbauer A, Burger KN et al. Glucosylceramide is synthesized at the cytosolic surface of various Golgi subfractions. J Cell Biol 1992; 117(2):259-67.
51. D'Angelo G, Polishchuk E, Di Tullio G et al. Glycosphingolipid synthesis requires FAPP2 transfer of glucosylceramide. Nature 2007; 449(7158):62-7.
52. Yamashita T, Wada R, Sasaki T et al. A vital role for glycosphingolipid synthesis during development and differentiation. Proc Natl Acad Sci USA 1999; 96(16):9142-7.
53. Jennemann R, Sandhoff R, Langbein L et al. Integrity and barrier function of the epidermis critically depend on glucosylceramide synthesis. J Biol Chem 2007; 282(5):3083-94.
54. Jennemann R, Sandhoff R, Wang S et al. Cell-specific deletion of glucosylceramide synthase in brain leads to severe neural defects after birth. Proc Natl Acad Sci USA 2005; 102(35):12459-64.
55. Yamashita T, Allende ML, Kalkofen DN et al. Conditional LoxP-flanked glucosylceramide synthase allele controlling glycosphingolipid synthesis. Genesis 2005; 43(4):175-80.

56. Huitema K, van den Dikkenberg J, Brouwers JF et al. Identification of a family of animal sphingomyelin synthases. Embo J 2004; 23(1):33-44.
57. Tafesse FG, Ternes P, Holthuis JC. The multigenic sphingomyelin synthase family. J Biol Chem 2006; 281(40):29421-5.
58. Villani M, Subathra M, Im YB et al. Sphingomyelin synthases regulate production of diacylglycerol at the Golgi. Biochem J 2008; 414(1):31-41.
59. Graf C, Niwa S, Müller M et al. Wild-type levels of ceramide and ceramide-1-phosphate in the retina of ceramide kinase-like-deficient mice. Biochem Biophys Res Commun 2008; 373(1):159-63.
60. Wijesinghe DS, Massiello A, Subramanian P et al. Substrate specificity of human ceramide kinase. J Lipid Res 2005; 46(12):2706-16.
61. Lamour NF, Stahelin RV, Wijesinghe DS et al. Ceramide kinase uses ceramide provided by ceramide transport protein: localization to organelles of eicosanoid synthesis. J Lipid Res 2007; 48(6):1293-304.
62. Kumagai K, Yasuda S, Okemoto K et al. CERT mediates intermembrane transfer of various molecular species of ceramides. J Biol Chem 2005; 280(8):6488-95.
63. Boath A, Graf C, Lidome E et al. Regulation and traffic of ceramide 1-phosphate produced by ceramide kinase: comparative analysis to glucosylceramide and sphingomyelin. J Biol Chem 2008; 283(13):8517-26.
64. Sugiura M, Kono K, Liu H et al. Ceramide kinase, a novel lipid kinase. Molecular cloning and functional characterization. J Biol Chem 2002; 277(26):23294-300.
65. Mitsutake S, Yokose U, Kato M et al. The generation and behavioral analysis of ceramide kinase-null mice, indicating a function in cerebellar Purkinje cells. Biochem Biophys Res Commun 2007; 363(3):519-24.
66. Graf C, Zemann B, Rovina P et al. Neutropenia with impaired immune response to Streptococcus pneumoniae in ceramide kinase-deficient mice. J Immunol 2008; 180(5):3457-66.
67. Bornancin F, Mechtcheriakova D, Stora S et al. Characterization of a ceramide kinase-like protein. Biochim Biophys Acta 2005; 1687(1-3):31-43.
68. Inagaki Y, Mitsutake S, Igarashi Y. Identification of a nuclear localization signal in the retinitis pigmentosa-mutated RP26 protein, ceramide kinase-like protein. Biochem Biophys Res Commun 2006; 343(3):982-7.
69. Boudker O, Futerman AH. Detection and characterization of ceramide-1-phosphate phosphatase activity in rat liver plasma membrane. J Biol Chem 1993; 268(29):22150-5.
70. Shinghal R, Scheller RH, Bajjalieh SM. Ceramide 1-phosphate phosphatase activity in brain. J Neurochem 1993; 61(6):2279-85.
71. Brindley DN, Xu J, Jasinska R et al. Analysis of ceramide 1-phosphate and sphingosine-1-phosphate phosphatase activities. Methods Enzymol 2000; 311:233-44.
72. Futerman AH, van Meer G. The cell biology of lysosomal storage disorders. Nat Rev Mol Cell Biol 2004; 5(7):554-65.
73. Sabourdy F, Kedjouar B, Sorli SC et al. Functions of sphingolipid metabolism in mammals—lessons from genetic defects. Biochim Biophys Acta 2008; 1781(4):145-83.
74. Duan RD, Cheng Y, Hansen G et al. Purification, localization and expression of human intestinal alkaline sphingomyelinase. J Lipid Res 2003; 44(6):1241-50.
75. Schissel SL, Keesler GA, Schuchman EH et al. The cellular trafficking and zinc dependence of secretory and lysosomal sphingomyelinase, two products of the acid sphingomyelinase gene. J Biol Chem 1998; 273(29):18250-9.
76. Schissel SL, Schuchman EH, Williams KJ et al. Zn2+-stimulated sphingomyelinase is secreted by many cell types and is a product of the acid sphingomyelinase gene. J Biol Chem 1996; 271(31):18431-6.
77. Brady RO, Kanfer JN, Mock MB et al. The metabolism of sphingomyelin. II. Evidence of an enzymatic deficiency in Niemann-Pick diseae. Proc Natl Acad Sci USA 1966; 55(2):366-9.
78. Horinouchi K, Erlich S, Perl DP et al. Acid sphingomyelinase deficient mice: a model of types A and B Niemann-Pick disease. Nat Genet 1995; 10(3):288-93.
79. Marathe S, Miranda SR, Devlin C et al. Creation of a mouse model for nonneurological (type B) Niemann-Pick disease by stable, low level expression of lysosomal sphingomyelinase in the absence of secretory sphingomyelinase: relationship between brain intra-lysosomal enzyme activity and central nervous system function. Hum Mol Genet 2000; 9(13):1967-76.
80. Sawai H, Domae N, Nagan N et al. Function of the cloned putative neutral sphingomyelinase as lyso-platelet activating factor-phospholipase C. J Biol Chem 1999; 274(53):38131-9.
81. Zumbansen M, Stoffel W. Neutral sphingomyelinase 1 deficiency in the mouse causes no lipid storage disease. Mol Cell Biol 2002; 22(11):3633-8.
82. Tani M, Hannun YA. Neutral sphingomyelinase 2 is palmitoylated on multiple cysteine residues. Role of palmitoylation in subcellular localization. J Biol Chem 2007; 282(13):10047-56.

83. Marchesini N, Osta W, Bielawski J et al. Role for mammalian neutral sphingomyelinase 2 in confluence-induced growth arrest of MCF7 cells. J Biol Chem 2004; 279(24):25101-11.
84. Stoffel W, Jenke B, Blöck B et al. Neutral sphingomyelinase 2 (smpd3) in the control of postnatal growth and development. Proc Natl Acad Sci USA 2005; 102(12):4554-9.
85. Stoffel W, Jenke B, Holz B et al. Neutral sphingomyelinase (SMPD3) deficiency causes a novel form of chondrodysplasia and dwarfism that is rescued by Col2A1-driven smpd3 transgene expression. Am J Pathol 2007; 171(1):153-61.
86. Krut O, Wiegmann K, Kashkar H et al. Novel tumor necrosis factor-responsive mammalian neutral sphingomyelinase-3 is a C-tail-anchored protein. J Biol Chem 2006; 281(19):13784-93.
87. Corcoran CA, He Q, Ponnusamy S et al. Neutral sphingomyelinase-3 is a DNA damage and nongenotoxic stress-regulated gene that is deregulated in human malignancies. Mol Cancer Res 2008; 6(5):795-807.
88. Momoi T, Ben-Yoseph Y, Nadler HL. Substrate-specificities of acid and alkaline ceramidases in fibroblasts from patients with Farber disease and controls. Biochem J 1982; 205(2):419-25.
89. Li CM, Park JH, Simonaro CM et al. Insertional mutagenesis of the mouse acid ceramidase gene leads to early embryonic lethality in homozygotes and progressive lipid storage disease in heterozygotes. Genomics 2002; 79(2):218-24.
90. Tani M, Iida H, Ito M. O-glycosylation of mucin-like domain retains the neutral ceramidase on the plasma membranes as a type II integral membrane protein. J Biol Chem 2003; 278(12):10523-30.
91. Tani M, Igarashi Y, Ito M. Involvement of neutral ceramidase in ceramide metabolism at the plasma membrane and in extracellular milieu. J Biol Chem 2005; 280(44):36592-600.
92. Kono M, Dreier JL, Ellis JM et al. Neutral ceramidase encoded by the Asah2 gene is essential for the intestinal degradation of sphingolipids. J Biol Chem 2006; 281(11):7324-31.
93. Ohlsson L, Palmberg C, Duan RD et al. Purification and characterization of human intestinal neutral ceramidase. Biochimie 2007; 89(8):950-60.
94. Sun W, Xu R, Hu W et al. Upregulation of the human alkaline ceramidase 1 and acid ceramidase mediates calcium-induced differentiation of epidermal keratinocytes. J Invest Dermatol 2008; 128(2):389-97.
95. Xu R, Jin J, Hu W et al. Golgi alkaline ceramidase regulates cell proliferation and survival by controlling levels of sphingosine and S1P. Faseb J 2006; 20(11):1813-25.
96. Mao C, Xu R, Szulc ZM et al. Cloning and characterization of a novel human alkaline ceramidase. A mammalian enzyme that hydrolyzes phytoceramide. J Biol Chem 2001; 276(28):26577-88.
97. Mao C, Obeid LM. Ceramidases: regulators of cellular responses mediated by ceramide, sphingosine and sphingosine-1-phosphate. Biochim Biophys Acta 2008; 1781(9):424-34.
98. Mao C, Xu R, Szulc ZM et al. Cloning and characterization of a mouse endoplasmic reticulum alkaline ceramidase: an enzyme that preferentially regulates metabolism of very long chain ceramides. J Biol Chem 2003; 278(33):31184-91.
99. Wattenberg BW, Pitson SM, Raben DM. The sphingosine and diacylglycerol kinase superfamily of signaling kinases: localization as a key to signaling function. J Lipid Res 2006; 47(6):1128-39.
100. Johnson KR, Becker KP, Facchinetti MM et al. PKC-dependent activation of sphingosine kinase 1 and translocation to the plasma membrane. Extracellular release of sphingosine-1-phosphate induced by phorbol 12-myristate 13-acetate (PMA). J Biol Chem 2002; 277(38):35257-62.
101. Pitson SM, Moretti PA, Zebol JR et al. Activation of sphingosine kinase 1 by ERK1/2-mediated phosphorylation. Embo J 2003; 22(20):5491-500.
102. Buehrer BM, Bardes ES, Bell RM. Protein kinase C-dependent regulation of human erythroleukemia (HEL) cell sphingosine kinase activity. Biochim Biophys Acta 1996; 1303(3):233-42.
103. Yamanaka M, Shegogue D, Pei H et al. Sphingosine kinase 1 (SPHK1) is induced by transforming growth factor-beta and mediates TIMP-1 up-regulation. J Biol Chem 2004; 279(52):53994-4001.
104. Nakade Y, Banno Y, T-Koizumi K et al. Regulation of sphingosine kinase 1 gene expression by protein kinase C in a human leukemia cell line, MEG-O1. Biochim Biophys Acta 2003; 1635(2-3):104-16.
105. Olivera A, Kohama T, Edsall L et al. Sphingosine kinase expression increases intracellular sphingosine-1-phosphate and promotes cell growth and survival. J Cell Biol 1999; 147(3):545-58.
106. Ancellin N, Colmont C, Su J et al. Extracellular export of sphingosine kinase-1 enzyme. Sphingosine-1-phosphate generation and the induction of angiogenic vascular maturation. J Biol Chem 2002; 277(8):6667-75.
107. Inagaki Y, Li PY, Wada A et al. Identification of functional nuclear export sequences in human sphingosine kinase 1. Biochem Biophys Res Commun 2003; 311(1):168-73.
108. Thompson CR, Iyer SS, Melrose N et al. Sphingosine kinase 1 (SK1) is recruited to nascent phagosomes in human macrophages: inhibition of SK1 translocation by Mycobacterium tuberculosis. J Immunol 2005; 174(6):3551-61.
109. Stahelin RV, Hwang JH, Kim JH et al. The mechanism of membrane targeting of human sphingosine kinase 1. J Biol Chem 2005.

110. Pitson SM, Xia P, Leclercq TM et al. Phosphorylation-dependent translocation of sphingosine kinase to the plasma membrane drives its oncogenic signalling. J Exp Med 2005; 201(1):49-54.
111. Olivera A, Spiegel S. Sphingosine-1-phosphate as second messenger in cell proliferation induced by PDGF and FCS mitogens. Nature 1993; 365(6446):557-60.
112. Edsall LC, Pirianov GG, Spiegel S. Involvement of sphingosine-1-phosphate in nerve growth factor-mediated neuronal survival and differentiation. J Neurosci 1997; 17(18):6952-60.
113. Shu X, Wu W, Mosteller RD et al. Sphingosine kinase mediates vascular endothelial growth factor-induced activation of ras and mitogen-activated protein kinases. Mol Cell Biol 2002; 22(22):7758-68.
114. Sarkar S, Maceyka M, Hait NC et al. Sphingosine kinase 1 is required for migration, proliferation and survival of MCF-7 human breast cancer cells. FEBS Lett 2005; 579(24):5313-7.
115. El-Shewy HM, Johnson KR, Lee MH et al. Insulin-like growth factors mediate heterotrimeric G protein-dependent ERK1/2 activation by transactivating sphingosine-1-phosphate receptors. J Biol Chem 2006.
116. Ma MM, Chen JL, Wang GG et al. Sphingosine kinase 1 participates in insulin signalling and regulates glucose metabolism and homeostasis in KK/Ay diabetic mice. Diabetologia 2007.
117. Spiegel S, Milstien S. Sphingosine-1-phosphate: an enigmatic signalling lipid. Nat Rev Mol Cell Biol 2003; 4(5):397-407.
118. Taha TA, Hannun YA, Obeid LM. Sphingosine kinase: biochemical and cellular regulation and role in disease. J Biochem Mol Biol 2006; 39(2):113-31.
119. Igarashi N, Okada T, Hayashi S et al. Sphingosine kinase 2 is a nuclear protein and inhibits DNA synthesis. J Biol Chem 2003; 278(47):46832-9.
120. Tani M, Ito M, Igarashi Y. Ceramide/sphingosine/sphingosine-1-phosphate metabolism on the cell surface and in the extracellular space. Cell Signal 2007; 19(2):229-37.
121. Venkataraman K, Thangada S, Michaud J et al. Extracellular export of sphingosine kinase-1a contributes to the vascular S1P gradient. Biochem J 2006; 397(3):461-71.
122. Cuvillier O. Sphingosine kinase-1—a potential therapeutic target in cancer. Anticancer Drugs 2007; 18(2):105-10.
123. Bonhoure E, Lauret A, Barnes DJ et al. Sphingosine kinase-1 is a downstream regulator of imatinib-induced apoptosis in chronic myeloid leukemia cells. Leukemia 2008.
124. Taha TA, Osta W, Kozhaya L et al. Down-regulation of sphingosine kinase-1 by DNA damage: dependence on proteases and p53. J Biol Chem 2004; 279(19):20546-54.
125. Taha TA, Kitatani K, Bielawski J et al. Tumor necrosis factor induces the loss of sphingosine kinase-1 by a cathepsin B-dependent mechanism. J Biol Chem 2005; 280(17):17196-202.
126. Taha TA, El-Alwani M, Hannun YA et al. Sphingosine kinase-1 is cleaved by cathepsin B in vitro: identification of the initial cleavage sites for the protease. FEBS Lett 2006; 580(26):6047-54.
127. Maceyka M, Sankala H, Hait NC et al. SphK1 and SphK2: Sphingosine kinase isoenzymes with opposing functions in sphingolipid metabolism. J Biol Chem 2005.
128. Mitra P, Oskeritzian CA, Payne SG et al. Role of ABCC1 in export of sphingosine-1-phosphate from mast cells. Proc Natl Acad Sci USA 2006; 103(44):16394-9.
129. Le Stunff H, Giussani P, Maceyka M et al. Recycling of sphingosine is regulated by the concerted actions of sphingosine-1-phosphate phosphohydrolase 1 and sphingosine kinase 2. J Biol Chem 2007; 282(47):34372-80.
130. Allende ML, Sasaki T, Kawai H et al. Mice deficient in sphingosine kinase 1 are rendered lymphopenic by FTY720. J Biol Chem 2004; 279(50):52487-92.
131. Kohno M, Momoi M, Oo ML et al. Intracellular role for sphingosine kinase 1 in intestinal adenoma cell proliferation. Mol Cell Biol 2006; 26(19):7211-23.
132. Kawamori T, Kaneshiro T, Okumura M et al. Role for sphingosine kinase 1 in colon carcinogenesis. Faseb J 2008.
133. Snider AJ, Kawamori T, Bradshaw SG et al. A role for sphingosine kinase 1 in dextran sulfate sodium-induced colitis. Faseb J 2008.
134. Mizugishi K, Yamashita T, Olivera A et al. Essential role for sphingosine kinases in neural and vascular development. Mol Cell Biol 2005; 25(24):11113-21.
135. Liu Y, Wada R, Yamashita T et al. Edg-1, the G protein-coupled receptor for sphingosine-1-phosphate, is essential for vascular maturation. J Clin Invest 2000; 106(8):951-61.
136. Mizugishi K, Li C, Olivera A et al. Maternal disturbance in activated sphingolipid metabolism causes pregnancy loss in mice. J Clin Invest 2007; 117(10):2993-3006.
137. Pyne S, Pyne NJ. Sphingosine-1-phosphate signalling and termination at lipid phosphate receptors. Biochim Biophys Acta 2002; 1582(1-3):121-31.
138. Pyne S, Long JS, Ktistakis NT et al. Lipid phosphate phosphatases and lipid phosphate signalling. Biochem Soc Trans 2005; 33(Pt 6):1370-4.

139. Alderton F, Darroch P, Sambi B et al. G-protein-coupled receptor stimulation of the p42/p44 mitogen-activated protein kinase pathway is attenuated by lipid phosphate phosphatases 1, 1a and 2 in human embryonic kidney 293 cells. J Biol Chem 2001; 276(16):13452-60.
140. Long J, Darroch P, Wan KF et al. Regulation of cell survival by lipid phosphate phosphatases involves the modulation of intracellular phosphatidic acid and sphingosine-1-phosphate pools. Biochem J 2005; 391(Pt 1):25-32.
141. Delon C, Manifava M, Wood E et al. Sphingosine kinase 1 is an intracellular effector of phosphatidic acid. J Biol Chem 2004; 279(43):44763-74.
142. Boujaoude LC, Bradshaw-Wilder C, Mao C et al. Cystic fibrosis transmembrane regulator regulates uptake of sphingoid base phosphates and lysophosphatidic acid: modulation of cellular activity of sphingosine-1-phosphate. J Biol Chem 2001; 276(38):35258-64.
143. Pyne S, Kong KC, Darroch PI. Lysophosphatidic acid and sphingosine-1-phosphate biology: the role of lipid phosphate phosphatases. Semin Cell Dev Biol 2004; 15(5):491-501.
144. Mandala SM, Thornton R, Tu Z et al. Sphingoid base 1-phosphate phosphatase: a key regulator of sphingolipid metabolism and stress response. Proc Natl Acad Sci USA 1998; 95(1):150-5.
145. Mandala SM, Thornton R, Galve-Roperh I et al. Molecular cloning and characterization of a lipid phosphohydrolase that degrades sphingosine-1-phosphate and induces cell death. Proc Natl Acad Sci USA 2000; 97(14):7859-64.
146. Ogawa C, Kihara A, Gokoh M et al. Identification and characterization of a novel human sphingosine-1-phosphate phosphohydrolase, hSPP2. J Biol Chem 2003; 278(2):1268-72.
147. Le Stunff H, Galve-Roperh I, Peterson C et al. Sphingosine-1-phosphate phosphohydrolase in regulation of sphingolipid metabolism and apoptosis. J Cell Biol 2002; 158(6):1039-49.
148. Kihara A, Sano T, Iwaki S et al. Transmembrane topology of sphingoid long-chain base-1-phosphate phosphatase, Lcb3p. Genes Cells 2003; 8(6):525-35.
149. Johnson KR, Johnson KY, Becker KP et al. Role of human sphingosine-1-phosphate phosphatase 1 in the regulation of intra- and extracellular sphingosine-1-phosphate levels and cell viability. J Biol Chem 2003; 278(36):34541-7.
150. Le Stunff H, Peterson C, Liu H et al. Sphingosine-1-phosphate and lipid phosphohydrolases. Biochim Biophys Acta 2002; 1582(1-3):8-17.
151. Mechtcheriakova D, Wlachos A, Sobanov J et al. Sphingosine-1-phosphate phosphatase 2 is induced during inflammatory responses. Cell Signal 2006.
152. Ikeda M, Kihara A, Igarashi Y. Sphingosine-1-phosphate lyase SPL is an endoplasmic reticulum-resident, integral membrane protein with the pyridoxal 5′-phosphate binding domain exposed to the cytosol. Biochem Biophys Res Commun 2004; 325(1):338-43.
153. Stoffel W, LeKim D, Sticht G. Distribution and properties of dihydrosphingosine-1-phosphate aldolase (sphinganine-1-phosphate alkanal-lyase). Hoppe Seylers Z Physiol Chem 1969; 350(10):1233-41.
154. Schwab SR, Pereira JP, Matloubian M et al. Lymphocyte sequestration through S1P lyase inhibition and disruption of S1P gradients. Science 2005; 309(5741):1735-9.
155. Fyrst H, Saba JD. Sphingosine-1-phosphate lyase in development and disease: sphingolipid metabolism takes flight. Biochim Biophys Acta 2008; 1781(9):448-58.
156. Mendel J, Heinecke K, Fyrst H et al. Sphingosine phosphate lyase expression is essential for normal development in Caenorhabditis elegans. J Biol Chem 2003; 278(25):22341-9.
157. Li G, Foote C, Alexander S et al. Sphingosine-1-phosphate lyase has a central role in the development of Dictyostelium discoideum. Development 2001; 128(18):3473-83.
158. Schmahl J, Raymond CS, Soriano P. PDGF signaling specificity is mediated through multiple immediate early genes. Nat Genet 2007; 39(1):52-60.

Chapter 2

Sphingolipid Transport

Laura Riboni,* Paola Giussani and Paola Viani

Abstract

Sphingolipids are a family of ubiquitous membrane components that exhibit multiple functional properties fundamental to cell properties. Sphingolipid transport represents a crucial aspect in the metabolism, signaling and biological role of sphingolipids. Different mechanisms of sphingolipid movements contribute to their selective localization in different membranes but also in different portions and sides of the same membrane, thus ensuring and regulating their interaction with different enzymes and target molecules.

In this chapter we will describe the knowledge of the different mechanisms of sphingolipid movements within and between membranes, focusing on the recent advances in this field and considering the role played by selective sphingolipid molecules in the regulation of these mechanisms.

Introduction

Sphingolipids are a family of amphipathic lipids that exhibit extraordinary structural and multiple functional properties fundamental to cell characteristics. They are ubiquitous membrane components, being present in different organelle membranes and particularly abundant in the plasma membrane (PM). In biological membranes, sphingomyelin (SM) and glycosphingolipids (GSLs) represent the major sphingolipids, these display an asymmetric or polarized distribution and play important roles in the regulation of membrane fluidity and sub-domain structure. Their physical-chemical properties enable them to fulfil and regulate a large spectrum of relevant biological functions such as molecular sorting, cell-cell interaction and intracellular transport. Moreover, different intermediates of sphingolipid metabolism, as sphingosine (Sph), sphingosine-1-phosphate (S1P), ceramide (Cer) and ceramide-1-phosphate (Cer1P), can act as bioactive molecules involved, as intra- or extracellular messengers, in the regulation of crucial processes as cell growth, death, adhesion, migration and senescence. A crucial and fundamental aspect in the spectrum of the biological functions of sphingolipids is represented by their intra- and extracellular transport. Indeed, the complex compartmentation of sphingolipid metabolism, the specific sphingolipid composition of different cells and organelles and the proper localization of selected sphingoid molecules, functional to the interaction with multiple targets in different locations, are strictly dependent from and regulated by different modalities of sphingolipid transport.

Sphingolipid transport has emerged as a sophisticated and complex tool crucial in moving, transferring and finally locating the different sphingolipids in a specific subcellular site or in the extracellular milieu. This transport involves short-distance movements within a membrane (intramembrane transport), including lateral and transmembrane motions and trafficking between different membranes (intermembrane transport), including that mediated by proteins and vesicle flow (Fig. 1).

*Corresponding Author: Laura Riboni—Department of Medical Chemistry, Biochemistry and Biotechnology, L.I.T.A.-Segrate, via Fratelli Cervi 93-20090 Segrate, Milan, Italy. Email: laura.riboni@unimi.it

Sphingolipids as Signaling and Regulatory Molecules, edited by Charles Chalfant and Maurizio Del Poeta.

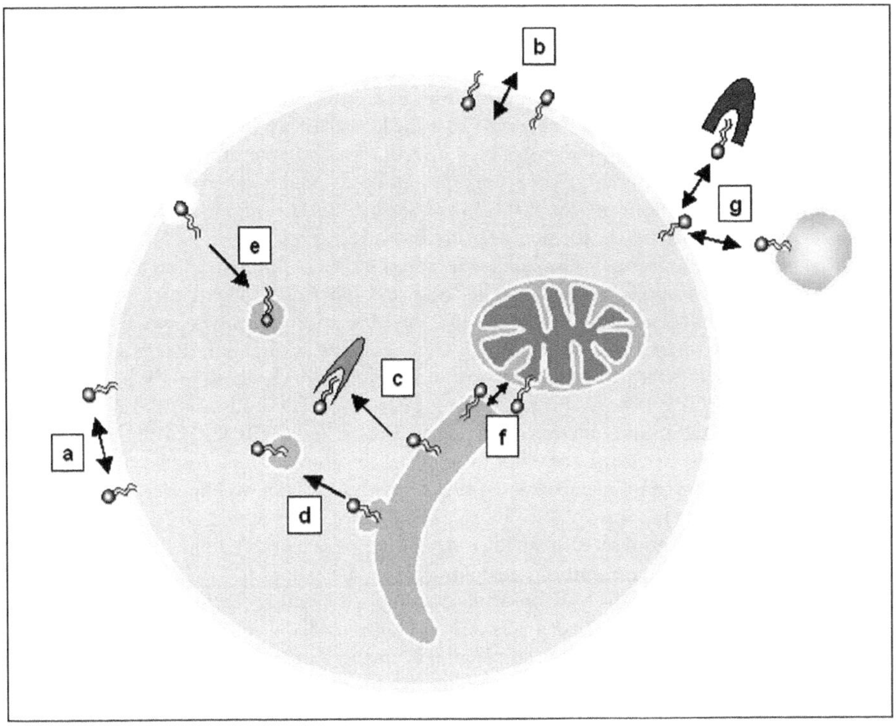

Figure 1. Different types of sphingolipid transport. Sphingolipids can display movements within a membrane, including lateral diffusion (a) and transbilayer motion (b). Sphingolipid transport between different cellular membranes can be protein-mediated (c) and vesicle-mediated, including vesicles of the biosynthetic route (d) and endocytotic vesicles (e). Sphingolipids could be exchanged as monomers between the cytosolic layer of organelle membranes (f). Plasma membrane sphingolipid may be exported through the interaction with specific extracellular protein/lipoproteins (g). For simplicity, the possible involvement of protein transporters in (b) and (g) is omitted. Most of these movements may occur at or between different cellular membranes. A color version of this figure is available at www.landesbioscience.com/curie.

Intramembrane Sphingolipid Movements

The behaviour of sphingolipids in a membrane represents a crucial aspect of their transport as well as biological properties. In this section we will first analyze the dynamics and organization of sphingolipids within a membrane layer and then the mechanisms underlying the translocation of sphingolipid molecules between the two leaflets of a biological membrane.

Lateral Diffusion and Lateral Phase Separation of Sphingolipids

Sphingolipids, as the other lipid components of cell membranes, are characterized by a high diffusion rate in a leaflet of the bilayer. The lateral diffusion of different sphingolipids in model and biological membranes has been evaluated by different techniques, which include spectroscopic methods such as fluorescence energy transfer, fluorescence quenching, including the monomer to excimer transition, diffusion measurements based on fluorescence recovery after photobleaching (FRAP), fluorescence correlation spectroscopy, single particle tracking (SPT) and NMR spectroscopy. The reported lateral diffusion coefficient of different sphingolipids ranges from 10^{-7} to 10^{-10} cm^2/sec, this variability is most likely due to the different methods employed. Studies performed with pyrene-labeled sphingolipids report that, in a fluid bilayer (i.e., in a liquid disordered phase), the diffusion coefficient of SM, glucosylceramide (GlcCer) and different gangliosides is similar (about $1.6 \times 10^{-7} cm^2/sec$)[1,2] and comparable to that of phosphatidylcholine (PC) obtained with the same spectroscopic approach.[3] Using gangliosides differently labelled at the polar head, the diffusion values determined with FRAP is in the range of 10^{-9} cm^2/sec.[4,5] From these data, it emerges that the lateral diffusion of a lipid in the bilayer is primarily determined by the hydrophobic membrane-bound portion of the molecule and does not significantly differ between glycero- and sphingolipids. Due to their high diffusion rate, sphingolipids should explore an area of 0.1-1 $\mu m^2/sec$ in a cell membrane and should display a homogeneous distribution in the membrane plane. However, a great amount of evidence has been presented indicating that lateral sphingolipid segregation in a membrane plane may occur. The existence in living cells of sphingolipid-enriched domains, functional to lipid sorting to the apical membrane was initially postulated by Simons and Van Meer.[6] Subsequently it was found that the GPI-anchored protein can be sorted through the same pathway.[7] The isolation from biological membranes at low temperature of detergent-insoluble sphingolipid-cholesterol (Chol) enriched domains containing GPI anchored proteins[8] was the basis for a large body of work addressing the composition, organization and function of these structures in living cells. Lateral phase separation of sphingolipids and the tendency of Chol to associate to sphingolipid-enriched domains have been demonstrated in model membranes by different studies (reviewed in refs. 9-11). Studies performed with ternary mixtures of PC, SM/GSL and Chol indicate that two distinct physical states can coexist in the fluid state in a wide range of temperatures and lipid composition. The physical state of sphingolipid-Chol enriched domains has been defined as a liquid ordered state (Lo) and differs from that of the bulk lipid bilayer described by the more fluid liquid disordered state ($L\alpha$).[12-14] In the Lo state the acyl chains exist in a highly ordered conformation which results in a reduction of the area per lipid, a tight lipid packing and a greater thickness of the bilayer. The intermolecular forces involved in the phase segregation of sphingolipids and Chol include lipid-lipid hydrogen bonds, in which sphingolipids can act both as acceptors and donors, weak dipolar interactions between sphingolipid polar heads and van der Waals interactions. The preferential partitioning of Chol in these domains can also involve the model of hydrophobic shielding or the 'umbrella effect' in which the steric hindrance of strongly hydrated sphingolipid polar heads may favour the Chol insertion in the bilayer. All these considerations attain to a structural and static description of sphingolipid-Chol enriched lipid domains, but how is the lateral diffusion of sphingolipids in such structures? In phospholipid/Chol systems, pulsed field gradient NMR technique[15] allowed to estimate the diffusion rate of a lipid in the Lo phase is about 3-5 fold slower than in a liquid disordered one, but not different enough to impede lipid exchange between the two phases.[1,15] When both SM and Chol were present in the membrane model, the diffusion coefficients were compatible with that determined in more simple systems, but in contrast to the latter, the lipid exchange between the Lo and the $L\alpha$ phases was slower.[17] These findings are in agreement with the results obtained in single particle tracking, which demonstrated a transient confinement of the 4-fold slower GM1 in a limited membrane portion.[18] Thus the physical state (i.e., Lo vs disordered state) is not the only determinant for the lipid dynamic in the domain. The hydrophobic mismatch between the thicker sphingolipid-enriched domain and the thinner surrounding bilayer, which contributes to the energy required to lipid domain formation and size,[19] can partially contribute to explain the confined diffusion of lipids in coexisting Lo and disordered phases.

Sphingolipid Raft Dynamics

The raft concept, originally hypothesized for the specific sorting of sphingolipids and proteins in the PM,[6,7] has been proposed as a key element in almost all the forms of vesicular traffic.[20,21] The majority of information about the composition and function of lipid rafts have been obtained on the basis of the operational definition of rafts as detergent-resistant membrane fraction at low temperature, which implies that stable lipid rafts do exist in biological membranes. Nevertheless, the existence, dimension, lifetime and ultimately the physiological function of sphingolipid-Chol enriched domains in living cells are a matter of great debate among the researchers in the field.

This controversy mainly comes from the difficulty to visualize rafts in living cells. Apart from caveolae, a subset of lipid rafts which can be easily detected by electron microscopy as about 100 nm-sized structures, the direct observation of rafts by optical microscopy has turned out to be impossible, thus suggesting that native rafts are smaller than the optical diffraction resolution limit of 300 nm. On the other hand, the sizing of these structure by indirect methods postulates raft dimensions ranging from 10 to 700 nm,[9,22] thus suggesting the possibility that the observed larger structures represent macrodomains derived from the aggregation/coalescence of nanometer scaled rafts. This implies that native membrane rafts are highly dynamic, heterogeneous and short lived structures. In the 2006 Keystone symposium on lipid rafts and cell functions, the biophysicists, biochemists and cell biologists converged on the following definition "Membrane rafts are small (10-200 nm), heterogeneous, highly dynamic, sterol- and sphingolipid-enriched domains that compartmentalize cellular processes. Small rafts can sometimes be stabilized to form larger platforms through protein-protein and protein-lipid interactions".[23] Different aspects can contribute to the metastability of small nanoscaled rafts in the membranes, as (i) the compartmentalization into small domains through actin-based membrane skeleton fence, with integral proteins as barrier palisade; and (ii) the lipid and protein heterogeneity of biological membranes, which limits the hydrophobic mismatch between thicker sphingolipid-enriched domains and the more thinner surrounding bilayer (reviewed in ref. 24).

The dynamic evolution of small membrane rafts into larger structures can be induced by ligand binding as, for example, the cholera toxin binding to GM1[25] and the EGF binding to its receptor, which induces the merging of two types of GM1-containing small domains.[26] In addition, Cer strongly associates with lipid rafts[27,28] and can induce the coalescence of large domains in PM, thus facilitating the raft-residing CD95/Fas death receptor clustering and the subsequent induction of apoptosis.[29] The Cer role in membrane lateral organization and dynamics has been extensively reviewed.[30-32] The formation of membrane domains highly enriched in Cer can be relevant to different physiological processes. First, it can contribute to the definition of signaling platforms generated by the activity of a signaling-related sphingomyelinase (SMase). Second, Cer can displace Chol from membrane rafts, as both compete in their association with lipid rafts due to the limited capacity of sphingolipids to shield the small polar groups of Cer and Chol from contact with water.[30,32] This can be at the basis of the Chol displacement from PMs upon Cer generation, which may affect membrane Chol homeostasis and may have implications for the association and activity of Chol-bound proteins. Finally, the formation of Cer-enriched microdomains can promote membrane vesiculation and vesicle budding. This has been observed in PC/SM giant unilamellar vesicles, in which Cer-enriched microdomains were rapidly formed after treatment of vesicles with SMase and can also occur in erythrocytes after loading with high concentrations of C18-Cer.[33] The Cer-mediated vesicle budding has been implicated in the sorting of subcellular membranes into different populations of intracellular vesicles[34] and in the ER-Golgi vesicular traffic of both sphingolipids and proteins.[35]

Transbilayer Transport

The transfer of sphingolipid molecules from one to the other leaflet of a biological membrane represents a further, intriguing aspect of sphingolipid movement. This movement may be of relevance to different processes, such as the formation and maintenance of a proper sphingolipid asymmetry, thus being important in the PM and organelle structure/functional properties and in the mechanisms of intracellular sphingolipid trafficking. It may also enable the sphingolipid molecule to interact with a protein/enzyme that resides on the opposite side of the membrane that must be crossed, thus being relevant in sphingolipid metabolism and signaling. In addition, it may be crucial to the interaction of a sphingolipid with an extracellular protein/lipoprotein and thus to its export into the extracellular milieu.

As for other amphipatic lipids, the transbilayer movement of sphingolipids can be from the outer to the inner monolayer of a membrane ("flip"), in the opposite direction ("flop"), or bi-directional (scrambling or "flip-flop"). All these motions can occur spontaneously or facilitated by the transport/interaction with a protein.

The spontaneous behaviour of a (sphingo)lipid molecule within a membrane is influenced by different factors, including the balance between its hydrophobic tail and polar headgroup, the composition of both the sphingolipid alkyl chain and the membrane. It is generally accepted that sphingolipids with a reduced or abundant polar headgroup will display fast or slow spontaneous transbilayer movements, respectively. Since most sphingolipids possess a bulky headgroup, they exhibit the tendency to reside in a membrane monolayer for long periods of time. In agreement, the studies on the spontaneous transbilayer diffusion of spin-labelled SMs and gangliosides in model membranes have shown that this exchange does not spontaneously occur or, if it does it, is very slow.[36-38] On the opposite, simple sphingolipids such as GlcCer and Cer can move between leaflets.[33,39,40]

Cellular membranes, particularly eukaryotic PMs, are equipped with specialised proteins that can either actively translocate lipids from one leaflet to the other or facilitate a passive equilibration of lipids between the two membrane halves. Protein transporters are referred as flippases (mediating transport from outer, noncytosolic, to inner, cytosolic, layer), floppases (from inner to outer layer) and scramblases (facilitating flip-flop, i.e., nonspecific redistribution across bilayers). Increasing evidence supports that different subfamilies of ATP binding cassette (ABC) transporter family contribute to, or are involved in, sphingolipid passage across a membrane (Table 1).[41] ABC proteins have been generally recognized as drug efflux pumps that protect the body from various toxic substances. However, they exhibit scarce selectivity and are able to move various molecules against their concentration gradients.[42,43] After its binding to the Nucleotide Binding Domain (NBD), ATP provides the energy required to actively transport substrates across the membrane. In eukaryotes, most ABC transporters contain two hydrophobic domains, each with 6-11 membrane-spanning α-helices and two NBDs exposed to the cytoplasm. Some ABCs contain only one transmembrane domain and one NBD, but can combine into homo- or hetero-dimers to form functional transporters. In mammals, ABC transporters are expressed in PM and different intracellular membranes, including those of endoplasmic reticulum (ER), Golgi, mitochondria, endosomes, lysosomes and exocytotic vesicles and almost all the eukaryotic ABC proteins studied to date are floppases, transporting substrates outwardly from cytosol into organelles or out of cells. On the bases of their sequence homology and NBD organization, human ABC transporter have been classified into seven subfamilies, designated A-G.

Notably, since the expression of ABC transporters may be restricted in a cell/tissue-specific fashion, different ABC expression may contribute to differences in sphingolipid transport and thus to cell/tissue sphingolipid pattern and functional specialization.

Transbilayer Transfer of Sphingomyelin and Complex Glycosphingolipids

The distribution of complex sphingolipids (SM and GSLs) across PMs is highly asymmetric, being present essentially in the outer monolayer. This asymmetry appears to be dictated by the exocytotic traffic of newly synthesized sphingolipids and maintained with low levels of flip-flop across the bilayer. In fact, SM and GSLs are synthesized from Cer and GlcCer on the luminal side of the Golgi, where they cannot translocate to the cytosolic face. Upon exocytosis, they reach the external leaflet of the PM, where little or no transport to the cytosolic face occurs under basal conditions. However, although in most cells the great majority (>80%) of SM is normally confined to the outer leaflet of the PM and to the luminal face (corresponding to the outer one) of organelle membranes, in some membranes the percentage of SM present in the outer leaflet is markedly reduced. For example, SM in the outer leaflet accounts for only 65% of intestinal brush border membranes and hepatocytes, possibly reflecting specific functional properties of these membranes.[44] In liver ER and brain microsomes, SM facing the luminal face accounts for 60%, being thus markedly lower than in other organelles.[44]

As biogenic (self-synthesizing) membrane, the ER membrane requires phospholipid scrambling for bilayer assembly and maintenance. Different studies reported that phospholipid scrambling in ER occurs very rapidly, bi-directionally, independently of the phospholipid head group and requires specific membrane proteins. Indeed, the transmembrane movement of SM in ER is significantly faster than in other membranes, although much slower than glycerophospholipids.[44]

Table 1. ABC transporters involved in sphingolipid transport

Transporter	Proposed Role	Membranes/Cells	ABC Localization	Notes	Deficiency-Associated Disease
ABCA1	SM floppase	Model membranes	PM	Transports mainly cholesterol and PC	HDL deficiency
ABCA1	S1P floppase	Astrocytes Blood brain barrier	PM		
ABC-like	S1P floppase	Platelets	PM	Inhibited by glyburide	
ABCA2	Transport of SM and gangliosides ??	Oligodendrocytes, some neurons	Lysosome, endosome, Golgi, PM	SL traffic during SL biosynthesis or degradation	Expression correlated with Alzheimer's disease
ABCA7	Cer transport ??	Keratinocytes	Lamellar granules	Facilitates Cer production at the ER and traffic from/to lamellar granules	
ABCA12	GlcCer floppase	Keratinocyte Lung	Golgi, lamellar granules	Mediates transport into lamellar granules	Harlequin ichthyosis
ABCB1 (P-gp, MDR1)	SM, GSL floppase	Blood brain barrier	PM		Defective multidrug resistance in cancer
ABCC1 (MRP1)	SM, GlcCer floppase	Pig kidney-derived polarized cells	PM		Defective multidrug resistance in cancer
ABCC1 (MRP1)	S1P floppase	Mast cells, HUVEC	PM		Defective multidrug resistance in cancer
ABCC7 (CFTR)	S1P flippase	Resistance arteries; epithelial cells	PM	Cl^- channel	Cystic fibrosis
ABCG1	SM floppase	Different cells	PM	Cholesterol and SM floppase	Pulmunary lipidosis
ABCG2 (BCRP)	Cer floppase	Placental trophoblasts	PM		Fetal growth restriction

Presumed roles of different members of the ABC transporter family in the transbilayer transfer of different sphingolipids. See text for details.

Of relevance, the transbilayer movement of complex sphingolipids (mainly shown for SM) can occur in and appears functional to, different processes. At the PM, the transbilayer asymmetry of SM can be compromised, along with that of glycerophospholipids, in several physiological and pathological conditions such as neutrophil activation, blood coagulation, apoptosis and suicidal death of erythrocytes (eryptosis).[45,46] Indeed, the asymmetric distribution of all the major phospholipid classes, including SM, is rapidly lost upon a persistent increase of cytoplasmic Ca^{2+}, caspase activation, or oxidative stress.[47,48] Increased cytosolic Ca^{2+} and enhanced Cer levels lead to membrane scrambling too and subsequent phosphatidylserine exposure.[49]

A phospholipid-scramblase, isolated from erythrocytes and cloned, has been implicated in the fast bi-directional phospholipids transport between the two monolayers.[50] This transporter, named phospholipid scramblase 1 (PLSCR1), is a 37-kDa membrane protein with a short C-terminal external sequence, a single transmembrane segment and a long cytoplasmic extension containing a calcium-binding segment and is expressed in multiple human tissues and cancer cell lines. Ca^{2+}-induced lipid scrambling is defective in cells from patients with Scott syndrome, a bleeding disorder originating from a reduced ability of activated platelets to expose procoagulant phospholipids on their surface.[51] Intriguingly, in lymphoid cells from a patient with Scott syndrome, scramblase is induced normally during apoptosis, but cannot be activated by Ca^{2+}, suggesting that apoptosis and Ca^{2+} operate through different pathways to activate the same scramblase.[52] More recent reports suggest that different proteins are responsible for scramblase activity and phospholipid scramblases emerged as a group of homologous proteins conserved in all eukaryotic organisms and exhibiting different cell expression and subcellular localization.[53] Although a certain amount of evidence supports the role of scramblase(s) in destroying PM phospholipid asymmetry at critical cellular events like activation, injury and apoptosis, the function of PLSCR1 as a phospholipids translocator has been recently challenged.[53] At present, the role of different scramblases in sphingolipid transport remains to be clarified.

A further mechanism, consisting in the SMase-induced asymmetric increase in Cer concentration has been proposed as a crucial one in to facilitate the scrambling of other lipids. The observation that a Cer asymmetric generation can promote lipid scrambling (see above) has led to postulate that endogenous SMase could be a lipid scramblase.[54]

In lipid vesicles, ganglioside GM3 can undergo spontaneous flip-flop as a consequence of SMase-induced Cer on the outer leaflet.[55] So far, no evidence has been provided on gangliosides scrambling during cell activation and apoptosis. Interestingly, PS exposure on the outer membrane leaflet during cell activation and apoptosis colocalizes with GM1, possibly at the level of lipid rafts.[49,56]

A further mechanism of complex sphingolipid transfer across the membrane bilayers involves ABC family proteins. Different members of the ABC family have been implicated in facilitating the transport of SM and gangliosides at the PM (Table 1). This transport might be functional not only in maintaining and restoring sphingolipid asymmetry, as example after cell activation, but also in the export of complex sphingolipids to acceptor proteins/lipoproteins.

ABCA1 and ABCG1, two ABC proteins required for lipoprotein generation, appear to be involved in SM outward transport. ABCA1 is able to transfer phospholipids and Chol to lipid-free apoA-I and lipidated apoE. In liposomes made of SM or PC, purified ABCA1 shows high ATPase activity, suggesting that it can recognize phospholipids with choline head groups.[57] However, when cellular SM content is reduced, apoA-I-dependent Chol efflux by ABCA1 increases, suggesting that ABCA1 preferentially transports PC and Chol.[57] In contrast, ABCG1, a floppase important in the efflux of excess Chol from peripheral tissues to preβ-HDL and HDL, appears to preferentially recognize Chol and SM as substrates to transport across the PM and secrete.[58] ABCG1 stimulates Chol and SM efflux to HDL2 or HDL-3.[58] Interestingly, Chol efflux by ABCG1 is strictly dependent on the cellular SM level and this level correlates with SM efflux.[59]

Recent data also implicate a role of ABCA2, a transporter involved in lipid movement to generate the myelin sheath, as a possible transporter of complex sphingolipids. In the *ABCA2* null mice, Sakai et al[60] reported several major alterations in the lipid composition of both whole brain and purified myelin fractions. Analysis of brain and myelin lipids revealed selective deficiencies of SM, whereas brain gangliosides and myelin GM1 were significantly increased.[60] Since ABCA2 is localized to late endosome/lysosomes and Golgi, it might be involved in the metabolism of neural SM and/or gangliosides.[60]

Some studies provided evidence that ABCB1 (P-gp, MDR1) and ABCC1 (MRP1) can transport short-chain fluorescent SM (and GlcCer, PC) analogs,[61,62] implicating these proteins in sphingolipid distribution between the leaflets of the PM. However, the role of these ABC transporters in the movement of natural sphingolipid molecules among biomembranes remains to be elucidated.

Transbilayer Transfer of Monohexosylsphingolipids

In contrast to more complex sphingolipids, the monohexosylsphingolipids GlcCer and galactosylceramide exhibit a spontaneous tendency to move between membrane bilayers. In model membranes, spin-labeled GlcCer and galactosylceramide analogues can spontaneously cross the bilayer by passive diffusion.[38] Of relevance, this movement was observed also in the PM of erythrocytes and hepatocyte-like cells, as well as in the ER and Golgi membranes.[38,63] In both model and plasma membranes, the spontaneous transbilayer movement exhibited a slow rate (half-time between 2 and 5 h at 20°C). However, in rat liver ER and Golgi membranes, the transverse diffusion of spin-labeled monohexosylsphingolipids was much faster (half-time of about 3 min at 20°C), displayed a saturable behaviour, was sensitive to proteases and occurred in both directions,[38] indicating that one or more specific scramblase(s) facilitates this rapid movement. This should promote a symmetric distribution of monohexosylsphingolipids across Golgi membranes, allowing their rapid access to the outer (noncytosolic) leaflet, where further glycosilations occurs. Using fluorescent short chain analogues, it has been reported that ABCB1 is involved in GlcCer translocation in the Golgi.[63] However, a recent study suggests that P-glycoprotein selectively translocates such analogues but not natural long chain lipids.[64]

Notably, the transbilayer movement of lactosylceramide has not been detectable in both artificial and natural membranes,[38] implying that once synthesised in the Golgi (or produced in the lysosome), it remains at the lumenal leaflet where it undergoes subsequent metabolic processing.

Recent studies have revealed a mechanism of GlcCer transport specific for epidermal keratinocytes and functional to the extracellular formation of Cer. This transport involves ABCA12 transporter in the delivery of GlcCer to secretory granules (termed lamellar granules) in the epidermal keratinocytes.[65] Notably, ABCA12 selectively transports long chain GlcCer, essential to the generation of a specific class of ceramides crucial to the correct formation of the epidermal permeability barrier.[66] Mutations in *ABCA12* result in the failure to deliver GlcCer to lamellar granules and consequent failure to form lamellar bodies and extracellular lamellar membranes[67] and underlie harlequin ichthyosis (HI) and lamellar ichthyosis, two devastating skin disorders.[65,67] Of relevance, corrective transfer of the *ABCA12* into HI keratinocytes restores normal GlcCer loading into lamellar granules.[65]

Transbilayer Transfer of Ceramide and Sphingoid Bases

Unlike other sphingolipids, Cer and the sphingoid bases Sph and sphinganine are characterised by the presence of a very small polar headgroup, suggesting they can easily undergo transbilayer diffusion. Indeed, different studies reported their rapid scrambling.[33,40,68] The reported half-times of Cer flip-flop are less than 1 min at 37°C for the natural C16-Cer in giant vesicles and in the erythrocyte PM.[33] This rapid scrambling should give Cer the ability to reside in both sides of the membrane, thus forming membrane spanning domains. Up to now, it is not known if the lipid/protein composition of different biological membranes might influence the Cer flip-flop. This could be of relevance in the metabolism and functional properties of Cer.

Cer acts as intracellular messenger of different stimuli in multiple cell types and Cer generation appears closely linked to scrambling in both apoptosis and eryptosis.[46,69] Intriguingly, recent studies reported that Cer itself can be responsible for the transbilayer motion of phospholipids.[33,54,70] The conversion of SM into Cer on only one side of a lipid vesicle appears to generate a surface tension that in turn can generate a surface asymmetry in the bilayer.[54] This asymmetry in turn promotes phospholipid scrambling across the PM in a nonspecific fashion. Notably, dihydroceramide, the Cer precursor unable to exert many of the physiological effects of Cer, does not induce phospholipid scrambling.[55]

Unexpectedly, some ABC proteins appear to be required also for the efficient translocation of Cer and long chain sphingoid bases (LCBs). It was found that during terminal keratinocyte differentiation, the upregulation of ABCA7 is paralleled by intracellular and surface Cer levels and overexpression of ABCA7 in HeLa cells results in increased levels of intracellular and surface Cer.[71] In addition, in placenta trophoblasts, ABCG2 has been proposed as transporter of Cer to the outer leaflet of the PM, protecting placenta from Cer-induced apoptosis.[72]

A novel transporter, referred as Rsb1p (resistant to sphingoid bases) and localized to the PM, was shown to be involved in the export of dihydrosphingosine and phytosphingosine in *Saccharomyces cerevisiae*.[73,74] Intriguingly, *Rsb1* expression is enhanced in cells with altered glycerophospholipid asymmetry due to the disruption of either the inward or outward movement of glycerophospholipids,[74] suggesting that altered membrane asymmetry can trigger *Rsb1* expression. Interestingly, Pdr3p, a transcription factor crucial to multidrug resistance, was shown to activate *Rsb1* transcription, suggesting that this is functional to LCBs detoxification in the retrograde response of *Saccharomyces cerevisiae*.[75] It will be interesting to evaluate whether a floppase exporting LCBs may exist in mammals as well.

Transbilayer Transfer of Sphingosine-1-Phosphate

Unique among sphingoid molecules, S1P exhibits the peculiar property of binding to specific cell surface receptors. Since most S1P is generated intracellularly, S1P transport across the PM is crucial for its action as first messenger. Different cells have been shown to release newly synthesized S1P into the extracellular milieu and such release could be important for the local action of S1P in the extracellular space.[76-78] Of note, the extracellular release of S1P appears cell specific, as not all cells exhibit the capacity to excrete S1P.

S1P release can occur constitutively, or can be induced by different stimuli, such as PKC activation, Ca^{2+} and growth factors.[76-78] However, the mechanism of S1P transmembrane traffic is still unclear. With its polar head group, S1P presumably cannot readily permeate the lipid bilayer, its export/secretion into the extracellular space most probably requiring specific transporters. Increasing evidence suggests that members of the ABC transporter family are of relevance for the S1P export out of cells (Table 1). It was shown that export of S1P from activated mast cells is mediated by ABCC1[79] and the member of the A subfamily ABCA1 is critical for S1P release from astrocytes and at the level of the blood brain barrier.[80] Similarly, it was suggested that S1P secretion from HUVEC is also mediated by ABCA1 and ABCC1.[81]

Some pieces of evidence indicate the involvement of ABC transporters also in the active S1P uptake from the extracellular milieu. In particular, the cystic fibrosis transmembrane regulator ABCC7 has been implicated in the uptake of extracellular S1P by smooth muscle cells,[82,83] suggesting that shuttling from the outside to the inside by ABCC7 participates to the regulation of S1P availability for its receptors.

Although all these studies suggest the involvement of different mechanisms in the transbilayer transport of sphingolipid and support the relevance of this movement, the mechanisms underlying it remain largely unknown and further intensive studies are necessary to elucidate the players, localisation and regulation of these mechanisms.

Intermembrane Sphingolipid Transport

Sphingolipids are unevenly distributed in different subcellular membranes and, as described above, in the single membrane as well. The heterogeneous distribution of complex sphingolipids in the different subcellular membranes, varying from 3-4% of total membrane lipids in the ER to about 20 mol % in the PM and lysosomes,[84] strongly reflects the main localization of these lipids in the PM, the topology of the enzymes involved in sphingolipid metabolism and the mode of sphingolipid transport between the different subcellular organelles. Sphingolipid

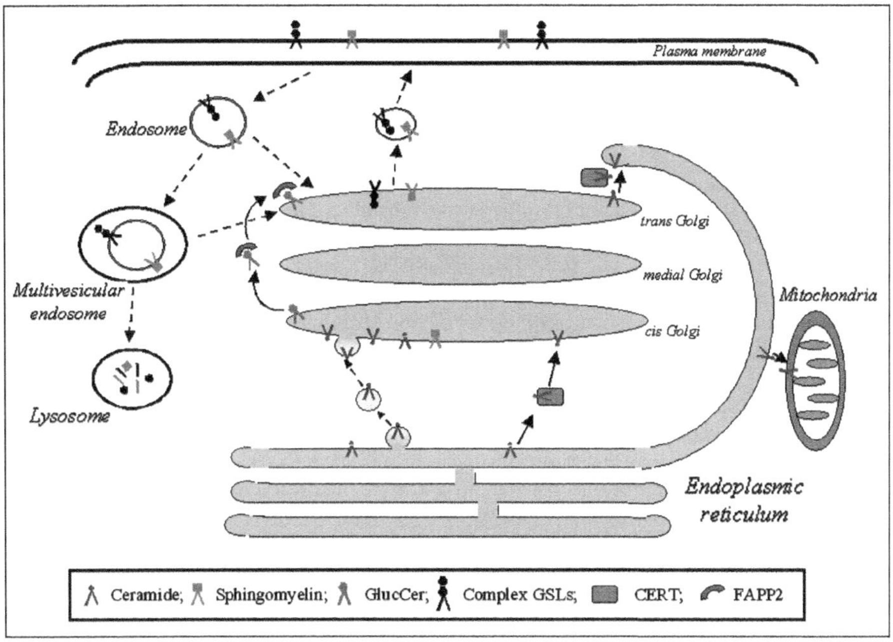

Figure 2. Mechanisms of intermembrane sphingolipid transport. Sphingolipids can move between different cellular membranes via protein-mediated (solid lines) and vesicle-mediated (dotted lines) trafficking routes. See text for details.

metabolism is indeed compartmentalized, their synthesis starting in the ER and continuing in the Golgi apparatus; degradation mainly occurs in the endosomal/lysosomal district. Due to their hydrophobic tail, sphingolipids are not free to diffuse in the aqueous cytosolic milieu. For this reason they must be transported to the different sites of synthesis and degradation.

Two major mechanisms of intermembrane (i.e., between different cell membranes) sphingolipid transport are operative in cells and include protein- and vesicular-mediated transports (Fig. 2).

Protein-Mediated Sphingolipid Transport

In spite the existence of proteins able to transfer sphingolipids in a nonvesicular manner has been postulated more than 20 years ago, in recent years the crucial role of two proteins in the transport of selective sphingolipids has been demonstrated. Indeed CERT and FAPP2 (four-phosphate-adaptor protein 2) have been identified as lipid transfer proteins, able to transfer Cer and GlcCer, respectively, to proper membranes in a nonvesicular manner (Fig. 2).

CERT-Mediated Transport of Ceramides

Cer synthesized at the cytosolic face of ER must be transported to the Golgi apparatus to be further metabolized to GlcCer, SM and Cer1P. Before the discovery of the specific Cer transfer protein CERT,[85] the studies addressed to understand the mode of Cer transport between ER and Golgi suggested that both vesicular and nonvesicular (possibly protein-mediated) mechanisms can contribute to this process.[84]

Hanada and coworkers[85] identified CERT in a mutant cell line (LY-A cells) characterized by a decrease of SM due to a defect in the transport of Cer to the Golgi.[86,87] This finding was pivotal to understand the mechanism of the nonvesicular delivery of Cer to the Golgi apparatus. Studies of Hanada's group demonstrated that in Chinese hamster ovary cells the biosynthesis of SM but not that of GlcCer is dependent on CERT.[85] This was confirmed in different cell lines by different studies[88-90] and was mainly attributed to the different localization of GCS and SMS in the cis and trans Golgi respectively. Moreover, down-regulation of CERT by RNAi strongly reduced the levels of C1P synthesized by Cer kinase (CerK) in the trans-Golgi.[91] However, this was not confirmed in other cell types by the use of pharmacological inhibition of CERT,[92] this discrepancy being mainly ascribed to the different experimental approach used to inhibit CERT and to the possible different subcellular compartmentalization of CERK.[92]

CERT is a 69 kDa protein identical to the splicing variant of the Goodpasture antigen binding protein (GPBP) that lacks a serine-rich domain of 26 amino acids. GPBP was previously identified as a Ser/Thr kinase which phosphorylates the noncollagenous-1 domain of the $\alpha3$ chain of type IV collagen.[93] Both the splicing variants can act as Cer-transfer proteins in vitro.[85] GPBP and CERT, coded by the *COL4A3BP* gene, are both highly conserved in the evolution[94] and widely expressed in various tissues,[95] thus suggesting a differentiated function for the two isoforms. The distinct functions of the two splicing variants GPBP and CERT are also supported by the evidence that they are differentially expressed during Zebrafish development[96] and they distinctly concur to the program of protein secretion and Cer traffic in the cell.[97]

CERT has in its structure a pleckstrin homology (PH) domain in the N-terminal region, a START domain in the C-terminal region, a FFAT motif and a serine repeat (SR) motif localized respectively at the C-terminal and N-terminal of in the middle region (MR) between the PH and START domains. All these regions specifically contribute the ER to Golgi Cer transport and its regulation. The START domain is responsible for the Cer binding activity of CERT.[85] It shows high structural similarities with START domains of other lipid binding proteins, such as PC transfer protein, the Chol binding protein MLN64 and STARD4, all containing a similarly structured hydrophobic pocket for lipid binding.[98] Notwithstanding these similarities, CERT specifically recognizes Cer but not other sphingolipids, phospholipids or Chol and recognizes only weakly diacylglycerol.[85] X-ray analysis of the crystal structures of CERT-START domain in its apo- form, or cocrystalized with Cer or diacylglycerol, allowed to identify the structural elements by which CERT can distinguish Cer from other lipids.[99] Among the different molecular species of Cer, CERT efficiently transfers those containing C14-C20 fatty acids, but is less efficient towards Cer containing longer chain fatty acids, with the exception of C24:1 species.[100,101] This suggests that the major determinant for CERT selectivity is the match between the depth of its hydrophobic pocket and the actual length of Cer acyl chains. CERT can also extract dihydroceramide, but with a relatively restricted fatty acid specificity.

The PH domain restricts the direction of transfer and destination of Cer through its specific binding to phosphatidylinositol 4-monophosphate (PI4P), a phosphoinositide mainly associated to the Golgi. This domain is essential for the correct transport of Cer for SM biosynthesis, since a mutation in this domain is responsible for the defect in Cer transport in LY-A cells.[85] The idea that CERT is recruited to the Golgi through the binding of the PH domain to PI4P and this is required for the efficient utilization of Cer for SM biosynthesis is enforced by evidence demonstrating the down regulation or inhibition of PI4 kinase IIIβ as a limiting factor for Cer transport and SM biosynthesis in COS-7 cells.[89] The FFAT motif is responsible for CERT binding to the ER-resident proteins, vesicle-associated membrane proteins (VAMP)-associated proteins (VAPs). A mutation in the FFAT motif impairs the ER-Golgi transport of Cer, thus demonstrating the crucial role of VAP-CERT interaction in this process.[102] These findings indicate that the START domain alone is not sufficient for the transport activity of CERT in vivo and the PH domain and FFAT motif spatially restrict the random Cer transfer activity of the START domain to efficiently direct Cer in the biosynthesis of SM. The molecular mechanism of CERT-mediated Cer transport is yet to be defined. At present two main hypotheses can be made (Fig. 3). The first implies a short distance shuttle activity of CERT (Fig. 3A). In this case in its apo form CERT, possibly in a conformation in which the PH domain is masked, binds to ER through its FFAT motif to extract Cer. Cer binding promotes the detachment of Cer-CERT from the ER, the unmasking of PH domain and the docking to PI4P in the Golgi with the subsequent release of Cer. The second model (Fig. 3B) implies that CERT operates in specific regions where ER and trans Golgi come into close apposition, also called membrane contact sites.[103] In this model the active form of CERT

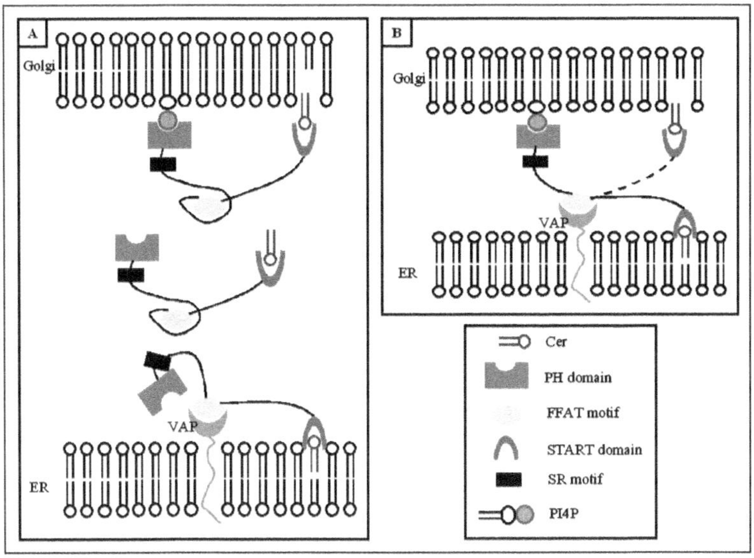

Figure 3. Proposed models of CERT-mediated transport of Cer between ER and Golgi. A) Hypothetical short distance shuttle activity of CERT; B) CERT may act as Cer transporter at ER-Golgi membrane contacts. See text for details.

is simultaneously bound to both VAP in the ER and PI4P in the Golgi. In both cases an open question exists about the signal/s and mechanisms responsible for the molecular dynamics of the process. A crucial point in this context is the phosphorylation state of CERT. In mammalian cells, CERT can be in a hyperphosphorylated state, that involve 7-9 Ser/Thr residues in the SR motif, close to the PH domain.[104] This represents an inactive state of CERT, as the treatment of human CERT with a bacterial protein phosphatase strongly increases its activity.[104] At present, two protein kinases are known to phosphorylate CERT: the casein kinase I[105] and the Golgi-associated protein kinase D.[106] In phosphorylated CERT, both the START and the PH domain binding capacity is inhibited, thus suggesting that phosphorylation induces a conformational change bringing the two domains to mutually interfere with each other. Recently it was demonstrated that an ER-resident protein phosphatase 2Cε (PP2Cε) dephosphorylates CERT in a VAP-dependent manner.[107] Over-expression of PP2Cε also increase the association of CERT to the Golgi, thus confirming dephospho-CERT as the active transfer protein. The control of CERT phosphorylation by the concerted, but compartmentalized, activities of both protein kinases and phosphatases seems to represent the main element in regulating CERT activity.

CERT activity in SM biosynthesis can be also regulated by the oxysterol binding protein OSBP[108] and by the SM/Chol ratio in the PM trough the stimulation of CERT dephosphorylation,[107] thus indicating a role for CERT in the maintenance of SM/Chol homeostasis in the PM.

Different lines of evidence indicate that CERT can be involved in the control of cellular functions other than sphingolipid metabolism. Down regulation of CERT increase the sensitivity to oxidative stress in *Drosophila*, determining a reduction in fly lifespan.[109] CERT is essential for mouse development and embryonic survival, since embryonic death associated to mitochondrial degeneration was observed in CERT-null mouse embryos.[110] CERT is up-regulated in drug-resistant cancer cells and down regulation of CERT sensitizes cancer cells to multiple cytotoxic agents.[111] Notwithstanding all these effects have been related to the Cer transfer capacity of CERT, the mechanisms involved are still not really clarified and future studies will be addressed to define the role of CERT in the control of cell fate.

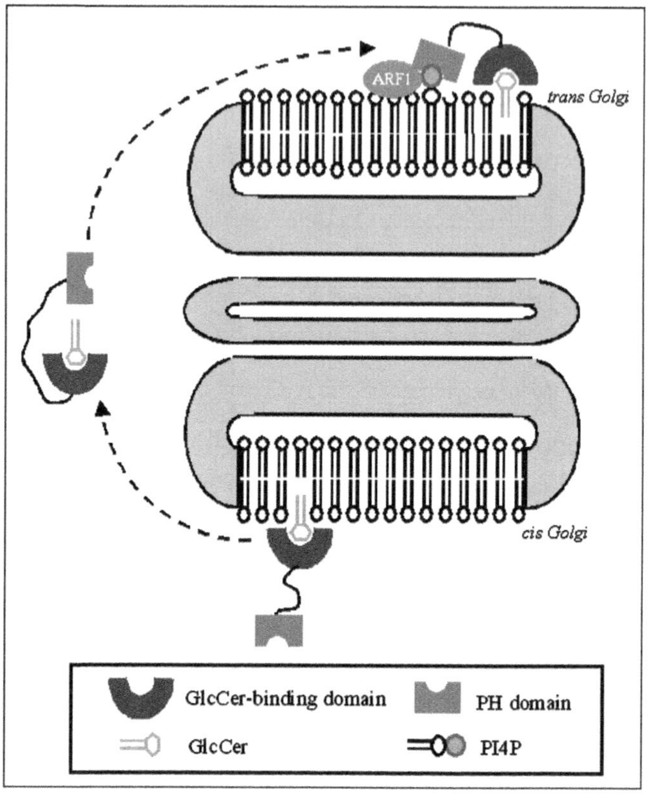

Figure 4. Proposed role of FAPP2 in the transport of GlcCer. According to the actual view, FAPP2 mediates GlcCer transport between cis- and trans-Golgi cisternae. See text for details.

FAPP2-Mediated Transport of Glucosylceramide

It has been recently demonstrated that FAPP2 mediates a nonvesicular transport GlcCer from cis-Golgi (its site of synthesis) to the trans-Golgi compartments,[90] where the enzymes of more complex GSLs the synthesis are localized (Fig. 4). Like CERT, FAPP2 contains a PH domain, PI4P binding to this domain being required for FAPP2 localization to Golgi. Besides this mechanism, the small GTPase ARF1 is also involved in the recruitment of FAPP2 to the trans Golgi network (Fig. 4). FAPP2 also possesses a glycolipid-transfer protein homology domain that is highly specifc for GlcCer, but does not transfer SM, PC or Cer.[90] All these observations suggest that GlcCer is transported from the cis side of Golgi to the *trans* side by FAPP2 in a nonvesicular manner. On the other hand, van Meer's group presented a different hypothesis[64] for this transport: GlcCer synthesized at the Golgi should be retrogradely transported by FAPP2 to the ER, here translocated to the lumen and then transported to the Golgi again to be further glycosylated. So far, whether and how FAPP2 can interact with the ER remains unknown.

Glycolipid Transfer Proteins

Glycolipid transfer proteins (GLTPs) are small, soluble and ubiquitous proteins able to accelerate the intermembrane transfer of glycolipids in vitro.[112] GLTP specificity encompasses both sphingoid- and glycerol-based glycolipids, but with a strict requirement that the initial sugar residue be β-linked to the hydrophobic lipid backbone.[113,114] Even if early studies with porcine and bovine GLTP established the ability of the protein to selectively accelerate the intermembrane transfer of GSLs in vitro, the mechanism of protein action remained poorly defined.[112]

Recently, Mattjus proposed that GLTP is not likely involved in the de novo synthesis of GSLs, but could rather play a role as a glycolipid sensor for the cellular levels of GlcCer.[115] Further studies are required to clarify GLTPs role in sphingolipid transport.

Membrane Contacts

Several organelles make close contacts with each other at zones of apposition called membrane contact sites, which might provide an alternative means of sphingolipid transport. For example, Cer synthesized in the ER has been proposed to diffuse to mitochondria at contact sites (Fig. 2).[116] This could be of relevance, as mitochondrial Cer has been implicated in the regulation of apoptosis and no pathway of vesicular transport to mitochondria is known. However, up to know, the transport of different sphingolipids to mitochondria (if any) remains largely unknown.

Sphingolipid Vesicular Transport

Vesicular transport represents a crucial mechanism of sphingolipid intracellular traffic. By this transport different sphingolipids travel as membrane components of small carrier vesicles from a donor compartment to a specific target membrane, where the vesicle fuses. The vesicular transport of sphingolipids is believed to be operated by molecular machineries similar to those of proteins, cytoskeleton proteins and associated molecular motors playing an important role.[117]

Sphingolipid vesicular flow is divided into two major routes depending on its direction in relation to the cell membrane: the biosynthetic and the endocytic pathways. Notably these routes may be interconnected at different levels.

Biosynthetic Vesicular Pathway

The biosynthetic route starts on membranes of the ER where the enzymes for Cer biosynthesis are localised and implies Cer transport to Golgi, where the biosynthesis of SM and GlcCer occurs. Several pieces of evidence support that, besides the CERT-mediated transport (see above), neosynthesized Cer at the ER appear to move to Golgi, through a vesicle-mediated route (Fig. 2).[85,118,119] Indeed, protein- and vesicular-dependent routes have been shown to coexist in mammalian cells and yeast[120,121] and to separately contribute to the control of Cer metabolism and levels.[88]

Vesicular transport appears crucial for Cer delivery to GlcCer synthase. In fact, CERT downregulation inhibits SM biosynthesis, but not that of GlcCer in different cells.[85,88,90] Although ER-Cer sorting into vesicles appears dictated by the specificity of CERT for specific SMs and functional to the transport of Cer for the biosynthesis of GlcCer, the biosynthesis of SM is, at least in part, dependent on the vesicular transport of a pool of Cer to SM synthase.[88,120]

The property of Cer of promoting membrane vesiculation and budding appears involved in the ER-Golgi vesicular transport of Cer.[35] In agreement, ongoing Cer synthesis is required for the efficient transport of GPI-anchored proteins from the ER,[122,123] suggesting that Cer has a function in the sorting of specific, ER-derived vesicles. Intriguingly, PI3K/Akt has been shown to regulate the ER-Golgi vesicular transport of Cer and consequently Cer levels in C6 glioma cells,[124] implying this traffic in Cer-mediated signaling too.

The transport of intermediate and final products of sphingolipid biosynthesis through the Golgi apparatus to the cell surface requires vesicular transport (Fig. 2). In fact, complex sphingolipids are synthesized on the luminal leaflet of the Golgi and in order to reach the PM, they must be sorted from the Golgi into anterograde-moving vesicles.[101,119,125] Even though the mechanisms underlying sphingolipid sorting remain unknown, sphingolipid and Chol segregation from coatomer protein I (COPI) vesicles has been shown to occur at the Golgi apparatus.[126]

In polarised cells (such as epithelial cells and hepatocytes) basolateral and apical membranes maintain a different sphingolipid composition, GSLs being enriched at the apical membrane, whereas the basolateral one displays a normal content.[119] These differences imply the sorting of distinct molecules and a specialised sphingolipid trafficking en route to the PM. Whereas basolateral sorting relies on the recognition of specific targeting proteins by adaptor-protein coats,[127] apical protein sorting is stimulated by PKA and appears dependent on the interactions between GSLs and specific glycoproteins (as glycosylphosphatidylinositol (GPI)-anchored proteins) in the trans-Golgi.[128] The selective sorting of GSLs in the apical transport remains poorly understood, but it must involve GSL lateral enrichment in budding apical precursor vesicles,[125,129] their hydrogen bonding tendency might possibly favour their segregation from other lipids. In hepatocytes, the displacement of type II PKA from its anchoring in the Golgi selectively excludes newly synthesized sphingolipid analogues from the direct Golgi-to-apical surface transport and results in their vesicular transport to the basolateral surface.[130] Moreover, GlcCer synthesis inhibition delays Golgi-to-apical surface trafficking of a specific set of protein,[130] suggesting that GSLs at the Golgi are sorted into specific vesicles, which are stimulated by type II-PKA and crucial to the delivery of specific proteins to the apical surface.

Endocytic Pathway

In the endocytic pathway, parts of the PM containing sphingolipids bud and are incorporated into early endosomes (Fig. 2). Endocytosis can occur through clathrin-dependent, caveolae-dependent or clathrin/caveolae-independent mechanisms, either constitutively or ligand-stimulated.[131] In a single cell type multiple endocytic mechanisms with different cargos and protein machinery, may exist.[131]

In early endosomes different and specific steps of sorting gradually occur. This sorting results in the formation of intraluminar vesicles and gives origin to multivesicular endosomes (MVEs or late endosomes). Different components of the Endosomal Sorting Complex Required for Transport (ESCRT) play a role in the formation of intralumenal vesicles, although their precise role remains to be clarified.[132]

It is believed that endocytosis occurs at sphingolipid-enriched regions of the PM and recent studies indicate that variations of specific sphingolipids influence endocytototic flow. For example, addition (or depletion) of GSLs, but not SM, selectively stimulates (or/inhibits) caveolar endocytosis in different cells.[133,134] In addition, SM accumulation at the PM, impairs the membrane targeting and activation of Rho A and leads to defects on RhoA-regulated endocytosis.[134,135] Thus sphingolipids appears of relevance and differentially required in the regulation of different endocytic mechanisms.

Sphingolipids can be delivered to different membranes as components of both the endosomal- or of intralumenal-vesicle membranes. After internalization via endocytosis, different vesicles transport sphingolipids to three major destinations, including the lysosomes (degradation pathway), the Golgi apparatus where additional glycosylations of GSLs may take place, or the PM (direct recycling and transcytosis).

The current view is that sphingolipids destined for degradation reach the lysosomes as membrane components of MVE internal vesicle (Fig. 2). After fusion with lysosomes, these vesicles become intralysosomal vesicles, their membranes (internal membranes) providing the platform of sphingolipid (and other membrane components) degradation.[136,137]

During the maturation of MVEs destined to lysosomes, lipid-sorting occurs and leads to the formation of internal vesicles enriched in bis(monoacylglycero)phosphate (BMP) (incorrectly also called lysobisphosphatidic) and depleted of Chol.[137] BMP-containing membranes play a role in GSL degradation[136] and regulate the export of Chol from internal membranes[138] and lipid sorting/export from late endosome intralumenal membranes to other cellular destinations.[132]

Chol removal from internal membrane vesicles is crucial for degradation processes. Schulze et al[137] proposed that during endocytosis the decrease in luminal pH is paralleled by SM degradation to Cer by acidic SMase and increased Cer may displace Chol from sphingolipid-enriched domains,[139] facilitating its transport out of the lysosomes.

Sphingolipid digestion by lysosomal hydrolases produces several molecules, such as fatty acids, sphingoid bases and monosaccharides, which are not further degraded in the lysosome, but exported and mainly recycled for biosynthetic purposes. These salvage pathways represent an important saving of energy in the cell economy and constitute a relevant event in sphingolipid turnover.[140] Despite Sph is efficiently recycled in different cells, the mechanism of its egress from lysosomes is unknown. Since Sph would be expected to partition in lysosomes by virtue of its ionizable positive charge, its exit from the lysosome should be protein-facilitated.

The direct recycling of endocytosed vesicles back to the PM can occur via early endosomes or MVEs. In early endosomes, some sphingolipids together with other lipids and proteins that need to be recycled back to the cell surface, such as receptors, are collected into tubular domains that eventually detach and form recycling endosomes.[141] The direct recycling of both fluorescent and natural SMs has been reported to be different, short-chain SMs being more efficiently recycled than the long chain SMs.[142] Furthermore, Chol loading or depletion of proteins involved in MVE formation or late endosomal trafficking altered the recycling and/or lysosomal degradation of SMs in a chain length-dependent manner.[142] Thus sphingolipid recycling appears a protein- and Chol-regulated process and might involve the specific sorting of sphingolipid molecular species.

At least in some cell types, recycling of MVEs can occur. Upon fusion with the PM, MVEs extracellularly release their intralumenal vesicles as exosomes, which play key role in multiple processes as intercellular communication and antigen presentation.In a recent study performed in an oligodendrocyte cell line, Trajkovic et al[34] reported that exosome formation is ESCRT-independent but requires neutral SMase 2-mediated Cer formation on the cytosolic side of endosomes and purified exosomes were found Cer-enriched. The Authors suggested that Cer-induced aggregation of lipid microdomains leads to domain induced inward budding of intraluminal vesicles.[34]

Besides the direct route from the Golgi, polarised cells can employ an indirect route (transcytosis) to transport GSLs (and specific proteins) to the apical membrane, involving prior sphingolipid delivery to the basolateral surface followed by rerouting to the apical membrane. For example, in polarized Madin-Darby-canine-kidney cells, newly synthesized GSLs and GPI-anchored proteins are first delivered to the basolateral membranes and then internalized by a clathrin-independent pathway that leads to the apical membrane.[145] Thus in some polarised cells the sorting of newly synthesized GSLs and GPI-anchored proteins might not occur at the trans-Golgi, but during internalization through caveolae from the basolateral membrane. Transcytosis is stimulated by cAMP, which requires dihydroCer synthase activation, probably functional to sphinganine removal.[146] In agreement, the accumulation of sphinganine results in frustrated cAMP-stimulated transcytosis and impaired apical membrane biogenesis.[146]

Conclusion

Cells regulate the localization of sphingolipids through a complex compartmentation of sphingolipid metabolism as well as a spectrum of intramembrane and intermembrane transport processes. These are sophisticated processes, controlled by local or extracellular interactions and stimuli and strictly interconnected, endowing the cells to fully integrate different sphingolipid molecules as key components and modulators of cell structure and functional properties. This complexity appears fundamental in sphingolipid transport, making it efficient, allowing it to move single or pools of sphingolipid molecules even in opposite directions and to proceed at different rates and adapting sphingolipid localization to specific inputs. In cells, this complexity should allow different sphingolipids to behave differently under different conditions, to respond to different cellular requirements and finally to display multiple functional properties.

In spite of extensive work, in most cases the precise nature of the molecular mechanisms that control the interaction between sphingolipids and their membrane environment and govern all the specific processes of sphingolipid transport in the living cells remains limited. Several important questions about its complexity remain also unanswered, such as the mechanisms underlying the interactions of multiple systems involved in sphingolipid transport, the control of sphingolipid transport by molecular interactions and signaling mechanism, as well as the pathological consequences of its derangement.

Solving these questions will represent an important challenge in the future. This will require the development of improved, new methods for manipulation and detection of specific molecular species or pools of sphingolipids at distinct intracellular sites, as well as an integrated approach that also includes the transport of membrane proteins.

References

1. Hresko RC, Sugar P, Barenholz Y et al. The lateral distribution of pyrene-labeled sphingomyelin and glucosylceramide in phosphatidylcholine bilayers. Biophys J 1987; 51:725-33.
2. Ollmann M, Schwarzmann G, Sandhoff K et al. Pyrene-labeled gangliosides: micelle formation in aqueous solution, lateral diffusion and thermotropic behavior in phosphatidylcholine bilayers. Biochemistry 1987; 26:5943-52.
3. Galla HJ, Hartmann W. Excimer-forming lipids in membrane research. Chem Phys Lipids 1980; 27:199-219.
4. Goins B, Masserini M, Barisas BG et al. Lateral diffusion of ganglioside GM1 in phospholipid bilayer membranes. Biophys J 1986; 49:849-56.
5. Spiegel S, Schlessinger J, Fishman PH. Incorporation of fluorescent gangliosides into human fibroblasts: mobility, fate and interaction with fibronectin. J Cell Biol 1984; 99:699-704.
6. van Meer G, Stelzer EH, Wijnaendts-van Resandt RW et al. Sorting of sphingolipids in epithelial (Madin-Darby canine kidney) cells. J Cell Biol 1987; 105:1623-35.
7. Lisanti MP, Sargiacomo M, Graeve L et al. Polarized apical distribution of glycosyl-phosphatidylinositol anchored proteins in a renal epithelial cell line. Proc Natl Acad Sci USA 1988; 85:9557-61.
8. Brown DA, Rose JK. Sorting of GPI anchored proteins to glycolipid-enriched membrane subdomains during transport to the apical cell surface. Cell 1992; 68:533-44.
9. Jacobson K, Dietrich C. Looking at lipid rafts? Trends Cell Biol 1999; 9:87-91.
10. Herman B, Krishnan RV, Centonze VE. Microscopic analysis of fluorescence resonance energy transfer (FRET). Methods Mol Biol 2004; 261:351-70.
11. Subczynski WK, Kusumi A. Dynamics of raft molecules in the cell and artificial membranes: approaches by pulse EPR spin labeling and single molecule optical microscopy. Biochim Biophys Acta 2003; 1610:231-43.
12. Almeida PF, Pkorny A, Hinderliter A. Thermodynamics of membrane domains. Biochim Biophys Acta 2005; 1720:1-13.
13. Brown DA, London E. Structure and origin or ordered lipid domains in biological membranes. J Membrane Biol 1998; 164:103-14.
14. Garcia-Sáez AJ, Chiantia S, Salgano J et al. Pore formation by a Bax-derived peptide: effect on the line tension of the membrane probed by AFM. Biophys J 2007; 93:103-12.
15. Filippov A, Orrad G, Lindblom G. The effect of cholesterol on the lateral diffusion of phospholipids in oriented bilayers. Biophys J 2003; 84:3079-86.
16. Dietrich C, Bagatolli LA, Volovyk ZN et al. Lipid rafts reconstituted in model membranes. Biophys J 2001; 80:1417-28.
17. Filippov A, Orrad G, Lindblom G. Lipid lateral diffusion in ordered and disordered phases in raft mixtures. Biophys J 2004; 86:891-96.
18. Sheets ED, Lee GM, Simson R et al. Transient confinement of a glycosylphosphatidylinositol-anchored protein in the plasma membrane. Biochemistry 1997; 36:12449-58.
19. Pike LJ. The challenge of lipid rafts. J Lipid Res 2009; 50:S232-8.
20. Marks DL, Pagano RE. Endocytosis and sorting of glycosphingolipids in sphingolipid storge disease. Trends Cell Biol 2002; 12:605-13.
21. Hanzal-Bayer MF, Hancock JF. Lipid rafts and membrane traffic. FEBS Lett 2007; 581:2098-104.
22. Mishra S, Joshi PG. Lipid raft heterogeneity: an enigma. J Neurochem 2007; 103(Supp 1):135-42.
23. Pike LJ. Rafts defined: a report on the Keystone symposium on lipid rafts and cell function. J Lipid Res 2006; 47:1597-98.
24. Kusumi A, Suzuki K. Towards understanding the dynamics of membrane-raft-based molecular interactions. Biochim Biophys Acta 2005; 1746:234-51.
25. Lingwood D, Ries J, Schwille P et al. Proc Natl Acad Sci USA 2008; 105:10005-10.
26. Hofman EG, Ruonala MO, Bader AN et al. EGF Induces coalescence of different lipid rafts. J Cell Sci 2008; 121:2519-28.
27. Wang TY, Silvius JR. Sphingolipid partitioning into ordered domains in cholesterol-free and cholesterol-containing lipid bilayers. Biophys J 2003; 84:367-78.
28. Xu X, Bittman R, Duportail G et al. Effect of the structure of natural sterols and sphingolipids on the formation of ordered sphingolipid/sterol domains (rafts). Comparison of cholesterol to plant, fungal and disease-associated sterols and comparison of sphingomyelin, cerebrosides and ceramide. J Biol Chem 2001; 276:33540-6.
29. Grassmé H, Cremesti A, Kolesnick R et al. Ceramide-mediated clustering is required for CD95-DISC formation. Oncogene 2003; 22:5457-70.
30. Staneva G, Chachaty C, Wolf C et al. The role of sphingomyelin in regulating phase coexistence in complex lipid model membranes: competition between ceramide and cholesterol. Biochim Biophys Acta 2008; 1778:2727-39.

31. Goñi FM, Alonso A. Biophysics of sphingolipids I. Membrane properties of sphingosine, ceramides and other simple sphingolipids. Biochim Biophys Acta 2006; 1758:1902-21.
32. Goñi FM, Alonso A. Effects of ceramide and other simple sphingolipids on membrane lateral structure. Biochim Biophys Acta 2009; 1788:169-77.
33. López-Montero I, Rodriguez N, Cribier S et al. Rapid transbilayer movement of ceramides in phospholipid vesicles and in human erythrocytes. J Biol Chem 2005; 280:25811-19.
34. Trajkovic K, Hsu C, Chiantia S et al. Ceramide triggers budding of exosome vesicles into multivesicular endosomes. Science 2008; 319:1244-7.
35. Giussani P, Maceyka M, Le Stunff H et al. Sphingosine-1-phosphate phosphohydrolase regulates endoplasmic reticulum-to-Golgi trafficking of ceramide. Mol Cell Biol 2006; 26:5055-69.
36. Kornberg RD, McConnell HM. Inside-outside transitions of phospholipids in vesicle membranes. Biochemistry 1971; 10:1111-20.
37. Sharom FJ, Grant CW. A model for ganglioside behaviour in cell membranes. Biochim Biophys Acta 1978; 507:280-93.
38. Buton X, Hervé P, Kubelt J et al. Transbilayer movement of monohexosylsphingolipids in endoplasmic reticulum and Golgi membranes. Biochemistry 2002; 41:13106-15.
39. Rosenwald AG, Pagano RE. Intracellular transport of ceramide and its metabolites at the Golgi complex: insights from short-chain analogs. Adv Lipid Res 1993; 26:101-18.
40. Bai J, Pagano RE. Measurements of spontaneous transfer and transbilayer movement of Bodipy-labeled lipids in lipid vesicles. Biochemistry 1997; 36:8840-48.
41. van Meer G, Halter D, Sprong H et al. ABC lipid transporters: extruders, flippases, or flopless activators? FEBS Lett 2006; 580:1171-77.
42. Borst P, Elferink RO. Mammalian ABC transporters in health and disease. Annu Rev Biochem 2002; 71:537-92.
43. Takahashi K, Kimura Y, Nagata K et al. ABC proteins: key molecules for lipid homeostasis. Med Mol Morphol 2005; 38:2-12.
44. Zachowski A. Phospholipids in animal eukaryotic membranes: transverse asymmetry and movement. Biochem J 1993; 294:1-14.
45. Bevers EM, Comfurius P, Dekkers DWC et al. Lipid translocation across the plasma membrane of mammalian cells. Biochim Biophys Acta 1999; 1439:317-30.
46. Lang F, Lang KS, Lang PA et al. Mechanisms and significance of eryptosis. Antioxidants and Redox Signaling 2006; 8:1183-92.
47. Bratton DL, Fadok VA, Richter DA et al. Appearance of phosphatidylserine on apoptotic cells requires calcium-mediated nonspecific flip-flop and is enhanced by loss of the aminophospholipid translocase. J Biol Chem 1997; 272:26159-65.
48. Daleke DL, Lyles JV. Identification and purification of aminophospholipid flippases. Biochim Biophys Acta 2000; 1486:108-27.
49. Zwaal RF, Comfurius P, Bevers EM. Surface exposure of phosphatidylserine in pathological cells. Cell Mol Life Sci 2005; 62:971-88.
50. Zhou Q, Zhao J, Stout JG et al. Molecular cloning of human plasma membrane phospholipid scramblase: a protein mediating transbilayer movement of plasma membrane phospholipids. J Biol Chem 1997; 272:18240-44.
51. Toti F, Satta N, Fressinaud E et al. Scott syndrome, characterized by impaired transmembrane migration of procoagulant phosphatidylserine and hemorrhagic complications, is an inherited disorder. Blood 1996; 87:1409-15.
52. Williamson P, Christie A, Kohlin T et al. Phospholipid scramblase activation pathways in lymphocytes. Biochemistry 2001; 40:8065-72.
53. Sahu SK, Gummadi SN, Manoj N et al. Phospholipid scramblases: An overview. Arch Biochem Biophys 2007; 462:103-14.
54. López-Montero I, Vélez M, Devaux PF. Surface tension induced by sphingomyelin to ceramide conversion in lipid membranes. Biochim Biophys Acta 2007; 1768:553-61.
55. Contreras FX, Villar AV, Alonso A et al. Sphingomyelinase activity causes transbilayer lipid translocation in model and cell membranes. J Biol Chem 2003; 278:37169-74.
56. Frasch SC, Henson PM, Nagaosa K et al. Phospholipid flip-flop and phospholipid scramblase 1 (PLSCR1) colocalize to uropod rafts in formylated Met-Leu-Phe-stimulated neutrophils. J Biol Chem 2004; 279:17625-33.
57. Nagao K, Takahashi K, Hanada K et al. Enhanced apoA-1-dependent cholesterol efflux by ABCA1 from sphingomyelin-deficient Chinese hamster ovary cells. J Biol Chem 2007; 282:14868-74.
58. Kobayashi A, Takanezawa Y, Hirata T et al. Efflux of sphingomyelin, cholesterol and phosphatidylcholine by ABCG1. J Lipid Res 2006; 47:1791-802.

59. Sano O, Kobayashi A, Nagao K et al. Sphingomyelin-dependend cholesterol efflux by ABCG1. J Lipid Res 2007; 48:2377-84.
60. Sakai H, Tanaka Y, Tanaka M et al. ABCA2 deficiency results in abnormal sphingolipid metabolism in mouse brain. J Biol Chem 2007; 282:19692-99.
61. Raggers RJ, van Helvoort A, Evers R et al. The human multidrug resistance protein MRP1 translocates sphingolipid analogs across the plasma membrane. J Cell Sci 1999; 112:415-22.
62. Dekkers DW, Comfurius P, van Gool RG et al. Multidrug resistance protein 1 regulates lipid asymmetry in erythrocyte membranes. Biochem J 2000; 350:531-35.
63. Burger KNJ, van der Bijl P, van Meer G. Transbilayer movement of monohexosylsphingolipids in endoplasmic reticulum and Golgi membranes. J Cell Biol 1996; 133:15-28.
64. Halter D, Neumann S, van Dijk SM et al. Pre- and post-Golgi translocation of glucosylceramide in glycosphingolipid synthesis. J Cell Biol 2007; 179:101-15.
65. Akiyama M, Sugiyama-Nakagiri Y, Sakai K et al. Mutations in lipid transporter ABCA12 in harlequin ichthyosis and functional recovery by corrective gene transfer. J Clin Invest 2005; 115:1777-84.
66. Zuo Y, Zhuang DZ, Han R et al. ABCA12 maintains the epidermal lipid permeability barrier by facilitating formation of ceramide linoleic esters. J Biol Chem 2008; 283:36624-35.
67. Elias PM, Williams ML, Holleran WM et al. Pathogenesis of permeability barrier abnormalities in the ichthyoses: inherited disorders of lipid metabolism. J Lipid Res 2008; 49:697-714.
68. Contreras FX, Basañez G, Alonso A et al. Asymmetric addition of ceramides but not dihydroceramides promotes transbilayer (flip-flop) lipid motion in membranes. Biophys J 2005; 88:348-59.
69. Tepper AD, Ruurs P, Wiedmer T et al. Sphingomyelin hydrolysis to ceramide during the execution phase of apoptosis results from phospholipid scrambling and alters cell-surface morphology. J Cell Biol 2000; 150:155-64.
70. Goñi FM, Contreras FX, Montes LR et al. Biophysics (and sociology) of ceramides. Biochem Soc Symp 2005; 72:177-88.
71. Kielar D, Kaminski WE, Liebisch G et al. Adenosine triphosphate binding cassette (ABC) transporters are expressed and regulated during terminal keratinocyte differentiation: a potential role for ABCA7 in epidermal lipid reorganization. J Invest Dermatol 2003; 121:465-474.
72. Evseenko DA, Murthi P, Paxton JW et al. The ABC transporter BCRP/ABCG2 is a placental survival factor and its expression is reduced in idiopathic human fetal growth restriction. FASEB J 2007; 21:3592-605.
73. Kihara A, Igarashi Y. Identification and characterization of a Saccharomyces cerevisiae gene, RSB1, involved in sphingoid long-chain base release. J Biol Chem 2002; 277:30048-54.
74. Kihara A, Igarashi Y. Cross talk between sphingolipids and glycerophospholipids in the establishment of plasma membrane asymmetry. Mol Biol Cell 2004; 15:4949-59.
75. Panwar SL, Moye-Rowley WS. Long chain base tolerance in Saccharomyces cerevisiae is induced by retrograde signals from the mitochondria. J Biol Chem 2006; 281:6376-84.
76. Yatomi Y. Plasma sphingosine 1-phosphate metabolism and analysis. Biochim Biophys Acta 2008; 1780:606-11.
77. Maceyka MW, Milstien S, Spiegel S. Sphingosine-1-phosphate: the Swiss army knife of sphingolipid signaling. J Lipid Res 2009; 50:S272-6.
78. Hannun YA, Obeid LM. Principles of bioactive lipid signalling: lessons from sphingolipids. Nat Rev Mol Cell Biol 2008; 9:139-50.
79. Mitra P, Oskeritzian CA, Payne SG et al. Role of ABCC1 in export of sphingosine-1-phosphate from mast cells. Proc Natl Acad Sci USA 2006; 103:16394-99.
80. Sato K, Malchinkhuu E, Horiuchi Y et al. Critical role of ABCA1 transporter in sphingosine 1-phosphate release from astrocytes. J Neurochem 2007; 103:2610-19.
81. Lee Y-M, Venkataraman K, Hwang S-I et al. A novel method to quantify sphingosine 1-phosphate by immobilized metal affinity chromatography (IMAC). Prostaglandins Other Lipid Mediat 2007; 84:154-62.
82. Boujaoude LC, Bradshaw-Wilder C, Mao C et al. Cystic fibrosis transmembrane regulator regulates uptake of sphingoid base phosphates and lysophosphatidic acid: modulation of cellular activity of sphingosine 1-phosphate. J Biol Chem 2001; 276:35258-64.
83. Peter BF, Lidington D, Harada A et al. Role of sphingosine-1-phosphate phosphohydrolase 1 in the regulation of resistance artery tone. Circ Res 2008; 103:315-24.
84. van Meer G, Holthuis JC. Sphingolipid transport in eukaryotic cells. Biochim Biophys Acta 2000; 1486:145-70.
85. Hanada K, Kumagai K, Yasuda S et al. Molecular machinery for nonvesicular trafficking of ceramide. Nature 2003; 426:803-9.

86. Hanada K, Hara T, Fukasawa M et al. Mammalian cell mutants resistant to a sphingomyelin-directed cytolysin. Genetic and biochemical evidence for complex formation of the LCB1 protein with the LCB2 protein for serine palmitoyltransferase. J Biol Chem 1998; 273:33787-94.
87. Fukasawa M, Nishijima M, Hanada K. Genetic evidence for ATP-dependent endoplasmic reticulum-to-Golgi apparatus trafficking of ceramide for sphingomyelin synthesis in Chinese hamster ovary cells. J Cell Biol 1999; 144:637-85.
88. Giussani P, Colleoni T, Brioschi L et al. Ceramide traffic in C6 glioma cells: evidence for CERT-dependent and independent transport from ER to the Golgi apparatus. Biochim Biophys Acta 2008; 1781:40-51.
89. Toth B, Balla A, Ma H et al. Phosphatidylinositol 4-kinase IIIb regulates the transport of ceramide between the endoplasmic reticulum and Golgi. J Biol Chem 2006; 281:36369-77.
90. D'Angelo G, Polishchuk E, Di Tullio G et al. Glycosphingolipid synthesis requires FAPP2 transfer of glucosylceramide. Nature 2007; 449:62-7.
91. Lamour NF, Stahelin RV, Wijesinghe DS et al. Ceramide kinase uses ceramide provided by ceramide transport protein: localization to organelles of eicosanoid synthesis. J Lipid Res 2007; 48:1293-304.
92. Boath A, Graf C, Lidome E et al. Regulation and traffic of ceramide 1-phosphate produced by ceramide kinase comparative analysis to glucosylceramide and sphingomyelin. J Biol Chem 2008; 283:8517-26.
93. Raya A, Revert F, Navarro S et al. Characterization of a novel type of serine/threonine kinase that specifically phosphorylates the human goodpasture antigen. J Biol Chem 1999; 274:12642-9.
94. Hanada K, Kumagai K, Tomishige N et al. CERT and intracellular trafficking of ceramide. Biochim Biophys Acta 2007; 1771:644-53.
95. Raya A, Revert-Ros F, Martínez-Martínez P et al. Goodpasture antigen-binding protein the kinase that phosphorylates the Goodpasture antigen, is an alternatively spliced variant implicated in autoimmune pathogenesis. J Biol Chem 2000; 276:40392-9.
96. Granero-Moltó F, Sarmah S, O'Rear L et al. Goodpasture antigen-binding protein and its spliced variant, ceramide transfer protein, have different functions in the modulation of apoptosis during zebrafish development. J Biol Chem 2008; 283:20495-504.
97. Revert-Ros F, Ventura I, Martínez-Martínez P et al. Goodpasture antigen-binding protein is a soluble exportable protein which interacts with type IV collagen: identification of novel membrane-bound isoforms. J Biol Chem 2008; 283:30246-55.
98. Yamaji T, Kumagai K, Tomishige N et al. Two sphingolipid transfer proteins, CERT and FAPP2: their roles in sphingolipid metabolism. IUBMB Life 2008; 60:511-8.
99. Kudo N, Kumagai K, Tomishige N et al. Structural basis for specific lipid recognition by CERT responsible for nonvesicular trafficking of ceramide. Proc Natl Acad Sci USA 2008; 105:488-93.
100. Kumagai K, Yasuda S, Okemoto K et al. CERT mediates intermembrane transfer of various molecular species of ceramides. J Biol Chem 2005; 280:64488-95.
101. Futerman AH. Intracellular trafficking of sphingolipids; relationship to biosynthesis. Biochim Biophys Acta 2006; 1758:18885-92.
102. Kawano M, Kumagai K, Nishijima M et al. Efficient trafficking of ceramide from the endoplasmic reticulum to the Golgi apparatus requires a VAMP-associated protein-interacting FFAT motif of CERT. J Biol Chem 2006; 281:30279-88.
103. Ladinsky MS, Mastronarde DN, McIntosh JR et al. Golgi structure in three dimensions: functional insights from the normal rat kidney cell. J Cell Biol 1999; 144:1135-1149.
104. Kumagai K, Kawano M, Shinkai-Ouchi F et al. Interorganelle trafficking of ceramide is regulated by phosphorylation-dependent cooperativity between the PH and START domains of CERT. J Biol Chem 2007; 282:14868-74.
105. Tomishige N, Kumagai K, Kusuda J et al. Casein kinase Ig2 down-regulates trafficking of ceramide in the synthesis of sphingomyelin. Mol Biol Cell 2009; 20:348-57.
106. Fugmann T, Hausser A, Schöffler et al. Regulation of secretory transport by protein kinase D-mediated phosphorylation of the ceramide transfer protein. J Cell Biol 2007; 178:15-22.
107. Saito S, Matsui H, Kawano M et al. Protein phosphatase 2Ce is an endoplasmic reticulum integral membrane protein that dephosphorylates the ceramide transport protein CERT to enhance its association with organelle membranes. J Biol Chem 2008; 283:6584-93.
108. Perry RJ, Ridgway ND. Oxysterol-binding protein and vesicle-associated membrane protein-associated protein are required for sterol-dependent activation of the ceramide transport protein. Mol Biol Cell 2006; 17:2604-16.
109. Rao RP, Yuan C, Allegood JC et al. Ceramide transfer protein function is essential for normal oxidative stress response and lifespan. Proc Natl Acad Sci USA 2007; 104:1364-9.
110. Wang X, Rao RP, Kasakowska-Cholody T et al. Mitochondrial degeneration and not apoptosis is the primary cause of embryonic lethality in ceramide transfer protein mutant mice. J Cell Biol 2009; 184:143-58.

111. Swanton C, Marani M, Pardo O et al. Regulators of mitotic arrest and ceramide metabolism are determinants of sensitivity to paclitaxel and other chemotherapeutic drugs. Cancer Cell 2007; 11:498-512.
112. Brown RE, Mattjus P. Glycolipid transfer proteins. Biochim Biophys Acta 2007; 1771:746-60.
113. Yamada K, Abe A, Sasaki T. Specificity of the glycolipid transfer protein from pig brain. J Biol Chem 1985; 260:4615-21.
114. Yamada K, Abe A, Sasaki T. Glycolipid transfer protein from pig brain transfers glycolipids with beta-linked sugars but not with alpha-linked sugars at the sugar-lipid linkage. Biochim Biophys Acta 1986; 879:345-349.
115. Mattjus P. Glycolipid transfer proteins and membrane interaction. Biochim Biophys Acta 2009; 1788:267-72.
116. Siskind LJ. Mitochondrial ceramide and the induction of apoptosis. J Bioenerg Biomembranes 2005; 37:143-53.
117. Gissen P, Maher ER. Cargos and genes: insights into vesicular transport from inherited human disease. J Med Genet 2007; 44:545-55.
118. Perry RJ, Ridgway ND. Molecular mechanisms and regulation of ceramide transport. Biochim Biophys Acta 2005; 1734:220-34.
119. van Meer G, Voelker DR, Feigenson GW. Membrane lipids: where they are and how they behave. Nat Rev Mol Cell Biol 2008; 9:112-24.
120. Viani P, Giussani P, Brioschi L et al. Ceramide in nitric oxide inhibition of glioma cell growth. Evidence for the involvement of ceramide traffic. J Biol Chem 2003; 278:9592-601.
121. Funato K, Riezman H. Vesicular and nonvesicular transport of ceramide from ER to the Golgi apparatus in yeast. J Cell Biol 2001; 55:949-59.
122. Horvath A, Sutterlin C, Manning-Krieg U et al. Ceramide synthesis enhances transport of GPI-anchored proteins to the Golgi apparatus in yeast. EMBO J 1994; 13:3687-95.
123. Kajiwara K, Watanabe R, Pichler H. Yeast ARV1 is required for efficient delivery of an early GPI intermediate to the first mannosyltransferase during GPI assembly and controls lipid flow from the endoplasmic reticulum. Mol Biol Cell 2008; 19:2069-82.
124. Giussani P, Brioschi L, Bassi R et al. Phosphatidylinositol 3-kinase/Akt pathway regulates the endoplasmic reticulum to Golgi traffic of ceramide in glioma cells: a link between lipid signaling pathways involved in the control of cell survival. J Biol Chem 2009; 284:5088-96.
125. Holthuis JCM, Levine TP. Lipid traffic: floppy drives and a superhighway. Nature 2005; 6:209-20.
126. Brügger B, Sandhoff R, Wegehingel S et al. Evidence for segregation of sphingomyelin and cholesterol during formation of COPI-coated vesicles. J Cell Biol 2000; 151:507-17.
127. Folsch H, Ohno H, Bonifacino JS et al. A novel clathrin adaptor complex mediates basolateral targeting in polarized epithelial cells. Cell 1999; 99:189-98.
128. Simons K, Ikonen E. Functional rafts in cell membranes. Nature 1997; 387:569-72.
129. Zegers MM, Hoekstra D. Mechanisms and functional features of polarized membrane traffic in epithelial and hepatic cells. Biochem J 1998; 336:257-69.
130. Wojtal KA, de Vries E, Hoekstra D et al. Efficient trafficking of MDR1/P-Glycoprotein to apical canalicular plasma membranes in HepG2 cells requires PKA-RIIa anchoring and glucosylceramide. Mol Biol Cell 2006; 17:3638-50.
131. Mayor S, Pagano RE. Pathways of clathrin-independent endocytosis. Nat Rev Mol Cell Biol 2007; 8:603-12.
132. Falguières T, Luyet P-P, Gruenberg J. Molecular assemblies and membrane domains in multivesicular endosome dynamics. Exp Cell Res 2009; 15:315:1567-73.
133. Sharma DK, Brown JC, Choudhury A et al. Selective stimulation of caveolar endocytosis by glycosphingolipids and cholesterol. Mol Biol Cell 2004; 15:3114-22.
134. Cheng ZJ, Singh RD, Sharma DK et al. Distinct mechanisms of clathrin-independent endocytosis have unique sphingolipid requirements. Mol Biol Cell 2006; 17:3197-210.
135. Galvan C, Camoletto PG, Cristofani F. Anomalous surface distribution of glycosyl phosphatidyl inositol—anchored proteins in neurons lacking acid sphingomyelinase. Mol Biol Cell 2008; 19:509-22.
136. Kolter T, Sandhoff K. Principles of lysosomal membrane digestion-stimulation of sphingolipid degradation by sphingolipid activator proteins and anionic lysosomal lipids. Annu Rev Cell Dev Biol 2005; 21:81-103.
137. Schulze H, Kolter T, Sandhoff K. Principles of lysosomal membrane degradation. Cellular topology and biochemistry of lysosomal lipid degradation. Biochim Biophys Acta 2009; 1793:674-83.
138. Chevallier J, Chamoun Z, Jiang G et al. Lysobisphosphatidic acid controls endosomal cholesterol levels. J Biol Chem 2008; 283:27871-80.
139. London ME. Ceramide selectively displaces cholesterol from ordered lipid domains (rafts): implications for lipid raft structure and function. J Biol Chem 2004; 279:9997-10004.

140. Tettamanti G, Bassi R, Viani P et al. Salvage pathways in glycosphingolipid metabolism. Biochimie 2003; 85:423-37.
141. Maxfield FR, McGraw TE. Endocytic recycling. Nat Rev Mol Cell Biol 2004; 5:121-32.
142. Koivusalo M, Jansen M, Somerharju P et al. Endocytic trafficking of sphingomyelin depends on its acyl chain length. Mol Biol Cell 2007; 18:5113-23.
143. Stoorvogel W, Kleijmeer MJ, Geuze HJ et al. The biogenesis and functions of exosomes. Traffic 2002; 3:321-30.
144. van Niel G, Porto-Carreiro I, Simoes S et al. Exosomes: a common pathway for a specialized function. J Biochem 2006; 140:13-21.
145. Polishchuk R, Di Pentima A, Lippincott-Schwartz J. Delivery of raft-associated, GPI-anchored proteins to the apical surface of polarized MDCK cells by a transcytotic pathway. Nat Cell Biol 2004; 6:297-307.
146. van IJzendoorn SCD, Van Der Wouden JM, Liebisch G et al. Polarized membrane traffic and cell polarity development is dependent on dihydroceramide synthase-regulated sphinganine turnover. Mol Biol Cell 2004; 15:4115-24.

Chapter 3

Sphingolipid Analysis by High Performance Liquid Chromatography-Tandem Mass Spectrometry (HPLC-MS/MS)

Jacek Bielawski,* Jason S. Pierce, Justin Snider, Barbara Rembiesa, Zdzislaw M. Szulc and Alicja Bielawska

Abstract

Sphingolipid (SPL) metabolism (Fig. 1) serves a key role in the complex mechanisms regulating cellular stress responses to environment. Several SPL metabolites, especially ceramide (Cer), sphingosine (Sph) and sphingosine1-phosphate (S1P) act as key bioactive molecules governing cell growth and programmed cell death (Fig. 2). Perturbations in sphingolipids of one type may enhance or interfere with the action of another. To monitor changes in SPL composition therefore, reliable analytical methods are necessary.

Here we present the liquid chromatography tandem mass spectrometry (LC-MS/MS) approach for simultaneous qualitative and quantitative monitoring of SPL components (classes and molecular species) in biological material as an effective tool to study sphingolipid signaling events. The LC-MS/MS methodology is the only available technique that provides high specificity and sensitivity, along with a wealth of structural identification information.

Introduction

Although sphingolipids (SPLs) have long been thought to function exclusively as structural constituents of the plasma membranes, in the past two decades research into the sphingolipids has progressed along two areas. First, SPLs have been shown to influence membrane structure, where they have been proposed to exist in clusters and form microdomains containing cholesterol at the plasma membrane, the so-called "lipid rafts".[1] These lipid microdomains are thought to function as platforms for effective signal transduction and correct protein sorting. Second, many SPLs have been shown to act as both first and second messengers, as well as bioactive mediators, in a variety of signaling pathways. Thus, the SPL metabolites—ceramide (Cer), ceramide-1-phosphate (Cer1P), sphingosine (Sph) and sphingosine 1-phosphate (S1P)—have emerged as a new class of lipid biomodulators for various cell functions and through participation in and influencing of multiple signaling pathways.[2-5] During the last 25 years there has been a dramatic increase in the studying of sphingolipid signaling in many patho-biological disorders but only recently new tools and approaches became available to examine these processes out, such as highly sensitive mass

*Corresponding Author: Jacek Bielawski—Department of Biochemistry and Molecular Biology, Medical University of South Carolina, 173 Ashley Ave., P.O. Box 250509, Charleston, South Carolina 29425, USA. Email: bielawsj@musc.edu

Sphingolipids as Signaling and Regulatory Molecules, edited by Charles Chalfant and Maurizio Del Poeta.

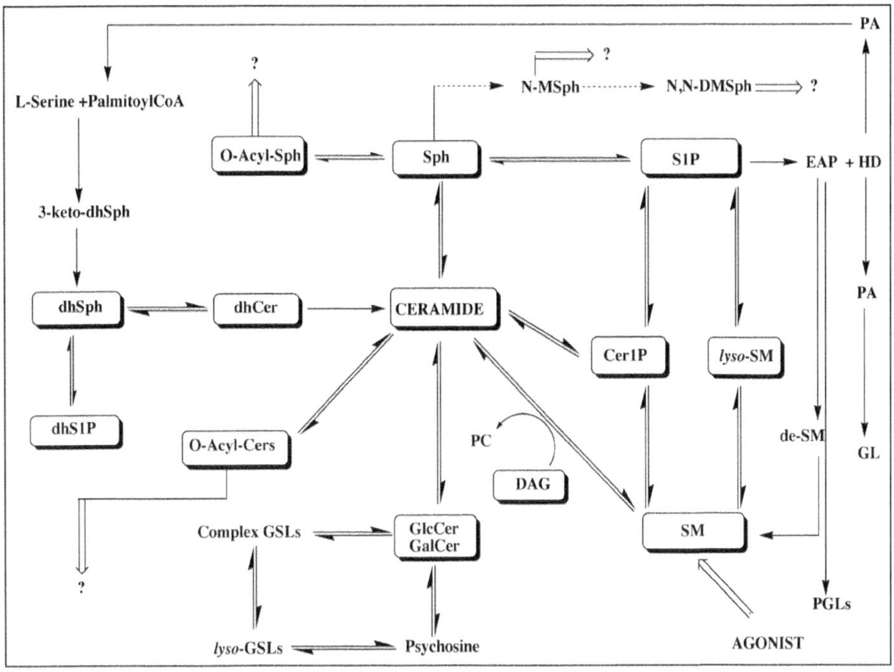

Figure 1. Sphingolipid biosynthesis and metabolic pathways; metabolomic profiling of sphingolipids. Abbreviations used in the figure: 3-keto-dhSph, 3-keto-dihydrosphingosine; dhSph, dihydrosphingosine; dhS1P, dihydrosphingosine 1-phosphate; dhCer, dihydroceramide; Sph, sphingosine; S1P, sphingosine 1-phosphate; Cer1P, ceramide-1-phosphate; SM, sphingomyelin; *lyso*-SM, lyso-sphingomyelin; DAG, diacylglycerol; O-Acyl-Sph. O-acyl-sphingosine, O-Acyl-Cer, O-acyl-ceramide; N-Me-Sph, N-methyl-sphingosine; N.N-DMSph, N,N-dimethyl-sphingosine; PA, palmitic acid; EAP, ethanolamine phosphate; HD, hexadecenal; GL, glycerolipids; de-SM, demethylated sphingomyelin; PGLs, phosphoglycerolipids; GlcCer, glucosylceramide; GalCer, galactosylceramide; GSLs, glycosphingolipids.

spectrometry methods for sphingolipid analysis. This chapter describes the high-performance liquid chromatography-tandem mass spectrometry (LC-MS/MS) for sphingolipid analysis in biological samples.

Sphingolipids: Structure and Composition

Prevalent complex SPLs: phosphosphingolipids (PSLs) and glycosphingolipids (GSLs) are found in all eukaryotes, some prokaryotes and viruses, mainly as components of the plasma membrane and related organelles. SPLs constitute about 30% of the total lipid of plasma membranes.

SPLs constitute one of the most structurally diversified classes of amphipathic lipids abundant in all living organisms. Variations in the nature of the head group attached to the primary hydroxyl group (carbohydrates, phosphocholine, phosphate or phosphoinositol), N-acyl group and sphingoid base (SB) backbone result in a great number of chemically distinct SPLs, where Sph, sphinganine (dhSph) or phytosphingosine (phytoSph) are the core structural moieties. Thousands of natural complex SPLs have been isolated based on almost 60 distinct species of sphingoid bases, although most of them are very minor components. SBs, the backbone of all SPLs, encompass a wide array of (2S, 3R, 4E)-2-amino-1,3-dihydroxyalkenes (Sphs), (2S, 3R)-2-amino-1,3-dihydroxy-alkanes (dhSphs) and (2S, 3S, 4R)-2-amino-1,3,4-trihydroxyalkanes (phytoSphs) with alkyl chain lengths from 14 to 22 carbon atoms and variations in the number and

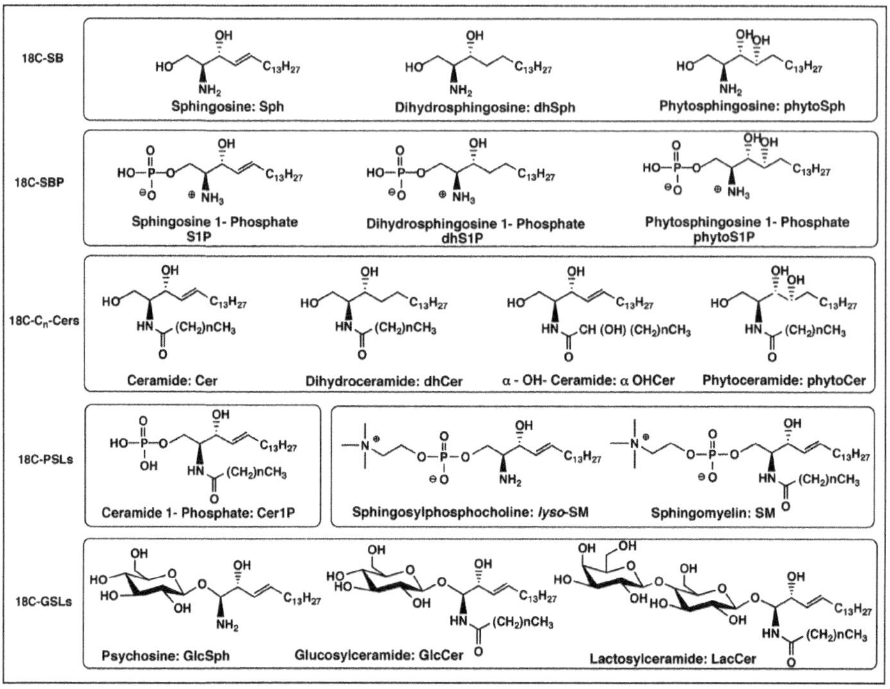

Figure 2. Natural sphingolipids are a highly heterogenous system related to the sphingoid bases and derivatization made on the amino- and hydroxy-functions. Structures shown in this figure represent derivatives of 18C-SB (sphingoid bases containing C18-backbone chain) indicating SPLs containing 2-amino-1,3-dihydroxy-octadecene-4E, 2-amino-1,3-dihydroxy-octadecane and 2-amino-1,3,4-trihydroxy-octadecane. General structures, nomenclature and abbreviations for SPLs are cited and described in this presentation. Cn -indicates the chain length of N-acyl part of SPLs.

position of the double bonds, hydroxyl groups and branching methyl groups. Mammalian SPLs are predominantly composed of 2-amino-1,3-dihydroxyoctadecene (18CSph, abbreviated here as Sph) and 2-amino-1,3-dihydroxyoctadecane (18CdhSph, abbreviated here as dhSph) (Fig. 2). Yeast and plant SBs are mainly composed of 2-amino-1,3,4-trihydroxyoctadecane (18CphytoSph), 18CdhSph and their eicosa-homologs (20CphytoSph and 20CdhSph). Additionally, some SPLs' SBs may contain a double bond in position 8 or have double bonds in positions 4 and 8 or/and have a methyl group in position 9 of the sphingosine backbone (which can be found in plant and fungi SPLs).

Ceramides are N-acyl-derivatives of SBs. Combinations of different SBs with different fatty acids (including their hydroxy-analogs) generates a huge variety of Cers, dhCers and phytoCers.

These basic SPLs are modified at the 1-hydroxyl group to: (i) phosphates (e.g., S1P and Cer1P), (ii) phosphocholine-analogs (e.g., sphingomyelin, SM and lysosphingomyelin, *lyso*-SM) and (iii) glucosyl- and galactosyl-analogs (e.g., glucosylceramide, GlcCer and galactosylceramide, GalCer, known as cerebrosides and their *lyso*-form: psychosine). Members of the latter group also serve as precursors to hundreds of different species of complex GSPLs.

SPLs constitute the second major category of polar lipids and for example they represent approximately 5-10% of total lipid mass in mammalian brain (6-8). Abnormal SPL metabolism could lead to their accumulation and deposition in multiple tissues, especially neural tissues, that results in potentially severe clinical manifestations, known as the sphingolipidoses.[6]

The structural diversity of SPLs dictates that every step in analysis of these natural products must be carefully evaluated.

LC-MS Methods for Detection and Analysis of Bioactive Sphingolipids

Technological advances in lipid detection, analysis and quantitation have played a key role in promoting the development of the sphingolipid research field. Traditional lipid analytical methods, such as thin-layer chromatography, are hampered by limited sensitivity, selectivity and resolution. Metabolic labeling using lipid precursors (such as serine or palmitic acid) have been wildly used for selective labeling of certain classes of lipids, which are then typically separated using thin-layer chromatography and visualized by autoradiography. However, this method is affected by incorporation of radioactive substrates and this does not always reflect the primary lipid contents in cells. Furthermore, thin-layer chromatography has low resolution and low sensitivity; thus it is difficult to identify the subspecies of individual SPLs.

The development of advanced mass-spectrometry-based methodologies has allowed the simultaneous assessment of several SPL subgroups as well as the probing of individual molecular subspecies such as various chain-length Cers. To understand the physiological function of sphingolipid metabolites, it has become important to know the metabolic change of particular SPLs and their individual subspecies from one sample. Here, we review methods for simultaneous analysis of SPLs using liquid chromatography tandem mass spectrometry techniques.

Lipidomic Approach

The term "lipidomics" has recently emerged[9-13] in relation to genomics and proteomics. Thus, lipidomics can be defined as the full characterization of lipid molecules in the studied biological material "Sphingolipidomics" will define the field of sphingolipids.

A variety of sample preparation, ionization modes and instrumental designs have been developed to analyze particular SPL classes by MS technology.[14] Design for this methodology has been based on the fact that different SPL subclasses dissociate into structurally distinctive patterns corresponding to their sphingoid bases, N-acyl chains and polar headgroups.[15-25] Recent advances in electrospray ionization (ESI) have provided a new approach to successfully examine total SPL components in crude lipid extracts.[19,22,23,25] Electrospray ionization (ESI) methodology allows generation of intact molecular ions of molecules from solution, delivered by direct infusion or by coupling high performance liquid chromatography (HPLC) column directly to the mass spectrometer. Further improvements in instrumentation, such as the triple quadruple with robust ion sources, fast scanning mass analyzers and reduced chemical noise (mainly in MS/MS technique) allow the identification and quantitation of SPLs with great sensitivity (sub-picomol detection limit) in a highly reproducible manner. SPL identification is accomplished by tandem mass spectrometry (MS/MS) with precursor ion scans to distinguish various molecular species in crude lipid extract by taking advantage of the unique molecular decomposition pattern[19,25] for each SPL class (Fig. 3). SPL quantitation is performed by using positive ionization and multiple reaction monitoring (MRM) in conjunction with HPLC separation.[25] Liquid chromatography/tandem mass spectrometry (LC-MS/MS) is the only technology available that provides structural specificity, quantitative precision and relatively high-throughput for analysis of complex SPLs in small samples.

Sample Preparation

The extraction process is one of the most important steps in pretreatment of solid (cell pellets, tissue) and liquid (plasma, serum, whole blood, biological fluids) samples.

Chloroform: methanol 2:1 (v/v) extraction, developed in 1956 for fish tissue,[26] further improved in 1959 by Bligh and Dyer,[27] became a golden standard procedure, known as the "Bligh & Dyer" (B & D) method and it is still commonly used for lipid extraction from all biological matrices. It involves a two step extraction employing chloroform: methanol: water at well-defined ratios of 1:2:0.8 and 2:2:1.8, respectively. According to the originators, the upper (methanol: water) phase contains all the "nonlipid" substances, while most lipids remain in the lower (chloroform) phase. This virtually unchanged procedure is commonly applied to most SPL sample preparation, regardless of the analytical procedure subsequently used e.g., TLC, HPLC

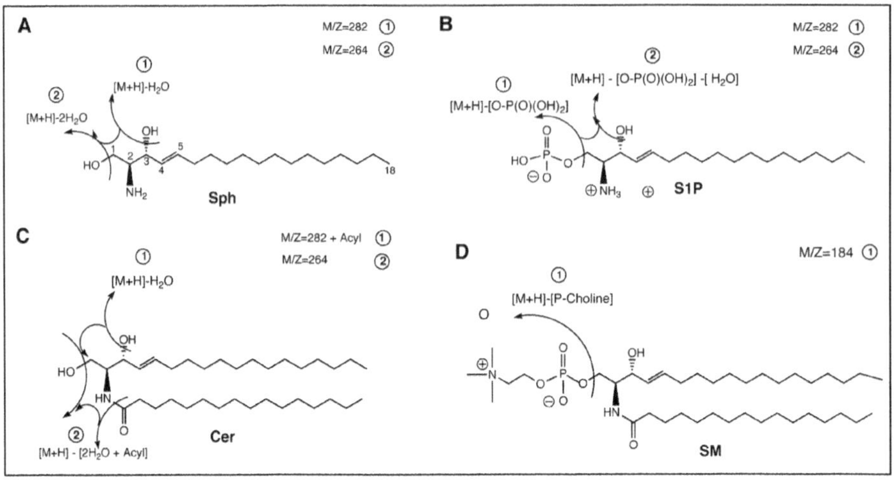

Figure 3. Typical MS/MS fragmentation patterns of sphingosine (A), sphingosine 1-phosphate (B), ceramide (C) and sphingomyelin (D). The specific common fragment ion of m/z = 264 (2, panel C) for Cers and m/z = 184 (1, panel D) for SM are used in the Parent Ion Scan experiments for determination of molecular species composition prior to quantitative analysis with the MRM experiments.

or MS, although the extraction efficiency, particularly for the most polar SPL components as S1P or *lyso*-SM seems to be questionable.

Over time, some modifications to the B & D have been incorporated in the isolation of SPLs, mostly intended to remove the bulk of the major co-extracted components, especially the glycerolipids, by subjecting the initial chloroform extract to a mild alkaline hydrolysis that cleaves ester linkage.[25,28-36] However, 1-O-acyl-ceramides (O-AcylCer) and related compounds[37] will also be hydrolyzed, thus artificially increasing the level of Cers. Comparison of the Cer level calculated from lipid extracts that were prepared with and without the base hydrolysis step can provide important data about the level of O-AcylCers. Our results showed some (20-40%) increase in Cers after this treatment. Nevertheless, this simple approach is recommended in the preparation of samples for SM analysis to allow elimination of phosphatidylcholine (PC) from the lipid extract, which may interfere with SM determination, even at highly specific LC-MS/MS analysis, due to close masses and fragmentation pattern.

Several attempts to further separate the initial total lipid extracts into particular lipid classes by a set of solid phase extraction (SPE) cartridges[38-40] proved to be very time consuming and not reproducible. Moreover, it may not be necessary when selective LC/MS technology is employed for analysis.

We developed[41] a one-phase extraction, using ethyl acetate: iso-propanol: water system at 60:30:10; (v/v/v) and 85:15:0 (v/v/v) for cell pellets and tissue homogenates and aqueous samples, respectively. The protocol describes lipid extraction under a safe and neutral condition to avoid destruction of the parent "soft" SPLs (e.g., SPLs containing O-acyl group), assuring efficient and quantitative extraction of the SB-1Ps from biological material since the latter are notoriously difficult to recover quantitatively.[39,42]

Readers interested in developing and/or improving existing methods of sample preparation are referred to an excellent review by McDowall.[43]

Analysis of Intact Sphingolipids by Mass Spectrometry

Mass spectrometry is a powerful detection technique that enables separation and characterization of compounds according to their mass-to-charge ratio (m/z). Its essential components include a sample inlet, ion source, mass analyzer, detector and data handling system. The combination of sensitivity, selectivity, speed and ability to provide invaluable structural information makes MS an ideal method for analysis of intact lipid molecular species.

The interest in the analysis of lipids in general especially SBs, SB1Ps, Cers, Cer1Ps, Glc/GalCers and SMs has continued to evolve due to the importance of these molecules in various biological transformations. SPL molecular species exist in nature as a complex mixture of closely related components which differ in the fatty acid chain length, degree on unsaturation and hydroxylation. These species differ greatly in their chemical and biological properties. Various analytical methods have been employed to separate and analyze individual species from intact (underivatized) form, out of which ESI/MS is the method of choice providing the following advantages:[44]

- Elimination of time consuming derivatization steps
- Making possible the study and follow-up of biosynthesis, metabolism, turnover and transport of the molecular species
- Protect possible rearrangement of the fatty acid chain during derivatization.

The sample introduction can be either by direct infusion or through preceding separation devices such as liquid chromatography (LC).

Mechanism of Electrospray Ionization Mass Spectrometry (ESI/MS)

ESI/MS, invented in the 1960s, was put into practice by Fenn et al.[45,46] It involves transformation of ions from the liquid to the gas phase. It is a method that operates at atmospheric pressure and ambient temperature. Initially, a solution containing the analytes of interest is introduced to the ESI ion source through capillary tubing. The narrow orifice at the end of the capillary and the dynamic forces facilitate formation of sprayed small droplets in the ionization chamber. Application of electric potential (approximately 2-5 kV) causes ionization, consequently the droplets carry a net charge. The charged droplets are then directed into the mass analyzer by the applied electric field. The applied potential may be positive or negative depending on physicochemical properties of the analytes. Passing through the ionization chamber, the droplets dissolve and this effect dramatically increases the columbic forces between the ions. Once this force exceeds the surface tension of the solvent, the droplets explode to form a fine mist of smaller droplets. This cycle is repeated until molecular ions are generated prior to their entrance into the mass analyzer.[47]

The soft ionization can generate lipid molecular ions without causing extensive fragmentation.[18,25,48-50]

MS Scan Modes

A number of mass analyzers are available, e.g., quadruple, ion trap, time of flight, ion cyclotron resonance, or sector instruments, which separate charged molecules in vacuum depending on their m/z ratio.

In so called full scan (FS) mode, a spectrum of primary, mostly molecular, ions is identified. This is the least specific mode with low sensitivity and it is mostly used for a rough assessment of major components of biological material when no or very limited information about SPL composition exist. However, interferences from other compounds present can either suppress ionization or cause a high chemical noise making such detection of SPL virtually impossible.

Moreover, the mass analyzers can also be used for fragmentation, predominantly in the triple quadruple instrument. In this instrument, the middle (Q2) field free quadruple either focuses and transmits all ions, or can be used as a collision cell for controlled fragmentation, called collision induced dissociation (CID). As results of a collision with an inert gas, introduced into the collision cell, the internal energy of the ions increases through conversion of kinetic energy breaking out specific bonds, depending on the collision energy applied.[51-52] The fragment ions are then analyzed in the second mass analyzer (third quadruple Q3). The choice of collision gas, its pressure and particularly the applied collision polarization and energy, affect the degree of fragmentation.

When a single quadruple instrument is used, partial fragmentation can be induced in the source by elevating the cone-to-skimmer potential difference. Protonated molecules desorbed from the

ESI droplets are accelerated between the cone and skimmer, undergoing CID upon collision with residual carrier gas molecules.[52]

Initial "soft" ionization of extracts prepared from biological samples results in numerous SPL molecular ions either positive $(M + H)^+$ or negative $(M\text{-}H)^-$. When the precursor ion fragments, it generates secondary (called daughter) distinctive pattern of ions related to the head group, SBs and fatty acids. This provides a wealth of structural information, enabling identification of SPLs in particular biological material. The positive ionization fragmentation can be enforced by incorporation of alkali metal ions $(M + Me)^+$ where Me = Li^+, Na^+, K^+, Rb^+, or Cs^+.[18,53] In addition to structural information, tandem MS provides a higher sensitivity, specificity and greatly reduced chemical background, thanks to very selected mode of monitored masses.

Specific Scan Modes for MS/MS Instrumentation

Product Ion Scan

In product ion scan, the first mass analyzer (Q1) allows a single ion with a set m/z value to pass and this is then further fragmented by CID in the second quadruple (Q2), the secondary (daughter) ions are then scanned over a defined mass range by the third quadruple (Q3) and passed to the detector. The relative abundance of the product ions depends on the dissociation dynamic; therefore, changing the CID collision energy, a fragmentation pattern is observed which is specific for each SPL class of compounds.

Neutral Loss (NL)

In a neutral loss scan, Q3 is offset from Q1 by fixed m/z, corresponding to specific neutral loss, e.g., 18 Dalton for loss of a water molecule. Both Q1 and Q3 scan over specified ranges of m/z values. In this mode, the detector records only those precursor ions that decompose, losing the specified neutral fragment. This type of MS experiments highly decreases chemical noise and is very helpful in identification of unknown SPLs.

Precursor Ion Scan (PI)

In a PI scan mode, the Q3 is set to pass specific m/z value, characteristic of a defined secondary ion. The Q1 scans across m/z range, recording only those primary ions which decompose to the specified product ion of interest. This highly specific scan mode eliminates or at least greatly reduces chemical noise and it constitutes a very useful identification tool since each class of SPLs yields at least one common product ion. Thus, setting Q3 to this specific daughter ion and scanning Q1 over the expected parent ion mass ranges, a spectrum of molecular species for an unknown biological sample may be identified.

Multiple Reaction Monitoring (MRM)

In a MRM experiment, the Q1 is set to pass specific precursor ion m/z and Q3 specific daughter ion m/z only.

This makes the MRM the most specific and sensitive MS/MS experiment allowing the analysis of even very minor components of a complex mixture with great precision and sensitivity. Such experiment practically eliminates chemical noise, thus makes it an ideal tool for quantitative analyses, particularly if coupled with HPLC physical separation. Multiple mass transitions, specific for particular compound, may be monitored sequentially; therefore, a large number of compounds may be analyzed together. Optimization of CID parameters for each compound of interest results in best sensitivity and specificity.

Sphingolipid Identification

Due to the complexity of sphingolipids, which usually constitute minor components of a crude lipid extracts, identification of individual molecular species is necessary before attempting any quantitative determination. This task can only be achieved with application of MS, particularly with precursor ion scan (PI) experiments. Although direct infusion full scan MS have been attempted,[30,36] reliable results may be obtained only for negative mode, in which a limited number of SPLs, such as free fatty acids, are ionized. In full scan positive mode, high chemical background makes any identification virtually impossible.

Qualitative analysis of SPLs from crude extracts is best accomplished by analysis of their unique molecular decomposition products using a PI scan of common fragment ions, characteristic for the particular class of SPLs (Fig. 3).[32,33,48] Readers are directed to the comprehensive studies on fragmentation patterns for mammalian and yeast SPLs, presented by Sullard[18] and Shevchenko,[54] respectively. Briefly, for mammalian SPLs, the m/z 264 and m/z 266 are the common fragment ions used for identification of Sph and its saturated counterpart, dhSph derivatives, respectively.

Considering the complexity of SPL composition, as well as the presence of many other lipid related compounds in biological material extracts, it is advisable to confirm initial identification, derived from PI scan, in order to avoid false identification. This may be accomplished by other, more compound-specific MS experiments, such as product ion scan of the newly identified molecular ion, or in a MRM experiment, with mass transition unique for the particular molecule e.g., single or double dehydration for Cers or NL of sugar moiety for GlcCers.

Ionization conditions and collision energy are optimized for individual molecular species to achieve maximum sensitivity and quantitative accuracy. SPLs composition has to be established for every new matrix.

Cer, Cer1P and GlcCers's molecular species (C18-SB) is established by the Precursor Ion scan, performed for the common Product Ion (m/z) 264.2 and 266.1 for Sph and dhSph derivatives, respectively at the high collision energy (35-55 eV), operating in positive ionization mode (Fig. 3). A representative sample extract is infused directly into ESI source and it is then scanned for molecular ions of the potential SPLs. Further confirmation of identity is achieved through MRM analysis with "soft" fragmentation (15-30 eV). Running sample through the HPLC system also confirms a reasonable retention time. Only SPLs that satisfy identification criteria in both analyses should be considered truly present in the sample.

SM and dhSM molecular species (18C-SB). Identification of the SM and dhSM components is performed similarly, employing common Product Ion (m/z 183.9) at 40 eV collision energy (Fig. 3).

Note: It is important to optimize the ionization conditions for each class of SPLs and collision energy for each individual molecular subspecies to be applied for quantitative MRM analysis.

HPLC-MS/MS Methodology

High Performance Liquid Chromatography (HPLC) is often employed for the separation of intact lipid molecules using various detectors. SPLs lack chromophores that would have enabled direct specific spectrophotometric detection. Some attempts have been made with UV[55-59] and evaporative light scattering (ELSD).[60-62] Both detectors, however, lack specificity and impose additional limitations. With UV detection, it is very difficult to select a working mobile phase (MB) since underivatized SPLs absorb close to the 200-210 nm range, depending upon the degree of unsaturation of the FA moiety and most of the commonly used solvents strongly absorb in this region.[59,63]

In the ELSD detector, the HPLC column effluent is evaporated, leaving the solute components as fine droplets, which are illuminated by laser and the scattered light is measured. This is an indiscriminatory detection since any compound that does not evaporate may be detected.[63]

An alternative technique that overcomes most of the above problems and provides both compound specificity and quantitative sensitivity is the use of HPLC coupled with ESI/MS. It is one of most powerful technologies for analysis of intact polar lipid molecules. The physical separation power of HPLC into either various lipid subclasses and/or individual molecular species within the class, together with MS highly selective detection, makes possible simultaneous determination of either protonated or deprotonated molecules, providing also invaluable structural information.[64]

Both Normal (NP) and Reverse Phase (RP) HPLC have been employed and it is important to select a solvent system that renders chromatographic resolution and ESI/MS compatibility to achieve maximum sensitivity.[18,25,32,33,41,65-67]

Recently, the most powerful technique applied for SPL molecular species is the HPLC-MS/MS instrumentation with the MRM scanning mode where each target analyte is uniquely identified by the Precursor-Product ion mass transition and the specific retention time.

Quantitation

Quantitation of SPLs in biological extracts has been a least developed segment of the HPLC/MS/MS analysis due to very limited supply of commercially available individual standards. Only laboratories that have access to custom made synthetic standards[41] were able to set up a reliable comprehensive quantitative protocols for various SPL classes. Recently, however, Avanti Polar Lipids Inc. (Alabaster, AL) and Matreya Inc. (Pleasant Gap, PA) significantly increased their offer of synthetic standards, so those major difficulties can be gradually overcome.

To achieve reliable quantitation of all molecular species, calibration curves should be generated for as many representative components of SPL as possible, due to diversified MS responses, as reflected by calibration curve slopes (Fig. 4).

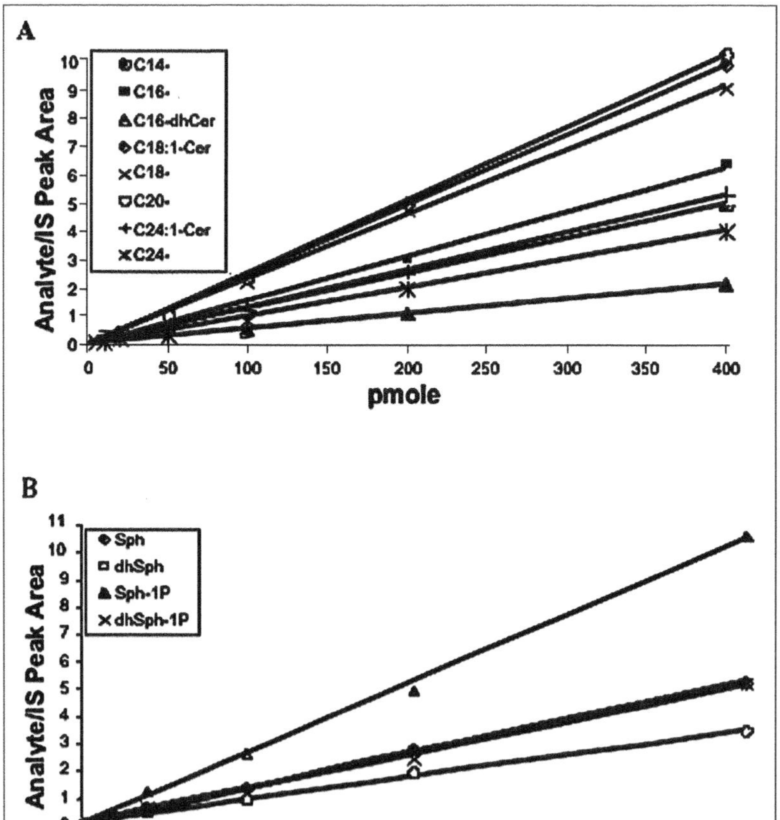

Figure 4. Calibration curves of sphingoid bases and ceramides. The MS response varies for molecular species even within the particular SPL class, as indicated by the calibration curve slops; therefore, individual calibrations are generated for as many target analytes as possible. Linear instrument response (R^2 value of .99) is obtained for the typical calibration ranges: 1.0-400.0 pmoles for SBs and SB 1Ps (lower panel), as well as for all ceramide species, Cn-Cers (upper panel).

Selection of Internal Standards (ISs)

Selection of a representative set of internal standards, which serve as a reference for both identification and quantitation, is critical for the analysis of a complex mixture of SPLs. Internal standards should be as close as possible to the target analytes, presenting similar MS fragmentation pattern as well as physicochemical properties reflected by similar solubility, extraction efficiency and mobile-stationary phase relationship during the HPLC separation. The best IS would be a compound which is chemically identical to target analyte labeled with a stable isotope, usually ^{2}H or ^{13}C. However, considering the large number of SPL molecular species, such approach is impractical; therefore, some compromise has to be applied. Often one IS per SPL class is used, mostly a sphingolipid with unnatural, usually lower, number of carbon atoms in the FA moiety.[18,25,31-34,66,67]

Bielawski et al.[41] have introduced ISs for particular SPL classes synthesized from C17- SB as the closest "unnatural" sphingoid base to the natural C18-SB counterpart. This selection gives a confidence that physicochemical properties such as the elution order and mass fragmentation pattern accurately reflect natural SPLs, but are not present in the analyzed sample. Moreover, since they are introduced to the samples prior to extraction, incomplete extraction efficiencies are compensated for, rendering quantitation of the target SPLs more precise.

Quantitative Calibration

Generating a calibration mechanism for each target SPL in a class greatly improves the quality of the obtained quantitative results. However, due to the limited availability of authentic standards (see above), calibration for, as many as is practical, representative standards should be generated, so that calibration devised for the synthetic standard can be also used for few structurally closely related analytes.

Sometimes[18,25,31-34,48,66,67] quantitation is performed using one IS as a single point calibration. This is not a very good practice since it assumes equal MS response to all molecular species in the class. Unfortunately, MS responses varies widely, depending on both structural features (number of carbon atoms, branching, unsaturation etc.) as well as mobile phase composition which changes over time, particularly when gradient elution is employed.

Based on the above considerations, we have adopted the following approach to SPL quantitative analysis.[41] In this approach, quantitative analyses of SPLs are based on the eight-point calibration curves generated for each target analyte. The synthetic standards along with a set of ISs are spiked into an artificial matrix, then subjected to the identical extraction procedures as the test samples and then analyzed by the HPLC-MS/MS system operating in positive MRM mode, employing HPLC with gradient elution. Peaks for the target analytes and IS are recorded and processed using the instrument software system. Plotting the analyte/IS peak area ratios against analyte concentration generates the analyte-specific calibration curves. Any SPL for which no standard is available is quantitated using the calibration curve of the closest counterpart.

Data Handling

Results from the MS analysis represent the mass level of particular SPLs (in pmols) per total sample used for lipid extract preparation and quantitative analysis. In general, treatment with exogenous agents causes changes in SPL levels and compositions. For the final data presentation, MS results should be normalized to some stable parameters (which are considered not affected by that particular treatment). Total protein (mg), or phospholipid contents Pi (nmol) present in the Bligh & Dyer extract,[27] which corresponds to the amount of the biological material used for MS analysis, can be used as the normalization parameters.[29,68,69] Also normalization to the total cell number is used.[70] Final results should be shown as changes in the relation to the control (% control). From our experience, data normalized to the protein or to the Pi (shown as % of the control) are not exactly the same. It is critical that once the user selects the normalization parameter, it carry it out consequently throughout the total study for consistency of the generated quantitative results.

Alternative Methodology

A variety of different techniques (mostly radio-labeling, HPLC analysis of fluorescent analogs and enzymatic methods) in addition to MS methodology are used for SPLs measurement. Up to now, the enzymatic method employing diacylglycerol kinase and (^{32}P) ATP has been the most commonly used procedure for total sphingolipids quantitation in the range of 25 pmols to 2 nmols.[69] Cellular SBs are most often analyzed by the HPLC technique developed for their fluorescent derivatives.[71] Cellular SB-1Ps are analyzed via their derivatization to (^{3}H) C2-ceramide phosphate, by an enzymatic method (employs alkaline phosphatase), followed by action of recombinant sphingosine kinase and (^{32}P ATP)) after TLC separation of S1P from the cellular Sph, or by employing HPLC analysis of OPA-derivatised S1P.[72-76] These procedures require less expensive equipment than mass spectrometry but are not as informative. SM may be determined by several different approaches including TLC analysis, GC analysis of silylated derivatives and MS techniques.[49,50,68,70] Total Cer and SM can be determined following hydrolysis and analysis of the liberated and derivatised SBs by means of HPLC[71,74] and the liberated fatty acids by means of GC[77] or GC/MS.[78]

Conclusion

This chapter describes quantitative analysis of virtually all compounds involved in sphingolipid metabolism and turnover (signaling) such as sphingoid bases, sphingoid base-1-phosphates, lysosphingolipids, ceramides, ceramide-1-phosphates, sphingomyelin and cerebrocides. The major emphasis was put on the most versatile LC/MS technology, that provides a wealth of structural information and specificity, essential in developing reliable analytical protocols, due to complexity of "sphingolipidome".

The LC/MS/MS methods have been successfully applied to a large number of different mammalian, and yeast cell lines, as well as various tissue samples, that typically contain many different sphingolipid subspecies, but constitute only a small fraction of crude lipid extracts.

So far there are not very good, mass spectrometry based analyses for some part of "sphingolipidome", namely complex gangliosides e.g GD3, due to lack of authentic, synthetic standards. Once this problem will overcome new analytical methodology will follow.

References

1. Simons K, Ikonen E, Functional rafts in cell membranes. Nature 1997; 387:569-572.
2. Pettus BJ, Chalfant CE, Hannun YA. Ceramide in apoptosis: an overview and current perspectives. Biochim Biophys Acta 2002; 1585:114-125.
3. Cuvillier O. Sphingosine in apoptosis signaling. Biochem Biophys Acta 2002; 1585:153-162.
4. Spiegel S, Kolesnick R. Sphingosine 1-phosphate as a therapeutic agent. Leukemia 2002; 16:1596-1602.
5. Lamour N F, Chalfant C E, Ceramide-1-phosphate: The "missing" link in eicosanoid biosynthesis and inflammation. Mol Interv 2005; 5:358-367.
6. Sastry PS. Lipids of nervous tissue: composition and metabolism. Prog Lipid Res 1985; 24(2):69-176.
7. Vos JP, Lopes-Cardozo M, Gadella BM. Metabolic and functional aspects of sulfogalactolipids. Biochim Biophys Acta 1994; 1211:125 -149
8. Merrill AH, Schmelz EM, Dillehay DL et al. Sphingolipids--the enigmatic lipid class: biochemistry, physiology, and pathophysiology. Toxicol Appl Pharmacol 1997; 142(1):208-225.
9. Han XL. Gross RW. Global analyses of cellular lipidomes directly from crude extracts of biological samples by ESI mass spectrometry: a bridge to lipidomics. J Lipid Res 2003; 44:1071-1079.
10. Han XL, Yang JY, Cheng H et al. Toward fingerprinting cellular lipidomes directly from biological samples by two-dimensional electrospray ionization mass spectrometry. Anal Biochem 2004; 330:317-331.
11. Spener F, Lagarde M, Géloën A et al. What is lipidomics? Eur J Lipid Sci Technol 2003; 105:481-482.
12. Lagarde M, Geloen A, Record M et al. Lipidomics is emerging. Biochim Biophys Acta 2003; 1634:61-69.
13. Hannun YA. Functional Lipidomics: Lessons and Examples from the Sphingolipids. In: Feng L, Prestwich G eds. Functional Lipidomics. Boca Raton: CRC press, 2005:147-158.
14. Adams J, Ann Q. Structure determination of sphingolipids by mass spectrometry. Mass Spectrom Rev 1993; 12:51-85.
15. Adams J, Ann Q. Structure determination of ceramides and neutral glycosphingolipids by collisional activation of (M + Li)+ ions. J Am Soc Mass Spectrom 1992; 3:260-263.

16. Ann Q, Adams J. Structure-specific collision-induced fragmentation of ceramides cationized with alkali-metal ions. Anal Chem 1993; 22:7-13.
17. Ann Q, Adams J. Collision-induced decomposition of sphingomyelins for structural elucidation. Biol Mass Spectrom 1993; 22:285-294.
18. Sullard MC. Sphingolipid Metabolism and Cell Signaling. Methods Enzymol 2000; 312:32-45.
19. Gu M, Kerwin JL, Watts JD et al. Ceramide profiling of complex lipid mixtures by electrospray ionization mass spectrometry. Anal Biochem 1997; 244:347-356
20. Van Veldhoven PP, De Ceuster P, Rozenberg R et al. On the presence of phosphorylated sphingoid bases in rat tissues. A mass-spectrometric approach. FEBS Lett 1994; 350(1):91-95.
21. Couch LH, Churchwell MI, Doerge DR et al. Identification of ceramides in human cells using liquid chromatography with detection by atmospheric pressure chemical ionization-mass spectrometry. Rapid Commun Mass Spectrom 1997;11:504-512.
22. Mano N, Oda Y, Yamada K et al. Simultaneous quantitative determination method for sphingolipid metabolites by liquid chromatography/ion spray ionization tandem mass spectrometry, Anal Biochem 1997; 244:291-300.
23. Liebisch G, Derobnik W, Reil M et al. Quantitative measurement of different ceramide spiecies from crude cellular lipid extracts by electrospray ionization tandem mass spectrometry (ESI-MS/MS). Lipid Res 1999; 40:1539-1546.
24. Pettus BJ, Kroesen BJ, Szulc ZM et al. Quantitative measurement of different ceramide species from crude cellular extracts by normal-phase high-performance liquid chromatography coupled to atmospheric pressure ionization mass spectrometry. Rapid Commun Mass Spectrom 2004; 18:577-583.
25. Sullard MC, Merrill AH. Analysis of sphingosine-1-phosphate, ceramides and other bioactive sphingolipids by high-performance liquid chromatography-tandem mass spectrometry. Sci STKE 2001; 67: 1-11.
26. Folch J, Lees M, Sloane HS. A simple method for the isolation and purification of total lipids from animal tissues. J Biol Chem. 1956; 196:497-509.
27. Bligh EG, Dyer WJ. A rapid method of total lipid extraction and purification, Can J Biochem and Physiol 1959; 37:911-917.
28. Lin T, GenestierL , Pinkoski MJ. Role of acidic sphingomyelinase in Fas/CD95-mediated cell death. J Biol Chem, 2000: 275(12):8657-8663.
29. Bose R, Verheij R, Haimovitz-Friedman A et al. Ceramide synthase mediates daunorubicin-induced apoptosis; an alternative mechanism for generating death signals. Cells 1995; 82:405-414.
30. Prasad, T, Saini P, Guar NA et al. Functional Analysis of CaIPT1, a Sphingolipid Biosynthetic Gene Involved in Multidrug Resistance and Morphogenesis of Candida albicans. Chemother 2005; 49:3442-3452.
31. Monick MM, Mallampalli RK, Bradford M et al. Cooperative prosurvival activity by ERK and Akt in human alveolar macrophages is dependent on high level of acid ceramidase activity. J Immunology. 2004; 173:123-135.
32. Adams JM, Pratipanawatr T, Berria R et al. Ceramide content is increased in skeletal muscle from obese insulin-resistant humans, Diabetes 2004; 53:25-31.
33. Zheng W, Kollmeyer J, Symolon H et al. Ceramides and other bioactive sphingolipids backbone in health and disease: Lipidomic analysis, metabolism and roles in membrane structure, dynamic, signalic and autopaphy, Biochim Biophys Acta 2006; 1758(12):1864-1884.
34. Maceyka M, Sankala H, Hait NC et al. SphK1 and SphK2, sphingosine kinase isoenzymes with opposition functions in sphingolipid metabolism. J Biol Chem 2005; 280:37118-37129.
35. VanDer Luit AH, Budde M, Zerp S et al. Resistance to alkyl-lysophospholipid-induced apoptosis due to downregulated sphingomyelin synthase 1 expression with consequent sphingomyelin- and cholesterol-deficiency in lipid rafts. Biochem J 2007; 401:541-549.
36. Guan XL, He X, Ong WY et al. Non-targeted profiling of lipids during kainate-induced neuronal injury. FASEB J 2006; 20:1152-1161.
37. Abe A, Shayman JA, Purification and characterization of 1-O-Acylceramide synthase, a novel phospholipase A2 with transacylase activity. J Biol Chem 1998; 273:8467-8474.
38. Brodennec J, Koul O, Aguado I et al. A procedure for fractionation of sphingolipid classes by solid-phase extraction on aminopropyl cartridges. J Lipid Res 2000; 41:1524-1531.
39. Bodennec J, Brichon G, Zwingelstein G et al. Purification of Sphingoid Classes by Solid-Phase Extraction with Aminopropyl and weak cation Exchange Cartridges. Methods Enzymol 2000; 312:101-114.
40. Rizov I, Doulis A. Separation of plant membrane lipids by multiple solid-phase extraction. J Chromatogr 2001; A922:347-354.
41. Bielawski J, Szulc ZM, Hannun YA et al. Simultaneous quantitative analysis of bioactive sphingolipids by high-performance liquid chromatography-tandem mass spectrometry. Methods 2006; 39:82-91.
42. Kralik SF, Du X, Patel C et al. A method for quantitative extraction of sphingosine 1-phosphate into organic solvent. Anal Biochem 2001; 294:190-193.
43. McDowall RD. Sample preparation for biomedical analysis. J Chromatogr 1989; 492:3-58.

44. Olsson NU, Salem N. Molecular species analysis of phospholipids. J Chromatogr B. 1997; 692:245-256.
45. Fenn JB, Mann M, Meng CK et al, Electrospray ionization for mass spectrometry of large biomolecules. Science 1989; 246:64 -72.
46. Fenn JB, Mann M, Meng CK et al. Electrospray ionization-principles and practice. Mass Spectrom 1990; 9:37-70.
47. Cech NB, Enke CG. Practical implications of some recent studies in electrospray ionization fundamentals. Mass Specrom Rev 2001; 20:362-387.
48. Gu M, Kerwin JL, Watts JD et al. Ceramide profiling of complex lipid mixtures by electrospray ionization mass spectrometry. Anal Biochem 1997; 244:347-356.
49. Vieu C, Chevy, F, Rolland C et al. Coupled assay of sphingomyelin and ceramide molecular species by gas liquid chromatography, J Lipid Res 2002; 43:510-522.
50. Isaac G, Bylund D, Masson JE et al. Analysis of phosphatidylcholine and sphingomyelin molecular species from brain extracts using capillary liquid chromatography electrospray ionization mass spectrometry. J Neurosci Methods. 2003;128(1-2):111-9.
51. Hoffman ED. Tandem mass spectrometry: a primer. J Mass Spectrom 1996; 31:129-137.
52. Kuksis A. Myher JJ. Application of tandem mass spectrometry for the analysis of long-chain carboxylic acids. J Chromatogr B 1995; 671:35-70.
53. Samuelsson K, Samuelsson B. Gas chromatographic and mass spectrometric studies of synthetic and naturally occurring ceramides. Chem Phys Lipids 1970; 5:44-49.
54. Ejsing CS, Moehring T, Bahr U et al. Collision-induced dissociation pathways of yeast sphingolipids and their molecular profiling in total lipid extracts: a study by quadrupole TOF and linear ion trap-orbitrap mass spectrometry. J Mass Spectrom 2006; 41:372-389.
55. Junganwala, FB, Hayssen V, Pasquini JM et al. Separation of molecular species of sphingomyelin by reversed-phase high-performance liquid chromatography. J Lipid Res 1979; 20:579-587.
56. Smith M, Jungalwala FB. Reversed-phase high performance liquid chromatography of phosphatidylcholine: a simple method for determining relative hydrophobic interaction of various molecular species J Lipid Res 1981; 22:697-704.
57. Patton GM, Fasulo JM, Robins SJ. Separation of phospholipids and individual molecular species of phospholipids by high-performance liquid chromatography J Lipid Res 1982; 23:190-196.
58. Abidi SL, Mounts TL. High-performance liquid chromatographic separation of molecular species of neutral phospholipids. J Chromatogr A 1992; 598:209-218.
59. McHowat J, Jones JH, Creer MH. Quantitation of individual phospholipid molecular species by UV absorption measurements. J Lipid Res. 1996; 37:2450-2460.
60. Christie WW. Separation of molecular species of triacylglycerols by high-performance liquid chromatography with a silver ion column. J Chromatogr 1988; 454:273-284.
61. Brouwers JF, Gadella BM, Golde LM et al. Quantitative analysis of phosphatidylcholine molecular species using HPLC and light scattering detection. J Lipid Res 1998; 9(2):344-353.
62. Brouwers JF, Vernooij EA, Tielens AG et al. Rapid separation and identification of phosphatidylethanolamine molecular species. Journal of lipid research. 1999; 40(1):164-169.
63. ChristieWW. Detectors for high-performance liquid chromatography of lipids with special reference to evaporative light-scattering detection. In: Christie WW, ed. Advances in Lipid Methodology-One. Bridgewater:Oily Press, 1992:239-271.
64. Karlsson AA, Michelsen P, Odham G. Molecular species of sphingomyelin: Determination by high-performance liquid chromatography mass spectrometry with electrospray and high-performance liquid chromatography tandem mass spectrometry with atmospheric pressure chemical ionization. J Mass Spectrom 1998; 33:1192-1198.
65. Cole RB. Electrospray Ionization Mass Spectrometry. New York: John-Willey, 1997.
66. Fox TE. Diabetes alters sphingolipid metabolism in the retina. Diabetes 2006; 55:3573-3580
67. Dragusin M, WehnerS , Kelly S et al. Effects of sphingosine-1-phosphate and ceramide-1-phosphate on rat intestinal smooth muscle cells: implications for postoperative ileus. FASEB 2006; 20:1930-1932.
68. Luberto C, Hannun YA. Sphingomyelin synthase, a potential regulator of intracellular levels of ceramide and diacylglycerol during SV40 transformation- Does sphingomyelin synthase account for the putative, phosphatidylcholine-specific phospholipase. C J Biol Chem 1998; 273:14550-14559.
69. Bielawska A, Perry DK, Hannun YA. Determination of Ceramides and Diglycerides by the Diglyceride Kinase Assay. Anal Biochem 2001; 298:141-150.
70. Sullard MC, Wang E, Peng Q et al. Metabolomic Profiling of Sphingolipids in Human Glioma Cell Lines by Liquid Chromatography Tandem Mass Spectrometry, Cell Mol Biol 2003; 49:789-797.
71. Merrill AH, Jr, Wang E, Mullins RE, Jamison WC, Nimkar S, Liotta DC. Quantitation of free sphingosine in liver by high-performance liquid chromatography. Anal Biochem 1988;171:373-381.
72. Edsall LC, Spiegel S. Enzymatic Measurement of Sphingosine 1-Phosphate. Anal Biochem 1999; 272:80-86.

73. Caligan TB, Peters K, Ou J, et al. A high-performance liquid chromatographic method to measure sphingosine-1-phosphate and related compounds from sphingosine kinase assays and other biological samples, Anal Biochem 2000; 281:36-44.
74. Quanren H, Hirofumi S, Neelesh S, et al. Ceramide Synthase Inhibition by Fumonisin B1 Treatment Activates Sphingolipid-Metabolizing Systems in Mouse. Liver ToxSci Advance Access Toxicol Sci 2006; 94:388-397.
75. Tani M, Kihara A, Igarashi Y. Rescue of cell growth by sphingosine with disruption of lipid microdomain formation in Saccharomyces cerevisiae deficient in sphingolipid biosynthesis Biochem J 2006; 394: 237-242.
76. James L. Carroll, J, Diann M. McCoy SE, et al. Pulmonary-specific expression of tumor necrosis factor-alters surfactant lipid metabolism. Am J Physiol Lung Cell Mol Physiol 2002; 282: 735-742.
77. Gaver RC, Sweeley CC. Methods for methanolysis of sphingolipids and direct determination of long-chain bases by gas chromatography. J Am Oil Chem Soc 1965; 42:294-298.
78. Samuelsson K, Samuelsson B. Gas chromatographic and mass spectrometric studies of synthetic and naturally occurring ceramides. Chem Phys Lipids 1970; 5:44-79.

Chapter 4

Ceramide Synthases:
Roles in Cell Physiology and Signaling

Johnny Stiban, Rotem Tidhar and Anthony H. Futerman*

Abstract

Ceramide synthases (CerS) are integral membrane proteins of the endoplasmic reticulum. Six mammalian CerS have been described, with each utilizing fatty acyl CoAs of relatively defined chain lengths for *N*-acylation of the sphingoid long chain base. In this chapter, we review the main functional features of the CerS proteins, discuss their fatty acid specificity, kinetics, tissue distribution and mode of inhibition, as well as possible posttranslational modifications. We then address the reason that mammals contain six distinct CerS, whereas most other enzymes in the sphingolipid biosynthetic pathway only occur in one or two isoforms. Finally, we discuss the putative roles of CerS and the ceramide derived from the CerS, in signaling pathways and in development of disease.

Introduction

Ceramide (Cer) is an important bioactive lipid that has been implicated in a variety of cell biological processes ranging from regulation of cell growth to cell death and senescence.[1-3] The biochemical pathways by which Cer is generated are highly conserved between mammals and yeast.[4] Cer is composed of a long chain base (LCB), sphingosine (or sphinganine in the case of dihydroceramide (Fig. 1A)), which is acylated at the free amine nitrogen to form an amide bond. The *N*-acylation reaction is catalyzed by ceramide synthases (CerS).

The first molecular characterization of genes involved in de novo Cer synthesis was made about 6-7 years ago with the observation that two genes in yeast, Lag1p and Lac1p, were required for C26-Cer production[5,6] (C26 indicates an acyl chain of 26 carbon atoms), the main Cer species found in yeast. Around this time, an earlier version of this book was published.[7] Strikingly, in the earlier version there was no discussion of the molecular identification of CerS; thus, the current chapter serves not only to update the earlier book, but also demonstrates the remarkable progress made in the study of CerS over the past 5-6 years.

After the discovery of Lag1p and Lac1p, database searches over the next couple of years[8,9] revealed six mammalian homologs, which were initially named Lass (Longevity Assurance) genes, but were recently renamed CerS (Ceramide Synthase)[10] due to the assignment of their function as genuine ceramide synthases.[11] Figure 1B shows the alignment of the sequences of the human CerS family compared to the two yeast proteins. Phylogenetically, CerS1 is more closely related to yeast Lag1p and Lac1p than to the other mammalian homologs (Fig. 1C).[10,12] Moreover, all mammalian CerS, except CerS1, contain a Hox-like domain;[13] however, this Hox-like domain is unlikely to act as a transcription factor since the first 15 amino acid residues of the Hox domain are missing as are key residues involved in DNA binding. In addition, most of the Hox-like domain can be deleted without affecting catalytic activity.[13] Thus, the function of the Hox-like domain in

*Corresponding Author: Anthony H. Futerman—Department of Biological Chemistry, Weizmann Institute of Science, Rehovot 76100, Israel. Email: tony.futerman@weizmann.ac.il

Sphingolipids as Signaling and Regulatory Molecules, edited by Charles Chalfant and Maurizio Del Poeta.

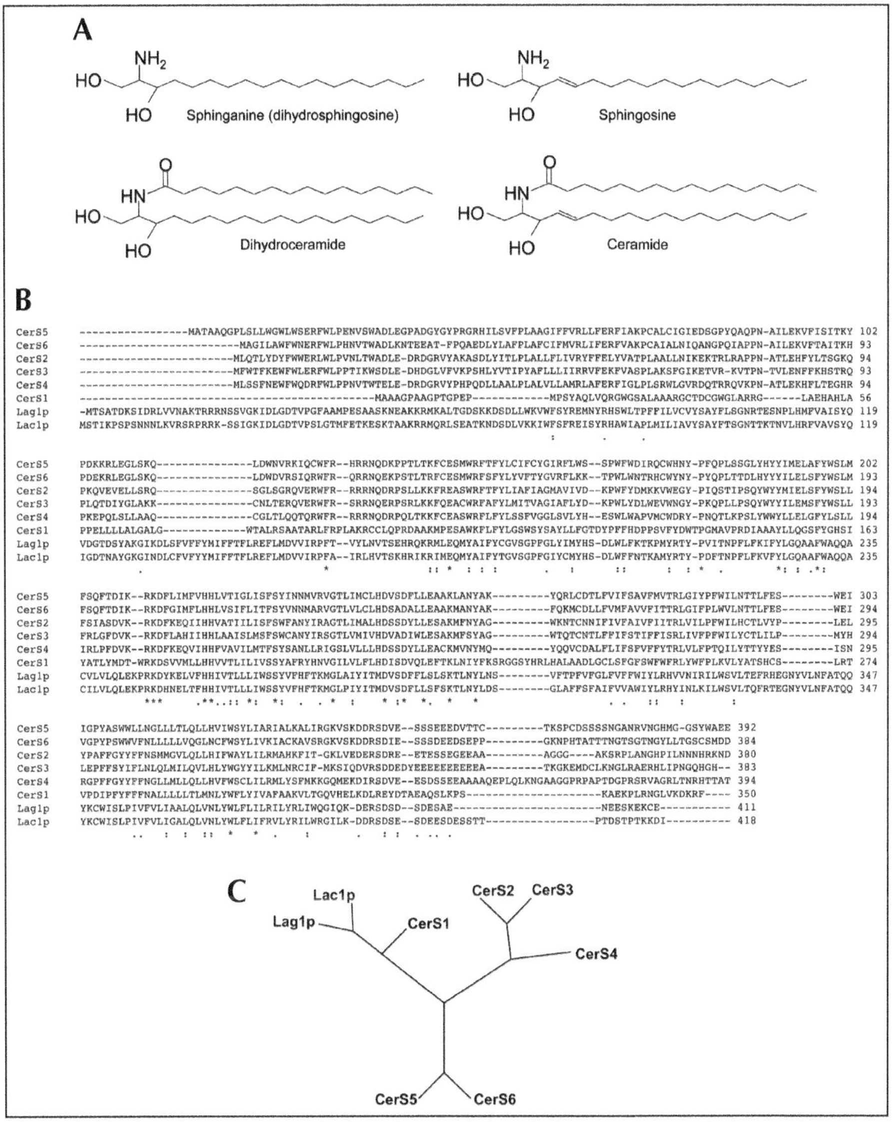

Figure 1. An overview of CerS biochemistry. A) Structures of CerS substrates and products. B) Alignments (performed using ClustalW2 (http://www.ebi.ac.uk/Tools/clustalw2/index.html)). C) Phylogenetic tree comparing human and yeast CerS.

(Hox)CerS (i.e., those CerS which contain a Hox-like domain) is currently unknown. In yeast, an additional subunit, Lip1, is required for ceramide synthesis,[14] but mammalian CerS do not appear to require any accessory proteins for activity.[11]

The main functional region of the CerS is the TLC (Tram-Lag-CLN8) domain,[9,10] a region of ~200 residues also found in other proteins.[9] This definition is based on two additional proteins, Tram1 and CLN8; Tram1 was found in a search for homologs of Lag1 in humans. The CerS active site is located in the TLC domain; however, other members of the TLC domain-containing family do not appear to have CerS activity although they may modulate CerS activity.[15]

Since their discovery, great interest has been shown in characterizing the CerS, with a view to understanding their biochemistry and their biology. It is these aspects that we will now review.

Fatty Acid Specificity, Kinetics and Tissue Distribution

Each mammalian CerS utilizes fatty acyl CoAs of relatively defined chain lengths for Cer synthesis (Table 1), using either sphinganine (derived from the biosynthetic pathway) or sphingosine (derived from SL recycling) as the LCB. Thus, CerS1 uses mostly C18-CoA,[16] CerS4 uses C18- and C20-CoAs,[17] CerS5 and CerS6 use mostly C16-CoA[12,17] and CerS3 uses very long chain acyl CoAs (C26 and higher).[18] CerS2 can utilize a wider range of fatty acyl CoAs, from C20 to C26, but does not use C16- or C18-CoAs.[19] The CerS produce 2-hydroxy (α-hydroxy) Cer with a chain length similar to that of the respective nonhydroxy-Cer.[20]

A recent study examined the kinetics of Cer formation. Reactions proceeded according to classical Michaelis-Menten kinetics[21] and the K_m values of all CerS towards sphinganine were in the low μM range (Table 1). It should be noted that the K_m values reported in this study were considerably lower than some reported earlier, in which values as high as 300 μM were obtained (reviewed in ref. 22). This may be due to the use of detergents in some of these earlier assays, rather than the detergent-free method described in reference 21. Interestingly, CerS4, which can use either C18-CoA or C20-CoA,[17] has an identical K_m value towards sphinganine irrespective of the acyl CoA chain length. This strongly supports the notion that the main biochemical difference between the CerS proteins is in their use of acyl CoAs and that the LCB binding site may be similar between CerS, although this cannot exclude the possibility that different CerS have different affinities towards different LCBs, as appears to be the case with CerS5.[17]

The CerS are differentially distributed in various tissues, such that subsets of Cer differing in acyl chain length could be made in specific tissues, presumably to meet the different physiological needs of each tissue. Analysis of levels of CerS mRNA in 14 mouse tissues[19] demonstrated that CerS2 was the most ubiquitously expressed, with the highest expression in kidney and liver (30-40 molecules RNA/ng total RNA) (Table 1). CerS5 and CerS6 are also expressed in most tissues, but

Table 1. Biochemical and physiological features of CerS

CerS	Acyl CoA Specificity[a,b,c]	Tissue Distribution[d]	mRNA Level in Predominant Tissue[d] (Molecules/ ng RNA)	K_m Towards Sphinganine[e] (μM)	Mitochondrial Localization[f]
1	C18, C18:1	Brain/skeletal muscles	3.3 ± 0.25	2.5 ± 0.7	Yes
2	C20, C22, C24, C24:1, C26	Kidney/liver	25-35	4.8 ± 0.4	Yes
3	C26 and above	Testes	22 ± 1.3	1.7 ± 0.4	Not measured
4	C18, C20	Heart/leucocytes/ liver/skin	6-8	1.8 ± 0.4	No
5	C14, C16, C18, C18:1	All tissues—high in lung and epithelia[g]	0.15-1.7	1.8 ± 0.4	No
6	C14, C16, C18	Intestine/kidney	2.5-3	2.0 ± 0.6	Yes

[a]Venkataraman et al, 2002; Riebeling et al, 2003; Lahiri et al, 2005; Laviad et al, 2008; [b]Mizutani et al, 2005; 2006; [c]Spassieva et al, 2006; [d]Riebeling et al, 2003; Laviad et al, 2008; [e]Lahiri et al, 2007; [f]Yu et al, 2007; [g]Xu et al, 2005.

the expression levels (~1-3 molecules RNA/ng total RNA) are much lower than those of CerS2; in another study, CerS5 was found to be highly expressed in mouse lung epithelial cells.[23] CerS4 is expressed in most tissues with the skin, leukocytes, heart and liver showing highest expression. CerS3 is exclusively expressed in testes[18,19] and skin,[19] specifically keratinocytes,[20] whereas CerS1 is highly expressed in brain and skeletal muscles.

To determine if there is a correlation between CerS mRNA levels and Cer acyl chain lengths in different tissues, the distribution of Cer subspecies was compared to those of the relative expression levels of CerS mRNA.[19] The two tissues with the highest CerS2 mRNA levels, kidney and liver, also had the highest proportions of C22 to C24-Cer. Kidney also has high proportions of C22-24 acyl chains in sphingomyelin (SM) and hexosylceramides (HexCer) as does liver, although the *N*-acyl chain composition of HexCer differs from that of Cer and SM. For the other three tissues examined (brain, testis and skeletal muscle), CerS2 mRNA was less prevalent than the mRNAs of the other CerS and the proportions of C22-, C24- and C24:1-Cer and -SMs are correspondingly lower. Interestingly, for two of these tissues (brain and skeletal muscle), HexCer contains surprisingly high proportions of C22-24-Cer, suggesting that there are factors other than the relative amounts of the CerS mRNA that affect the Cer subspecies distribution, particularly in downstream complex SLs and glycosphingolipids.

Most studies, at least those using immunofluorescence to examine the localization of ectopically-expressed proteins, suggest that CerS are located exclusively to the endoplasmic reticulum with no colocalization with mitochondrial markers;[16,17,19] earlier biochemical studies demonstrated that the CerS are found on the cytoplasmic leaflet of the ER.[24,25] Nevertheless, there is some evidence, based on biochemical isolation of sub-cellular fractions followed by Western blotting, that CerS can be detected in mitochondria and/or mitochondrial-associated ER membranes[26,27] (Table 1). Determination of the precise intracellular localization of the CerS awaits the generation of high quality specific antibodies.

Inhibitors

A number of specific inhibitors of CerS have been described, the most notable of which is Fumonisin, a mycotoxin derived from *Fusarium*. Another fungal-derived inhibitor is Australifungin. A third compound, the immunomodulator, FTY720, was recently shown to inhibit CerS activity.

Fumonisins

Fumonisin (FB) was first shown to inhibit Cer synthesis in 1991,[28] before the molecular identification of the CerS. Fumonisins are isolated from *Fusarium* moulds (which occur mainly in maize) and they bear considerable structural similarity to the LCB backbone of SLs. Two derivatives, FB1 and FB2, are potent inhibitors of CerS; FB1 inhibits CerS activity in rat liver microsomes and in isolated hepatocytes with an IC_{50} value of ~0.1 μM.[28] Further early studies demonstrated that inhibition occurs via competitive-like inhibition towards both sphinganine and C18-CoA[29] and that FB1 and FB2 block the proliferation of LLC-PK1 cells at concentrations between 10 and 35 μM; concentrations over 35 μM were cytotoxic.[30] FB1 is a specific inhibitor of the CerS, as no inhibitory effects were observed on the activities of other enzymes in the SL biosynthetic pathway.[31] In addition, FB1 was shown to block Cer synthesis in cultured hippocampal neurons, which led to significant changes in the rates of neuronal growth.[32,33]

Since these early studies, FB1 has been extensively used as an inhibitor of the de novo SL biosynthetic pathway in cell cultures as well as in animals and is routinely used to distinguish the effects of Cer generated via the action of sphingomyelinases compared to Cer generated de novo. However, since FB1 inhibits not only Cer synthesis, but also the synthesis of all subsequent down-stream SLs (i.e., SM and glycosphingolipids), care must be taken in the interpretation of results obtained after relatively long times of inhibition with FB1 (i.e., more than one hour). Moreover, upon over-expression of CerS activity in some cultured cells, FB1 elevates Cer levels.[8,17] The reason for this is unclear and to date there is no evidence that Cer is elevated by treatment with FB1 in cells that do not overexpress CerS.

Australifungin

Australifungin is a broad-range antifungal agent that acts against human pathogenic fungi at ranges from 0.015 to 1.0 μg/ml. Australifungin specifically inhibits SL synthesis by blocking CerS activity.[34] In crude membranes derived from yeast, the inhibitory concentration of Australifungin was 10 μM.[35] However, Australifungin is much less widely used as a CerS inhibitor than FB1.

FTY720

FTY720, a sphingosine analog, is in clinical trials as an immunomodulator. The biological effects of FTY720 are believed to occur mainly after its metabolism to FTY720 phosphate (FTY720-P). However, until recently, it was not known if FTY720 itself could interact with and modulate the activity of other enzymes of SL metabolism. Recently, we demonstrated that FTY720 inhibits CerS activity in vitro by noncompetitive inhibition towards acyl CoAs and uncompetitive inhibition towards sphinganine; the EC_{50} of inhibition varied from ~12-66 μM depending on the CerS and on the acyl chain length.[36] In cultured cells, FTY720 had a more complex effect, with Cer synthesis inhibited at high (500 nM to 5 μM) but not low (<200 nM) sphinganine concentrations, consistent with FTY720 acting as an uncompetitive inhibitor towards sphinganine. Finally and unexpectedly, elevated levels of Cer, SM and HexCers were observed after short times of incubation with FTY720. These data suggest that some of the effects of FTY720 observed in vivo might need to be re-evaluated in light of its ability to modulate CerS activity.[36]

Posttranslational Modifications

Indirect evidence from over a decade ago, based on rapid changes in CerS activity after various stimuli, suggested that CerS are likely to be modified posttranslationally.[37,38] Accumulating evidence supports a role for phosphorylation in the posttranslational modification of CerS. Activation of protein kinase C (PKC) increases de novo Cer synthesis,[39] which was attributed to up-regulation of CerS5 activity.[40] Deletion of the α'-catalytic subunit of casein kinase II in yeast reduced levels of Cer produced in vitro[35] and yeast lacking TOR (Target of Rapamycin) could not synthesize Cer. Ypk2, a kinase activated by TOR2, induces CerS activity and this step is antagonized by the Ca^{2+}/calmodulin-dependent phosphatase, calcineurin.[41] Calcineurin has also been shown to negatively regulate the formation of complex SLs[42,43] and the overexpression of calcineurin B subunit enhanced the oncogenic potential of HEK 293T cells.[44] Thus, indirect evidence suggests that CerS are modulated by phosphorylation, a notion supported by data from high performance mass spectrometry suggesting that mouse liver CerS2 and CerS5 are phosphorylated.[45] Moreover, CerS1 turnover after various drug treatments[46] appears to be regulated by the opposing actions of p38 MAP kinase and protein kinase C (PKC); p38 MAP kinase is a positive regulator of turnover, while PKC is a negative regulator of turnover. Pulse-chase labeling experiments demonstrated that CerS1 is phosphorylated in vivo and activation of PKC increases the phosphorylation of the protein.[47] The possible role of phosphorylation in regulating CerS activity is illustrated in Figure 2.

Membrane Topology

Early proteolytic digestion experiments implied that the active site of CerS faces the cytosol.[25] More recently, Igarashi and colleagues suggested that CerS2, 5 and 6 have five transmembrane domains; moreover, the N-terminus of these CerS is inside the lumen of the endoplasmic reticulum and the C-terminus, at least of CerS6, is in the cytosol.[12] However, another study suggested that the yeast CerS, Lag1p and Lac1p, have eight putative trans-membrane domains, with both the N- and C-termini facing the cytosol;[48] the conserved Lag motif, which contains the potential active site, was suggested to be embedded in the membrane.[48] We have tried to resolve this issue using the PHD predicted protein server,[49] which suggests that mammalian CerS have six putative trans-membrane domains with both N- and C-termini facing the cytosol (Fig. 3). Verification of the trans-membrane topology of the CerS awaits more detailed structural analyses and ultimately, resolution of their crystal structures.

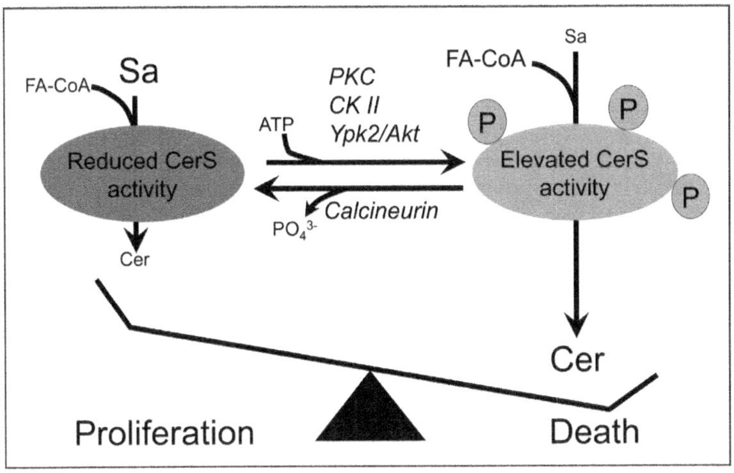

Figure 2. A putative role for CerS in regulating cell death. There have been suggestions that CerS can be phosphorylated (see text). Phosphorylation could occur via PKC (protein kinase C), CK II (casein kinase 2) or Ypk2 (or its mammalian orthologs, SGK and/or Akt/PKB) resulting in activated CerS and generation of pro-apoptotic Cer. Calcineurin could act in the opposite manner, by dephosphorylating CerS. The phosphates shown on the CerS are for illustration purposes only since there are currently no reports on CerS phosphorylation sites.

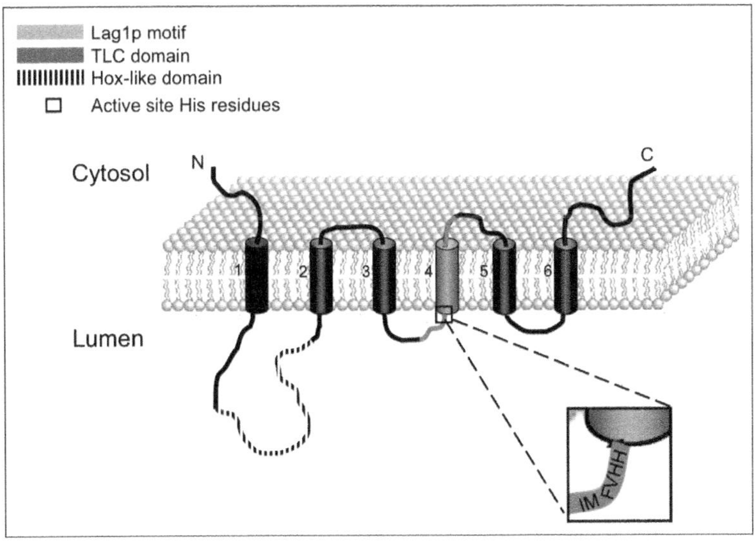

Figure 3. Predicted membrane topology of CerS. The prediction was performed using the PHD predict protein server (http://cubic.bioc.columbia.edu), which suggested six membrane spanning domains with both the N- and C-termini on the cytosolic leaflet (of the endoplasmic reticulum). The Hox-like domain (found in CerS2-6), shown as a dashed line, is located in the 1st lumenal loop. The TLC domain, shown in dark gray, begins at the end of the 1st lumenal loop and continues to the end of the 6th transmembrane domain. The Lag1p motif (which is part of the TLC domain) is shown in light gray and begins in the middle of the 2nd lumenal loop and continues through to the 2nd cytosolic loop. The two conserved histidine residues, which are proposed to be active site residues, are also shown; the enlargement shows the sequence of human CerS5.

Why Are There So Many Mammalian CerS?

The key question related to the physiology of CerS is why there are six distinct enzymes that essentially carry out the same reaction, namely *N*-acylation of the LCB, albeit with strict acyl CoA chain length specificity. A relatively straightforward answer is that ceramides containing specific fatty acids play more vital roles in cell physiology than once thought,[10] though what these roles are have not been fully delineated. It is known that ceramides with different acyl chain lengths have distinct biophysical properties[50,51] and the ceramides could themselves influence the biophysical properties of the membranes in which they are found, by for instance, differentially interacting with other membrane components. Evidence is also accumulating that ceramides with specific acyl chain lengths are generated in different signaling pathways and that these ceramides can differentially interact with downstream components in such pathways.[10,52] Furthermore, the multiple levels of regulation of CerS expression[17,19,23] and activity[37-45] support a vital role for the acyl chain length of ceramide in key events of cell physiology.

Although an integrated picture of how the CerS function together is currently lacking, considerable progress has been made in understanding the roles of individual CerS. CerS1, the first mammalian CerS to be described,[16] specifically synthesizes C18-Cer and is mainly expressed in the brain and in skeletal muscle and is almost undetectable in other tissues[12,19] (Table 1). CerS1 appears to be involved in cancer regulation and in modulating drug sensitivity. Overexpression of CerS1 inhibited the growth of human head and neck squamous cell carcinoma cells (HNSCC) and also increased the chemotherapy-induced apoptosis of these cells.[53,54] Moreover, there was a correlation between attenuated C18-Cer levels and the state of clinical disease, suggesting that C18-Cer is an important player in the regulation of HNSCC growth and/or pathogenesis.[55] CerS1 also has a unique role in regulating sensitivity to chemotherapeutic drugs;[46,56] CerS1 expression led to an increased sensitivity to cisplatin, which is widely used to treat a variety of solid tumors. In response to cisplatin, CerS1 expression increased the activation of the p38 mitogen-activated protein (MAP) kinase and concomitantly CerS1 was translocated from the ER to the Golgi apparatus.[46] This data suggests a potential role for CerS1 as a target for improving the efficacy of cisplatin therapy. In summary, CerS1 is regulated by mechanisms that involve PKC, ubiquitination and ER-to-Golgi translocation leading to its eventual proteasomal degradation.[47]

CerS2, which synthesizes C20-C26-Cer (Table 1) is the most ubiquitously expressed of all the CerS and has the broadest tissue distribution.[17,19] In a study on the expression of CerS in the brain, CerS2 was found to have the highest expression of all CerS in oligodendrocytes and Schwann cells and its up-regulation during myelination suggests it is responsible for the synthesis of the majority of SLs in myelin.[57] CerS2 is also regulated by a unique mechanism, namely via sphingosine 1-phosphate (S1P). S1P, but not lyso-phosphatidic acid[36] interacts with and inhibits CerS2 via two residues that are part of an S1P receptor-like motif, which is found only in CerS2.[19] The opposing functions that Cer and S1P play in signaling pathways suggests that this mode of regulation might be of significance in cell physiology and signaling. CerS2 also displays genomic features characteristic of a 'housekeeping' gene, although no other CerS genes display these characteristics.[19]

CerS3 is expressed at high levels in the skin,[17] which contains very long acyl chain ceramides that are involved in maintaining the water permeability barrier function[58] and in the testes, but is almost undetectable in other tissues.[12,19] CerS3 has been postulated to be involved in sperm formation and androgen production;[18] indeed, levels of germinal SLs containing very long acyl chains (synthesized by CerS3) increase during postnatal testicular maturation and are important for completion of spermatogenesis.[59]

CerS4, which uses C18- and C20-CoAs[17] is expressed mainly in skin, leukocytes, heart and liver.[19] CerS5, which synthesizes C16-Cer, is expressed in most tissues.[12,19] C16-Cer is the most abundant short-chain Cer in fibroblasts, endothelial cells and cells of the immune system[12,60] and has been shown to be of particular importance in apoptosis.[52,61,62] CerS6 is also expressed in most tissues[12,19] and produces short acyl chain ceramides (Table 1). However, little is known about the roles of CerS4, 5 and 6 in specific events in cell physiology.

One interesting difference between CerS1 and CerS4 and 5 emerged from a study looking at the role of each of these CerS in mediating drug sensitivity. CerS1 expression rendered cells more sensitive to cisplatin, carboplatin, doxorubicin and vincristine, but in contrast, expression of

CerS4 did not have any effect on the cellular sensitivity to any of the agents tested, while CerS5 expression increased the sensitivity only to doxorubicin and vincristine, but not to cisplatin and carboplatin.[46] These results strongly support the idea that the CerS genes are not equivalent in function. Similarly in yeast, the response to stress can vary from one isoform to another. Lac1p, but not Lag1p, is regulated by the pleiotropic drug resistance (Pdr) regulatory pathway, with Lac1p expression ~3 times higher than that of Lag1p.[63]

As summarized in this section, evidence is currently emerging that different CerS play different roles in mediating specific biochemical events and thus specific roles in cell physiology and it is to be hoped that their precise functions will have been clarified by the time of publication of the next edition of this book.

Roles of CerS in Signal Transduction and Disease

While study of the roles of individual CerS and their modes of regulation is currently in its infancy, much more is known about the roles of Cer and in particular about the roles of Cer containing specific acyl chain lengths. Much of this data has emerged from study of the generation of Cer from SM hydrolysis (via both neutral- and acid-sphingomyelinase), a research area that is somewhat more advanced than study of the generation of ceramides via the biosynthetic pathway.[64-66] The cross-talk between the generation of these two pathways is not very well understood; likewise, the coordinate regulation between CerS and other enzymes in the biosynthetic pathway (i.e., serine palmitoyl transferase, glucosylceramide synthase etc), needs to be further studied.

However, a number of studies have shown that Cer generated via CerS can influence cell fate, with up-regulation of CerS activity causing apoptosis and down-regulation inducing tumor formation. For instance, arsenic trioxide induces the production of cytotoxic levels of Cer by up-regulating de novo synthesis,[67] Cer production is stimulated in hypoxia/reoxygenation in NT-2 neuronal precursor cells by the concerted actions of acid sphingomyelinase and CerS5,[68] Cer levels are increased via the de novo pathway following p53 up-regulation in leukemia and colon cancer cells and CerS5 transcriptional up-regulation increases C16-Cer levels, ultimately causing cell death.[69] A variety of stress stimuli are known to increase Cer levels and some of these act via CerS.[70] A list of agents that induce stress is given in Table 2, which also summarizes the putative role of CerS in these processes.

Table 2. Some stress stimuli reported to cause ceramide elevation and apoptosis. For a number of these stimuli, ceramide synthesis via activation of CerS has been implicated, as indicated.

Stressor	CerS Activation	Reference
UVB[a] radiation	Yes	87
Heat	Yes	88-90
Etoposide	No	91
Hypoxia/reoxygenation	Yes	68, 92
Ischemia/reperfusion	Yes	27, 93
TNFα[b]	Yes	94
INFγ[c]	No available data	
IL-1β[d]	No	95
Gemcitabine/doxorubicin	Yes	54
Daunorubicin	Yes	96
4-HPR[e]	Yes	15

[a]Ultraviolet B; [b]Tumor necrosis factor α; [c]Interferon γ; [d]Interleukin-1 β; [e]4-(N-hydroxyphenyl) retinamide.

De novo Cer synthesis has also been implicated in a number of diseases,[71] such as diabetes,[72] cystic fibrosis (CF)[73,74] and chronic obstructive pulmonary disease (COPD). In the case of the latter disease, which is characterized by alveolar cell apoptosis, Cer up-regulation via the de novo pathway was shown to be directly involved, since inhibition of de novo synthesis by FB1 and myriocin prevented disease onset.[75,76] Cer is also involved in several liver conditions[77] such as hepatic ischemia/reperfusion,[78] steatohepatitis[79] and Wilson disease.[80] Cer may also play a role in Alzheimer's disease,[77] with long acyl chain Cer enriched in regions of the brain which are vulnerable to Alzheimer's disease[81] with concurrent elevation of CerS2 and CerS4 gene expression.[82] Cer is involved in neuronal death[83] and white matter dysfunction[84,85] and increased Cer levels were suggested to be involved in cerebral ischemia and stroke.[77,86] The relationship between Cer generated via de novo synthesis and that generated via sphingomyelinase action is an area of active study in each of these diseases and thus the role of regulation of CerS activity remains to be established.

Conclusion

In this chapter, we have summarized the remarkable progress made in study of the CerS genes and proteins since their identification in the early 2000s. Much still remains to be understood, not least whether these genes are coordinately regulated, whether the CerS proteins somehow interact with each other (or with other proteins) and thus modulate their activity and finally, the precise roles of the different Cer species made by each CerS in the various tissues where they are made. The discovery of the CerS has added a new dimension to SL research and the coming years are sure to yield many more unexpected findings.

References

1. Hannun YA, Obeid LM, Wolff RA. The novel second messenger ceramide: identification, mechanism of action and cellular activity. Adv Lipid Res 1993; 25:43-64.
2. Hannun YA. Functions of ceramide in coordinating cellular responses to stress. Science 1996; 274:1855-1859.
3. Futerman AH, Hannun YA. The complex life of simple sphingolipids. EMBO Rep 2004; 5:777-782.
4. Futerman AH, Riezman H. The ins and outs of sphingolipid synthesis. Trends Cell Biol 2005; 15:312-318.
5. Guillas I, Kirchman PA, Chuard R et al. C26-CoA-dependent ceramide synthesis of Saccharomyces cerevisiae is operated by Lag1p and Lac1p. Embo J 2001; 20:2655-2665.
6. Schorling S, Vallee B, Barz WP et al. Lag1p and Lac1p are essential for the Acyl-CoA-dependent ceramide synthase reaction in Saccharomyces cerevisae. Mol Biol Cell 2001; 12:3417-3427.
7. Futerman AH, editor. Ceramide Signaling: Kluwer Academic/Plenum Publishers 2002.
8. Venkataraman K, Futerman A. Do longevity assurance genes containing Hox domains regulate cell development via ceramide synthesis? FEBS Lett 2002; 528:3-4.
9. Winter E, Ponting CP. TRAM, LAG1 and CLN8: members of a novel family of lipid-sensing domains? Trends Biochem Sci 2002; 27:381-383.
10. Pewzner-Jung Y, Ben-Dor S, Futerman AH. When do Lasses (longevity assurance genes) become CerS (ceramide synthases)?: Insights into the regulation of ceramide synthesis. J Biol Chem 2006; 281:25001-25005.
11. Lahiri S, Futerman AH. LASS5 is a bona fide dihydroceramide synthase that selectively utilizes palmitoyl-CoA as acyl donor. J Biol Chem 2005; 280:33735-33738.
12. Mizutani Y, Kihara A, Igarashi Y. Mammalian Lass6 and its related family members regulate synthesis of specific ceramides. Biochem J 2005; 390:263-271.
13. Mesika A, Ben-Dor S, Laviad EL et al. A new functional motif in Hox domain-containing ceramide synthases: identification of a novel region flanking the Hox and TLC domains essential for activity. J Biol Chem 2007; 282:27366-27373.
14. Vallee B, Riezman H. Lip1p: a novel subunit of acyl-CoA ceramide synthase. Embo J 2005; 24:730-741.
15. Schulz A, Mousallem T, Venkataramani M et al. The CLN9 protein, a regulator of dihydroceramide synthase. J Biol Chem 2006; 281:2784-2794.
16. Venkataraman K, Riebeling C, Bodennec J et al. Upstream of growth and differentiation factor 1 (uog1), a mammalian homolog of the yeast longevity assurance gene 1 (LAG1), regulates N-stearoyl-sphinganine (C18-(dihydro)ceramide) synthesis in a fumonisin B1-independent manner in mammalian cells. J Biol Chem 2002; 277:35642-35649.

17. Riebeling C, Allegood JC, Wang E et al. Two mammalian longevity assurance gene (LAG1) family members, trh1 and trh4, regulate dihydroceramide synthesis using different fatty acyl-CoA donors. J Biol Chem 2003; 278:43452-43459.
18. Mizutani Y, Kihara A, Igarashi Y. LASS3 (longevity assurance homologue 3) is a mainly testis-specific (dihydro)ceramide synthase with relatively broad substrate specificity. Biochem J 2006; 398:531-538.
19. Laviad EL, Albee L, Pankova-Kholmyansky I et al. Characterization of ceramide synthase 2: tissue distribution, substrate specificity and inhibition by sphingosine 1-phosphate. J Biol Chem 2008; 283:5677-5684.
20. Mizutani Y, Kihara A, Chiba H et al. 2-Hydroxy-ceramide synthesis by ceramide synthase family: enzymatic basis for the preference of FA chain length. J Lipid Res 2008; 49:2356-2364.
21. Lahiri S, Lee H, Mesicek J et al. Kinetic characterization of mammalian ceramide synthases: determination of K(m) values towards sphinganine. FEBS Lett 2007; 581:5289-5294.
22. Wang E, Merrill AH. Ceramide Synthase. Methods Enzymol 1999; 311:15-21.
23. Xu Z, Zhou J, McCoy DM et al. LASS5 is the predominant ceramide synthase isoform involved in de novo sphingolipid synthesis in lung epithelia. J Lipid Res 2005; 46:1229-1238.
24. Mandon EC, Ehses I, Rother J et al. Subcellular localization and membrane topology of serine palmitoyltransferase, 3-dehydrosphinganine reductase and sphinganine N-acyltransferase in mouse liver. J Biol Chem 1992; 267:11144-11148.
25. Hirschberg K, Rodger J, Futerman AH. The long-chain sphingoid base of sphingolipids is acylated at the cytosolic surface of the endoplasmic reticulum in rat liver. Biochem J 1993; 290:751-757.
26. Bionda C, Portoukalian J, Schmitt D et al. Subcellular compartmentalization of ceramide metabolism: MAM (mitochondria-associated membrane) and/or mitochondria? Biochem J 2004; 382:527-533.
27. Yu J, Novgorodov SA, Chudakova D et al. JNK3 signaling pathway activates ceramide synthase leading to mitochondrial dysfunction. J Biol Chem 2007; 282:25940-25949.
28. Wang E, Norred WP, Bacon CW et al. Inhibition of sphingolipid biosynthesis by fumonisins. Implications for diseases associated with Fusarium moniliforme. J Biol Chem 1991; 266:14486-14490.
29. Merrill AH Jr, van Echten G, Wang E et al. Fumonisin B1 inhibits sphingosine (sphinganine) N-acyltransferase and de novo sphingolipid biosynthesis in cultured neurons in situ. J Biol Chem 1993; 268:27299-27306.
30. Yoo HS, Norred WP, Wang E et al. Fumonisin inhibition of de novo sphingolipid biosynthesis and cytotoxicity are correlated in LLC-PK1 cells. Toxicol Appl Pharmacol 1992; 114:9-15.
31. Wu WI, McDonough VM, Nickels JT et al. Regulation of lipid biosynthesis in Saccharomyces cerevisiae by fumonisin B1. J Biol Chem 1995; 270:13171-13178.
32. Harel R, Futerman AH. Inhibition of sphingolipid synthesis affects axonal outgrowth in cultured hippocampal neurons. J Biol Chem 1993; 268:14476-14481.
33. Schwarz A, Futerman AH. Inhibition of sphingolipid synthesis, but not degradation, alters the rate of dendrite growth in cultured hippocampal neurons. Brain Res Dev Brain Res 1998; 108:125-130.
34. Mandala SM, Thornton RA, Frommer BR et al. The discovery of australifungin, a novel inhibitor of sphinganine N-acyltransferase from Sporormiella australis. Producing organism, fermentation, isolation and biological activity. J Antibiot (Tokyo) 1995; 48:349-356.
35. Kobayashi SD, Nagiec MM. Ceramide/long-chain base phosphate rheostat in Saccharomyces cerevisiae: regulation of ceramide synthesis by Elo3p and Cka2p. Eukaryot Cell 2003; 2:284-294.
36. Lahiri S, Park H, Laviad EL et al. Ceramide synthesis is modulated by the sphingosine analog FTY720 via a mixture of uncompetitive and non-competitive inhibition in an acyl CoA chain length-dependent manner. J Biol Chem 2009; 284:16090-16098.
37. Bose R, Verheij M, Haimovitz-Friedman A et al. Ceramide synthase mediates daunorubicin-induced apoptosis: an alternative mechanism for generating death signals. Cell 1995; 82:405-414.
38. Yano M, Kishida E, Muneyuki Y et al. Quantitative analysis of ceramide molecular species by high performance liquid chromatography. J Lipid Res 1998; 39:2091-2098.
39. Becker KP, Kitatani K, Idkowiak-Baldys J et al. Selective inhibition of juxtanuclear translocation of protein kinase C betaII by a negative feedback mechanism involving ceramide formed from the salvage pathway. J Biol Chem 2005; 280:2606-2612.
40. Kitatani K, Idkowiak-Baldys J, Bielawski J et al. Protein kinase C-induced activation of a ceramide/protein phosphatase 1 pathway leading to dephosphorylation of p38 MAPK. J Biol Chem 2006; 281:36793-36802.
41. Aronova S, Wedaman K, Aronov PA et al. Regulation of ceramide biosynthesis by TOR complex 2. Cell Metab 2008; 7:148-158.
42. Mulet JM, Martin DE, Loewith R et al. Mutual antagonism of target of rapamycin and calcineurin signaling. J Biol Chem 2006; 281:33000-33007.
43. Tabuchi M, Audhya A, Parsons AB et al. The phosphatidylinositol 4,5-biphosphate and TORC2 binding proteins Slm1 and Slm2 function in sphingolipid regulation. Mol Cell Biol 2006; 26:5861-5875.
44. Wang YL, Wang Y, Tong L et al. Overexpression of calcineurin B subunit (CnB) enhances the oncogenic potential of HEK293 cells. Cancer Sci 2008; 99:1100-1108.

45. Villen J, Beausoleil SA, Gerber SA et al. Large-scale phosphorylation analysis of mouse liver. Proc Natl Acad Sci USA 2007; 104:1488-1493.
46. Min J, Mesika A, Sivaguru M et al. (Dihydro)ceramide synthase 1 regulated sensitivity to cisplatin is associated with the activation of p38 mitogen-activated protein kinase and is abrogated by sphingosine kinase 1. Mol Cancer Res 2007; 5:801-812.
47. Sridevi P, Alexander H, Laviad EL et al. Ceramide synthase 1 is regulated by proteasomal mediated turnover. Biochim Biophys Acta 2009; 1793:1218-1227.
48. Kageyama-Yahara N, Riezman H. Transmembrane topology of ceramide synthase in yeast. Biochem J 2006; 398:585-593.
49. Rost B. PHD: predicting one-dimensional protein structure by profile-based neural networks. Methods Enzymol 1996; 266:525-539.
50. Sot J, Aranda FJ, Collado MI et al. Different effects of long- and short-chain ceramides on the gel-fluid and lamellar-hexagonal transitions of phospholipids: a calorimetric, NMR and x-ray diffraction study. Biophys J 2005; 88:3368-3380.
51. Pinto SN, Silva LC, de Almeida RF et al. Membrane domain formation, interdigitation and morphological alterations induced by the very long chain asymmetric C24:1 ceramide. Biophys J 2008; 95:2867-2879.
52. Kroesen BJ, Jacobs S, Pettus BJ et al. BcR-induced apoptosis involves differential regulation of C16 and C24-ceramide formation and sphingolipid-dependent activation of the proteasome. J Biol Chem 2003; 278:14723-14731.
53. Koybasi S, Senkal CE, Sundararaj K et al. Defects in cell growth regulation by C18:0-ceramide and longevity assurance gene 1 in human head and neck squamous cell carcinomas. J Biol Chem 2004; 279:44311-44319.
54. Senkal CE, Ponnusamy S, Rossi MJ et al. Role of human longevity assurance gene 1 and C18-ceramide in chemotherapy-induced cell death in human head and neck squamous cell carcinomas. Mol Cancer Ther 2007; 6:712-722.
55. Karahatay S, Thomas K, Koybasi S et al. Clinical relevance of ceramide metabolism in the pathogenesis of human head and neck squamous cell carcinoma (HNSCC): attenuation of C(18)-ceramide in HNSCC tumors correlates with lymphovascular invasion and nodal metastasis. Cancer Lett 2007; 256:101-111.
56. Baran Y, Salas A, Senkal CE et al. Alterations of ceramide/sphingosine 1-phosphate rheostat involved in the regulation of resistance to imatinib-induced apoptosis in K562 human chronic myeloid leukemia cells. J Biol Chem 2007; 282:10922-10934.
57. Becker I, Wang-Eckhardt L, Yaghootfam A et al. Differential expression of (dihydro)ceramide synthases in mouse brain: oligodendrocyte-specific expression of CerS2/Lass2. Histochem Cell Biol 2008; 129:233-241.
58. Coderch L, Lopez O, de la Maza A et al. Ceramides and skin function. Am J Clin Dermatol 2003; 4:107-129.
59. Rabionet M, van der Spoel AC, Chuang CC et al. Male germ cells require polyenoic sphingolipids with complex glycosylation for completion of meiosis: a link to ceramide synthase-3. J Biol Chem 2008; 283:13357-13369.
60. Pettus BJ, Baes M, Busman M et al. Mass spectrometric analysis of ceramide perturbations in brain and fibroblasts of mice and human patients with peroxisomal disorders. Rapid Commun Mass Spectrom 2004; 18:1569-1574.
61. Eto M, Bennouna J, Hunter OC et al. C16 ceramide accumulates following androgen ablation in LNCaP prostate cancer cells. Prostate 2003; 57:66-79.
62. Osawa Y, Uchinami H, Bielawski J et al. Roles for C16-ceramide and sphingosine 1-phosphate in regulating hepatocyte apoptosis in response to tumor necrosis factor-alpha. J Biol Chem 2005; 280:27879-27887.
63. Kolaczkowski M, Kolaczkowska A, Gaigg B et al. Differential regulation of ceramide synthase components LAC1 and LAG1 in Saccharomyces cerevisiae. Eukaryot Cell 2004; 3:880-892.
64. Dickson RC. Thematic review series: sphingolipids. New insights into sphingolipid metabolism and function in budding yeast. J Lipid Res 2008; 49:909-921.
65. Bartke N, Hannun YA. Bioactive sphingolipids: metabolism and function. J Lipid Res 2008.
66. Reynolds CP, Maurer BJ, Kolesnick RN. Ceramide synthesis and metabolism as a target for cancer therapy. Cancer Lett 2004; 206:169-180.
67. Dbaibo GS, Kfoury Y, Darwiche N et al. Arsenic trioxide induces accumulation of cytotoxic levels of ceramide in acute promyelocytic leukemia and adult T-cell leukemia/lymphoma cells through de novo ceramide synthesis and inhibition of glucosylceramide synthase activity. Haematologica 2007; 92:753-762.
68. Jin J, Hou Q, Mullen TD et al. Ceramide generated by sphingomyelin hydrolysis and the salvage pathway is involved in hypoxia/reoxygenation-induced Bax redistribution to mitochondria in NT-2 cells. J Biol Chem 2008; 283:26509-26517.
69. Panjarian S, Kozhaya L, Arayssi S et al. De novo N-palmitoylsphingosine synthesis is the major biochemical mechanism of ceramide accumulation following p53 up-regulation. Prostaglandins Other Lipid Mediat 2008; 86:41-48.

70. Pandey S, Murphy RF, Agrawal DK. Recent advances in the immunobiology of ceramide. Exp Mol Pathol 2007; 82:298-309.
71. Lahiri S, Futerman AH. The metabolism and function of sphingolipids and glycosphingolipids. Cell Mol Life Sci 2007; 64:2270-2284.
72. Summers SA. Ceramides in insulin resistance and lipotoxicity. Prog Lipid Res 2006; 45:42-72.
73. Ito Y, Sato S, Ohashi T et al. Reduction of airway anion secretion via CFTR in sphingomyelin pathway. Biochem Biophys Res Commun 2004; 324:901-908.
74. Vilela RM, Lands LC, Meehan B et al. Inhibition of IL-8 release from CFTR-deficient lung epithelial cells following pretreatment with fenretinide. Int Immunopharmacol 2006; 6:1651-1664.
75. Petrache I, Natarajan V, Zhen L et al. Ceramide upregulation causes pulmonary cell apoptosis and emphysema-like disease in mice. Nat Med 2005; 11:491-498.
76. Petrache I, Natarajan V, Zhen L et al. Ceramide causes pulmonary cell apoptosis and emphysema: a role for sphingolipid homeostasis in the maintenance of alveolar cells. Proc Am Thorac Soc 2006; 3:510.
77. Morales A, Lee H, Goni FM et al. Sphingolipids and cell death. Apoptosis 2007; 12:923-939.
78. Llacuna L, Mari M, Garcia-Ruiz C et al. Critical role of acidic sphingomyelinase in murine hepatic ischemia-reperfusion injury. Hepatology 2006; 44:561-572.
79. Mari M, Caballero F, Colell A et al. Mitochondrial free cholesterol loading sensitizes to TNF- and Fas-mediated steatohepatitis. Cell Metab 2006; 4:185-198.
80. Lang PA, Schenck M, Nicolay JP et al. Liver cell death and anemia in Wilson disease involve acid sphingomyelinase and ceramide. Nat Med 2007; 13:164-170.
81. Cutler RG, Kelly J, Storie K et al. Involvement of oxidative stress-induced abnormalities in ceramide and cholesterol metabolism in brain aging and Alzheimer's disease. Proc Natl Acad Sci USA 2004; 101:2070-2075.
82. Wang G, Silva J, Dasgupta S et al. Long-chain ceramide is elevated in presenilin 1 (PS1M146V) mouse brain and induces apoptosis in PS1 astrocytes. Glia 2008; 56:449-456.
83. Schwarz A, Futerman AH. Distinct roles for ceramide and glucosylceramide at different stages of neuronal growth. J Neurosci 1997; 17:2929-2938.
84. Lee JT, Xu J, Lee JM et al. Amyloid-beta peptide induces oligodendrocyte death by activating the neutral sphingomyelinase-ceramide pathway. J Cell Biol 2004; 164:123-131.
85. Jana A, Pahan K. Fibrillar amyloid-beta peptides kill human primary neurons via NADPH oxidase-mediated activation of neutral sphingomyelinase. Implications for Alzheimer's disease. J Biol Chem 2004; 279:51451-51459.
86. Yu ZF, Nikolova-Karakashian M, Zhou D et al. Pivotal role for acidic sphingomyelinase in cerebral ischemia-induced ceramide and cytokine production and neuronal apoptosis. J Mol Neurosci 2000; 15:85-97.
87. Uchida Y, Nardo AD, Collins V et al. De novo ceramide synthesis participates in the ultraviolet B irradiation-induced apoptosis in undifferentiated cultured human keratinocytes. J Invest Dermatol 2003; 120:662-669.
88. Jenkins GM, Richards A, Wahl T et al. Involvement of yeast sphingolipids in the heat stress response of Saccharomyces cerevisiae. J Biol Chem 1997; 272:32566-32572.
89. Wells GB, Dickson RC, Lester RL. Heat-induced elevation of ceramide in Saccharomyces cerevisiae via de novo synthesis. J Biol Chem 1998; 273:7235-7243.
90. Jenkins GM, Cowart LA, Signorelli P et al. Acute activation of de novo sphingolipid biosynthesis upon heat shock causes an accumulation of ceramide and subsequent dephosphorylation of SR proteins. J Biol Chem 2002; 277:42572-42578.
91. Perry DK, Carton J, Shah AK et al. Serine palmitoyltransferase regulates de novo ceramide generation during etoposide-induced apoptosis. J Biol Chem 2000; 275:9078-9084.
92. Basnakian AG, Ueda N, Hong X et al. Ceramide synthase is essential for endonuclease-mediated death of renal tubular epithelial cells induced by hypoxia-reoxygenation. Am J Physiol Renal Physiol 2005; 288:F308-314.
93. Cuzzocrea S, Di Paola R, Genovese T et al. Anti-inflammatory and anti-apoptotic effects of fumonisin B1, an inhibitor of ceramide synthase, in a rodent model of splanchnic ischemia and reperfusion injury. J Pharmacol Exp Ther 2008; 327:45-57.
94. Meyer SG, de Groot H. Cycloserine and threo-dihydrosphingosine inhibit TNF-alpha-induced cytotoxicity: evidence for the importance of de novo ceramide synthesis in TNF-alpha signaling. Biochim Biophys Acta 2003; 1643:1-4.
95. Veluthakal R, Jangati GR, Kowluru A. IL-1beta-induced iNOS expression, NO release and loss in metabolic cell viability are resistant to inhibitors of ceramide synthase and sphingomyelinase in INS 832/13 cells. Jop 2006; 7:593-601.
96. Turnbull KJ, Brown BL, Dobson PR. Caspase-3-like activity is necessary but not sufficient for daunorubicin-induced apoptosis in Jurkat human lymphoblastic leukemia cells. Leukemia 1999; 13:1056-1061.

Chapter 5

Tales and Mysteries of the Enigmatic Sphingomyelin Synthase Family

Joost C.M. Holthuis and Chiara Luberto*

Abstract

In the last five years tremendous progress has been made toward the understanding of the mechanisms that govern sphingomyelin (SM) synthesis in animal cells. In line with the complexity of most biological processes, also in the case of SM biosynthesis, the more we learn the more enigmatic and finely tuned the system appears. Therefore with this review we aim first, at highlighting the most significant discoveries that advanced our knowledge and understanding of SM biosynthesis, starting from the discovery of SM to the identification of the enzymes responsible for its production; and second, at discussing old and new riddles that such discoveries pose to current investigators.

Sphingomyelin Biosynthesis: An Historical Perspective

Initial Milestones

Sphingomyelin (SM) was first isolated by the German biochemist Thudichum in 1884 and its name derived from both the enigmatic and novel nature of its chemical structure (in the Greek mythology the sphinx is a monster that posed a riddle) and the tissue where it was isolated from (myelin).[1] In spite of the initial suggestion that SM might have a specific role in neural function, later studies showed that SM is present in all mammalian tissues as well as lipoproteins. SM is indeed one of the most abundant phospholipids and, in cells, it forms a concentration gradient along the secretory pathway with the highest concentration in the plasma membrane (where it accumulates in the exoplasmic leaflet).

SM is composed of a ceramide module and a phosphocholine (P-choline) moiety bound to the primary hydroxyl group (Fig. 1). It was first proposed by Sribney and Kennedy[2] that CDP-choline was the donor of the P-choline headgroup of SM, similarly to the reaction that leads to the biosynthesis of phasphatidylcholine(PC), known as the "Kennedy pathway".[3] In their experimental conditions, it was found that ceramide with the nonnatural threo configuration was a better substrate as compared to the naturally occurring erythro. In 1965, an alternative reaction for SM biosynthesis was proposed by Roscoe Brady and colleagues.[4] In this case, evidence was provided to show that, in rat brain preparations, the synthesis of SM could also occur from acylation of lyso-SM by stearoyl-CoA and later found that both erythro or threo-lyso-SM were equally active as lipid acceptors in the reaction.[5] Shortly after, Fujino et al[6] demonstrated that, by twitching the protocol for the preparation of the incubation mixture, also erythro- as well as threo-ceramide was recognized as acceptor of the P-choline group from CDP-choline to form SM according to the Kennedy reaction, even if these results were later questioned.[7] In 1972 a novel reaction for the synthesis of SM was proposed

*Corresponding Author: Chiara Luberto—Department of Biochemistry and Molecular Biology, Medical University of South Carolina, Charleston, South Carolina, USA. Email: luberto@musc.edu

Sphingolipids as Signaling and Regulatory Molecules, edited by Charles Chalfant and Maurizio Del Poeta. ©2010 Landes Bioscience and Springer Science+Business Media.

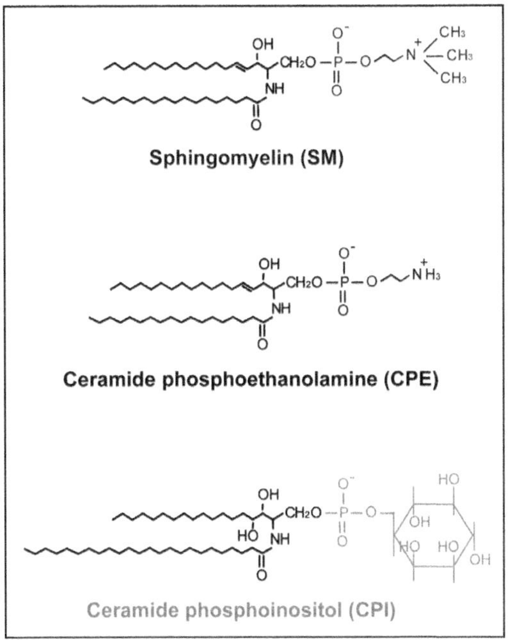

Figure 1. Schematic of the biochemical structures of SM, CPE and CPI. In black is represented the basic ceramide module common to the three molecules and in color is represented each specific head-group. The ceramide module of CPI is more often characterized by long and very long chain fatty acid and the presence of an hydroxy group in position 4 of the sphingoid backbone. A color version of this image is available at www.landesbioscience.com/curie.

by Diringer et al which identified PC as the donor of the P-choline, by using pulse-chase labeling techniques with ^{32}P-labeled phosphorus and ^{3}H-choline and looking at the kinetics by which the radioactivity associated with PC or SM was changing in growing SV40 transformed mouse fibroblasts.[8] Finally, in 1974 Ullman and Radin conclusively demonstrated in vitro that PC is indeed the donor of the P-choline for synthesis of SM in mouse liver preparations.[9] They found that natural erythro-ceramide was the preferred substrate, that the reaction did not need addition of cations and that it was characterized by an optimal neutral pH. In addition to liver, the reaction occurred in preparations from kidney, lung, spleen and heart but it was strikingly absent in brain, suggesting the existence of an alternative route for SM synthesis in this organ. Ever since, the scientific community settled on PC as the donor of the P-choline moiety of SM, with few exceptions,[10] such that SMS is also referred to as the phosphatidylcholine:ceramide cholinephosphotransferase (Fig. 2).

Localization of SM Synthase Activity in Cells

In the early studies addressing the biochemical characterization of the P-choline donor for SMS activity, the source of the enzyme was predominantly microsomal preparations, thus no major information on the subcellular localization of SMS activity was provided, except for the fact that it was tightly membrane-bound. Later on it was suggested that, in mouse SV40-transformed fibroblasts, rat liver, or 3T3-L1 fibroblasts most of the activity resided in the plasma membrane.[11-13] SMS activity at the Golgi was later detected also in the same mouse SV40-transformed fibroblasts and in Chinese hamster lung fibroblasts and epithelial Madin-Darby canine kidney cells.[14-17] In order to clarify the issue of subcellular localization of SM synthesis, a thorough study was performed in 1990 by Futerman and coworkers using the well-characterized protocol for subcellular fractionation of rat liver and by paying particular attention to inhibit the activity of the SM-metabolizing enzyme

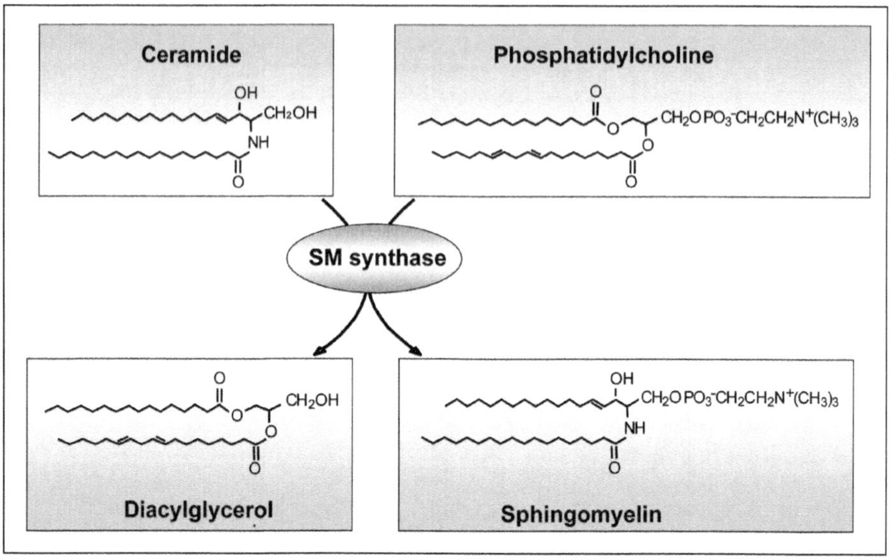

Figure 2. Schematic of the biochemical reaction catalyzed by SM synthase.

neutral sphingomyelinase by addition of metal chelators.[18] In this study it was shown that most of the SMS activity resided in the cis-medial enriched Golgi (75%) (with a luminal orientation of the catalytic site) and minor activity was detected in the trans enriched Golgi (15%) and plasma membrane (~10%) fractions. Similarly, Jeckel and coworkers confirmed enrichment of the SMS in enriched cis Golgi apparatus from rat liver.[19,20] Confirmation of the Golgi as the main site of SM biosynthesis came also from studies using inhibitors of vescicular trafficking (low temperatures or pharmacological agents). In this case, no appearance of newly synthesized SM on the plasma membrane could be detected, suggesting that the main site for SM synthesis resided in the cell and that vescicular trafficking was required for SM delivery to the plasma membrane.[16,17] To complicate matters, the idea that, in BHK cells, SM destined to the plasma membrane is synthesized in recycling endosomes was also put forward[21-23] but later unequivocally disproved.[24] Also, SMS activity seemed absent from an intermediate ER-Golgi compartment (ERGIC) and, also in the Vero cells employed in this study, it was mostly enriched in the Golgi fraction.[25] In 1998, the use of preparative free-solution isotachophoresis (FS-ITP) allowed separation of enzymatically active Golgi subfractions from rat liver isolated Golgi apparatus.[26] SMS activity was concentrated in the cis/medial fractions where the specific activity was the highest and where the ceramide was mostly concentrated. On the other hand, newly synthesized SM seemed to be transferred to the trans Golgi fractions where SM mass was most abundant. Shortly after, the question whether significant SMS activity could be present in the trans Golgi network (TGN) was raised by Allan and coworkers[27] using BHK cells infected with Semiliki Forest virus. It was postulated that this viral infection promoted incorporation of viral proteins into the TGN causing a significant increase in the density of these membranes. It was shown that, after infection, the shifted peak, which most likely contained trans Golgi network membranes, carried most of the SMS activity and it was enriched in SM whereas the profile of glucosylceramide synthase activity remained unchanged. The confusion about the sub-cellular localization of SMS within the Golgi apparatus found partial resolution when a study by Sadeghlar et al[28] showed that different cells might present a different distribution of SMS activity along this compartment. In fact, in primary neurons and neuroblastoma cells SM synthesis seemed to be primarily localized in late compartments of the Golgi apparatus, probably the TGN whereas in fibroblasts most of the activity resided in the early/mid Golgi. Finally, in rat

Sertoli cells most of SMS activity was found associated with the trans-Golgi cisternae and only a minor fraction was present in the early Golgi and plasma membrane.

Discovery of a Ceramide Transfer Protein with a Key Role in SM Biosynthesis

Additional clues on the localization and mechanism of SM synthesis came more recently with the identification of a ceramide transfer protein (CERT). In an attempt to identify the enzyme(s) responsible for SM biosynthesis, Hanada and coworkers set up a screen for CHO mutant cells searching for clones resistant to lysenin, a SM-directed cytolysin.[29] One of the mutant clones, LY-A showed a defect in SM biosynthesis without any appreciable impairment in the SMS enzymatic activity per se or in the synthesis of the precursors of the reaction, namely ceramide and PC.[30] By exploiting the hypersensitivity of LY-A cells to agents that extract cholesterol such as methyl-β-cyclodextrin (MCD), the authors searched for cDNAs that were able to revert the MCD-sensitivity and restore SM biosynthesis. This approach yielded a cDNA coding for CERT. Two splice variants of CERT exist, a long less represented variant, identical to the Goodpasture antigen-binding protein (GPBP) and a shorter more represented variant, missing 26 amino acids and identical to GPBPΔ26.[31] Both were found to mediate intermembrane transfer of ceramide.[30] CERT is characterized by three different domains: (i) a FFAT motif which favors CERT approximation to the ER, the site of ceramide synthesis, through the binding to the ER resident membrane protein vamp associated protein (VAP)[32]; (ii) a pleckstrin homology (PH) domain that binds to phosphatidylinositol-4-monophosphate (PI4P) and targets CERT to the Golgi/trans Golgi network[30]; (iii) a START domain that facilitates the extraction of ceramide from a donor membrane and its subsequent transfer to an acceptor membrane.[33]

The nonvescicular transfer of ceramide from the ER to the Golgi requires ATP.[34,35] The ATP-dependent step for ceramide trafficking might be attributable to the synthesis of PI4P in the Golgi. Alternatively, ATP may allow the recycling of CERT after its binding to PI4P or VAP or it could energize an ATP-dependent translocase that moves ceramide from the cytosolic side of the Golgi to the luminal side where the catalytic domain of SMS is located.[36] The spatial organization of the CERT-mediated transfer of ceramide between the ER and Golgi is not clear. The over-expression of a green fluorescent protein (GFP) tagged-CERT in CHO cells showed its localization at the cis/medial Golgi apparatus pointing to the early Golgi as the site for ceramide transfer.[30] Interestingly though, ER-trans Golgi contact sites exist[37,38] and therefore it is plausible that the transfer of ceramide from the ER to the Golgi could also occur at these sites of contact. This would also imply that a pool of SMS should reside at the trans Golgi.

Even though CERT regulates a main route for the transport of ceramide from the ER to the Golgi, it has been recently demonstrated that, in glioma cells, an additional vescicular mechanism of ceramide transport, sensitive to nitric oxide, is active alongside the CERT-mediated transport.[39] Whether the ceramide transported through this route can reach alternative pools of SMS in the Golgi apparatus (i.e., early Golgi) is a fascinating still open question.

Alternative Pathways of SM Biosynthesis and Analogous Reactions

The observation that brain, in spite of containing large amounts of SM, lacked any obvious SMS activity led to speculations on the existence of an alternative pathway of SM biosynthesis.[9] A transferase reaction similar to the one catalyzed by SMS was described in which phosphatidylethanolamine (PE) rather than PC is used as head group donor, yielding the SM analogue ceramide phosphoethanolamine (CPE)[40] (Fig. 2). Subsequent stepwise methylation reactions would then add three additional methyl groups to the phosphoethanolamine head group, producing SM. Indeed Malgat et al provided evidence for the existence of such a pathway in brain and liver microsomes and plasma membranes.[41] Interestingly, it was shown that in addition to CPE, both tissues were also able to synthesize SM when endogenous PC (and not dipalmitoylcholine) was utilized as substrate and the active site of CPE synthase was oriented versus the luminal side of the membrane bilayer.[42] CPE synthase activity was also observed in ram spermatozoa, mice synaptic vesicles and sciatic nerve.[43,44] Although the existence of CPE synthase has been proven, its contribution to SM formation or the function of CPE itself are still unresolved questions.

Another complex sphingolipid analogue of SM is the ceramide phosphoinositol (CPI). CPI is found in yeast and plants where no SM is present and the reaction leading to its synthesis is virtually identical to the one catalyzed by SMS, except that the head group donor molecule is phosphatydylinositol (PI) instead of PC.[45,46] Importantly, neither can mammalian SMS catalyze the CPI synthase reaction nor can CPI synthase produce SM. On the other hand, some parasitic protozoa have the ability to synthesize CPI, SM and CPE, depending on the phase of the life cycle.[47] Indeed, it has been recently demonstrated that during the procyclic phase of *Trypanosoma brucei* both CPI and SM can be detected whereas during the bloodstream stage, SM and CPE are produced instead by the *TbSLS1-4* gene product. This raises the question of whether, even in mammalian cells, a single transferase can catalyze the production of both SM and CPE.

Physicochemical Properties of SM

SM comprises a heterogeneous group of molecules because of the different fatty acyl chains linked with amide bond to the sphingoid backbone and for the nature of the sphingoid backbone itself.[48] The large majority of SM carries saturated fatty acids (ranging from C14 to C26 in length) and sphingosine as the sphingoid backbone. Alternatively, dihydrosphingosine (sphinganine) or 4-hydroxysphinganine (phytosphingosine) could account for the sphingoid backbone and unsaturated and/or branched fatty acids could also be found.[49,50]

Even though structurally similar, SM and phosphatidylcholine (PC) differ significantly for a number of features that determine the higher melting temperature of the first. In fact, the predominant presence of saturated fatty acids and the potential intermolecular hydrogen binding, via the amide bond and the free hydroxyl group at position 3, favor a high-density packing of SM, which increases the compactness and impermeability of the membrane.[51-53] According to the umbrella model,[54] the preferential mixing of sterols with SM is caused by shielding of the nonpolar sterol molecule by the phosphocholine head group of SM. Collectively these physicochemical features are believed to influence the lateral organization of cellular membranes. The lipid raft hypothesis postulates the existence of SM/sterol-enriched microdomains in the plasma membrane that serve as dynamic platforms for the clustering of membrane proteins with a role in signal transduction, membrane trafficking and cell adhesion.[55]

The Multigenic Sphingomyelin Synthase (SMS) Family

SMS Cloning Strategies

Initial studies indicated that mammalian SM synthases are membrane-bound enzymes that readily lose activity following solubilization with detergent.[12] This feature complicated their identification by classical biochemical approaches. Purification of a soluble SM synthase released by *Pseudomonas aeruginosa*[56] provided no clues on the identity of the mammalian enzyme. Efforts to isolate SM synthase mutants by screening CHO cells for resistance to a SM-directed cytolysin led to the identification of CERT (see above).[29,30]

A complementary approach was based on structural information available for the enzyme catalyzing synthesis of CPI in yeast. CPI production requires the product of the *AUR1* gene,[45] a protein containing the C2 and C3 active site motifs characteristic for members of the lipid phosphate phosphatase (LPP) superfamily.[57] A database search for novel sequences encoding integral membrane proteins containing the C2 and C3 domains common to Aur1p and LPPs identified three families of candidate SM synthase (*CSS*) genes with homologues in multiple animal species[58] (Fig. 3). Several members of each family were cloned and analyzed for their ability to mediate SM synthesis upon heterologous expression in yeast. One family, the CSS3 family, was found to contain multiple members with SM synthase activity. A detailed analysis of two human members, SMS1 and SMS2, revealed that they met all criteria previously assigned to mammalian SM synthase. First, their expression proved sufficient to support SM synthesis in yeast, an organism lacking endogenous SM synthase activity. Second, both proteins function as bidirectional lipid cholinephosphotransferases capable of converting PC and ceramide into SM and diacylglycerol (DAG) and vice versa. Third, the proteins reside in cellular organelles where SM synthesis is known to occur, namely in

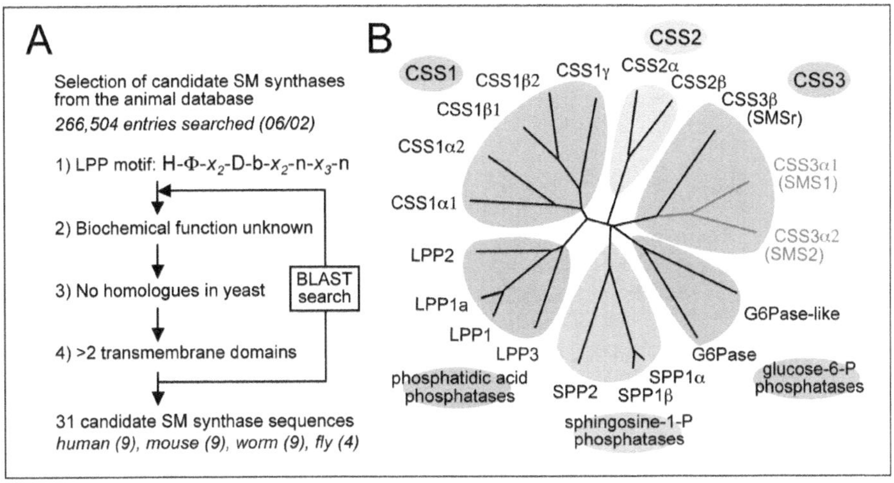

Figure 3. Selection and phylogenetic analysis of candidate SM synthases. A) Animal entries in SwissProt/TrEMBL were searched for the presence of a sequence motif shared by LPPs and Aur1p proteins and then further selected on the basis of three additional criteria, as indicated. B) Phylogenetic tree of human candidate SM synthases (CSS) and previously characterized members of the human LPP superfamily. Figure reproduced from Huitema et al, 2004.

the trans-Golgi (SMS1 and SMS2) and plasma membrane (SMS2). Finally, SMS1 and SMS2 share sequence motifs containing putative active site residues with Aur1p and LPPs that are facing the exoplasmic leaflet, the side of the membrane where SM synthesis is known to occur.

Taking advantage of a mouse lymphoid cell line with strongly reduced SM synthase activity and SM levels, a subsequent study reported the expression cloning of *SMS1* as a gene able to restore SM levels in these cells.[59] Due to a decreased SM content of the plasma membrane, these lymphoid cells are highly susceptible to the toxic effects of methyl-β-cyclodextrin (MβCD), a compound used to deplete cholesterol from cellular membranes. *SMS1* was recovered from a human cDNA expression library on the basis of its ability to restore MβCD resistance. Importantly, this work showed that SMS1 corresponds to a major SM synthase activity in mammalian cells with a critical role in sustaining cell growth (see also below).

Interestingly, at the time of the identification of SMS1 as a major SMS, *SMS1* was already annotated as Hmob33 (human medulla oblongata 33) or MOB, a gene with a predominant brain expression mapping on chromosome 10.[60] Structural and functional analysis of its transcripts predicted regulation at both transcriptional and translational level.[60,61] Subsequently, several alternative spliced products of *SMS1* were found both in human cerebellum and in the mouse[62] suggesting complex regulation for *SMS1* expression.

Structural Organization and Reaction Chemistry of SMS Family Members

Members of the SMS family share a common membrane topology with LPPs.[57,58] Both groups of enzymes contain a six times membrane-spanning core domain with the termini facing the cytosol and the putative C2 and C3 active site residues facing the exoplasmic leaflet (Fig. 4). The active site includes a catalytic triad of histidine and aspartate residues previously implicated in LPP-mediated hydrolysis of lipid phosphate esters.[63] Thus, it is likely that SMS family members utilize a reaction chemistry similar to that described for LPPs to catalyze the choline phosphotransferase reaction. This is predicted to occur via a two-step process, involving: (i) a nucleophilic attack on the lipid-phosphate ester bond in PC by the histidine in C3 assisted by the conserved aspartate in this motif, resulting in formation of a choline phosphohistidine intermediate and the release of

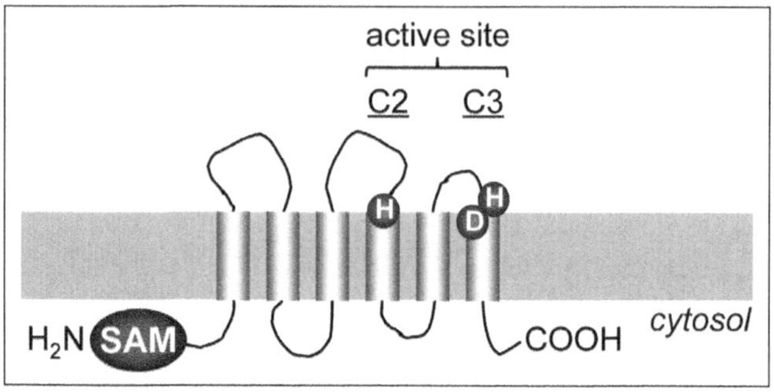

Figure 4. Predicted membrane topology of SMS family members. Putative active site residues in LPP sequence motifs C2 and C3 are highlighted in red and the sterile alpha motif or SAM domain in blue. A color version of this image is available at www.landesbioscience.com/curie.

DAG; (ii) a nucleophilic attack on the choline phosphohistidine intermediate by the oxygen of the ceramide hydroxyl group assisted by the histidine in C2, resulting in transfer to the sphingoid base and the release of SM. Consistent with this model, mutation of one of the histidine or aspartate residues that make up the catalytic triad is sufficient to abolish sphingomyelin synthase activity of SMS1 and SMS2 without affecting their subcellular distribution.[64]

A remarkable difference between the two SMS isoforms is that SMS1, but not SMS2, contains a N-terminal Sterile Alpha Motif or SAM domain.[58,60] SAM domains have been shown to homo- and hetero-oligomerize, forming multiple self-association architectures. They can also associate with various non-SAM domain-containing proteins and appear to possess the ability to bind RNA.[65] The function of the SAM domain in SMS1 is still unknown. Fusion of this domain to the N-terminus of SMS2 does not affect the subcellular distribution of SMS2, nor does its removal from SMS1 lead to redistribution of that enzyme.[64] Hence, it appears that the SAM domain is not involved in targeting SMS1 to the Golgi apparatus.

SMS Family Members Display Striking Variations in Substrate Specificity

The SMS family displays a multiplicity of SMS genes in essentially all organisms generating SM.[58,66] In addition to SMS1 and SMS2 genes, the mammalian genome contains a third, SMS-related (SMSr) gene that is highly conserved, from humans and worms to the fruit fly, *Drosophila melanogaster*. Interestingly, *Drosophila* lacks SMS1 and SMS2 homologues and does not synthesize SM. Instead, this organism produces the SM analogue CPE as a major membrane constituent.[67,68] Mammals also produce small amounts of CPE. Two CPE synthase activities have been described in mammalian cells, one associated with the ER and the other one associated with the plasma membrane.[41,42] As PE serves as the headgroup donor for both activities, the enzymes involved can be classified as PE:ceramide ethanolaminephosphotransferases. Consequently, SM and CPE biosynthesis in mammals share common reaction chemistry. As SMS1, SMS2 and SMSr are structurally related and share the C2 and C3 active site residues, SMSr provides an attractive candidate for the elusive CPE synthase.[58] Indeed, we recently demonstrated that *Drosophila* and human SMSr proteins catalyze EPC biosynthesis and, contrary to SMS1 and SMS2, localize to the ER.[69] SMSr thus qualifies as the ER-resident CPE synthase originally described by Malgat et al.[41,42]

Remarkably, we have recently found that SMSr produces only trace amounts of CPE and that bulk production of CPE in *Drosophila* S2 cells requires a different enzyme. This second, insect-specific CPE synthase uses CDP-ethanolamine instead of PE as headgroup donor,[69] analogous to the ethanolaminephosphotransferases of the Kennedy pathway.[70] This implies that, contrary to SM synthesis in mammals, bulk production of CPE in insects occurs in the cytosolic leaflet of the membrane. As insects require CERT for efficient CPE production,[68] the CDP-ethanolamine-dependent CPE synthase likely resides in the Golgi apparatus. Its identity remains to be established.

Since the plasma membrane-associated CPE synthase activity in mammalian cells shares the same subcellular distribution as SMS2, we recently reinvestigated the substrate specificities of all three mammalian SMS family members. This revealed that, while SMS1 and SMSr are monofunctional SM and CPE synthases, respectively, SMS2 is a bifunctional enzyme producing both SM and CPE.[71] Thus, SMS2 likely accounts for the plasma membrane-resident CPE synthase described previously.[41,42] These recent findings demonstrate an unexpected diversity in substrate specificity among mammalian SMS family members. Interestingly, characterization of a trypanosome sphingolipid synthase family that is orthologous to the *Leishmania* CPI synthase revealed an enzyme with dual SM and EPC synthase activity.[47]

Differential Expression of SMS1 and SMS2

Whereas SMS1 and SMS2 are expressed to a similar level in most human and murine tissues,[58,62] they seem to be differentially expressed in various cell lines. Interestingly, expression of both SMS1 and SMS2 has been observed in all adherent cell lines analyzed so far (breast cancer MCF-7 cells, cervical cancer HeLa cells, hepatocellular carcinoma HepG2 cells, colon cancer $CaCo_2$ cells, human lung fibroblasts, human epithelial HEK-293 cells, mouse melanoma MEB4, mouse fibroblasts) whereas expression of SMS2 seems to be very low or absent in the majority of suspension cell lines (S49, WR19L/Fas, Ramos and U937 lymphoma cells, HL-60, Molt-4 and K562 leukemia cells and primary human B, T and monocytic cells)[58,62,72-74] and Luberto, unpublished observations). Since SMS2 is in part localized at the plasma membrane, with its catalytic site oriented toward the outside of the cell and it is expressed mainly in adherent cells, it may serve a specialized function in cell-cell or cell to matrix interactions.

Cellular Functions of SMS Family Members

SMS1 and SMS2 as Regulators of SM Homeostasis and Receptor-Mediated Signaling

The accumulation of SM in the exoplasmic leaflet of the plasma membrane together with its high-density packing and affinity for cholesterol (see above) implies a vital role in the barrier function of the plasma membrane.[49-53] Recent evidence has further enriched this model supporting a role for SM in the homeostasis of lipid microdomains at the plasma membrane, often sites of receptor-mediated signaling. Taking advantage of the identification of SMS1 as bona fide SM synthase, the role of SM in the activation of Fas signaling was investigated by Miyaji et al.[75] Comparing SMS1 deficient cells (with over expressed Fas receptor) with cells in which expression of SMS1 was restored, the authors provide evidence for a critical role of SM in Fas-mediated signaling by enabling the formation of the death-inducing signaling complex (DISC) and consequently allowing caspase activation and production of ceramide at the plasma membrane. Subsequently, the involvement of SMS1 in the maintenance of raft homeostasis has been postulated in S49 mouse lymphoma cells.[73] A variant S49 cell line resistant to apoptosis induced by alkyl-lysophospholipids (ALP) turned out to be deficient in SM synthesis due to down-regulation of SMS1 expression. Since ALP appears to be internalized via raft-dependent endocytosis, it was suggested that SM depletion perturbs the internalization of ALP in the resistant S49 variant. Indeed, down-regulation of SMS1 in the parental S49 cells mimicked the resistant phenotype observed in the S49 variant cell line, i.e., a diminished raft-dependent uptake of ALP and ALP-induced apoptosis.

More recently, involvement of SMS2 in the maintenance of SM levels in plasma membrane lipid microdomains was determined in HEK 293 cells.[72] In these cells, it was shown that siRNA-mediated down-regulation of either SMS1 or SMS2 induced a significant reduction of SM content in detergent-resistant fractions. Confirmation of the involvement of SMS2 in the physiological maintenance of plasma membrane SM has come from studies using macrophages from SMS2 knock out (KO) mice.[76] Macrophages from SMS2 KO mice showed diminished recruitment of the Toll Like Receptor 4-MD2 complex on the cell surface in response to lypopolysaccharide (LPS) treatment, consistent with the reduction of LPS-mediated apoptosis observed in THP-1-derived macrophages after siRNA-mediated down-regulation of SMS1 or SMS2.[77] Likewise, siRNA-mediated down-regulation of SMS2 in HEK 293 cells partly inhibited recruitment of the tumor necrosis factor (TNF) α receptor 1 to detergent-resistant microdomains in response to stimulation with TNF, decreasing TNF-mediated down-stream signaling. These observations complement the fact that over-expression of SMS1 or SMS2 in Chinese hamster ovary (CHO) cells increased the number of detergent-insoluble microdomains and TNF-induced apoptosis.[77] In both the SMS2 KO macrophages and HEK 293 cells, NF-κB activation was one of the down-stream signaling events that were inhibited upon loss or down-regulation of SMS2, thus supporting an early report in which a link between stimulation of SM synthesis and nuclear translocation of NF-κB was established.[78] SMS1 has been also implicated in raft-dependent activation of T-cell receptor in Jurkat cells.[79] By using Jurkat cells stably expressing shRNA targeting human SMS1, the authors established a cell line with impaired SM synthesis and levels in the plasma membrane. Stimulation of T-cell activation with CD3 revealed impaired expression of CD69 (early marker of leukocyte activation), adhesion, proliferation and TCR clustering and translocation to lipid rafts.

Finally, regulation of cellular SM has been observed upon modulation of either SMS1 or SMS2 in HeLa cells.[74,80] Over-expression of either SMS1 or SMS2 resulted in a net increment of basal SM mass.[74] Moreover, expression of either SMS1 or SMS2 favored comparable SM resynthesis after hydrolysis of plasma membrane SM induced by treatment with bacterial sphingomyelinase.[74] Since in HeLa cells, SMS1 localizes exclusively at the Golgi and SMS2 localizes at the Golgi and plasma membrane, these observations indicate that the unique plasma membrane localization of SMS2 did not specifically facilitate SM resynthesis from plasma membrane derived-ceramide, thus suggesting that the pool of SMS2 in the Golgi may be of significant activity. In support of this conclusion, over-expression of either SMS1 or SMS2 in HeLa cells comparably increased de novo SM biosynthesis, known to occur in the Golgi,[74] whereas down-regulation of either one significantly inhibited it in HeLa,[74,80] Huh,[72] HEK 293 cells and macrophages isolated from KO mice.[76] In HeLa cells down-regulation of either SMS1 or SMS2 inhibited TNF-mediated NF-κB activation by altering plasma membrane receptor activation, similarly to HEK 293 cells (Luberto and Marimuthu, unpublished observations).

SMS1 and SMS2 as Regulators of Lipid-Based Signaling

Given the enzymatic activity of SMS and the biological relevance of its substrates and products of the reaction, similarly to what discussed for SM, significant effort has been directed at determining the contribution of SMS1 and SMS2 to the homeostasis of ceramide and DAG levels in mammalian cells and their role in the cellular functions mediated by these bioactive lipids. Modulation of SMS1 and/or SMS2 expression has been achieved by gene over-expression or by their down-regulation using silencing RNA or by *SMS2* gene KO.

The effect of over expression of SMS1 on ceramide levels was first reported in Jurkat cells after stable transfection.[81] In these cells, enhanced SMS1 levels caused a general increase of sphingolipids in resting cells. On the other hand, over expression of SMS1 prevented accumulation of ceramide and dihydroceramides following photodamage, as compared to vector control cells, even though no SM accumulation could be observed. In this system over expression of SMS1 prevented apoptosis associated to photodamage. On the other hand, in wild type Jurkat cells, siRNA-mediated down regulation of SMS1 enhanced accumulation of ceramides, dihydroceramides and sphingosine following photodamage, with a concomitant enhancement of apoptosis.[82] Similarly, increased apoptosis was also observed after photodamage when SMS2 was down-regulated. A comparable correlation between SMS1 expression and cell death was observed in yeast.[83] In fact mouse *SMS1* was identified as a gene able to rescue yeast cells from cell death induced by expression of the pro-apoptotic Bcl-2 family member, Bax and other cytotoxic stimuli such as hydrogen peroxide, osmotic stress and elevated temperature. Since expression of *SMS1* would favor growth of yeast cells in the presence of either short-chain ceramide analogues or phytosphingosine and a mouse splice variant of *SMS1* missing the key catalytic residues did not rescue bax-induced cytotoxicity,[62] it was speculated that the SMS1-mediated effect was due to metabolism of stress-inducing intermediates of sphingolipid metabolism, even though no lipid analysis was reported in these studies.

Similar to Jurkat cells,[81] over-expression of SMS1 or SMS2 in CHO cells caused an increase of ceramide levels.[77] An increase of ceramide levels was also observed upon down-regulation of either SMS1 or SMS2 in HeLa and Huh cells,[66,72,74] suggesting that changes in ceramide levels due to modulation of SMSs might be differently regulated depending on the specific cellular context.

Whereas the levels of PC do not seem to change following modulation of SMSs,[66,72,74,77] altering the expression of SMS1 or SMS2 exerts diverse responses on DAG levels. In CHO cells over-expression of SMS1 or SMS2 induced a significant accumulation of total DAG.[77] On the other hand, in HeLa cells down-regulation of either SMS1 or SMS2 did not alter total DAG levels[66,74] and in Huh7, SMSs down-regulation caused a tendency to decrease that was not significant.[72] Therefore, also in the case of DAG as for ceramide, cells react differently to modulation of DAG levels by SMS activity and that may have to do with the bioactivity of these lipids. Importantly, even though HeLa cells did not show a change in total DAG levels, a local decrease of DAG could be probed at the level of the Golgi when SMSs were down-regulated and SM synthesis stimulated.[74] These observations suggest that first, SMSs at the Golgi are able to produce DAG and that in resting cells, there are in place systems that effectively prevent large oscillations of DAG due to alteration of SMS activities. When the equilibrium of the system is tilted by enhanced availability of ceramide for SMS1 or SMS2 in the Golgi, then the metabolism of DAG produced in this organelle is not adequate and DAG might accumulate in this compartment. As also eloquently discussed by Richard Pagano,[84] a number of metabolic pathways could hypothetically regulate the level of SMS-produced DAG at the Golgi: (1) DAG produced at the Golgi could recycle back to the endoplasmic reticulum where it is utilized for a new round of PC biosynthesis; (2) DAG could be utilized in the Golgi itself to produce a new molecule of PC; and (3) DAG could be readily degraded through the action of cytoplasmic lipases or metabolized into other molecules, such as phosphatydic acid. Interestingly, stimulation of sphingolipid synthesis in HeLa cells led to translocation of the DAG-binding protein protein kinase D (PKD) to the Golgi, possibly through a SMS-mediated mechanism.[85] Indeed, with the identification of SMSs, it was later shown that down-regulation of SMS1 or SMS2 inhibited translocation of PKD to the Golgi induced by stimulation of SM synthesis, thus reinforcing the notion that if accumulated, the DAG produced by SMSs at the Golgi is biologically functional.[74] Since PKD is involved in regulation of the secretory pathway from the trans Golgi network, these observations may implicate SMS-derived DAG as potential regulator of the budding of secretory vesicles from the Golgi apparatus to the plasma membrane.

Conclusion

Over the last few years a wealth of novel insight into the dynamics of SM biosynthesis has emerged and yet a large number of still unresolved questions are ahead of us. For example, the enzymes responsible for SM biosynthesis in mammals belong to a conserved protein family whose members display striking differences in substrate specificity. This SMS family contains both single and dual activity enzymes, with SMS1 being a monofunctional SM synthase, SMSr a monofunctional CPE synthase and SMS2 a bifunctional enzyme with both SM and CPE synthase activity. These results, combined with the observation that each enzyme mainly resides in a different organelle along the secretory pathway, establishes an unexpected level of complexity in the organization of sphingolipid biosynthesis in mammals. What is the significance of SMSr-mediated CPE biosynthesis in the ER? Does this enzyme play a role in controlling ceramide-induced stress pathways that originate from the ER? The relative extent of cell surface-associated SM and CPE synthase activity mediated by the bifunctional enzyme SMS2 will be controlled, at least in part, by the availability of its substrates PC and PE. In healthy mammalian cells, PE is largely confined to the cytosolic leaflet of the PM whereas PC in the exoplasmic leaflet should be plentiful. However, when PM lipid asymmetry is dissipated, for example during apoptosis, the higher proportion of PE in the exoplasmic leaflet would stimulate SMS2-mediated CPE production. This raises the intriguing possibility that transbilayer lipid arrangement may have a direct impact on the activity of SMS2 as a negative regulator of ceramide signaling at the PM. Moreover, SMS1 and a pool of SMS2 are both localized to the Golgi.[58,66,72,74,76] Why is this organelle equipped with two different SMS enzymes? What is the relevance of the reverse activity of SMS enzymes? Does it really take place in cells? Is the activity of SMS enzymes merely regulated by the reciprocal local concentrations of their substrates and products or are there other levels of regulation? And what is the fate and function of DAG molecules produced by either SMS1 or SMS2?

Several reports have described a complex regulation of SMS activity and SM, ceramide and DAG in the nucleus.[86-88] SMS activity has been detected in the nuclear envelope and associated with the chromatin.[89] Interestingly though, it seems that these two activities might have opposite regulation during proliferation and apoptosis. In fact, during proliferation, chromatin SMS decreases whereas the one localized in the nuclear envelope increases, while an opposite profile has been observed during apoptosis.[90,91] The changes in chromatin-associated SMS might regulate accessibility of DNA for transcription by regulation of DNA associated SM, similarly to what observed in the case of double stranded RNA.[92] Surprisingly, neither SMS1 nor SMS2 localize to the nucleus and so the identities of the nuclear SMS activities remain to be established.

Finally, it has been proposed that SMS activity might be responsible for the activity of the elusive PC-specific phospholipase C (PC-PLC), an enzyme potentially involved in the regulation of DAG but still of unknown identity.[78] Whether SMS1 or SMS2 are also responsible for carrying out the reaction proposed for PC-PLC, perhaps under specific conditions such as poor ceramide availability, remains to be established. In the near future, we hope to find answers to some of these questions.

Acknowledgements

This work was supported in part by grants from the Dutch Organization of Sciences (NWO-CW), the Utrecht University High Potential Program (to J.H.) and the by NIH Grant # P20 RR-017677 from the National Center for Research Resources and Institutional Research Grant from the American Cancer Society IRG 97-219-08 (to C.L.).

References

1. Thudichum JLW. A treatise on the chemical constitution of the brain. 1884 Baillière, Tindall and Cox, London. Republished in 1962 by Archon Books, Hamden, Connecticut.
2. Sribney M, Kennedy EP. The enzymatic synthesis of sphingomyelin. J Biol Chem 1958; 233(6):1315-1322.
3. Weiss SB, Smith SW, Kennedy EP. The enzymatic formation of lecithin from cytidine diphosphate choline and D-1,2-diglyceride. J Biol Chem 1958; 231(1):53-64.
4. Brady RO, Bradley RM, Young OM et al. An alternative pathway for the enzymatic synthesis of sphingomyelin. J Biol Chem 1965; 240(9):3693-3694.
5. Fujino Y, Negishi T. Investigation of the enzymatic synthesis of sphingomyelin. Biochim Biophys Acta 1968; 152(2):428-430.
6. Fujino Y, Nakano M, Negishi T et al. Substrate specificity for ceramide in the enzymatic formation of sphingomyelin. J Biol Chem 1968; 243(17):4650-4651.
7. Sribney M. Stimulation of sphingomyelin synthetase by sulfhydryl reagents. Can J Biochem 1971; 49(3):306-310.
8. Diringer H, Marggraf WD, Koch MA et al. Evidence for a new biosynthetic pathway of sphingomyelin in SV 40 transformed mouse cells. Biochem Biophys Res Commun 1972; 47(6):1345-1352.
9. Ullman MD, Radin NS. The enzymatic formation of sphingomyelin from ceramide and lecithin in mouse liver. J Biol Chem 1974; 249(5):1506-1512.
10. Stoffel W, Melzner I. Studies in vitro on the biosynthesis of ceramide and sphingomyelin. A reevaluation of proposed pathways. Hoppe Seylers Z Physiol Chem 1980; 361(5):755-771.
11. Marggraf WD, Anderer FA, Kanfer JN. The formation of sphingomyelin from phosphatidylcholine in plasma membrane preparations from mouse fibroblasts. Biochim Biophys Acta 1981; 664(1):61-73.
12. Voelker DR, Kennedy EP. Cellular and enzymic synthesis of sphingomyelin. Biochemistry 1982; 21(11):2753-2759.
13. Nelson DH, Murray DK. Dexamethasone increases the synthesis of sphingomyelin in 3T3-L1 cell membranes. Proc Natl Acad Sci USA 1982; 79(21):6690-6692.
14. Marggraf WD, Zertani R, Anderer FA et al. The role of endogenous phosphatidylcholine and ceramide in the biosynthesis of sphingomyelin in mouse fibroblasts. Biochim Biophys Acta 1982; 710(3):314-323.

15. Lipsky NG, Pagano RE. Sphingolipid metabolism in cultured fibroblasts: microscopic and biochemical studies employing a fluorescent ceramide analogue. Proc Natl Acad Sci USA 1983; 80(9):2608-2612.
16. Lipsky NG, Pagano RE. Intracellular translocation of fluorescent sphingolipids in cultured fibroblasts: endogenously synthesized sphingomyelin and glucocerebroside analogues pass through the Golgi apparatus en route to the plasma membrane. J Cell Biol 1985; 100(1):27-34.
17. van Meer G, Stelzer EH, Wijnaendts-van-Resandt RW et al. Sorting of sphingolipids in epithelial (Madin-Darby canine kidney) cells. J Cell Biol 1987; 105(4):1623-1635.
18. Futerman AH, Stieger B, Hubbard AL et al. Sphingomyelin synthesis in rat liver occurs predominantly at the cis and medial cisternae of the Golgi apparatus. J Biol Chem 1990; 265(15):8650-8657.
19. Jeckel D, Karrenbauer A, Birk R et al. Sphingomyelin is synthesized in the cis Golgi. FEBS Lett 1990; 261(1):155-157.
20. Jeckel D, Karrenbauer A, Burger KN et al. Glucosylceramide is synthesized at the cytosolic surface of various Golgi subfractions. J Cell Biol 1992; 117(2):259-267.
21. Kallen KJ, Quinn P, Allan D. Effects of brefeldin A on sphingomyelin transport and lipid synthesis in BHK21 cells. Biochem J 1993; 289(Pt 1):307-312.
22. Allan D, Kallen KJ, Quinn P. Biosynthesis of sphingomyelin and its delivery to the surface of baby hamster kidney (BHK) cells. Biochem Soc Trans 1993; 21(2):240-244.
23. Kallen KJ, Allan D, Whatmore J et al. Synthesis of surface sphingomyelin in the plasma membrane recycling pathway of BHK cells. Biochim Biophys Acta 1994; 1191(1):52-58.
24. van Helvoort A, Stoorvogel W, van Meer G et al. Sphingomyelin synthase is absent from endosomes. J Cell Sci 1997; 110(Pt 6):781-788.
25. Schweizer A, Clausen H, van Meer G et al. Localization of O-glycan initiation, sphingomyelin synthesis and glucosylceramide synthesis in Vero cells with respect to the endoplasmic reticulum-Golgi intermediate compartment. J Biol Chem 1994; 269(6):4035-4041.
26. Hofmeister R, Bottcher A, Schmitz G. Preparation of Golgi subfractions with free-solution isotachophoresis: analysis of sphingomyelin synthesis in Golgi subfractions from rat liver. Electrophoresis 1998; 19(7):1185-1194.
27. Allan D, Obradors MJ. Enzyme distributions in subcellular fractions of BHK cells infected with Semliki forest virus: evidence for a major fraction of sphingomyelin synthase in the trans-golgi network. Biochim Biophys Acta 1999; 1450(3):277-287.
28. Sadeghlar F, Sandhoff K, van Echten-Deckert G. Cell type specific localization of sphingomyelin biosynthesis. FEBS Lett 2000; 478(1-2):9-12.
29. Hanada K, Hara T, Fukasawa M et al. Mammalian cell mutants resistant to a sphingomyelin-directed cytolysin. Genetic and biochemical evidence for complex formation of the LCB1 protein with the LCB2 protein for serine palmitoyltransferase. J Biol Chem 1998; 273(50):33787-33794.
30. Hanada K, Kumagai K, Yasuda S et al. Molecular machinery for nonvesicular trafficking of ceramide. Nature 2003; 426(6968):803-809.
31. Raya A, Revert-Ros F, Martinez-Martinez P et al. Goodpasture antigen-binding protein, the kinase that phosphorylates the goodpasture antigen, is an alternatively spliced variant implicated in autoimmune pathogenesis. J Biol Chem 2000; 275(51):40392-40399.
32. Kawano M, Kumagai K, Nishijima M et al. Efficient trafficking of ceramide from the endoplasmic reticulum to the Golgi apparatus requires a VAMP-associated protein-interacting FFAT motif of CERT. J Biol Chem 2006; 281(40):30279-30288.
33. Kumagai K, Yasuda S, Okemoto K et al. CERT mediates intermembrane transfer of various molecular species of ceramides. J Biol Chem 2005; 280(8):6488-6495.
34. Fukasawa M, Nishijima M, Hanada K. Genetic evidence for ATP-dependent endoplasmic reticulum-to-Golgi apparatus trafficking of ceramide for sphingomyelin synthesis in Chinese hamster ovary cells. J Cell Biol 1999; 144(4):673-685.
35. Funakoshi T, Yasuda S, Fukasawa M et al. Reconstitution of ATP- and cytosol-dependent transport of de novo synthesized ceramide to the site of sphingomyelin synthesis in semi-intact cells. J Biol Chem 2000; 275(39):29938-29945.
36. Hanada K, Kumagai K, Tomishige N et al. CERT and intracellular trafficking of ceramide. Biochim Biophys Acta 2007; 1771(6):644-653.
37. Ladinsky MS, Mastronarde DN, McIntosh JR et al. Golgi structure in three dimensions: functional insights from the normal rat kidney cell. J Cell Biol 1999; 144(6):1135-1149.
38. Mogelsvang S, Marsh BJ, Ladinsky MS et al. Predicting function from structure: 3D structure studies of the mammalian Golgi complex. Traffic 2004; 5(5):338-345.
39. Giussani P, Colleoni T, Brioschi L et al. Ceramide traffic in C6 glioma cells: evidence for CERT-dependent and independent transport from ER to the Golgi apparatus. Biochim Biophys Acta 2008; 1781(1-2):40-51.

40. Broad TE, Dawson RM. Formation of ceramide phosphorylethanolamine from phosphatidylethanolamine in the rumen protozoon Entodinium caudatum (Short Communication). Biochem J 1973; 134(2):659-662.
41. Malgat M, Maurice A, Baraud J. Sphingomyelin and ceramide-phosphoethanolamine synthesis by microsomes and plasma membranes from rat liver and brain. J Lipid Res 1986; 27(3):251-260.
42. Malgat M, Maurice A, Baraud J. Sidedness of ceramide-phosphoethanolamine synthesis on rat liver and brain microsomal membranes. J Lipid Res 1987; 28(2):138-143.
43. Hinkovska-Galcheva V, Petkova DH, Nikolova MN. Sphingomyelin and ceramide: phosphoethanolamine synthesis in ram spermatozoa plasma membrane. Int J Biochem. 1989; 21(10):1153-1156.
44. Maurice A, Malgat M. Evidence for the biosynthesis of ceramide-phosphoethanolamine in brain synaptic plasma membrane vesicles and in sciatic nerve microsomes from normal and Trembler mice. Neurosci Lett 1990; 118(2):177-180.
45. Nagiec MM, Nagiec EE, Baltisberger JA et al. Sphingolipid synthesis as a target for antifungal drugs. Complementation of the inositol phosphorylceramide synthase defect in a mutant strain of Saccharomyces cerevisiae by the AUR1 gene. J Biol Chem 1997; 272(15):9809-9817.
46. Dickson RC, Sumanasekera C, Lester RL. Functions and metabolism of sphingolipids in Saccharomyces cerevisiae. Prog Lipid Res 2006; 45(6):447-465.
47. Sutterwala SS, Hsu FF, Sevova ES et al. Developmentally regulated sphingolipid synthesis in African trypanosomes. Mol Microbiol 2008; 70(2):281-296.
48. Barenholz Y, Thompson TE. Sphingomyelins in bilayers and biological membranes. Biochim Biophys Acta 1980; 604(2):129-158.
49. Merrill AH, Jr., Sullards MC, Allegood JC et al. Sphingolipidomics: high-throughput, structure-specific and quantitative analysis of sphingolipids by liquid chromatography tandem mass spectrometry. Methods 2005; 36(2):207-224.
50. Yang K, Zhao Z, Gross RW et al. Systematic analysis of choline-containing phospholipids using multi-dimensional mass spectrometry-based shotgun lipidomics. J Chromatogr B Analyt Technol Biomed Life Sci 2009.
51. Guo W, Kurze V, Huber T et al. A solid-state NMR study of phospholipid-cholesterol interactions: sphingomyelin-cholesterol binary systems. Biophys J 2002; 83(3):1465-1478.
52. Chiu SW, Vasudevan S, Jakobsson E et al. Structure of sphingomyelin bilayers: a simulation study. Biophys J 2003; 85(6):3624-3635.
53. Aittoniemi J, Niemela PS, Hyvonen MT et al. Insight into the putative specific interactions between cholesterol, sphingomyelin and palmitoyl-oleoyl phosphatidylcholine. Biophys J 2007; 92(4):1125-1137.
54. Ikonen E. Cellular cholesterol trafficking and compartmentalization. Nat Rev Mol Cell Biol 2008; 9(2):125-138.
55. van Meer G, Voelker DR, Feigenson GW. Membrane lipids: where they are and how they behave. Nat Rev Mol Cell Biol 2008; 9(2):112-124.
56. Luberto C, Stonehouse MJ, Collins EA et al. Purification, characterization and identification of a sphingomyelin synthase from Pseudomonas aeruginosa. PlcH is a multifunctional enzyme. J Biol Chem 2003; 278(35):32733-32743.
57. Heidler SA, Radding JA. Inositol phosphoryl transferases from human pathogenic fungi. Biochim Biophys Acta 2000; 1500(1):147-152.
58. Huitema K, van den Dikkenberg J, Brouwers JF et al. Identification of a family of animal sphingomyelin synthases. Embo J 2004; 23(1):33-44.
59. Yamaoka S, Miyaji M, Kitano T et al. Expression cloning of a human cDNA restoring sphingomyelin synthesis and cell growth in sphingomyelin synthase-defective lymphoid cells. J Biol Chem 2004; 279(18):18688-18693.
60. Vladychenskaya IP, Dergunova LV, Limborska SA. In vitro and in silico analysis of the predicted human MOB gene encoding a phylogenetically conserved transmembrane protein. Biomol Eng 2002; 18(6):263-268.
61. Vladychenskaya IP, Dergunova LV, Dmitrieva VG et al. Human gene MOB: structure specification and aspects of transcriptional activity. Gene 2004; 338(2):257-265.
62. Yang Z, Jean-Baptiste G, Khoury C et al. The mouse sphingomyelin synthase 1 (SMS1) gene is alternatively spliced to yield multiple transcripts and proteins. Gene 2005; 363:123-132.
63. Neuwald AF. An unexpected structural relationship between integral membrane phosphatases and soluble haloperoxidases. Protein Sci 1997; 6(8):1764-1767.
64. Yeang C, Varshney S, Wang R et al. The domain responsible for sphingomyelin synthase (SMS) activity. Biochim Biophys Acta 2008; 1781(10):610-617.
65. Kim CA, Bowie JU. SAM domains: uniform structure, diversity of function. Trends Biochem Sci 2003; 28(12):625-628.

66. Tafesse FG, Huitema K, Hermansson M et al. Both sphingomyelin synthases SMS1 and SMS2 are required for sphingomyelin homeostasis and growth in human HeLa cells. J Biol Chem 2007; 282(24):17537-17547.
67. Rietveld A, Neutz S, Simons K et al. Association of sterol- and glycosylphosphatidylinositol-linked proteins with Drosophila raft lipid microdomains. J Biol Chem 1999; 274(17):12049-12054.
68. Rao RP, Yuan C, Allegood JC et al. Ceramide transfer protein function is essential for normal oxidative stress response and lifespan. Proc Natl Acad Sci USA 2007; 104(27):11364-11369.
69. Vacaru AM, Tafesse FG, Ternes P et al. Sphingomyelin synthase-related protein SMSr controls ceramide homeostasis in the ER. J Cell Biol 2009; 185(6):1013-1027.
70. Vance JE. Phosphatidylserine and phosphatidylethanolamine in mammalian cells: two metabolically related aminophospholipids. J Lipid Res 2008; 49(7):1377-1387.
71. Ternes P, Brouwers JF, van den Dikkenberg J et al. Sphingomyelin synthase SMS2 displays dual activity as ceramide phosphoethanolamine synthase. J Lipid Res 2009.
72. Li Z, Hailemariam TK, Zhou H et al. Inhibition of sphingomyelin synthase (SMS) affects intracellular sphingomyelin accumulation and plasma membrane lipid organization. Biochim Biophys Acta 2007; 1771(9):1186-1194.
73. Van der Luit AH, Budde M, Zerp S et al. Resistance to alkyl-lysophospholipid-induced apoptosis due to downregulated sphingomyelin synthase 1 expression with consequent sphingomyelin- and cholesterol-deficiency in lipid rafts. Biochem J 2007; 401(2):541-549.
74. Villani M, Subathra M, Im YB et al. Sphingomyelin synthases regulate production of diacylglycerol at the Golgi. Biochem J 2008; 414(1):31-41.
75. Miyaji M, Jin ZX, Yamaoka S et al. Role of membrane sphingomyelin and ceramide in platform formation for Fas-mediated apoptosis. J Exp Med 2005; 202(2):249-259.
76. Hailemariam TK, Huan C, Liu J et al. Sphingomyelin synthase 2 deficiency attenuates NFkappaB activation. Arterioscler Thromb Vasc Biol 2008; 28(8):1519-1526.
77. Ding T, Li Z, Hailemariam T et al. SMS over expression and knockdown: impact on cellular sphingomyelin and diacylglycerol metabolism and cell apoptosis. J Lipid Res 2008; 49(2):376-385.
78. Luberto C, Hannun YA. Sphingomyelin synthase, a potential regulator of intracellular levels of ceramide and diacylglycerol during SV40 transformation. Does sphingomyelin synthase account for the putative phosphatidylcholine-specific phospholipase C? J Biol Chem 1998; 273(23):14550-14559.
79. Jin ZX, Huang CR, Dong L et al. Impaired TCR signaling through dysfunction of lipid rafts in sphingomyelin synthase 1 (SMS1)-knockdown T-cells. Int Immunol 2008; 20(11):1427-1437.
80. Tafesse FG, Ternes P, Holthuis JC. The multigenic sphingomyelin synthase family. J Biol Chem 2006; 281(40):29421-29425.
81. Separovic D, Hanada K, Maitah MY et al. Sphingomyelin synthase 1 suppresses ceramide production and apoptosis postphotodamage. Biochem Biophys Res Commun 2007; 358(1):196-202.
82. Separovic D, Semaan L, Tarca AL et al. Suppression of sphingomyelin synthase 1 by small interference RNA is associated with enhanced ceramide production and apoptosis after photodamage. Exp Cell Res 2008; 314(8):1860-1868.
83. Yang Z, Khoury C, Jean-Baptiste G et al. Identification of mouse sphingomyelin synthase 1 as a suppressor of Bax-mediated cell death in yeast. FEMS Yeast Res 2006; 6(5):751-762.
84. Pagano RE. What is the fate of diacylglycerol produced at the Golgi apparatus? Trends Biochem Sci 1988; 13(6):202-205.
85. Baron CL, Malhotra V. Role of diacylglycerol in PKD recruitment to the TGN and protein transport to the plasma membrane. Science 2002; 295(5553):325-328.
86. Keenan TW, Berezney R, Crane FL. Lipid composition of further purified bovine liver nuclear membranes. Lipids 1972; 7(3):212-215.
87. James JL, Clawson GA, Chan CH et al. Analysis of the phospholipid of the nuclear envelope and endoplasmic reticulum of liver cells by high pressure liquid chromatography. Lipids 1981; 16(7):541-545.
88. Albi E, Mersel M, Leray C et al. Rat liver chromatin phospholipids. Lipids 1994; 29(10):715-719.
89. Albi E, Magni MV. Sphingomyelin synthase in rat liver nuclear membrane and chromatin. FEBS Lett 1999; 460(2):369-372.
90. Albi E, Pieroni S, Viola Magni MP et al. Chromatin sphingomyelin changes in cell proliferation and/or apoptosis induced by ciprofibrate. J Cell Physiol 2003; 196(2):354-361.
91. Albi E, Cataldi S, Bartoccini E et al. Nuclear sphingomyelin pathway in serum deprivation-induced apoptosis of embryonic hippocampal cells. J Cell Physiol 2006; 206(1):189-195.
92. Micheli M, Albi E, Leray C et al. Nuclear sphingomyelin protects RNA from RNase action. FEBS Lett 1998; 431(3):443-447.

Chapter 6

Ceramide in Stress Response

Mariana N. Nikolova-Karakashian* and Krassimira A. Rozenova

Abstract

Evidence has consistently indicated that activation of sphingomyelinases and/or ceramide synthases and the resulting accumulation of ceramide mediate cellular responses to stressors such as lipopolysaccharide, interleukin 1β, tumor necrosis factor α, serum deprivation, irradiation and various antitumor treatments. Recent studies had identified the genes encoding most of the enzymes responsible for the generation of ceramide and ongoing research is aimed at characterizing their individual functions in cellular response to stress. This chapter discusses the seminal and more recent discoveries in regards to the pathways responsible for the accumulation of ceramide during stress and the mechanisms by which ceramide affects cell functions. The former group includes the roles of neutral sphingomyelinase 2, serine palmitoyltransferase, ceramide synthases, as well as the secretory and endosomal/lysosomal forms of acid sphingomyelinase. The latter summarizes the mechanisms by which ceramide activate its direct targets, PKCζ, PP2A and cathepsin D. The ability of ceramide to affect membrane organization is discussed in the light of its relevance to cell signaling. Emerging evidence to support the previously assumed notion that ceramide acts in a strictly structure-specific manner are also included. These findings are described in the context of several physiological and pathophysiological conditions, namely septic shock, obesity-induced insulin resistance, aging and apoptosis of tumor cells in response to radiation and chemotherapy.

Introduction

Cells and organisms have developed various strategies to deal with adverse changes in their environment. Cellular insult by infectious agents, toxins, nutrient deprivation, or genotoxic stress produces a coordinated systemic response (generally referred to as inflammation), which is aimed at neutralization of the insult and initiation of tissue repair. The first line of defense at systemic level is stimulation of the innate immune response, which consists of regulated production of inflammatory mediators, including cytokines like IL-1β, TNFα and IL-6. The typical cellular response to environmental stressors includes the induction of cellular apoptosis (in response to either genotoxic stress or cytokines like TNFα), growth arrest (during nutrient deprivation), increased eicosanoid production, cell migration and adhesion (in the presence of infectious agent). While acute inflammation is protective for the organisms, excessive and long-standing inflammation is harmful and underlies diseases like septic shock, atherosclerosis, asthma, rheumatoid arthritis and inflammatory bowel disease.

Diverse signaling pathways mediate cellular response to stress. The sphingolipid second messengers ceramide, ceramide-1-phosphate, sphingosine and sphingosine-1-phosphate play important role as mediators in many of these pathways. This chapter is focused on the role of ceramide in cellular stress response and the mechanisms of its generation and action.

*Corresponding Author: Mariana N. Nikolova-Karakashian—Department of Physiology, University of Kentucky College of Medicine, Lexington, Kentucky 40536.
Email: mnikolo@uky.edu

Sphingolipids as Signaling and Regulatory Molecules, edited by Charles Chalfant and Maurizio Del Poeta.

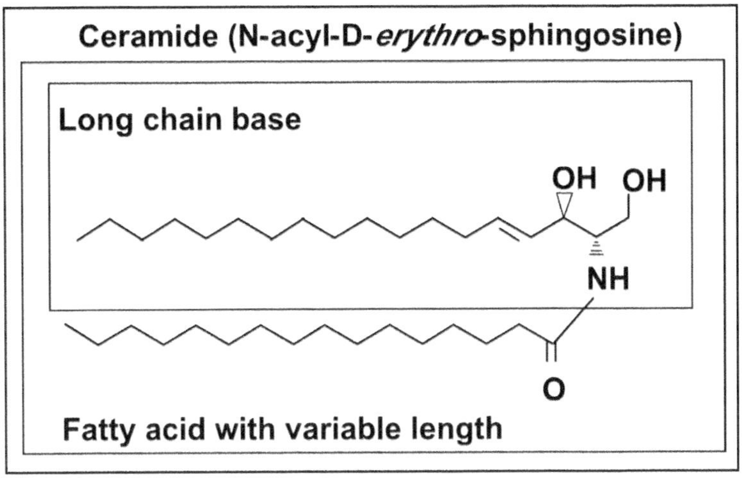

Figure 1. Structure of ceramide.

Chemical Structure and Biophysical Properties of Ceramide

Ceramides (Fig. 1) form the hydrophobic backbone of all complex sphingolipids[1] and consist of a long chain sphingoid base and amid linked fatty acid which is either saturated or unsaturated and vary in length from two to 28 carbon atoms. In mammalian cells, the most commonly found ceramides have D-*erythro*-sphingosine and a saturated fatty acyl chain of 16 carbon atoms and are among the most hydrophobic lipids in the membrane. Free ceramide has a very low critical micellar concentration (cmc $< 10^{-10}$M) and cannot exist in aqueous solutions.[2] Nevertheless, ceramides are still considered as amphiphiles because the hydroxyl group at the first carbon and the amide bond are hydrophilic moieties. Dihydroceramide differs from ceramide inasmuch as the latter contains a trans 4, 5 double bond, which is essential for some of the bioactive roles of ceramide.

The structural features in the ceramide molecule that are required for its biological properties are not well understood, however the two hydroxyl groups, amid group protons and the trans-double bond seems to be involved.[3] A network of intramolecular hydrogen bonds involving the OH and NH groups establishes a unique conformation arrangement of ceramide molecule.

Changes in Ceramide Mass during Stress

The basal ceramide concentrations in the cells are low and may change during cellular differentiation or progression through the cell cycle. Various inducers of cellular stress however lead to an accumulation of ceramide that promotes apoptotic, inflammatory and growth inhibitory signals and also mediates the onset of a specific response. The data in Table 1 illustrate what is the magnitude of changes in ceramide levels under various conditions of stress. These reported differences however, are likely to be an underestimation since they are measured in total cell preparations and not in the specific membrane fractions where ceramide was generated. Furthermore, with the exception of mass spectrometry—based assays, the most commonly used methods of ceramide quantification do not separate the individual ceramide species.[4] This might be important as the impact ceramide has on cell functions seemingly depends on the type of fatty acid attached to the sphingoid base. Recent data indeed support the notion that the different biological effects caused by ceramide may be mediated by distinct molecular species of the lipid, underscoring the necessity to evaluate changes in ceramide content in a structure-specific manner.

A comparison of the magnitudes of ceramide accumulation observed under various treatments reveals that the changes are more robust in response to chemotherapeutics or irradiation. As the outcome of these treatments has been the induction of cell death via apoptosis, some investigators

Table 1. Changes in ceramide content in response to stress and during some pathophysiological conditions

	Control	Treatment	Assay	Tissue/Cell Type	Ref.
Hypoxia/ Re-oxygenation	4.5 pmol/nmol LP	8.5 pmol/nmol LP	Mass Spec	NT-2	6
	100%	200% of control	DAGK	Cardiac myocytes	7
	22 pmol/nmol LP[#]	55 pmol/nmol LP	DAGK	NRK-52E	8
	250 pmol/mg Pr	900 pmol/mg Pr	DAGK	PC-12	9
	100%	300% of control	DAGK	HUVEC	10
	100%	50% of control	DAGK	A7r5	11
Ischemia/ Reperfusion	110% of sham	175% of sham	Mass Spec	Brain	12
	100%	200% of control	DAGK	Rat cardiomyocytes	13
	50 pmol/nmol	90 pmol/nmol LP	DAGK	Renal cortex	14
	190 pmol/mg Pr	290 pmol/mg Pr	DAGK	Liver	15
IL-1β	100%	390% of control	DAGK	Rat astrocytes	16
	100%	170% of control	DAGK	CG-4	17
	18 pmol/nmol LP	45 pmol/nmol LP	DAGK	Rat myocytes	18
	75 cpm/10^6cells	160 cpm/10^6 cells	Labeling	Rat mesangial cells	19
	1 nmol/mg Pr	1.5 nmol/mg Pr	HPLC	Rat hepatocytes	20
	1.1 nmol/mg Pr	1.48 nmol/mg Pr	HPLC	Rat hepatocytes	21
	250 pmol/10^6cells	500 pmol/10^6 cells	DAGK	HUVEC	22
	450 pmol/10^6cells	800 pmol/10^6 cells	DAGK	Dendritic cells	23
	100%	170% of control	Labeling	Thyroid cells	24
TNF-α	100%	150% of control	HPLC	Cortical neurons	25
	100%	330% of control	DAGK	Rat astrocytes	16
	200 pmol/mg Pr	850 pmol/mg Pr	DAGK	U-87 MG cells	26
	22 pmol/10^6 cells	37 pmol/10^6cell	DAGK	Mesangial cells	27
	100%	300% of control	HPLC	Hepatocytes	28
	0.8 nmol/mg Pr	1.4 nmol/mg Pr	DAGK	Lung microsomes	29
	5.0 nmol/mg Pr	10.9 nmol/mg Pr	HPLC	Alveolar lavage	30
	0.5 nmol/nmol LP	1.1 nmol/nmol LP	Mass Spec	Endothelial cells	31
	100%	460% of control	DAGK	MIN6 cells	32
	100%	700% of control	DAGK	MCF-7 cells	33
	100%	175% of control	DAGK	MCF-7 cells	34
	100%	250% of control	DAGK	L929	35
	5 pmol/nmol LP	13 pmol/nmol LP	DAGK	U937 cells	36
	5 pmol/nmol LP	12 pmol/nmol LP	DAGK	SK2	36
Serum starvation	100%	200% of control	DAGK	Mouse keratinocytes	37
LPS	2.8 pmol/nmol LP	4.3 pmol/nmol LP	HPLC	Brain	38
	4 nmol/ml	8 nmol/ml	HPLC	Plasma	38
	3.2 nmol/ml	4.8 nmol/ml	HPLC	Mouse serum,	39
	100%	200-500% of contr.	HPLC	Human plasma	39
	3.5 nM/mg Pr	16 nM/mg Pr	HPLC	THP-1 cells	40
	258 pmol/mg Pr	1634 pmol/mg Pr	HPLC	Mouse Mø	41
	8 nM/mg Pr	22 nM/mg Pr	HPLC	Human alveolar Mø	42
	1.1 nmol/mg Pr	2.1 nmol/mg Pr	DAGK	Intestinal mucosa	43

continued on next page

Table 1. Continued

	Control	Treatment	Assay	Tissue/Cell Type	Ref.
ER stress	5 pmol/mg Pr	9 pmol/mg Pr	Mass Spec	INS-1 cells	44
Ionizing radiation	100%	700% of control	DAGK	OLG	45
	100%	320% of control	Labeling	CHK cells	46
	0.95 nmol/10^6 cells	1.25 nmol/10^6 cells	DAGK	BAEC	47
	6-12 nmol/ml	9-18 nmol/ml	HPLC	Human serum	48
	100%	250% of control	Labeling	Keratinocytes	49
	10 pmol/10^6 cells	50 pmol/10^6 cells	Mass Spec	Jurkat cells	50
	100%	130% of control	DAGK	HEK293	51
	100%	140% of control	DAGK	Jurkat T-cells	51
	10 pmol/nmol LP	17 pmol/nmol LP	Mass Spec	MCF-7	52
	100%	320% of control	DAGK	HeLa	53
	100%	140% of control	Labeling	U937 cells	54
	100%	900% of control	DAGK	Human platelets	54
Diabetes/ obesity	0.5 pmol/µg Pr	1.0 pmol/µg Pr	DAGK	Rat heart tissue	55
	12 pmol/mg tissue	23 pmol/mg tissue	Mass Spec	Muscle	56
	25 pmol/mg tissue	40 pmol/mg tissue	DAGK	Muscle	57
	0.2 nmol/mg tissue	0.3 nmol/mg tissue	DAGK	Liver	57
	15 pmol/mg	23 pmol/mg	Mass Spec	Muscle	58
	100%	No change	Mass Spec	Liver	58
	3.5 pmol/µl	9 pmol/µl	Mass Spec	Serum	58
	0.5 nmol/mg Pr	1.5 nmol/mg Pr	HPLC	Liver	59
	2.4 nmol/ml	3.1 nmol/ml	Mass Spec	Human plasma	60
	120 pmol/150µl	190 pmol/150µl	Mass Spec	Mouse Serum	61
	380 pmol/mg Pr	580 pmol/mg Pr	Mass Spec	Epididymal fat	61
	300 pmol/mg Pr	550 pmol/mg Pr	Mass Spec	Subcutaneous fat	61
	650 pmol/ml	1400 pmol/ml	Mass Spec	Mouse plasma	62
	310 pmol/mg Pr	210 pmol/mg Pr	Mass Spec	Epididymal fat	62
	5 nmol/g	8 nmol/g	Mass Spec	Subcutaneous fat	63
	0.6 nmol/mg Pr	1.2 nmol/mg Pr	DAGK	Pancreatic islets	64
	250 pmol/mg Pr	180 pmol/mg Pr	Mass Spec	Retina	65
Aging	100%	170% of control	Mass Spec	Cerebral cortex	66
	100%	220% of control	DAGK	Heart	67
	175 pmol/µg DNA	250 pmol/µg DNA	Mass Spec	Adipose tissue	68
	0.45% of total LP	0.8% of total LP	HPLC	Liver	69
	1.3 nmol/mg Pr	1.7 nmol/mg Pr	HPLC	Rat hepatocytes	70
	0.5 pmol/mg Pr	1.1 pmol/mg Pr	DAGK	Endothelium	71
	3.5 pmol/nmol LP	6.0 pmol/nmol LP	DAGK	WI-38 HDF	72
Alzheimer's disease	27 nM/ml	52 nM/ml	DAGK	CSF	73
	100%	175% of control	Mass Spec	Brain	74
	2.5 nmol/mg Pr	9 nmol/mg Pr	Mass Spec	Temporal cortex	75
	2.0 nmol/mg Pr	8 nmol/mg Pr	Mass Spec	Cerebellum	75
Heat stress	1.5 pmol/nmol LP	3.8 pmol/nmol LP	DAGK	Molt-4	76

continued on next page

Table 1. Continued

	Control	Treatment	Assay	Tissue/Cell Type	Ref.
Oxidative stress	100%	800% of control	DAGK	Human primary	77
	100%	200% of control	DAGK	OLG	78
	100%	170% of control	DAGK	Cerebral cortex	79
	100%	200% of control	DAGK	PC12	80
	100%	350% of control	Mass Spec	Airway epithelium	81
	100%	140% of control	Labeling	APRE-19	82
	100%	180% of control	DAGK	SMC HAEC	83
Cardiovascular disease	210.5 pmol/mg Pr	305.9 pmol/mg Pr	DAGK	Mouse heart	84
	100%	140% of control	Mass Spec	Mouse heart	85
	135 nmol/ml	220 nmol/ml	Mass Spec	Plasma	86
Anti-cancer therapies	100%	1000% of control	Labeling	LnCaP	87
	100%	200% of control	Labeling	PC-3	87
	7 pmol/nmol LP	70 pmol/nmol LP	DAGK	MDA-MB 231	88
	100%	220% of control	Labeling	MDA-MB 468	89
	100%	350% of control	Labeling	MCF-7	89
	1% of total LP	10% of total LP	Labeling	MCF-7	90
	100%	900% of control	Labeling	MCF-7	91
	100%	180% of control	Labeling	BT-20	92
	100%	300% of control	Labeling	MDA-MB 231	92
	100%	650% of control	Labeling	MDA-MB 468	92
	100%	300% of control	Labeling	Hs-578T	92
	100%	380% of control	Labeling	T47D	92
	100%	620% of control	Labeling	MCF-7	92
	100%	700% of control	Labeling	HL-60/VCR	93
	100%	220% of control	Labeling	U-937	93
	100%	220% of control	Labeling	CHLA-90	94
	0.6 ng/10^6 cells	1.8 ng/10^6 cells	DAGK	JHU-022	95

In some cases numerical values were taken from graphically represented data and were rounded to the nearest whole number. Abbreviations: General: CSF, cerebrospinal fluid; DAGK, diacylglycerol Kinase; LP, lipid phosphate; Mø, primary macrophages; OLG, oligodendrocytes; Pr, Protein. Cells lines: A7r5, rat embryonic thoracic aorta smooth muscle; APRE-19, human retinal pigment epithelium; BAEC, bovine aortic endothelium; BT-20—human breast carcinoma; CG4, oligodendrocyte progenitor; CHK, Chinese hamster kidney; CHLA-90, human neuroblastoma; HAEC, human airway epithelial; HDF, human diploid fibroblasts; HEK 293, human embryonic kidney; HeLa, cervical carcinoma; HL60/VCR, drug resistant human leukemia; Hs-578T, breast carcinoma; HUVEC, human umbilical vein endothelium; INS-1, rat insulinoma; JHU-022, squamous cell carcinoma; Jurkat, T-lymphocytes; L929, mouse fibroblasts; MDA-MB 231, human Caucasian breast adenocarcinoma; MDA-MB 468, human Black breast adenocarcinoma; MCF-7, human breast adenocarcinoma; MIN6, murine beta cells; Molt-4; human acute lymphoblastic leukemia; LnCaP, human prostate carcinoma; NRK-52E, rat kidney epithelium; NT-2, human neuronal precursor; PC3, human prostate carcinoma, PC12, rat adrenal medulla carcinoma; SK2, hybridoma; SP-1, mammary intraductal adenocarcinoma; T47D, human ductal breast epithelial tumor; TNP-1, human acute monocytic leukemia; U87, human astrocytoma; U937, human leukemic monocyte lymphoma.

suggest that ceramide plays the role of a gauge that senses the level of cell injury and depending on downstream factors determines specific biological outcome.[5]

Mechanisms for Ceramide Generation during Stress

Ceramide does not move spontaneously between cellular membranes and is transported either during normal membrane biogenesis, or with the aid of CERT, a protein that transfers ceramides synthesized at the endoplasmic reticulum to the Golgi apparatus. This insolubility of ceramide suggests that once generated, ceramide is likely to remain localized at the place of its synthesis until metabolized to other sphingolipids. Consequently, the increases in ceramide concentration during stress response are compartmentalized in distinct locations within the cells and might affect diverse sets of down stream targets and responses.

The two main metabolic pathways for generation of excess ceramide during stress response are (i) the de novo synthesis in the endoplasmic reticulum and (ii) the turnover of sphingomyelin (SM) either at the plasma membrane or in the endosomal/lysosomal compartment (Fig. 2). A number of agonists have been shown to activate these pathways leading to transient elevation in ceramide. The magnitude and temporal pattern of ceramide accumulation is further influenced by the activity of ceramidases, SM synthases, ceramide kinase and glucosyl/galactosyl ceramide synthases. These enzymes catalyze the conversion of ceramide to other sphingolipids and some agonists seem to coordinately regulate both, the ceramide production and turnover.[20,96]

Role of the De Novo Pathway for Ceramide Generation in Cellular Stress Response

The de novo pathway for synthesis of ceramide consists of 4 reactions: the serine palmitoyl-transferase (SPT), which condenses palmitoyl-CoA and serine into 3-ketosphinganine, the 3-ketosphinganine reductase that generates sphinganine, the (dihydro)ceramide synthase, which

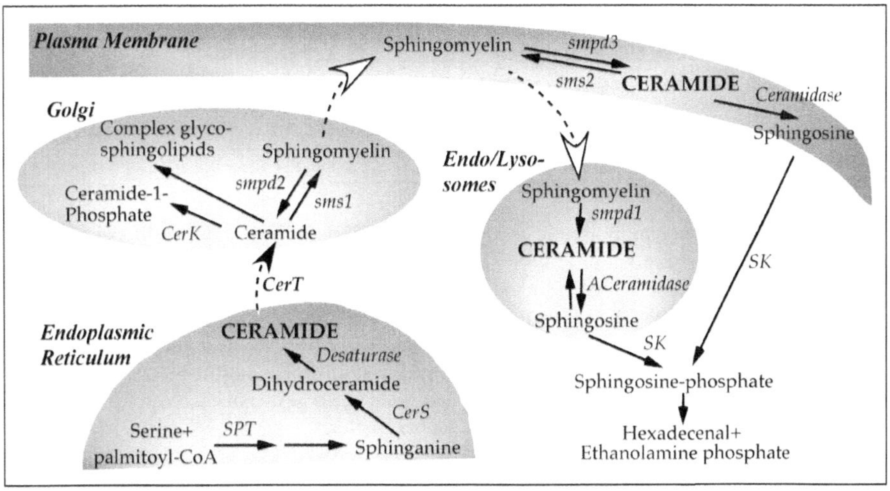

Figure 2. Metabolic pathways responsible for ceramide synthesis and degradation. The pools of ceramide involved in cellular response to stress are depicted in capital letters. Name of the enzymes are in *plain italic*. The names of relevant subcellular organelles are shown in ***bold italic.*** Black solid arrows are used to depict metabolic conversions. Black dashed arrow indicates protein-mediated transfer. White dashed arrow indicates vesicular transport. Abbreviations: SPT: Serine Palmitoyltransferase; Cers: Ceramide synthases; CerT: ceramide transfer protein. SMS1 & 2: sphingomyelin synthase 1 & 2; smpd1: Acid Sphingomyelinase, smpd2: Neutral sphingomyelinase1, smpd3: Neutral sphingomyelinase 2. CerK1, Ceramide Kinase1; SK: sphingosine Kinase. ACeramidase: Acid Ceramidase.

acylates sphinganine to dihydroceramide and the dihydroceramide desaturase, which converts relatively inactive dihydroceramides to ceramides. Stimulation of the de novo pathway during cellular stress response happens through up-regulation of the activity of SPT and/or (dihydro) ceramide synthase. The enzymes of the de novo synthesis of ceramide are in the endoplasmic reticulum. The newly generated ceramide is actively transported to the Golgi apparatus, where it serves as a rate-limiting substrate in the synthesis of complex sphingolipids, like SM and glycosphingolipids. It is noteworthy that some of the known ceramidases may also act as ceramide synthases in a CoA-dependent and independent manner as they exhibit reverse activity. This alternative route for ceramide synthesis is often referred to as the salvage pathway and its role in stress response is not well understood.

The activation of the de novo pathway apparently leads not only to an elevation in ceramide but also to increases in the concentration of SM and glycosphingolipids. Such parallel accumulation is documented in response to LPS, palmitate treatment and heat shock, among others and implies that a wide range of cell functions might be affected through specific and nonspecific mechanisms. Inhibitors of glucosylceramide synthase and labeling with radioactive precursors have been successfully used to elucidate the specific role of ceramide in each case.

De novo synthesis of ceramide is target of a number of fungal inhibitors like fumonisin B1, which inhibits dihydroceramide synthase and the reverse reaction of some ceramidases[97] and myriocin/ISP-1,[98] cycloserine[99] and beta-chloroalanine,[100] all of which inhibit SPT. These widely used inhibitors remain a critical test of establishing whether the de novo pathway is involved in a particular cellular response (reviewed in Merrill[101] and in Perry[96]).

The activation of the de novo pathway during conditions of stress was initially discovered in yeast, where it regulates growth and the response to heat or osmotic stress (reviewed in Dickson[102] and in Meier et al[103]). However, in yeast the signaling mediator is not ceramide but the free long chain bases, phytosphingosine and dihydrosphingosine, most likely reflecting the specifics of sphingolipid synthesis in yeast. Later, the activation of the de novo pathway in C. Elegans in response to ionizing radiation was described and found to be involved in activation of the CED-3 caspase[104] and the resulting apoptosis of germ cells.

In mammalian systems, the de novo pathway seems to play a prominent role in cellular response to heat or chemical stress,[105] septic shock,[106] lipo-apoptosis[107] and insulin resistance,[108] as well as in receptor-dependent and -independent induction of apoptosis by variety of chemotherapeutic agents like etoposides[109] and doxorubicin.[110]

Heat Stress

In the human acute lymphoblastic leukemia cell line Molt-4, heat shock induces more than twofold increase in total ceramide levels with C16 ceramide being the major species affected. This accumulation of ceramide has been linked to the induction of c-jun and apoptosis.[105] Labeling with tritiated palmitate has shown an accumulation of ceramides, but not sphingoid bases, thus confirming that in contrast to yeast, where sphingoid bases mediate heat shock, in mammals, ceramide is apparently involved. Both, myriocin and fumonisin B1 inhibit the increase in total ceramide mass thus confirming that ceramides produced upon heat shock were products of SPT and ceramide synthase activity.

Septic Shock

Similar observations delineate a role of the de novo synthesized ceramide in septic shock in vitro as well as in vivo.[106] Administration of LPS or cytokines to rabbits increases hepatic sphingolipid biosynthesis leading to the accumulation of ceramide, SM and glycosphingolipids. Studies in cell lines link the stimulation of the de novo pathway to the activation of the MAP kinases and NFκB and respectively to the innate immune response. In nonhepatic tissues the elevation in ceramide biosynthesis during septic shock is correlated with the elevated glycosphingolipid synthesis that seems to play a role in pathogen recognition.

Activation of ceramide synthesis in the liver has been linked to production and secretion of lipoproteins enriched in ceramides, sphingomyelins and glycosphingolipids. The functions of these lipoproteins with "altered" sphingolipid content was not well understood; it became clear however that the activation of the de novo synthesis during septic shock is paralleled by activation of another ceramide-generating enzyme, the Zn^{2+}-dependent secretory form of ASMase, termed SSMase. This enzyme hydrolyses SM in the Low Density Lipoproteins (LDL), leading to a robust increase in LDL ceramide content. Several lines of evidence suggest that LDL that are rich in ceramide might mediate injury to the arterial wall during inflammation. LDL extracted from atherosclerotic plaques contain higher ceramide content as compared to LDL isolated from the plasma.[111] Experimentally-induced elevation in LDL ceramide content has been further linked to higher rate of LDL aggregation[112] and oxidation, as well as with an increased potential to induce apoptosis in microvascular endothelial cell.[113]

Lipotoxicity and Insulin Desensitization

Consumption of diet rich in saturated fats (also known as Western diet) is the main risk factor for the development of insulin resistance, hyperglycemia and atherosclerosis. Palmitic acid is the main component of the Western diet and is linked to excessive accumulation of lipids in lean tissues, mainly muscle and liver, lipotoxicity and insulin resistance. Because SPT has almost exclusive preference for the CoA-thiol ester of palmitate, numerous studies have investigated whether Western diet affects de novo synthesis of ceramide. These studies have shown that excess palmitate (delivered via the consumption of Western diet, i.v. infusion, or directly added to cells in culture) stimulates the flux through the de novo pathway resulting in accumulation of ceramide, SM and glycosphingolipids.[58,114] Inhibition of SPT by myriocin prevents not only the palmitate flux through the pathway but also inhibits lipotoxicity, improves insulin response and leads to better glucose regulation.

The exact mechanism by which ceramide affects insulin response is not completely understood. Schmitz-Peiffer et al[115] observed that the accumulation of ceramide in myotubes exposed to palmitate was paralleled by inhibition of Akt/PKB. Cotreatment with myriocin, cycloserine, or fumonisin B1 restored insulin-stimulated Akt phosphorylation, even in the presence of excess palmitate, suggesting that palmitate-induced stimulation of de novo synthesis is required for inhibition of insulin responsiveness. Recent data however imply that in vivo, the diet-induced ceramide increases have to be accompanied by increased triacylglycerides synthesis and accumulation in order to affect the overall insulin response. Furthermore, stimulation of the de novo synthesis of sphingolipids and that of triacylglycerides seems to be correlated in liver.[59]

Programmed Cell Death

Perhaps, the most extensively studied cellular response to ceramide is the induction of programmed cell death or apoptosis. Activation of (dihydro)ceramide synthase in particular is implicated in endothelial cell death induced by TNFα,[116] in daunorubicin,[110] doxorubicin and gemcitabine-induced apoptosis and may account for some aspects of the toxicity of phorbol esters,[117] angiotensin II,[118] cannabinoids[119] and ischemia-reperfusion. In the latter case, investigation of intracellular sites of ceramide accumulation reveals that the elevation of ceramide is in mitochondria and is caused by the activation of a mitochondrial ceramide synthase via posttranslational mechanisms. Furthermore, ceramide accumulation appears to cause mitochondrial respiratory chain damage that could be mimicked by exogenously added natural ceramide to mitochondria.[12]

The recent cloning and characterization of 6 members of the ceramide synthase family CerS1-6 (also known as longevity assurance gene 1-6 (LASS1-6)) provided the opportunity to finally begin studying the role of these enzymes in apoptosis in more details. CerS1 but not CerS2-6 for example, was found to be sufficient and required for apoptosis in response to gemcitabine/doxorubicin treatment.[120] Notably, the gemcitabine/doxorubicin combination treatment was discovered to increase only the levels of C18-ceramide, the preferable substrate for CerS1. These and other studies have provided experimental evidence to the earlier assumptions that not only the type of sphingoid base (i.e., sphingosine vs sphinganine) but also the fatty acid length and degree of saturation influences the biological effectiveness of ceramide.

The role of the de novo pathway in apoptosis had also been investigated in different animal models. Inhibition of the enzymes controlling the de novo pathway prevents alveolar cell apoptosis, oxidative stress and emphysema (the prevalent disease caused by cigarette smoking) in both rats and mice,[121] delays the progression of atherosclerotic plaques in mice[122] and ameliorates some of the pathological consequences of spinal cord injury.[123]

Autophagy

The final paradigm to be discussed is the role of the de novo synthesis of ceramide in autophagy, an evolutionary conserved cytoprotective mechanism that sustains cells during periods of nutrient limitation, but under certain conditions may lead to mammalian cell death. Ceramide addition is sufficient to induce autophagy in some cells and a correlation between the rate of autophagy and ceramide synthesis has been observed in same pathological conditions suggesting a mechanistic link between ceramide, autophagy and disease (reviewed in Zheng et al[2]). Activation of de novo ceramide synthesis mediates autophagy in response to a bioenergetic crisis resulting in the rapid and profound down regulation of nutrient transporter proteins.[124] Ceramide is also shown to mediate the tamoxifen-dependent accumulation of autophagic vacuoles in the human breast cancer MCF-7 cells and to counteract interleukin 13-dependent inhibition of macroautophagy in HT29 cells.[125] The mechanism seems to involve the pro-autophagic protein beclin 1.

Mechanisms of Activation of De Novo Synthesis of Ceramide during Stress

Regulation of SPT at a transcriptional level has been seen with a number of agents, including endotoxin and cytokines,[106] UVB irradiation[126] and others.[127] Induction of both form of SPT, SPT1 and SPT2, occurs in balloon-injured rat carotid artery.[128] In contrast, long-term consumption of food rich in palmitate is correlated with increases in SPT1 protein but not mRNA level.[59] Activation of SPT occurs also posttranslationally in response to etoposide[33] and to heat shock in yeast.[34] Mitochondrial injury in cerebral ischemia/reperfusion activates ceramide synthases via posttranslational mechanism which is dependent of JNK.[12] Cers1 mRNA transcription is up-regulated in response to a gemcitabine/doxorubicin combination treatment,[120] while cisplatin is shown to cause a specific translocation of Cers1 from the endoplasmic reticulum to the Golgi apparatus.[129] The de novo pathway is also modulated through a negative feedback mechanism determined by the rate of sphingolipid degradation in the lysosomes since lipoproteins, sphingosine phosphate[130] and ASMase activity[59] seems to inhibit the flux through it. This might be a mechanism to prevent excessive synthesis of ceramide during normal healthy state of the cell.

Role of the Sphingomyelinases in Cellular Stress Response

The SMase family is a group of biochemically and genetically different enzymes all of which hydrolyze SM to ceramide and phosphorylcholine. SMase activities with neutral and acidic pH optima are found in most mammalian cells and an enzyme active in alkaline pH is localized in the intestinal wall. Currently, research is focused on 4 genes encoding different mammalian SMases: *smpd1* encodes two forms of acidic SMase, one associated with the endosomal/lysosomal compartment (ASMase) and a second one found in the plasma and the conditioned medium of stimulated cells (SSMase). *smpd2* and *smpd3* encode the Neutral SMase 1 (nSMase1) and 2 (nSMase2), both of which are Mg^{2+}-dependent but differ in their subcellular localization and role in signaling. Data from several labs have shown that in mammalian cells, nSMase1 is a housekeeping enzyme with no particular function in signaling. The recently cloned zebra fish nSMase1, however seems to mediate heat-induced apoptosis in zebra fish embryonic cells.[131] In mammalian cells, nSMase2 is regulated by cytokines like IL-1β and TNFα and mediates some of the cytokine effects.[132-134] The recently cloned *smpd4*, is suggested to encode a novel form of NSMase, nSMase3 that is found predominantly in skeletal muscle and heart.

Neutral Sphingomyelinase

Hepatic Acute Phase Response

The acute phase response of liver is an essential component of the systemic host response to bacterial infection and injury and requires IL-1β, a prototypic inflammatory cytokine. Activation of NSMase by IL-1β resulting in transient elevation in ceramide concentration is observed in number of cells including hepatocytes, mesangial cells, EL-4 cells and it has been related to activation of TAK-1, JNK and NF-κB, all of which have important roles in the IL-1β cascade.[135-137]

In hepatocytes, specific silencing of nSMase2 with siRNA has only a minimal effect on the basal cellular NSMase activity, however it results in a complete inhibition of the IL-1β-stimulated NSMase activity.[138] Therefore, nSMase2 is probably an inducible enzyme that contributes little to the basal turnover of SM, but at the same time it is also the only neutral SMase activated by IL-1β. The role of NSMase2 in the IL-1β signaling cascade is rather complex: seemingly, the activation of nSMase2 modulates the pattern of phosphorylation of JNK by IL-1β and respectively the magnitude of transcriptional induction of the hepatic acute phase proteins like C-Reactive Protein, α1 Acid Glycoprotein and Insulin-like Growth Factor Binding Protein-1. nSMase2 activation in hepatocytes leads to potentiation of JNK phosphorylation due to stabilization of the IL-1β receptor-associated kinase-1. The latter most likely involves a phosphatase, like the ceramide-activated protein phosphatase 2A (PP2A).[139]

Vascular Inflammation

Activation of nSMase2 by another pro-inflammatory cytokine, TNFα is well documented in cells of the vasculature and various cancer cells.[132,140,141] TNFα-induced nSMase2 activation is a prerequisite for endothelial nitric oxide synthase activation in HUVEC cells,[134] as well as for the up-regulation in A549 lung epithelial cells of vascular cell adhesion molecule and intracellular adhesion molecule 1,[141] all of which have prominent roles in vascular inflammatory responses.

Apoptosis

Studies into the activation of NSMase during apoptosis have focused mainly on the role of NSMase-mediated ceramide production in the apoptosis induced by 55 kDa receptor for TNFα.[142] A protein factor associated with NSMase activation, FAN, which interacts with the membrane-proximal domain of the p55 receptor and couples stimulation of the receptor to neutral SMase activation, is required for TNFα-induced ceramide generation, caspase processing and apoptosis.[143] Subsequent work in MCF-7 cells found that activation of NSMase in TNFα-stimulated cell death is upstream of mitochondrial changes, cytochrome C release and caspase-9 activation.[144]

Oxidative stress-induced cell death may also be attributed to activation of NSMase. Free oxygen (H_2O_2), but not nitrogen ($ONOO^-$) radicals specifically activate nSMase2, while silencing nSMase2 prevents H_2O_2-induced apoptosis, but had no effect on $ONOO^-$-induced apoptosis.[83]

Very recent studies had began to suggest that a novel form of neutral SMase, nSMase3 is a DNA damage and nongenotoxic stress-regulated gene that is deregulated in human malignancies and modulate the sensitivity of cancer cells to adriamycin-induced cell killing.[145] Early studies seem to suggest that at least the overexpressed nSMase3 can be activated by TNFα within seconds of stimulation.

Growth Arrest

A number of studies have established a role for NSMase-generated ceramide in regulating the cell cycle and mediating growth arrest, possibly through dephosphorylation of retinoblastoma protein and/or regulation of cyclin dependent kinases. Serum withdrawal causes activation of NSMase and cell cycle arrest at G0/G1 in Molt-4 cells.[146] Interestingly, a study focusing on genes up-regulated during confluence-induced arrest had initially identified nSMase2 as a confluence-arrest gene, CCA1, in rat 3Y1 cells.[147] In line with that, nSMase2 seems to mediate confluence-induced growth arrest of MCF-7 cells. The latter was preceded by confluence-induced translocation of nSMase2 to the plasma membrane.[148]

Aging and Cancer

Constitutive up-regulation of nSMase2 and elevation of ceramide concentration during aging has been observed in liver, brain, macrophages and other tissues.[149] This aging-associated elevation in ceramide seems important for the onset of aging process since it has been causatively linked to hyperresponsiveness to IL-1β[70] and LPS[41] and to deregulation of nitric oxide production in endothelium.[71] A substantial decline in hepatic GSH content, which is characteristic for the aging process is responsible for this constitutive increase in nSMase2 activity.[150] nSMase2 apparently follows a pattern of regulation consistent with "developmental-aging" continuum, since in animal models of delayed aging, like calorie-restricted animals, the aging-associated changes in NSMase activity and function are reversed.[150] In cellular model of senescence, similar induction of NSMase activity and accumulation of ceramide has been observed in senescent cells and linked to the decline in proliferative capacity and onset of senescence.[72]

In contrast, a somatic homozygous deletion specifically targeting nSMase2 (Smpd3) is found in a genomic screen for gene copy losses contributing to tumorigenesis in a mouse osteosarcoma model, while loss-of-function mutations in smpd3 gene were identified in 5% of acute myeloid leukemia and 6% of acute lymphoid leukemia cancers. It has been suggested that the mutation could be linked to a defect in plasma membrane translocation of nSMase2 and decreased responsiveness to TNFα. Reconstitution of smpd3 expression in mouse tumor cells lacking the endogenous gene enhances the TNFα-induced reduction of cell viability.[151]

Mechanisms of Activation of NSMase

(i) **Translocation**: Studies have shown that nSMase2 can be translocated to the plasma membrane when cell confluence is reached.[132] This translocation of nSMase2 is also required for the confluence-induced cell cycle arrest to ensue.[152] In oligodendroma-derived cells, a regulated translocation of NSMase2 to the caveolae, which are the signaling domains of the plasma membrane[153] has been observed, while studies with highly differentiated nonproliferating primary hepatocytes have shown that the overexpressed nSMase2 is localized constitutively at the plasma membrane.[133] These data suggest that translocation of nSMase2 to the plasma membrane might be important mechanisms for regulation of its activity *in situ* by bringing the enzyme into contact with its substrate. Pharmacological inhibitors and specific siRNA has implicated the novel PKC, specifically PKCδ, in TNF and PMA-stimulated nSMase2 translocation to the plasma membrane.[52]

(ii) **NSMase as a redox-sensitive enzyme.** The major scavenger of reactive oxygen species, GSH, has been found to be a reversible inhibitor of cellular NSMase activity.[33,154] This is noteworthy, because depletion of cellular GSH content is observed in conditions of increased oxidative stress. The modulation of NSMase activity by GSH was first established in the context of regulation of TNFα signaling and apoptosis.[33,154] Later, the ability of GSH to affect the sensitivity of T47D/H3 breast cancer cells to doxorubicin was attributed to the inhibitory effect GSH has on NSMase activity.[155] A correlation between oxidative stress and NSMase activity was also found in long-lived rats on vitamin Q10 enriched diet[156] and in astrocytes treated with vitamin E.[157] Finally, recent research has shown that specific downregulation of nSMase2 with siRNA blocks H_2O_2-induced apoptosis of human aortic endothelial cells, identifying nSMase2 as a redox-sensitive protein.[158] Detailed analyses of sensitivity of nSMase2 to GSH in hepatocytes suggest that GSH depletion exerts its effect on NSMase activity following a sigmoid dose dependent curve. A rapid and significant activation of NSMase is observed only when hepatic GSH concentration drops below 30% of its basal level, implying that there is a threshold required for NSMase activation during oxidative stress.[150] Interestingly, nSMase1 is also sensitive to changes in the GSH/GSSG ratio.[159] This suggests that redox sensitivity might be a common property of neutral SMases.

Acid Sphingomyelinase

Endotoxic Shock

Patients with severe sepsis exhibit an enhanced SMase activity in plasma. The increase is correlated with the severity of illness and the fatal outcome.[160] Studies in humans and mice administered with LPS confirm these observations.[39] Deletion of ASMase protects against LPS-induced elevation in the plasma SMase activity[161] and attenuates endothelial apoptosis and animal death[43] implicating a secretory isoform of ASMase (SSMase) in apoptosis and organ failure in sepsis. The physiological and biochemical properties of this SSMase are not well understood. It has been postulated that SSMase degrades SM in the secreting cell outer membrane leaflet, which contains almost ¾ of the cellular SM. This SM pool is seemingly inaccessible for other cellular SMases, since the active center of nSMase2, the sole plasma membrane-associated SMase, is facing the cytosolic leaflet of the membrane. These proposed autocrine functions of SSMase however had not been rigorously tested. In turn, substantial evidence had implicated SSMase in modifying the SM/ceramide content of circulating LDL as discussed earlier in the chapter.

Apoptosis (reviewed in ref. 162)

Irradiation of tumor cells with ionizing radiation[52,163-165] transiently activates ASMase with maximal activity detected between 1 to 10 min post irradiation. The activation of ASMase has been linked to the ability of radiation to induce apoptosis since ASMase deficiency causes apoptosis resistance in various tissues in ex vivo or in vivo experiments.[163-165] Genetic restoration of the activity also restores the radiation effects indicating that ASMase mediates apoptosis via ceramide, at least in some cells including B cells, endothelial cells or mesothelium, lung epithelial cells, MCF-7 breast cancer cells, etc. However, some other cell types, for example, thymocytes, remained sensitive to radiation in ASMase-deficient mice suggesting that radiation effects are mediated by different mechanisms depending on the cell type.

ASMase is also involved in death receptor-mediated apoptosis. Stimulation via the CD95 receptor leads to ASMase activation and ceramide accumulation that precedes the induction of cell death.[166-168] Furthermore, studies with fibroblasts from Niemann–Pick disease type A patients, who exhibit deficiency of ASMase and hepatocytes from ASMase null mice show that ASMase activation is required and sufficient for CD95 induced apoptosis.[168,169]

Viral and Bacterial Infections

Cellular ASMase activity seems to play an important role in susceptibility of mammalian organisms to microbial infections. ASMase-deficient mice were found to be more sensitive to infections with the Gram-negative bacteria *L. monocytogenes*, due to a defect in lyso-phagosomal fusion. In wild-type macrophages, the phagosome rapidly fuse with the lysosomes to form a phago-lysosome and to kill and digest bacteria, while in macrophages deficient in ASMase the process is slower and leads to inefficient transfer of lysosomal antibacterial hydrolyses into phagosomes.[170] Apparently, ASMase is required for the proper fusion of late phagosomes with lysosomes.[171] ASMase-deficient mice were also more susceptible to infection with Sindbis virus, an enveloped virus with a single-stranded RNA that is involved in fatal alpha virus encephalomyelitis, due to more rapid replication and spread of the virus in the nervous system.[172]

Mechanisms of Activation of ASMase

One mechanism for activation of ASMase involves its phosphorylation by PKCδ at Ser^{508} which mediates UV light-induced ASMase activation and cell death in MCF-7 breast cancer cells.[52] Phosphorylation of ASMase seems to be correlated with its translocation to the plasma membrane (see below) in irradiated cells.

Evidence for a Coordinated Regulation of Multiple Pathways for Ceramide Generation

A handful of studies suggest that the activity of the two SMases and the de novo pathway might be regulated in a coordinated fashion by the same agonist, resulting in the transient generation of distinct "waves" of ceramide increases that might serve to diversify the biological effects of ceramide. These studies also emphasize the significance of specific ceramide species in defined stages of cellular stress response.

For example, a transient increase of ceramide is observed within minutes after exposure to ionizing irradiation which is a consequence of DNA damage-independent acid SMase[173] or neutral SMase activation.[47] Several hours after irradiation, a second wave of ceramide accumulation is observed depending on the DNA damage-dependent activation of ceramide synthase. It seems that the late ceramide accumulation is also dependent on the first one and is rate limiting for the apoptotic process induced by irradiation.[173]

Kroesen et al[174] on the other hand, showed that cross linking of the B-cell receptor generates C16-ceramide upstream of the mitochondria in a caspase independent manner and that inhibition of C16 ceramide generation rescues from cell death.[56] C24 ceramide however was generated downstream of mitochondrial dysfunction in a caspase dependent manner. All of the increases are seemingly due to activation of de novo synthesis and apparently different members of the ceramide synthase family might be differentially regulated.

A classical example of coordinated regulation of neutral and acidic SMase is the early work of Kronke and colleagues.[175] In these studies, stimulation with TNFα lead to activation of both enzymes through apparently separate mechanisms since different domains of the cytosolic tail of the TNFα receptor were responsible. Furthermore, while the activation of ASMase was linked to NFκB activation, NSMase seemed to regulate the activation of proline-directed serine/threonine protein kinase(s).

Mechanisms of Ceramide Effects on Cellular Functions

Ceramide-Interacting Molecules

PKCζ (reviewed in ref. 114)

PKC ζ, atypical PKC isoform, was identified as a molecule that ceramide directly binds and activate. Notably, ceramide-induced activation of PKCζ is linked to inhibition of Akt-1. Akt-1 is a key regulatory molecule for various metabolic cellular functions, cell proliferation and cell death, which has been known for a long time to be inhibited by ceramide. Ceramide-activated PKCζ seems to phosphorylate Ser^{34} of the Akt-1 pleckstrin homology domain, thus preventing the interaction of Akt-1 with PIP_3 and respectively with the plasma membrane.[176-178] In addition to stimulating the kinase activity of PKCζ, ceramide binding seems to affect the ability of PKCζ to interact with other proteins. In vascular smooth muscle, ceramide stabilizes the interaction of Akt-1 and PKCζ within caveolin-enriched lipid microdomains to inactivate Akt and specifically reduces the association of PKCζ with 14-3-3, a scaffold protein localized to less structured regions within membranes.[179] In differentiating embryonic stem cells, ceramide binding to PKCζ similarly leads to activation of its kinase activity but also increases PKCζ binding to its inhibitor protein, prostate apoptosis response-4 (PAR-4), thus compromising the antiapoptotic activity of PKCζ and inducing apoptosis.[180]

PP2A (reviewed in ref. 181)

Long chain D-*erythro*-C18-ceramide has been shown to activate PP2A (more specifically the AB'C trimer), PP2AC and PP1-γC and -αC in vitro suggesting a direct and stereospecific effect of ceramide on the phosphatase activity. This ability of ceramide to activate PP2A was later found to mediate the ceramide effects on various substrates relevant to the induction of apoptosis, growth arrest or inflammation, including c-Jun, Bcl-2, Akt/PKB, Rb, PKCα, ERK1/2, IRAK-1, SR proteins and many others.

Cathepsin D

Ceramide specifically binds and activates the endosomal acidic aspartate protease cathepsin D. Direct interaction of ceramide with cathepsin D results in autocatalytic proteolysis of the 52 kDa procathepsin D to the enzymatically active 48/32 kDa isoforms that can subsequently cleave and activate the apoptotic regulator Bid.[182] Studies in ASMase deficient cells strongly suggest that ASMase activity is responsible for the generation of ceramide that can activate the protease.[183] This is in contrast to the ability of ceramide to activate PKCζ and PP2A, which seemingly requires activation of the de novo pathway or NSMase.

Indirect Targets of Ceramide

Modulators of Apoptosis (reviewed in ref. 184)

The members of the Bcl-2 family of proteins are essential modulators of apoptotic cell death following genotoxic and nongenotoxic stress. Strong evidence links ceramide to the regulation of two members of that family, the antiapoptotic Bcl-2 and the pro-apoptotic Bax. The mechanisms involved are far from understood and apparently quite complex. TNFα- and etoposide-induced activation of NSMase leads to apoptosis via a pathway where ceramide is upstream of the antiapoptotic member of the family, Bcl-2 and lead to its inhibition (5).[185] These effects seem to involve PP2A. Bcl-2, whose phosphorylation at Ser^{70} is required for its anti-apoptotic function, becomes dephosphorylated in response to ceramide in a PP2A-dependent manner and consequently is degraded in the proteasomes.[186] In turn, gemcitabine-induced ceramide generation is found to enhance the expression of proapoptotic Bcl-x.[187] Ceramide accumulation is also linked to the activation of the execution caspases, but it is downstream of the initiator caspases.[188,189]

C16-ceramide, which is generated by ASMase in response to irradiation is shown to induce a conformation change of the pro-apoptotic member Bax leading to its activation and cytochrome C release.[190] It was suggested that only the production of ceramide in the mitochondria can induce the oligomerization of Bax that drives cell death.[191] As Bcl-2, Bax is also regulated by ceramide via PP2A, since its dephosphorylation is associated with conformational change and release of cytochrome C from the mitochondria. Ceramide generation by ASMase has been suggested to mediate cell death by caspase- dependent and independent mechanisms depending on the death stimulus.[192]

Regulators of Cell Cycle

Dephosphorylation of the Retinoblastoma protein, activation of the cyclin dependent kinase inhibitor p21 and inhibition of the cyclin dependent kinase 2[193,194] are essential steps in the pathway leading to cell cycle arrest in response to ceramide accumulation. Ceramide induced-dephosphorylation of Retinoblastoma protein is mediated through PP2A, for which Retinoblastoma protein is a direct substrate. The hypophosphorylated Retinoblastoma protein binds and sequesters E2F, an essential factor for progression through the cell cycle. The senescent-associated growth may be attributed to a defect in the phospholipase D/protein kinase C (PLD/PKC) pathway and ceramide can inhibit both PLD and PKC.[195,196]

Regulators of Inflammation

Studies by Hannun and Brenner were the first to show that TNFα activates the stress-activated protein kinases JNKs, resulting in the stimulation of AP-1-transcription factor and induces the translocation of NFκB to the nucleus, resulting in the stimulation of NFκB-dependent gene transcription, through ceramide.[197] Ceramide induced JNK activation has also been reported in response to FAS activation,[198] irradiation[199] and many cytokines. The activation of JNK by ceramide involves Rac-1[200] in the case of radiation-induced apoptosis, PKCζ in response to IL-1β[201] and TAK-1 in response to TNF-α, IL-1β, or anti-Fas antibody.[202] In hepatocytes, nSMase2-generated ceramide has been shown to modulate the effects of IL-1β on JNK phosphorylation and activity in PP2A and IRAK-1-depndent manner.[133]

Ceramide-mediated JNK activation inhibits differentiation of skeletal muscle progenitor cells to myoblasts[203] and induces the expression of various acute phase proteins like Insulin-like Growth Factor Binding Protein-1 in liver.[204] JNK activation mediates apoptosis in neurons in response to ceramide and Amyloid β, the protein involved in the ethiology of Alzheimer's disease. In that system, an inhibition of NSMase was found to attenuate Amyloid β-induced JNK phosphorylation and AP-1 DNA binding activity suggesting the involvement of the plasma membrane-associated SMase activity.[205]

Ceramide Effects on Membrane Organization

Some of the biological responses to ceramide generation are probably due to the effects ceramide has on the structure of membrane rafts and caveolae, on the membrane curvature and the membrane permeability to aqueous solutes. The two hydroxyl groups, the amid group protons and the trans-double bond in ceramide molecule, together with the lack of large hydrophilic head group determine the ability of ceramide to affect the properties and organization of biological membrane. The long-chain base and long saturated N-acyl chains accompanied by a very small hydrophilic head promote the partitioning of ceramide into ordered membrane domains. Together with SM, which has a strong affinity for interacting with membrane cholesterol, ceramide participate in the formation of the caveolae, the signaling platforms on the cell surface. While SM is by far much more abundant than ceramide in the lipid rafts, the generation of ceramide within rafts dramatically alters the biophysical properties of these membrane domains. Ceramide molecules have the tendency to spontaneously self associate to small ceramide-enriched membrane microdomains.[206] Gulbins and colleagues had proposed that these ceramide microdomains spontaneously fuse to form large ceramide-enriched macrodomains or platforms. The formation of such platforms is proposed to underline the process of agonist induced receptor clustering and consequently, the reorganization of intracellular signaling molecules to transmit a signal into the cell. Among the receptors affected by ceramide-mediated aggregation/clustering are CD95, CD28, TNFα, CD40, FcgRII.[207] Intriguingly, the acidic form of SMase seems to be involved in the formation of these signaling platforms, since it has been shown to translocate to the plasma membrane under certain conditions. In further support to that extend, overexpression of the SM synthase, SMS-1, which enriches plasma membrane with SM also enhances the translocation of Fas into lipid rafts leading to subsequent Fas clustering, DISC formation, activation of caspases and apoptosis, suggesting that SM levels in the lipid rafts might also be a critical factor possible by acting as a source for ceramide.[208]

Another intriguing property of ceramide is its ability to spontaneously form channels into planer bilayers, liposomes and biological membranes indicating that ceramide might regulate membrane leakiness. Experiments with unilamellar vesicles show that ceramide forms large pores that can allow the efflux of proteins as large as 60 kDa. Studies by Colombini and colleagues had shown that at physiologically relevant levels, ceramides form stable channels in mitochondrial outer membranes capable of passing the largest proteins known to exit mitochondria during apoptosis including cytochrome C.[209] Ongoing efforts on characterizing the ceramide metabolism in this organelle will shed more light into the physiological function of ceramide in the mitochondria.

Finally, studies in cells isolated from ASMase-deficient mice or Niemann-Pick patients had shown defects in the phagolysosomal fusion and rab-4-mediated endocytosis[210] indicating that ASMase regulates select vesicular fusion processes by modifying the steric conformation of cellular membranes.[211] These conclusions are supported by series of studies on the infectivity of *Listeria monocytogenes* and *Mycobacterium avium* discussed earlier.

Conclusion

It was realized long ago that sphingolipid structure is highly diverse and encodes distinct chemical information that can be conveyed during cell signaling.[212] Research spanning through the last two decades established experimentally the role of this novel family of bioactive lipids in signaling. Ceramide emerged as the key metabolite in several sphingolipid-based signaling networks and was recognized to mediate conserved pathways of cellular stress response. With the genetic cloning and characterization of the majority of ceramide metabolizing enzymes, it also became clear that ceramide metabolism is finely tuned leading to spatial, temporal and species-specific accumulation during stress. A strict substrate specificity and distinct cellular localization of the enzymes that catalyze ceramide synthesis and degradation has brought diversity in the signaling molecules and cellular responses regulated by ceramide. The distinct ability of ceramide to alter the biophysical properties of cellular membrane adds additional layer of complexity in our understanding of the roles of ceramide in cellular stress response.

References

1. GoÒi FM, Alonso A. Biophysics of sphingolipids I. Membrane properties of sphingosine, ceramides and other simple sphingolipids. Biochimica et Biophysica Acta (BBA)—Biomembranes 2006; 1758:1902-1921.
2. Zheng W, Kollmeyer J, Symolon H et al. Ceramides and other bioactive sphingolipid backbones in health and disease: Lipidomic analysis, metabolism and roles in membrane structure, dynamics, signaling and autophagy. Biochimica et Biophysica Acta (BBA)—Biomembranes 2006; 1758:1864-1884.
3. Ramstedt B, Slotte JP. Sphingolipids and the formation of sterol-enriched ordered membrane domains. Biochimica et Biophysica Acta (BBA)—Biomembranes 2006; 1758:1945-1956.
4. Shaner RL, Allegood JC, Park H et al. Quantitative analysis of sphingolipids for lipidomics using triple quadrupole and quadrupole linear ion trap mass spectrometers. J Lipid Res 2008.
5. Zhang J, Alter N, Reed JC et al. Bcl-2 interrupts the ceramide-mediated pathway of cell death. Proc Natl Acad Sci USA 1996; 93:5325-5328.
6. Jin J, Hou Q, Mullen TD et al. Ceramide generated by sphingomyelin hydrolysis and the salvage pathway is involved in hypoxia/reoxygenation-induced Bax redistribution to mitochondria in NT-2 cells. J Biol Chem 2008; 283:26509-26517.
7. Hernandez OM, Discher DJ, Bishopric NH et al. Rapid activation of neutral sphingomyelinase by hypoxia-reoxygenation of cardiac myocytes. Circ Res 2000; 86:198-204.
8. Basnakian AG, Ueda N, Hong X et al. Ceramide synthase is essential for endonuclease-mediated death of renal tubular epithelial cells induced by hypoxia-reoxygenation. Am J Physiol Renal Physiol 2005; 288:F308-314.
9. Yoshimura S, Banno Y, Nakashima S et al. Ceramide formation leads to caspase-3 activation during hypoxic PC12 cell death. Inhibitory effects of Bcl-2 on ceramide formation and caspase-3 activation. J Biol Chem 1998; 273:6921-6927.
10. Therade-Matharan S, Laemmel E, Carpentier S et al. Vicaut. Reactive oxygen species production by mitochondria in endothelial cells exposed to reoxygenation after hypoxia and glucose depletion is mediated by ceramide. Am J Physiol Regul Integr Comp Physiol 2005; 289:R1756-1762.
11. Yun JK, Kester M. Regulatory role of sphingomyelin metabolites in hypoxia-induced vascular smooth muscle cell proliferation. Arch Biochem Biophys 2002; 408:78-86.
12. Yu J, Novgorodov SA, Chudakova D et al. JNK3 signaling pathway activates ceramide synthase leading to mitochondrial dysfunction. J Biol Chem 2007; 282:25940-25949.
13. Bielawska AE, Shapiro JP, Jiang L et al. Ceramide is involved in triggering of cardiomyocyte apoptosis induced by ischemia and reperfusion. Am J Pathol 1997; 151:1257-1263.
14. Zager RA, Iwata M, Conrad DS et al. Altered ceramide and sphingosine expression during the induction phase of ischemic acute renal failure. Kidney Int 1997; 52:60-70.
15. Llacuna L, Mari M, Garcia-Ruiz C et al. Critical role of acidic sphingomyelinase in murine hepatic ischemia-reperfusion injury. Hepatology 2006; 44:561-572.
16. Singh I, Pahan K, Khan M et al. Cytokine-mediated induction of ceramide production is redox-sensitive. Implications to proinflammatory cytokine-mediated apoptosis in demyelinating diseases. J Biol Chem 1998; 273:20354-20362.
17. Brogi A, Strazza M, Melli M et al. Induction of intracellular ceramide by interleukin-1 beta in oligodendrocytes. J Cell Biochem 1997; 66:532-541.
18. Schreur KD, Liu S. Involvement of ceramide in inhibitory effect of IL-1 beta on L-type Ca2+ current in adult rat ventricular myocytes. Am J Physiol 1997; 272:H2591-2598.
19. Kaszkin M, Huwiler A, Scholz K et al. Negative regulation of interleukin-1beta-activated neutral sphingomyelinase by protein kinase C in rat mesangial cells. FEBS Lett 1998; 440:163-166.
20. Nikolova-Karakashian M, Morgan ET, Alexander C et al. Bimodal regulation of ceramidase by interleukin-1beta. Implications for the regulation of cytochrome p450 2C11. J Biol Chem 1997; 272:18718-18724.
21. Chen J, Nikolova-Karakashian M, Merrill AH et al. Regulation of cytochrome P450 2C11 (CYP2C11) gene expression by interleukin-1, sphingomyelin hydrolysis and ceramides in rat hepatocytes. J Biol Chem 1995; 270:25233-25238.
22. Masamune A, Igarashi Y, Hakomori S. Regulatory role of ceramide in interleukin (IL)-1 beta-induced E-selectin expression in human umbilical vein endothelial cells. Ceramide enhances IL-1 beta action, but is not sufficient for E-selectin expression. J Biol Chem 1996; 271:9368-9375.
23. Sallusto F, Nicolo C, De Maria R et al. Ceramide inhibits antigen uptake and presentation by dendritic cells. J Exp Med 1996; 184:2411-2416.
24. Schneider C, Delorme N, El Btaouri H et al. Martiny. Interleukin 1 beta (IL-1 beta) action in porcine thyroid cells involves the ceramide signalling pathway. Cytokine 2001; 13:174-178.
25. Liu J, Ginis I, Spatz M et al. Hypoxic preconditioning protects cultured neurons against hypoxic stress via TNF-alpha and ceramide. Am J Physiol Cell Physiol 2000; 278: C144-153.

26. Sawada M, Kiyono T, Nakashima S et al. Molecular mechanisms of TNF-alpha-induced ceramide formation in human glioma cells: P53-mediated oxidant stress-dependent and -independent pathways. Cell Death Differ 2004; 11:997-1008.
27. Kitatani K, Akiba S, Sato T. Ceramide-induced enhancement of secretory phospholipase A2 expression via generation of reactive oxygen species in tumor necrosis factor-alpha-stimulated mesangial cells. Cell Signal 2004; 16:967-974.
28. Garcia-Ruiz C, Colell A, Mari M et al. Defective TNF-alpha-mediated hepatocellular apoptosis and liver damage in acidic sphingomyelinase knockout mice. J Clin Invest 2003; 111:197-208.
29. Mallampalli RK, Peterson EJ, Carter AB et al. TNF-alpha increases ceramide without inducing apoptosis in alveolar type II epithelial cells. Am J Physiol 1999; 276: L481-490.
30. Ryan AJ, McCoy DM, McGowan SE et al. Alveolar sphingolipids generated in response to TNF-alpha modifies surfactant biophysical activity. J Appl Physiol 2003; 94:253-258.
31. Medler TR, Petrusca DN, Lee PJ et al. Apoptotic sphingolipid signaling by ceramides in lung endothelial cells. Am J Respir Cell Mol Biol 2008; 38:639-646.
32. Ishizuka N, Yagui K, Tokuyama Y et al. Kanatsuka. Tumor necrosis factor alpha signaling pathway and apoptosis in pancreatic beta cells. Metabolism 1999; 48:1485-1492.
33. Liu B, Andrieu-Abadie N, Levade T et al. Glutathione regulation of neutral sphingomyelinase in tumor necrosis factor-alpha-induced cell death. J Biol Chem 1998; 273:11313-11320.
34. Luberto C, Hassler DF, Signorelli P et al. Inhibition of tumor necrosis factor-induced cell death in MCF7 by a novel inhibitor of neutral sphingomyelinase. J Biol Chem 2002; 277:41128-41139.
35. Strelow A, Bernardo K, Adam-Klages S et al. Overexpression of acid ceramidase protects from tumor necrosis factor-induced cell death. J Exp Med 2000; 192:601-612.
36. Monney L, Olivier R, Otter I et al. Role of an acidic compartment in tumor-necrosis-factor-alpha-induced production of ceramide, activation of caspase-3 and apoptosis. Eur J Biochem 1998; 251:295-303.
37. Tsuji K, Mitsutake S, Yokose U et al. Role of ceramide kinase in peroxisome proliferator-activated receptor beta-induced cell survival of mouse keratinocytes. Febs J 2008; 275:3815-3826.
38. Zimmermann C, Ginis I, Furuya K et al. M. Hallenbeck. Lipopolysaccharide-induced ischemic tolerance is associated with increased levels of ceramide in brain and in plasma. Brain Res 2001; 895:59-65.
39. Lightle S, Tosheva R, Lee A et al. Elevation of ceramide in serum lipoproteins during acute phase response in humans and mice: role of serine-palmitoyl transferase. Arch Biochem Biophys 2003; 419:120-128.
40. Cuschieri J, Billgren J, Maier RV. Phosphatidylcholine-specific phospholipase C (PC-PLC) is required for LPS-mediated macrophage activation through CD14. J Leukoc Biol 2006; 80:407-414.
41. Claycombe KJ, Wu D, Nikolova-Karakashian M et al. Ceramide mediates age-associated increase in macrophage cyclooxygenase-2 expression. J Biol Chem 2002; 277:30784-30791.
42. Monick MM, Mallampalli RK, Carter AB et al. Ceramide regulates lipopolysaccharide-induced phosphatidylinositol 3-kinase and Akt activity in human alveolar macrophages. J Immunol 2001; 167:5977-5985.
43. Haimovitz-Friedman A, Cordon-Cardo C, Bayoumy S et al. Lipopolysaccharide induces disseminated endothelial apoptosis requiring ceramide generation. J Exp Med 1997; 186:1831-1841.
44. Lei X, Zhang S, Bohrer A et al. The group VIA calcium-independent phospholipase A2 participates in ER stress-induced INS-1 insulinoma cell apoptosis by promoting ceramide generation via hydrolysis of sphingomyelins by neutral sphingomyelinase. Biochemistry 2007; 46:10170-10185.
45. Lu FG, Wong CS. Radiation-induced apoptosis of oligodendrocytes and its association with increased ceramide and down-regulated protein kinase B/Akt activity. Int J Radiat Biol 2004; 80:39-51.
46. Uchida Y, Nardo AD, Collins V et al. De novo ceramide synthesis participates in the ultraviolet B irradiation-induced apoptosis in undifferentiated cultured human keratinocytes. J Invest Dermatol 2003; 120:662-669.
47. Haimovitz-Friedman A, Kan CC, Ehleiter D et al. Ionizing radiation acts on cellular membranes to generate ceramide and initiate apoptosis. J Exp Med 1994; 180:525-535.
48. Sathishkumar S, Boyanovsky B, Karakashian AA et al. Elevated sphingomyelinase activity and ceramide concentration in serum of patients undergoing high dose spatially fractionated radiation treatment: implications for endothelial apoptosis. Cancer Biol Ther 2005; 4:979-986.
49. Grether-Beck S, Bonizzi G, Schmitt-Brenden H et al. Non-enzymatic triggering of the ceramide signalling cascade by solar UVA radiation. EMBO J 2000; 19:5793-5800.
50. Thomas RL Jr, Matsko CM, Lotze MT et al. Mass spectrometric identification of increased C16 ceramide levels during apoptosis. J Biol Chem 1999; 274:30580-30588.
51. Chatterjee M, Wu S. Cell line dependent involvement of ceramide in ultraviolet light-induced apoptosis. Mol Cell Biochem 2001; 219:21-27.
52. Zeidan YH, Wu BX, Jenkins RW et al. A novel role for protein kinase Cdelta-mediated phosphorylation of acid sphingomyelinase in UV light-induced mitochondrial injury. FASEB J 2008; 22:183-193.
53. Dai Q, Liu J, Chen J et al. Mitochondrial ceramide increases in UV-irradiated HeLa cells and is mainly derived from hydrolysis of sphingomyelin. Oncogene 2004; 23:3650-3658.

54. Charruyer A, Grazide S, Bezombes C et al. UV-C light induces raft-associated acid sphingomyelinase and JNK activation and translocation independently on a nuclear signal. J Biol Chem 2005; 280:19196-19204.
55. Zhou YT, Grayburn P, Karim A et al. Lipotoxic heart disease in obese rats: implications for human obesity. Proc Natl Acad Sci USA 2000; 97:1784-1789.
56. Adams JM 2nd, Pratipanawatr T, Berria R et al. Ceramide content is increased in skeletal muscle from obese insulin-resistant humans. Diabetes 2004; 53:25-31.
57. Turinsky J, O'Sullivan DM, Bayly BP. 1,2-Diacylglycerol and ceramide levels in insulin-resistant tissues of the rat in vivo. J Biol Chem 1990; 265:16880-16885.
58. Holland WL, Brozinick JT, Wang LP et al. Summers. Inhibition of ceramide synthesis ameliorates glucocorticoid-, saturated-fat- and obesity-induced insulin resistance. Cell Metab 2007; 5:167-179.
59. Deevska GM, Rozenova KA, Giltiay NV et al. Acid sphingomyelinase deficiency prevents diet-induced hepatic triacylglycerol accumulation and hyperglycemia in mice. J Biol Chem 2008.
60. Haus JM, Kashyap SR, Kasumov T et al. Plasma Ceramides Are Elevated in Obese Subjects with Type 2 Diabetes and Correlate with the Severity of Insulin Resistance. Diabetes 2008.
61. Shah C, Yang G, Lee I et al. Protection from high fat diet-induced increase in ceramide in mice lacking plasminogen activator inhibitor 1. J Biol Chem 2008; 283:13538-13548.
62. Samad F, Hester KD, Yang G et al. Altered adipose and plasma sphingolipid metabolism in obesity: a potential mechanism for cardiovascular and metabolic risk. Diabetes 2006; 55:2579-2587.
63. Kolak M, Westerbacka J, Velagapudi VR et al. Adipose tissue inflammation and increased ceramide content characterize subjects with high liver fat content independent of obesity. Diabetes 2007; 56:1960-1968.
64. Shimabukuro M, Zhou YT, Levi M et al. Fatty acid-induced beta cell apoptosis: a link between obesity and diabetes. Proc Natl Acad Sci USA 1998; 95:2498-2502.
65. Fox TE, Han X, Kelly S et al. Diabetes alters sphingolipid metabolism in the retina: a potential mechanism of cell death in diabetic retinopathy. Diabetes 2006; 55:3573-3580.
66. Costantini C, Scrable H, Puglielli L. An aging pathway controls the TrkA to p75NTR receptor switch and amyloid beta-peptide generation. EMBO J 2006; 25:1997-2006.
67. Rodriguez-Calvo R, Serrano L, Barroso E et al. Peroxisome proliferator-activated receptor alpha down-regulation is associated with enhanced ceramide levels in age-associated cardiac hypertrophy. J Gerontol A Biol Sci Med Sci 2007; 62:1326-1336.
68. Wu D, Ren Z, Pae M et al. Aging up-regulates expression of inflammatory mediators in mouse adipose tissue. J Immunol 2007; 179:4829-4839.
69. Lightle SA, Oakley JI, Nikolova-Karakashian MN. Activation of sphingolipid turnover and chronic generation of ceramide and sphingosine in liver during aging. Mech Ageing Dev 2000; 120:111-125.
70. Rutkute K, Karakashian AA, Giltiay NV et al. Aging in rat causes hepatic hyperresposiveness to interleukin-1beta which is mediated by neutral sphingomyelinase-2. Hepatology 2007; 46:1166-1176.
71. Smith AR, Visioli F, Frei B et al. Age-related changes in endothelial nitric oxide synthase phosphorylation and nitric oxide dependent vasodilation: evidence for a novel mechanism involving sphingomyelinase and ceramide-activated phosphatase 2A. Aging Cell 2006; 5:391-400.
72. Venable ME, Lee JY, Smyth MJ et al. Role of ceramide in cellular senescence. J Biol Chem 1995; 270:30701-30708.
73. Satoi H, Tomimoto H, Ohtani R et al. Astroglial expression of ceramide in Alzheimer's disease brains: a role during neuronal apoptosis. Neuroscience 2005; 130:657-666.
74. Cutler RG, Kelly J, Storie K et al. Involvement of oxidative stress-induced abnormalities in ceramide and cholesterol metabolism in brain aging and Alzheimer's disease. Proc Natl Acad Sci USA 2004; 101:2070-2075.
75. Han X, D MH, McKeel DW Jr et al. Substantial sulfatide deficiency and ceramide elevation in very early Alzheimer's disease: potential role in disease pathogenesis. J Neurochem 2002; 82:809-818.
76. Jenkins GM, Cowart LA, Signorelli P et al. Acute activation of de novo sphingolipid biosynthesis upon heat shock causes an accumulation of ceramide and subsequent dephosphorylation of SR proteins. J Biol Chem 2002; 277:42572-42578.
77. Jana A, Pahan K. Oxidative stress kills human primary oligodendrocytes via neutral sphingomyelinase: implications for multiple sclerosis. J Neuroimmune Pharmacol 2007; 2:184-193.
78. Song JH, Shin SH, Ross GM. Oxidative stress induced by ascorbate causes neuronal damage in an in vitro system. Brain Res 2001; 895:66-72.
79. Song JH, Shin SH, Wang W et al. Involvement of oxidative stress in ascorbate-induced proapoptotic death of PC12 cells. Exp Neurol 2001; 169:425-437.
80. Goldkorn T, Balaban N, Shannon M et al. H2O2 acts on cellular membranes to generate ceramide signaling and initiate apoptosis in tracheobronchial epithelial cells. J Cell Sci 1998; 111 (Pt 21):3209-3220.
81. Yamada Y, Tian J, Yang Y et al. Oxidized low density lipoproteins induce a pathologic response by retinal pigmented epithelial cells. J Neurochem 2008; 105:1187-1197.

82. Auge N, Nikolova-Karakashian M, Carpentier S et al. Role of sphingosine 1-phosphate in the mitogenesis induced by oxidized low density lipoprotein in smooth muscle cells via activation of sphingomyelinase, ceramidase and sphingosine kinase. J Biol Chem 1999; 274:21533-21538.
83. Castillo SS, Levy M, Thaikoottathil JV et al. Reactive nitrogen and oxygen species activate different sphingomyelinases to induce apoptosis in airway epithelial cells. Exp Cell Res 2007; 313:2680-2686.
84. Park TS, Hu Y, Noh HL et al. Ceramide is a cardiotoxin in lipotoxic cardiomyopathy. J Lipid Res 2008; 49:2101-2112.
85. Karpanen T, Bry M, Ollila HM et al. Overexpression of vascular endothelial growth factor-B in mouse heart alters cardiac lipid metabolism and induces myocardial hypertrophy. Circ Res 2008; 103:1018-1026.
86. Ichi I, Nakahara K, Miyashita Y et al. Association of ceramides in human plasma with risk factors of atherosclerosis. Lipids 2006; 41:859-863.
87. Wang H, Charles AG, Frankel AJ et al. Increasing intracellular ceramide: an approach that enhances the cytotoxic response in prostate cancer cells. Urology 2003; 61:1047-1052.
88. Scarlatti F, Sala G, Somenzi G et al. Resveratrol induces growth inhibition and apoptosis in metastatic breast cancer cells via de novo ceramide signaling. FASEB J 2003; 17:2339-2341.
89. Charles AG, Han TY, Liu YY et al. Taxol-induced ceramide generation and apoptosis in human breast cancer cells. Cancer Chemother Pharmacol 2001; 47:444-450.
90. Roberts CG, Gurisik E, Biden TJ et al. Synergistic cytotoxicity between tamoxifen and the plant toxin persin in human breast cancer cells is dependent on Bim expression and mediated by modulation of ceramide metabolism. Mol Cancer Ther 2007; 6:2777-2785.
91. Cabot MC, Han TY, Giuliano AE. The multidrug resistance modulator SDZ PSC 833 is a potent activator of cellular ceramide formation. FEBS Lett 1998; 431:185-188.
92. Wang H, Giuliano AE, Cabot MC. Enhanced de novo ceramide generation through activation of serine palmitoyltransferase by the P-glycoprotein antagonist SDZ PSC 833 in breast cancer cells. Mol Cancer Ther 2002; 1:719-726.
93. Senchenkov A, Han TY, Wang H et al. Enhanced ceramide generation and induction of apoptosis in human leukemia cells exposed to DT(388)-granulocyte-macrophage colony-stimulating factor (GM-CSF), a truncated diphtheria toxin fused to human GM-CSF. Blood 2001; 98:1927-1934.
94. Wang H, Maurer BJ, Reynolds CP et al. N-(4-hydroxyphenyl)retinamide elevates ceramide in neuroblastoma cell lines by coordinate activation of serine palmitoyltransferase and ceramide synthase. Cancer Res 2001; 61:5102-5105.
95. Gu X, Song X, Dong Y et al. Vitamin E succinate induces ceramide-mediated apoptosis in head and neck squamous cell carcinoma in vitro and in vivo. Clin Cancer Res 2008; 14:1840-1848.
96. Coroneos E, Wang Y, Panuska JR et al. Sphingolipid metabolites differentially regulate extracellular signal-regulated kinase and stress-activated protein kinase cascades. Biochem J 1996; 316 (Pt 1):13-17.
97. Merrill AH Jr, van Echten G, Wang E et al. Fumonisin B1 inhibits sphingosine (sphinganine) N-acyltransferase and de novo sphingolipid biosynthesis in cultured neurons in situ. J Biol Chem 1993; 268:27299-27306.
98. Paumen MB, Ishida Y, Muramatsu M et al. Inhibition of Carnitine Palmitoyltransferase I Augments Sphingolipid Synthesis and Palmitate-induced Apoptosis. J Biol Chem 1997; 272:3324-3329.
99. Blazquez C, Galve-Roperh I, Guzman M. De novo-synthesized ceramide signals apoptosis in astrocytes via extracellular signal-regulated kinase. FASEB J 2000; 14:2315-2322.
100. Lehtonen JYA, Horiuchi M, Daviet L et al. Activation of the de novo Biosynthesis of Sphingolipids Mediates Angiotensin II Type 2†Receptor-induced Apoptosis. J Biol Chem 1999; 274:16901-16906.
101. Merrill AH Jr. De Novo Sphingolipid Biosynthesis: A Necessary, but Dangerous, Pathway. J Biol Chem 2002; 277:25843-25846.
102. Dickson RC. Thematic Review Series: Sphingolipids. New insights into sphingolipid metabolism and function in budding yeast. J Lipid Res 2008; 49:909-921.
103. Meier KD, Deloche O, Kajiwara K et al. Sphingoid base is required for translation initiation during heat stress in Saccharomyces cerevisiae. Mol Biol Cell 2006; 17:1164-1175.
104. Deng X, Yin X, Allan R et al. Ceramide Biogenesis Is Required for Radiation-Induced Apoptosis in the Germ Line of C. elegans. Science 2008; 322:110-115.
105. Jenkins GM. The emerging role for sphingolipids in the eukaryotic heat shock response. Cell Mol Life Sci 2003; 60:701-710.
106. Memon RA, Holleran WM, Moser AH et al. Endotoxin and Cytokines Increase Hepatic Sphingolipid Biosynthesis and Produce Lipoproteins Enriched in Ceramides and Sphingomyelin. Arterioscler Thromb Vasc Biol 1998; 18:1257-1265.
107. Shimabukuro M, Higa M, Zhou YT et al. Lipoapoptosis in beta-cells of obese prediabetic fa/fa rats. Role of serine palmitoyltransferase overexpression. J Biol Chem 1998; 273:32487-32490.
108. Pickersgill L, Litherland GJ, Greenberg AS et al. Key role for ceramides in mediating insulin resistance in human muscle cells. J Biol Chem 2007; 282:12583-12589.

109. Perry DK, Carton J, Shah AK et al. Serine Palmitoyltransferase Regulates de Novo Ceramide Generation during Etoposide-induced Apoptosis. J Biol Chem 2000; 275:9078-9084.
110. Bose R, Verheij M, Haimovitz-Friedman A et al. Ceramide synthase mediates daunorubicin-induced apoptosis: An alternative mechanism for generating death signals. Cell 1995; 82:405-414.
111. Schissel SL, Tweedie-Hardman J, Rapp JH et al. Rabbit aorta and human atherosclerotic lesions hydrolyze the sphingomyelin of retained low-density lipoprotein. Proposed role for arterial-wall sphingomyelinase in subendothelial retention and aggregation of atherogenic lipoproteins. J Clin Invest 1996; 98:1455-1464.
112. Walters MJ, Wrenn SP. Effect of sphingomyelinase-mediated generation of ceramide on aggregation of low-density lipoprotein. Langmuir 2008; 24:9642-9647.
113. Boyanovsky B, Karakashian A, King K et al. Uptake and metabolism of low density lipoproteins with elevated ceramide content by human microvascular endothelial cells: implications for the regulation of apoptosis. J Biol Chem 2003; 278:26992-26999.
114. Holland WL, Summers SA. Sphingolipids, Insulin Resistance and Metabolic Disease: New Insights from in Vivo Manipulation of Sphingolipid Metabolism. Endocr Rev 2008; 29:381-402.
115. Schmitz-Peiffer C, Craig DL, Biden TJ. Ceramide Generation Is Sufficient to Account for the Inhibition of the Insulin-stimulated PKB Pathway in C2C12 Skeletal Muscle Cells Pretreated with Palmitate. J Biol Chem 1999; 274:24202-24210.
116. Xu J, Yeh CH, Chen S et al. Involvement of de Novo Ceramide Biosynthesis in Tumor Necrosis Factor-alpha/Cycloheximide-induced Cerebral Endothelial Cell Death. J Biol Chem 1998; 273:16521-16526.
117. Garzotto M, White-Jones M, Jiang Y et al. 12-O-tetradecanoylphorbol-13-acetate-induced apoptosis in LNCaP cells is mediated through ceramide synthase. Cancer Res 1998; 58:2260-2264.
118. Lehtonen JY, Horiuchi M, Daviet L et al. Activation of the de novo biosynthesis of sphingolipids mediates angiotensin II type 2 receptor-induced apoptosis. J Biol Chem 1999; 274:16901-16906.
119. Guzman M, Galve-Roperh I, Sanchez C. Ceramide: a new second messenger of cannabinoid action. Trends Pharmacol Sci 2001; 22:19-22.
120. Senkal CE, Ponnusamy S, Rossi MJ et al. Role of human longevity assurance gene 1 and C18-ceramide in chemotherapy-induced cell death in human head and neck squamous cell carcinomas. Mol Cancer Ther 2007; 6:712-722.
121. Petrache I, Natarajan V, Zhen L et al. Ceramide upregulation causes pulmonary cell apoptosis and emphysema-like disease in mice. Nat Med 2005; 11:491-498.
122. Park TS, Rosebury W, Kindt EK et al. Serine palmitoyltransferase inhibitor myriocin induces the regression of atherosclerotic plaques in hyperlipidemic ApoE-deficient mice. Pharmacol Res 2008; 58:45-51.
123. Cuzzocrea S, Deigner HP, Genovese T et al. Inhibition of Ceramide Biosynthesis Ameliorates Pathological Consequences of Spinal Cord Injury. Shock 2008.
124. Guenther GG, Peralta ER, Rosales KR et al. Ceramide starves cells to death by downregulating nutrient transporter proteins. Proc Natl Acad Sci USA 2008; 105:17402-17407.
125. Scarlatti F, Bauvy C, Ventruti A et al. Ceramide-mediated Macroautophagy Involves Inhibition of Protein Kinase B and Up-regulation of Beclin 1. J Biol Chem 2004; 279:18384-18391.
126. Farrell AM, Uchida Y, Nagiec MM et al. UVB irradiation up-regulates serine palmitoyltransferase in cultured human keratinocytes. J Lipid Res 1998; 39:2031-2038.
127. Linn SC, Kim HS, Keane EM et al. Regulation of de novo sphingolipid biosynthesis and the toxic consequences of its disruption. Biochem Soc Trans 2001; 29:831-835.
128. Uhlinger DJ, Carton JM, Argentieri DC et al. Increased expression of serine palmitoyltransferase (SPT) in balloon-injured rat carotid artery. Thromb Haemost 2001; 86:1320-1326.
129. Min J, Mesika A, Sivaguru M et al. (Dihydro)ceramide synthase 1 regulated sensitivity to cisplatin is associated with the activation of p38 mitogen-activated protein kinase and is abrogated by sphingosine kinase 1. Mol Cancer Res 2007; 5:801-812.
130. Laviad EL, Albee L, Pankova-Kholmyansky I et al. Characterization of ceramide synthase 2: tissue distribution, substrate specificity and inhibition by sphingosine 1-phosphate. J Biol Chem 2008; 283:5677-5684.
131. Yabu T, Imamura S, Yamashita M et al. Identification of Mg2+ -dependent neutral sphingomyelinase 1 as a mediator of heat stress-induced ceramide generation and apoptosis. J Biol Chem 2008; 283:29971-29982.
132. Marchesini N, Osta W, Bielawski J et al. Role for Mammalian Neutral Sphingomyelinase 2 in Confluence-induced Growth Arrest of MCF7 Cells. J Biol Chem 2004; 279:25101-25111.
133. Karakashian AA, Giltiay NV, Smith GM et al. Expression of neutral sphingomyelinase-2 (NSMase-2) in primary rat hepatocytes modulates IL-beta-induced JNK activation. FASEB J 2004; 18:968-970.
134. De Palma C, Meacci E, Perrotta C et al. Endothelial Nitric Oxide Synthase Activation by Tumor Necrosis Factor {alpha} Through Neutral Sphingomyelinase 2, Sphingosine Kinase 1 and Sphingosine 1 Phosphate Receptors: A Novel Pathway Relevant to the Pathophysiology of Endothelium. Arterioscler Thromb Vasc Biol 2006; 26:99-105.

135. Shirakabe K, Yamaguchi K, Shibuya H et al. TAK1 mediates the ceramide signaling to stress-activated protein kinase/c-Jun N-terminal kinase. J Biol Chem 1997; 272:8141-8144.
136. Verheij M, Bose R, Lin XH et al. Requirement for ceramide-initiated SAPK/JNK signalling in stress-induced apoptosis. Nature 1996; 380:75-79.
137. Westwick JK, Bielawska AE, Dbaibo G et al. Ceramide activates the stress-activated protein kinases. J Biol Chem 1995; 270:22689-22692.
138. Rutkute K, Karakashian AA, Giltiay NV et al. Aging in rat causes hepatic hyperresponsiveness to interleukin 1 b which is mediated by neutral sphingomyelinase-2. Hepatology (Baltimore, Md 2007; 46:1166-1176.
139. Chalfant CE, Kishikawa K, Mumby MC et al. Long chain ceramides activate protein phosphatase-1 and protein phosphatase-2A. Activation is stereospecific and regulated by phosphatidic acid. J Biol Chem 1999; 274:20313-20317.
140. Tellier E, Negre-Salvayre A, Bocquet B et al. Role for furin in tumor necrosis factor alpha-induced activation of the matrix metalloproteinase/sphingolipid mitogenic pathway. Mol Cell Biol 2007; 27:2997-3007.
141. Clarke CJ, Truong TG, Hannun YA. Role for neutral sphingomyelinase-2 in tumor necrosis factor alpha-stimulated expression of vascular cell adhesion molecule-1 (VCAM) and intercellular adhesion molecule-1 (ICAM) in lung epithelial cells: p38 MAPK is an upstream regulator of nSMase2. J Biol Chem 2007; 282:1384-1396.
142. Levade T, Malagarie-Cazenave S, Gouazé V et al. Ceramide in Apoptosis: A Revisited Role. Neurochemical Research 2002; 27:601-607.
143. Segui B, Cuvillier O, Adam-Klages S et al. Involvement of FAN in TNF-induced apoptosis. J Clin Invest 2001; 108:143-151.
144. Luberto C, Hassler DF, Signorelli P et al. Inhibition of Tumor Necrosis Factor-induced Cell Death in MCF7 by a Novel Inhibitor of Neutral Sphingomyelinase. J Biol Chem 2002; 277:41128-41139.
145. Corcoran CA, He Q, Ponnusamy S et al. Neutral sphingomyelinase-3 is a DNA damage and nongenotoxic stress-regulated gene that is deregulated in human malignancies. Mol Cancer Res 2008; 6:795-807.
146. Jayadev S, Liu B, Bielawska AE et al. Role for ceramide in cell cycle arrest. J Biol Chem 1995; 270:2047-2052.
147. Hayashi Y, Kiyono T, Fujita M et al. cca1 Is Required for Formation of Growth-arrested Confluent Monolayer of Rat 3Y1 Cells. J Biol Chem 1997; 272:18082-18086.
148. Marchesini N, Luberto C, Hannun YA. Biochemical Properties of Mammalian Neutral Sphingomyelinase2 and Its Role in Sphingolipid Metabolism. J Biol Chem 2003; 278:13775-13783.
149. Nikolova-Karakashian M, Karakashian A, Rutkute K. Role of neutral sphingomyelinases in aging and inflammation. Subcell biochem 2008; 49:469-486.
150. Rutkute K, Asmis RH, Nikolova-Karakashian MN. Regulation of neutral sphingomyelinase-2 by GSH: a new insight to the role of oxidative stress in aging-associated inflammation. J Lipid Res 2007; 48:2443-2452.
151. Kim WJ, Okimoto RA, Purton LE et al. Mutations in the neutral sphingomyelinase gene SMPD3 implicate the ceramide pathway in human leukemias. Blood 2008; 111:4716-4722.
152. Hayashi Y, Kiyono T, Fujita M et al. cca1 Is Required for Formation of Growth-arrested Confluent Monolayer of Rat 3Y1 Cells. J Biol Chem 1997; 272:18082-18086.
153. Goswami R, Ahmed M, Kilkus J et al. Differential regulation of ceramide in lipid-rich microdomains (rafts): Antagonistic role of palmitoyl:protein thioesterase and neutral sphingomyelinase 2. J Neurosci Res 2005; 81:208-217.
154. Liu B, Hannun YA. Inhibition of the neutral magnesium-dependent sphingomyelinase by glutathione. J Biol Chem 1997; 272:16281-16287.
155. Gouaze V, Mirault ME, Carpentier S et al. Andrieu-Abadie. Glutathione peroxidase-1 overexpression prevents ceramide production and partially inhibits apoptosis in doxorubicin-treated human breast carcinoma cells. Mol Pharmacol 2001; 60:488-496.
156. Bello RI, Gomez-Diaz C, Buron MI et al. Enhanced anti-oxidant protection of liver membranes in long-lived rats fed on a coenzyme Q10-supplemented diet. Exp Gerontol 2005; 40:694-706.
157. Ayasolla K, Khan M, Singh AK et al. Inflammatory mediator and beta-amyloid (25-35)-induced ceramide generation and iNOS expression are inhibited by vitamin E Free Radic Biol Med 2004; 37:325-338.
158. Castillo SS, Levy M, Thaikoottathil JV et al. Reactive nitrogen and oxygen species activate different sphingomyelinases to induce apoptosis in airway epithelial cells. Exp Cell Res 2007; 313:2680-2686.
159. Martin SF, Sawai H, Villalba JM et al. Redox regulation of neutral sphingomyelinase-1 activity in HEK293 cells through a GSH-dependent mechanism. Arch Biochem Biophys 2007; 459:295-300.
160. Claus RA, Bunck AC, Bockmeyer CL et al. Role of increased sphingomyelinase activity in apoptosis and organ failure of patients with severe sepsis. FASEB J 2005; 19:1719-1721.
161. Wong ML, Xie B, Beatini N et al. Acute systemic inflammation up-regulates secretory sphingomyelinase in vivo: a possible link between inflammatory cytokines and atherogenesis. Proc Natl Acad Sci USA 2000; 97:8681-8686.

162. Gulbins E, Kolesnick R. Raft ceramide in molecular medicine. Oncogene 22:7070-7077.
163. Santana P, Pena LA, Haimovitz-Friedman A et al. Acid sphingomyelinase-deficient human lymphoblasts and mice are defective in radiation-induced apoptosis. Cell 1996; 86:189-199.
164. Morita Y, Tilly JL. Sphingolipid regulation of female gonadal cell apoptosis. Ann N Y Acad Sci 2000; 905:209-220.
165. Pena LA, Fuks Z, Kolesnick RN. Radiation-induced apoptosis of endothelial cells in the murine central nervous system: protection by fibroblast growth factor and sphingomyelinase deficiency. Cancer Res 2000; 60:321-327.
166. Cifone MG, De Maria R, Roncaioli P et al. Apoptotic signaling through CD95 (Fas/Apo-1) activates an acidic sphingomyelinase. J Exp Med 1994; 180:1547-1552.
167. Gulbins E, Bissonnette R, Mahboubi A et al. FAS-induced apoptosis is mediated via a ceramide-initiated RAS signaling pathway. Immunity 1995; 2:341-351.
168. Grassme H, Jekle A, Riehle A et al. CD95 signaling via ceramide-rich membrane rafts. J Biol Chem 2001; 276:20589-20596.
169. Cremesti A, Paris F, Grassme H et al. Ceramide enables fas to cap and kill. J Biol Chem 2001; 276:23954-23961.
170. Utermohlen O, Herz J, Schramm M et al. Fusogenicity of membranes: the impact of acid sphingomyelinase on innate immune responses. Immunobiology 2008; 213:307-314.
171. Schramm M, Herz J, Haas A et al. Utermohlen. Acid sphingomyelinase is required for efficient phago-lysosomal fusion. Cell Microbiol 2008; 10:1839-1853.
172. Ng CG, Griffin DE. Acid sphingomyelinase deficiency increases susceptibility to fatal alphavirus encephalomyelitis. J Virol 2006; 80:10989-10999.
173. Vit JP, Rosselli F. Role of the ceramide-signaling pathways in ionizing radiation-induced apoptosis. Oncogene 2003; 22:8645-8652.
174. Kroesen BJ, Jacobs S, Pettus BJ et al. BcR-induced apoptosis involves differential regulation of C16 and C24-ceramide formation and sphingolipid-dependent activation of the proteasome. J Biol Chem 2003; 278:14723-14731.
175. Wiegmann K, Sch,tze S, Machleidt T et al. Functional dichotomy of neutral and acidic sphingomyelinases in tumor necrosis factor signaling. Cell 1994; 78:1005-1015.
176. Bourbon NA, Sandirasegarane L, Kester M. Ceramide-induced Inhibition of Akt Is Mediated through Protein Kinase Czeta . Implications for Growth Arrest. J Biol Chem, 2002; 277:3286-3292.
177. Summers SA, Garza LA, Zhou H et al. Regulation of insulin-stimulated glucose transporter GLUT4 translocation and Akt kinase activity by ceramide. Mol Cell Biol 1998; 18:5457-5464.
178. Powell DJ, Hajduch E, Kular G et al. Ceramide disables 3-phosphoinositide binding to the pleckstrin homology domain of protein kinase B (PKB)/Akt by a PKCzeta-dependent mechanism. Mol Cell Biol 2003; 23:7794-7808.
179. Fox TE, Houck KL, O'Neill SM et al. Ceramide recruits and activates protein kinase C zeta (PKC zeta) within structured membrane microdomains. J Biol Chem 2007; 282:12450-12457.
180. Wang G, Silva J, Krishnamurthy K et al. Direct binding to ceramide activates protein kinase Czeta before the formation of a pro-apoptotic complex with PAR-4 in differentiating stem cells. J Biol Chem 2005; 280:26415-26424.
181. Pettus BJ, Chalfant CE, Hannun YA. Ceramide in apoptosis: an overview and current perspectives. Biochim Biophys Acta 2002; 1585:114-125.
182. Heinrich M, Wickel M, Schneider-Brachert W et al. Cathepsin D targeted by acid sphingomyelinase-derived ceramide. EMBO J 1999; 18:5252-5263.
183. Heinrich M, Neumeyer J, Jakob M et al. Cathepsin D links TNF-induced acid sphingomyelinase to Bid-mediated caspase-9 and -3 activation. Cell Death Differ 2004; 11:550-563.
184. Taha TA, Mullen TD, Obeid LM. A house divided: ceramide, sphingosine and sphingosine-1-phosphate in programmed cell death. Biochim Biophys Acta 2006; 1758:2027-2036.
185. Sawada M, Nakashima S, Banno Y et al. Ordering of ceramide formation, caspase activation and Bax/Bcl-2 expression during etoposide-induced apoptosis in C6 glioma cells. Cell Death Differ 2000; 7:761-772.
186. Ito T, Deng X, Carr B et al. Bcl-2 phosphorylation required for anti-apoptosis function. J Biol Chem 1997; 272:11671-11673.
187. Chalfant CE, Rathman K, Pinkerman RL et al. De novo ceramide regulates the alternative splicing of caspase 9 and Bcl-x in A549 lung adenocarcinoma cells. Dependence on protein phosphatase-1. J Biol Chem 2002; 277:12587-12595.
188. Sawada M, Kiyono T, Nakashima S et al. Molecular mechanisms of TNF-[alpha]-induced ceramide formation in human glioma cells:P53-mediated oxidant stress-dependent and -independent pathways. Cell Death Differ 2004; 11:997-1008.

189. Tepper AD, de Vries E, van Blitterswijk WJ et al. Ordering of ceramide formation, caspase activation and mitochondrial changes during CD95- and DNA damage-induced apoptosis. J Clin Invest 1999; 103:971-978.
190. Kashkar H, Wiegmann K, Yazdanpanah B et al. Acid sphingomyelinase is indispensable for UV light-induced Bax conformational change at the mitochondrial membrane. J Biol Chem 2005; 280:20804-20813.
191. Birbes H, Luberto C, Hsu YT et al. A mitochondrial pool of sphingomyelin is involved in TNFalpha-induced Bax translocation to mitochondria. Biochem J 2005; 386:445-451.
192. Rotolo JA, Zhang J, Donepudi M et al. Caspase-dependent and -independent Activation of Acid Sphingomyelinase Signaling. J Biol Chem 2005; 280:26425-26434.
193. Dbaibo GS, Pushkareva MY, Jayadev S et al. Retinoblastoma gene product as a downstream target for a ceramide-dependent pathway of growth arrest. Proceedings of the National Academy of Sciences of the United States of America 1995; 92:1347-1351.
194. Lee JY, Bielawska AE, Obeid LM. Regulation of Cyclin-Dependent Kinase 2 Activity by Ceramide. Exp Cell Res 2000; 261:303-311.
195. Venable ME, Bielawska A, Obeid LM. Ceramide Inhibits Phospholipase D in a Cell-free System. J Biol Chem 1996; 271:24800-24805.
196. Lee JY, Hannun YA, Obeid LM. Ceramide Inactivates Cellular Protein Kinase Calpha. J Biol Chem 1996; 271:13169-13174.
197. Westwick JK, Bielawska AE, Dbaibo G et al. Ceramide activates the stress-activated protein kinases. J Biol Chem 1995; 270:22689-22692.
198. Latinis KM, Koretzky GA. Fas ligation induces apoptosis and Jun kinase activation independently of CD45 and Lck in human T-cells. Blood 1996; 87:871-875.
199. Verheij M, Bose R, Lin XH et al. Requirement for ceramide-initiated SAPK/JNK signalling in stress-induced apoptosis. Nature 1996; 380:75-79.
200. Brenner B, Koppenhoefer U, Weinstock C et al. Fas- or Ceramide-induced Apoptosis Is Mediated by a Rac1-regulated Activation of Jun N-terminal Kinase/p38 Kinases and GADD153. J Biol Chem 1997; 272:22173-22181.
201. Bourbon NA, Yun J, Kester M. Ceramide Directly Activates Protein Kinase C zeta to Regulate a Stress-activated Protein Kinase Signaling Complex. J Biol Chem 2000; 275:35617-35623.
202. Shirakabe K, Yamaguchi K, Shibuya H et al. TAK1 Mediates the Ceramide Signaling to Stress-activated Protein Kinase/c-Jun N-terminal Kinase. J Biol Chem 1997; 272:8141-8144.
203. Strle K, Broussard SR, McCusker RH et al. C-jun N-terminal kinase mediates tumor necrosis factor-alpha suppression of differentiation in myoblasts. Endocrinology 2006; 147:4363-4373.
204. Rutkute K, Nikolova-Karakashian MN. Regulation of insulin-like growth factor binding protein-1 expression during aging. Biochem Biophys Res Commun 2007; 361:263-269.
205. Chen S, Lee JM, Zeng C et al. Amyloid beta peptide increases DP5 expression via activation of neutral sphingomyelinase and JNK in oligodendrocytes. J Neurochem 2006; 97:631-640.
206. Kolesnick RN, Goñi FM, Alonso A. Compartmentalization of ceramide signaling: physical foundations and biological effects. J Cell Physiol 2000; 184:285-300.
207. Gulbins E, Li PL. Physiological and pathophysiological aspects of ceramide. AJP—Regulatory, Integrative and Comparative Physiology 2006; 290: R11-26.
208. Miyaji M, Jin ZX, Yamaoka S et al. Okazaki and H. Umehara. Role of membrane sphingomyelin and ceramide in platform formation for Fas-mediated apoptosis. J Exp Med 2005; 202:249-259.
209. Siskind LJ, Kolesnick RN, Colombini M. Ceramide Channels Increase the Permeability of the Mitochondrial Outer Membrane to Small Proteins. J Biol Chem 2002; 277:26796-26803.
210. Choudhury A, Sharma DK, Marks DL et al. Elevated endosomal cholesterol levels in Niemann-Pick cells inhibit rab4 and perturb membrane recycling. Mol Biol Cell 2004; 15:4500-4511.
211. Uterm^hlen O, Herz J, Schramm M et al. Fusogenicity of membranes: The impact of acid sphingomyelinase on innate immune responses. Immunobiology 2008; 213:307-314.
212. Merrill AH Jr, Jones DD. An update of the enzymology and regulation of sphingomyelin metabolism. Biochim Biophys Acta 1990; 1044:1-12.

Chapter 7

Animal Models for Studying the Pathophysiology of Ceramide

Toshihiko Kawamori*

Abstract

Bioactive sphingolipids play key roles in the regulation of several fundamental biological processes such as proliferation, apoptosis and transformation. The recent development of genetically engineered mouse (GEM) models has enabled the study of functional roles of sphingolipids in normal development and disease. In this chapter, we review the phenotypes of GEM models (knockout mice) that lack sphingolipid metabolism-related enzymes, discuss what we have learned from animal models and describe future directions of animal models in sphingolipid research.

Introduction

Sphingolipids constitute a class of lipids that share the presence of a sphingosine base in the backbone of their structures. In addition to their important structural functions, sphingolipids are currently considered key bioactive molecules that modulate cellular processes such as proliferation, differentiation, senescence, apoptosis and transformation. The recent development of genetically engineered mouse (GEM) models has contributed extensively to the field of sphingolipid research. In this review, we discuss biological functions that have been revealed using GEM for sphingolipid metabolism-related enzymes: sphingosine kinase 1 and 2, ceramidases, sphingomyelinases, sphingomyelin synthases and sphingosine-1-phosphate (S1P) lyase.

Sphingosine Kinase 1/2

Sphingosine kinases (SphKs), which are highly conserved enzymes found in mammals,[1-4] insects,[5] plants,[6] yeast,[7] worms[8] and slime molds,[9,10] catalyze the synthesis of sphingosine 1-phosphate (S1P) via the phosphorylation of sphingosine. To date, two distinct isoforms of SphK have been identified in mammals—SphK1 and SphK2. Two enzymes in mice contain five highly conserved regions (C1-C5) and an ATP binding site within a conserved lipid kinase catalytic domain.[1,2] Despite sharing two large conserved regions, these kinases have different kinetics of expression during development as well as different subcellular localizations.[2,11,12] Northern blot and quantitative PCR have revealed that SphK1 mRNA is high at embryonic day 7 (E7) and then decreases, but SphK2 mRNA expression remains high at later embryonic developmental stages.[2] SphK1 is predominantly localized to the cytoplasm but can be induced to localize to the inner leaflet of the plasma membrane.[13] SphK1 is distributed differently than SphK2 in the tissue, although both enzymes are widely expressed in most tissues.[2,14] Mice lacking the gene that encodes either *SphK1* or *SphK2* have no abnormal phenotypes, but SphK1/2 double knockout (KO) mice are lethal prior to

*Toshihiko Kawamori—Department of Pathology & Laboratory Medicine, Medical University of South Carolina, 165 Ashley Avenue, Suite 309, MSC 908, Charleston, South Carolina 29425, USA. Email: Kawamori@musc.edu

Sphingolipids as Signaling and Regulatory Molecules, edited by Charles Chalfant and Maurizio Del Poeta.

E13.5 with severe vascular and neural tube defects, resembling S1P 1 receptor-null mice.[15,16] These results indicate that S1P signaling during embryonic development is critical for neurogenesis and angiogenesis. Further analysis of SphK-deficient mice revealed that SphK1$^{-/-}$SphK2$^{+/-}$ females—but not mutant males nor any other mutant female combination—were infertile due to severe defects in decidual cells and decidual blood vessels, leading to early embryonic lethality.[17] A function of SphK1 and 2 in immunoregulatory system is well studied. FTY720, a potent immunosuppressive agent that, when phosphorylated, functions as an agonist for S1P receptors, S1P1, S1P3, S1P4 and S1P5, caused lymphopenia in SphK1-null mice, but not in SphK2-null mice, indicating that SphK2 and not SphK1, is the primary kinase responsible for phosphorylating FTY720 in vivo.[15,18] Conditional deletion of SphK1 and SphK2 in mice prevented lymphocyte egress from the thymus and peripheral lymphoid organs via the reduction of plasma and lymph S1P levels.[19]

Recently, roles of the SphK1/S1P pathway in inflammation and cancer have been revealed. The SphK1/S1P pathway has been reported to regulate the cyclooxygenase-2 (COX-2)/prostaglandin E_2 (PGE_2) pathway and pro-inflammatory products in several cell lines, such as murine fibroblasts and human colon and lung cancer cells.[20,21] In mast cells, macrophages and neutrophils, SphK activity is stimulated by the ligation of plasma membrane receptors for C5a, Fc and formyl peptide (fMLP).[22-25] Two reports of studies using SphK KO mice suggested a role for SphKs in inflammation and neutrophil functions.[26,27] SphK1 KO mice exhibit normal inflammatory cell recruitment during thioglycollate-induced peritonitis and SphK1-null neutrophils respond normally to formyl peptide. In a collagen-induced arthritis model of rheumatoid arthritis, SphK1 KO mice develop arthritis with a normal incidence and severity.[26] Neutrophils isolated from the bone marrow of SphK1 KO mice or SphK2 KO mice have normal increases in intracellular Ca^{2+} stimulated by fMLP, platelet-activating factor, the anaphylatoxin C5a, or ATP and normal migration towards fMLP and C5a.[27] These findings do not support the hypothesis that SphKs play a critical role in inflammation. Further analysis using a dextran sodium sulfate (DSS) colitis model has revealed that SphK1 KO mice have significantly less severe DSS-induced acute colitis.[28] Very recently, SphK1 KO mice have been shown to have significantly less aberrant crypt foci (ACF) formation, which is a preneoplastic lesion of colon cancer induced by the rodent colon carcinogen azoxymethane (AOM).[29] In addition, SphK1 deficiency significantly protects against colon cancer development induced by AOM/DSS treatment in a colitis-induced colon carcinogenesis model. An AOM-induced colon carcinogenesis mouse or rat model has a similar spectrum of colonic lesions as various human colon neoplastic diseases including the ACF-adenoma-carcinoma sequence. SphK1 has been reported to play a critical role in intestinal tumor formation in an *Apc*^{Min} (Min) mouse model.[30] The Min mouse model was generated by random ethylnitrosourea mutagenesis. It carries a nonsense mutation at codon 850 of the *Apc* gene leading to a truncated *Apc* polypeptide and is relevant to human familial adenomatous polyposis coli. Deletion of the SphK1 gene in Min mice resulted in profoundly suppressed adenoma size but not its incidence. These findings suggest that SphK1/S1P signaling may play a pivotal role in inflammation and cancer.

Ceramidases

In the sphingolipid degradative pathway, ceramidases (EC 3.5.1.23) catalyze the hydrolysis of the *N*-acyl group of ceramide to yield sphingosine and fatty acids. Ceramidases have traditionally been classified according to the pH range (acid, neutral, or alkaline) that supports their optimal activity.

Acid Ceramidase

An acid ceramidase is encoded by the *Asah1* gene.[31-33] Mutations in the corresponding human gene cause Farber disease, a lysosomal storage disorder that results in the accumulation of ceramide. To date, 17 different mutations in the acid ceramidase gene have been found in people with Farber disease.[31,32,34-36]

An acid ceramidase-null mouse model was created and homozygosity for the mutant allele led to an early, embryonic lethal phenotype before E8.5.[37] Acid ceramidase homozygous embryos could not survive beyond the 2-cell stage and underwent apoptotic death, suggesting that acid ceramidase is essential for embryo survival, removing ceramide from the newly formed embryos and preventing the default apoptosis pathway.[38]

Neutral Ceramidase

A neutral ceramidase in mice is encoded by the *Asah2* gene.[39-41] Orthologous genes have been identified in human,[42] rat,[43] zebrafish,[44] *Drosophila*[45] and bacteria.[46] The neutral ceramidase has been identified as a Type II integral membrane protein that can be cleaved to yield a soluble secreted protein.[47] The enzyme has also been found in the apical membranes of proximal and distal tubules, collecting ducts of kidney, endosome-like organelles of hepatocytes[43] and in the epithelia of the jejunum and ileum,[41,48] suggesting possible diverse physiological functions.

A complete null mouse for neutral ceramidase has been created by Dr. Proia's group.[49] It has been shown that neutral ceramidase-null mice are viable with a normal life span with no obvious abnormality. Neutral ceramidase-null mice are deficient in the intestinal degradation of ceramide, suggesting that neutral ceramidase is important for the catabolism of dietary sphingolipids and regulation of bioactive sphingolipid metabolites in the intestinal tract.

Sphingomyelinases (SMase) and Sphingomyelin Synthases (SMS)

Sphingomyelin is ubiquitously present in eukaryotic cells distributed in a gradient fashion from membranes of the endoplasmic reticulum, Golgi apparatus and lysosomes to the plasma membrane, which contains 70-90% of total cellular sphingomyelin. Sphingomyelinases (SMase, EC 3.1.4.12), which have been implicated in important and diverse cellular functions, catalyze the hydrolysis of sphingomyelin to ceramide and phosphocholine and are characterized by their optimal pH into acid, neutral and basic SMase species.[50,51]

On the other hand, sphingomyelin is synthesized by sphingomyelin synthase (SMS), which transfers the phosphorylcholine moiety from phosphatidylcholine onto ceramide, producing sphingomyelin and diacylglycerol.[52] SMS is implicated in NF-κB activation mediated by TNF-α and phorbol ester using cell culture systems.[53]

Acid Sphingomyelinase (ASMase)

The cellular glycoprotein acid sphingomyelinase (ASMase) has been shown to be located in the acidic lysosomal compartment and contributes to lysosomal sphingomyelin turnover.[54] The cDNA and gene encoding ASMase (designated *sphingomyelin phosphodiesterase 1, SMPD1*) were cloned in 1989 and 1992, respectively.[55,56] In humans, an inherited deficiency of ASMase activity results in Type A and B forms of Niemann-Pick disease.[57,58] More than 300 Niemann-Pick disease cases and a dozen distinct mutations in the ASMase gene have been reported. A mouse model of Niemann-Pick disease Type A has been generated by targeted disruption of the ASMase gene.[59] These ASMase-null (ASM KO) mice mimic the human disease phenotype inasmuch as they die by 8 months-of-age, display ataxia and tremors and show visceral symptoms similar to those observed in human patients. Recent evidence for the involvement of ASMase in membrane reorganization suggests that neurons from ASM KO mice have elevated sphingomyelin in detergent-resistant membrane microdomains, leading to an aberrant distribution of glycosyl phosphatidyl inositol-anchored proteins.[60] In addition, ASMase plays a role in cerebral ischemia. In wild-type, but not ASM KO mice, an experimental model of transient focal cerebral ischemia resulted in neuronal increases in ASMase, ceramide and the production of inflammatory cytokines. Wild type mice also had larger infarct size and worse behavioral outcomes than ASM KO mice.[61] Interestingly, increased lung ceramide was reported in cigarette-smoking patients with emphysema, so ceramide may be involved in disease prevention.[62] In addition, a recent paper described that membrane ceramide levels were elevated in respiratory tissue from two different cystic fibrosis mouse models and patients with cystic fibrosis. ASM heterozygous KO mice with a cystic fibrosis transgene or under pharmacological ASMase inhibition were more resistant to *Pseudomonas* and their baseline pulmonary inflammation was decreased,[63] suggesting reduced ASMase inhibited *Pseudomonas* infection and increased survival in cystic fibrosis mice. Interestingly, *Pseudomonas* infection has been shown to be more lethal in ASM homozygous KO mice than in wild type mice, likely due to reduced bacterial internalization caused by an inability to form ceramide-enriched microdomains.[64] ASMase plays a key role in stress-induced apoptosis. Among a variety of other functions, p53 is required for many cells to enter apoptosis after a lethal DNA-damaging dose of ionizing irradiation. In the thymus of p53-null mice, apoptotic cells were dramatically reduced in ASM KO mice. In the small intestine of p53-null mice, a significant amount of apoptotic endothelial cells were observed after irradiation, an event not observed in ASM KO mice, suggesting that ASM KO mice were protected from gastrointestinal tract syndrome (fatal bleeding into the gastrointestinal tract after irradiation).[65]

Neutral Sphingomyelinase (nSMase) 1/2

Two mammalian neutral sphingomyelinases have been identified: nSMase 1 and nSMase 2 (or in *unigene* nomenclature—SMPD2 and SMPD3, respectively). nSMase 1 is ubiquitously expressed with the mRNA and protein being greatest in the kidney.[66] nSMase 2, which has a different domain structure, is expressed mainly in brain.[67]

nSMase 1 deficient mice were generated in 2002.[68] Interestingly, nSMase homozygous KO mice are phenotypically normal and show neither lipid accumulation nor detectable changes in sphingomyelin, despite a gross reduction of nSMase activity in all organs except in the central nervous system.

Two lines of nSMase 2 (SMPD3)-deficient mice were generated: chemical-induced (fro/fro) and targeted (KO) mice. The mutation, fragilitas ossium (fro), was discovered in a random-bred stock of mice after treatment with the chemical mutagen tris(1-aziridinyl) phosphine-sulphine.[69] The mutation in the mouse has been demonstrated to have clinical, radiographic and morphologic manifestations similar to those which arise in autosomal recessive forms of osteogenesis imperfecta occurring in humans.[70] Recently, positional cloning results revealed that this mutation is a deletion in *Smpd3*, the gene encoding nSMase 2 and nSMase activity was abolished in fro/fro mice.[71] At birth, affected mice are smaller than normal with deformities and multiple fractures of ribs and long bones. Cartilage formation is normal, but the matrix of developing bones is severely undermineralized. Mortality is elevated in the perinatal period (up to 30%), but the condition stabilizes in mutant mice that survive to weaning. Most fro/fro adults breed and have normal behavior and lifespan. In mutant mice, blood calcium is normal, parathyroid hormone is elevated and bone osteonectin is decreased by 30%. Severe tooth and alveolar bone abnormalities linked to impaired mineralization were also observed. Another line of nSMase 2 deficient mice was generated with a gene-targeting method.[72] nSMase KO mice developed a dwarf phenotype with severe retardation of late embryonic and postnatal growth, a new hypothalamic form of combined pituitary hormone deficiency. These mice showed no sphingomyelin storage abnormalities, unlike nSMase 1 KO mice which exhibited massive sphingomyelin storage in lysosomes of the reticuloendothelial system.

Sphingomyelin Synthases (SMS)

Two SMS genes, SMS1 and SMS2, have been cloned and characterized for their cellular localizations.[73,74] SMS1 is found in the trans-Golgi apparatus, whereas SMS2 is predominantly found at the plasma membrane. Using cell culture systems, SMS1 has been implicated in the regulation of lipid raft sphingomyelin and raft functions, such as FAS receptor clustering,[75] endocytosis and apoptosis.[76] SMS2 homozygous KO mice were generated.[77] SMS2 KO mice display no obvious abnormalities, grow into adulthood and breed normally in a conventional environment. Using macrophages extracted from these mice, SMS2 deficiency significantly attenuated NF-κB activation, suggesting that SMS2 is a modulator of NF-κB activation and may be important in inflammation during atherogenesis.

S1P Lyase

S1P lyase catalyzes the final step of sphingolipid catabolism, namely the irreversible degradation of S1P. The human S1P lyase gene, S1P lyase 1 (*SGPL1*), encodes a protein of 568 amino acids with a molecular weight of 63.5 kDa.[78] The murine ortholog carries the same name and its amino-acid sequence is 84% identical and 91% similar to the human S1P lyase. In mice and rats, S1P lyase activity and expression are highest in the small intestines, colon, thymus and liver and lowest in the heart and brain, with the exception of the olfactory mucosa epithelium, where the enzyme is highly enriched.[79,80] Inhibition of S1P lyase was shown to prevent lymphocyte trafficking by disrupting S1P gradients in blood and tissues, demonstrating that S1P lyase can significantly effect S1P signaling.[81] In contrast, S1P lyase expression promotes apoptosis in human cells.[82] It has been also found that S1P lyase is down-regulated in human colorectal cancers and in Apc^{Min} mouse intestinal polyps, suggesting that S1P lyase loss of function may correlate with and/or contribute intestinal carcinogenesis.[83] Homozygous S1P lyase KO mice do not survive beyond 3-4 weeks after birth and they show significant growth failure and anemia.[84] Several congenital abnormalities were reported, including vascular abnormalities, which lead to fatal hemorrhage and anemia; skeletal defects, which involved improper palatal fusion; thoracic malformations of the sternum, ribs and vertebrae; and renal abnormalities. These results suggest that S1P lyase may have a role in the regulation of mammalian angiogenesis and other developmental processes.

The Other GEM for Sphingolipid-Related Enzymes

There are several GEM available for other important enzymes involved in the sphingolipid pathway, namely dihydroceramide desaturase 1, ceramide galactosyltransferase, ceramide glucosyltransferase and ceramide kinase (CEK).

Mice were generated lacking the gene for dihydroceramide desaturase 1 (*Des1*), which encodes the enzyme that converts metabolically inactive dihydroceramide into active ceramide.[85] Homozygous KO pups have no detectable DES1 protein and much less ceramide, but dramatically more dihydroceramide than wild type or heterozygous littermates. The homozygous null mice have incompletely penetrant lethality and surviving animals are small with a complex phenotype, including scaly skin and sparse hair, tremors, signs of growth retardation, including notably decreased mean body weight and length, total tissue mass, lean body mass, bone mineral content and density and liver dysfunctions (increased serum alkaline phosphatase, alkaline aminotransferase and total bilirubin).

Galactocerebroside is synthesized by addition of galactose to ceramide in a single step with UDP-galactose as the donor. The enzyme responsible for this reaction is UDP-galactose:ceramide galactosyltransferase. Mice lacking UDP-galactose:ceramide galactosyltransferase do not synthesize galactocerebroside or sulfatide, but they form myelin containing glucocerebroside, a lipid not previously identified in myelin.[86] These mice exhibit severe generalized tremoring and mild ataxia. With age, these mice develop progressive hindlimb paralysis and extensive vacuolation of the ventral region of the spinal cord.

Ceramide kinase (CEK) phosphorylates ceramide to form ceramide-1-phosphate. CER homozygous KO mice are viable and CER activity is completely abolished in these mice.[87] Neutrophils are strikingly reduced in the blood and spleen of CER homozygous KO mice and they develop more severe *Streptococcus pneumoniae* infections.

Glucosylceramide (GSL), the core structure of the majority of glycosphingolipids, is synthesized on the cytoplasmic face of the Golgi by glucosylceramide (GSL) synthase via the transfer of a glucose residue from UDP-glucose to ceramide.[88] Dr. Proia's group has disrupted the gene (*Ugcg*) encoding GSL synthase.[89] Disruption of GSL synthase causes embryonic lethality beginning about E7.5 with complete resorption of the embryo by E9.5 due to enhanced apoptosis in the ectoderm during gastrulation.

Conclusion

The past 20 years have witnessed tremendous progress in the field of sphingolipids, which has been made by the recent development of mass spectrometry coupled to synthesis of internal standards and development of gene targeting approaches, such as the KO mouse and RNA interfering techniques. KO mice phenotypes suggest multiple disease-related genetic and therapeutic possibilities involving the sphingolipid pathway and these mice are instrumental in analyzing biological functions. Further studies with stress-induced models such as carcinogenesis models with genetically engineered mice could better elucidate roles of sphingolipid pathways in disease and aid in identifying and validating involved enzymes and mechanisms for further therapies.

Acknowledgements

This work was supported in part by National Institutes of Health grants P01CA97132 and R01CA124687. We regret being unable to cite other important studies because of space limitations.

References

1. Kohama T, Olivera A, Edsall L et al. Molecular cloning and functional characterization of murine sphingosine kinase. J Biol Chem 1998; 273(37):23722-23728.
2. Liu H, Sugiura M, Nava VE et al. Molecular cloning and functional characterization of a novel mammalian sphingosine kinase type 2 isoform. J Biol Chem 2000; 275(26):19513-19520.
3. Nava VE, Lacana E, Poulton S et al. Functional characterization of human sphingosine kinase-1. FEBS Lett 2000; 473(1):81-84.
4. Melendez AJ, Carlos-Dias E, Gosink M et al. Human sphingosine kinase: molecular cloning, functional characterization and tissue distribution. Gene 2000; 251(1):19-26.
5. Herr DR, Fyrst H, Creason MB et al. Characterization of the Drosophila sphingosine kinases and requirement for Sk2 in normal reproductive function. J Biol Chem 2004; 279(13):12685-12694.
6. Nishiura H, Tamura K, Morimoto Y et al. Characterization of sphingolipid long-chain base kinase in Arabidopsis thaliana. Biochem Soc Trans 2000; 28(6):747-748.
7. Nagiec MM, Skrzypek M, Nagiec EE et al. The LCB4 (YOR171c) and LCB5 (YLR260w) genes of Saccharomyces encode sphingoid long chain base kinases. J Biol Chem 1998; 273(31):19437-19442.
8. Mendel J, Heinecke K, Fyrst H et al. Sphingosine phosphate lyase expression is essential for normal development in Caenorhabditis elegans. J Biol Chem 2003; 278(25):22341-22349.
9. Morio T, Urushihara H, Saito T et al. The Dictyostelium developmental cDNA project: generation and analysis of expressed sequence tags from the first-finger stage of development. DNA Res 1998; 5(6):335-340.
10. Li G, Foote C, Alexander S et al. Sphingosine-1-phosphate lyase has a central role in the development of Dictyostelium discoideum. Development (Cambridge, England) 2001; 128(18):3473-3483.
11. Olivera A, Kohama T, Edsall L et al. Sphingosine kinase expression increases intracellular sphingosine-1-phosphate and promotes cell growth and survival. J Cell Biol 1999; 147(3):545-558.
12. Igarashi N, Okada T, Hayashi S et al. Sphingosine kinase 2 is a nuclear protein and inhibits DNA synthesis. J Biol Chem 2003; 278(47):46832-46839.
13. Johnson KR, Becker KP, Facchinetti MM et al. PKC-dependent activation of sphingosine kinase 1 and translocation to the plasma membrane. Extracellular release of sphingosine-1-phosphate induced by phorbol 12-myristate 13-acetate (PMA). J Biol Chem 2002; 277(38):35257-35262.
14. Fukuda Y, Kihara A, Igarashi Y. Distribution of sphingosine kinase activity in mouse tissues: contribution of SPHK1. Biochemical and Biophysical Research Communications 2003; 309(1):155-160.
15. Allende ML, Sasaki T, Kawai H et al. Mice deficient in sphingosine kinase 1 are rendered lymphopenic by FTY720. J Biol Chem 2004; 279(50):52487-52492.
16. Mizugishi K, Yamashita T, Olivera A et al. Essential role for sphingosine kinases in neural and vascular development. Mol Cell Biol 2005; 25(24):11113-11121.
17. Mizugishi K, Li C, Olivera A et al. Maternal disturbance in activated sphingolipid metabolism causes pregnancy loss in mice. J Clin Invest 2007; 117(10):2993-3006.
18. Zemann B, Kinzel B, Muller M et al. Sphingosine kinase type 2 is essential for lymphopenia induced by the immunomodulatory drug FTY720. Blood 2006; 107(4):1454-1458.
19. Pappu R, Schwab SR, Cornelissen I et al. Promotion of lymphocyte egress into blood and lymph by distinct sources of sphingosine-1-phosphate. Science: 2007; 316(5822):295-298.
20. Pettus BJ, Bielawski J, Porcelli AM et al. The sphingosine kinase 1/sphingosine-1-phosphate pathway mediates COX-2 induction and PGE2 production in response to TNF-alpha. Faseb J 2003; 17(11):1411-1421.
21. Kawamori T, Osta W, Johnson KR et al. Sphingosine kinase 1 is up-regulated in colon carcinogenesis. Faseb J 2006; 20(2):386-388.
22. Alemany R, Meyer zu Heringdorf D, van Koppen CJ et al. Formyl peptide receptor signaling in HL-60 cells through sphingosine kinase. J Biol Chem 1999; 274(7):3994-3999.
23. Choi OH, Kim JH, Kinet JP. Calcium mobilization via sphingosine kinase in signalling by the Fc epsilon RI antigen receptor. Nature 1996; 380(6575):634-636.

24. Ibrahim FB, Pang SJ, Melendez AJ. Anaphylatoxin signaling in human neutrophils. A key role for sphingosine kinase. J Biol Chem 2004; 279(43):44802-44811.
25. Melendez AJ, Ibrahim FB. Antisense knockdown of sphingosine kinase 1 in human macrophages inhibits C5a receptor-dependent signal transduction, Ca2+ signals, enzyme release, cytokine production and chemotaxis. J Immunol 2004; 173(3):1596-1603.
26. Michaud J, Kohno M, Proia RL et al. Normal acute and chronic inflammatory responses in sphingosine kinase 1 knockout mice. FEBS Lett 2006; 580(19):4607-4612.
27. Zemann B, Urtz N, Reuschel R et al. Normal neutrophil functions in sphingosine kinase type 1 and 2 knockout mice. Immunol Lett 2007; 109(1):56-63.
28. Snider AJ, Kawamori T, Bradshaw SG et al. A role for sphingosine kinase 1 in dextran sulfate sodium-induced colitis. Faseb J 2009; 23(1):143-152.
29. Kawamori T, Kaneshiro T, Okumura M et al. Role for sphingosine kinase 1 in colon carcinogenesis. Faseb J 2009; 23(2):405-414.
30. Kohno M, Momoi M, Oo ML et al. Intracellular role for sphingosine kinase 1 in intestinal adenoma cell proliferation. Mol Cell Biol 2006; 26(19):7211-7223.
31. Koch J, Gartner S, Li CM et al. Molecular cloning and characterization of a full-length complementary DNA encoding human acid ceramidase. Identification of the first molecular lesion causing Farber disease. J Biol Chem 1996; 271(51):33110-33115.
32. Li CM, Park JH, He X et al. The human acid ceramidase gene (ASAH): structure, chromosomal location, mutation analysis and expression. Genomics 1999; 62(2):223-231.
33. Li CM, Hong SB, Kopal G et al. Cloning and characterization of the full-length cDNA and genomic sequences encoding murine acid ceramidase. Genomics 1998; 50(2):267-274.
34. Bar J, Linke T, Ferlinz K et al. Molecular analysis of acid ceramidase deficiency in patients with Farber disease. Hum Mutat 2001; 17(3):199-209.
35. Zhang Z, Mandal AK, Mital A et al. Human acid ceramidase gene: novel mutations in Farber disease. Mol Genet Metab 2000; 70(4):301-309.
36. Muramatsu T, Sakai N, Yanagihara I et al. Mutation analysis of the acid ceramidase gene in Japanese patients with Farber disease. J Inherit Metab Dis 2002; 25(7):585-592.
37. Li CM, Park JH, Simonaro CM et al. Insertional mutagenesis of the mouse acid ceramidase gene leads to early embryonic lethality in homozygotes and progressive lipid storage disease in heterozygotes. Genomics 2002; 79(2):218-224.
38. Eliyahu E, Park JH, Shtraizent N et al. Acid ceramidase is a novel factor required for early embryo survival. Faseb J 2007; 21(7):1403-1409.
39. Tani M, Okino N, Mori K et al. Molecular cloning of the full-length cDNA encoding mouse neutral ceramidase. A novel but highly conserved gene family of neutral/alkaline ceramidases. J Biol Chem 2000; 275(15):11229-11234.
40. Okino N, Mori K, Ito M. Genomic structure and promoter analysis of the mouse neutral ceramidase gene. Biochem Biophys Res Commun 2002; 299(1):160-166.
41. Choi MS, Anderson MA, Zhang Z et al. Neutral ceramidase gene: role in regulating ceramide-induced apoptosis. Gene 2003; 315:113-122.
42. El Bawab S, Roddy P, Qian T et al. Molecular cloning and characterization of a human mitochondrial ceramidase. J Biol Chem 2000; 275(28):21508-21513.
43. Mitsutake S, Tani M, Okino N et al. Purification, characterization, molecular cloning and subcellular distribution of neutral ceramidase of rat kidney. J Biol Chem 2001; 276(28):26249-26259.
44. Yoshimura Y, Tani M, Okino N et al. Molecular cloning and functional analysis of zebrafish neutral ceramidase. J Biol Chem 2004; 279(42):44012-44022.
45. Yoshimura Y, Okino N, Tani M et al. Molecular cloning and characterization of a secretory neutral ceramidase of Drosophila melanogaster. J Biochem 2002; 132(2):229-236.
46. Okino N, Ichinose S, Omori A et al. Molecular cloning, sequencing and expression of the gene encoding alkaline ceramidase from Pseudomonas aeruginosa. Cloning of a ceramidase homologue from Mycobacterium tuberculosis. J Biol Chem 1999; 274(51):36616-36622.
47. Tani M, Iida H, Ito M. O-glycosylation of mucin-like domain retains the neutral ceramidase on the plasma membranes as a type II integral membrane protein. J Biol Chem 2003; 278(12):10523-10530.
48. Lundgren P, Nilsson A, Duan RD. Distribution and properties of neutral ceramidase activity in rat intestinal tract. Dig Dis Sci 2001; 46(4):765-772.
49. Kono M, Dreier JL, Ellis JM et al. Neutral ceramidase encoded by the Asah2 gene is essential for the intestinal degradation of sphingolipids. J Biol Chem 2006; 281(11):7324-7331.
50. Obeid LM, Linardic CM, Karolak LA et al. Programmed cell death induced by ceramide. Science 1993; 259(5102):1769-1771.
51. Kolesnick RN, Kronke M. Regulation of ceramide production and apoptosis. Annu Rev Physiol 1998; 60:643-665.

52. Merrill AH Jr, Jones DD. An update of the enzymology and regulation of sphingomyelin metabolism. Biochimica et Biophysica Acta 1990; 1044(1):1-12.
53. Luberto C, Yoo DS, Suidan HS et al. Differential effects of sphingomyelin hydrolysis and resynthesis on the activation of NF-kappa B in normal and SV40-transformed human fibroblasts. J Biol Chem 2000; 275(19):14760-14766.
54. Sandhoff K, Klein A. Intracellular trafficking of glycosphingolipids: role of sphingolipid activator proteins in the topology of endocytosis and lysosomal digestion. FEBS Lett 1994; 346(1):103-107.
55. Quintern LE, Schuchman EH, Levran O et al. Isolation of cDNA clones encoding human acid sphingomyelinase: occurrence of alternatively processed transcripts. The EMBO Journal 1989; 8(9):2469-2473.
56. Schuchman EH, Levran O, Pereira LV et al. Structural organization and complete nucleotide sequence of the gene encoding human acid sphingomyelinase (SMPD1). Genomics 1992; 12(2):197-205.
57. Brady RO, Kanfer JN, Mock MB et al. The metabolism of sphingomyelin. II. Evidence of an enzymatic deficiency in Niemann-Pick diseae. Proc Natl Acad Sci USA 1966; 55(2):366-369.
58. Schneider PB, Kennedy EP. Sphingomyelinase in normal human spleens and in spleens from subjects with Niemann-Pick disease. J Lipid Res 1967; 8(3):202-209.
59. Horinouchi K, Erlich S, Perl DP et al. Acid sphingomyelinase deficient mice: a model of types A and B Niemann-Pick disease. Nat Genet 1995; 10(3):288-293.
60. Galvan C, Camoletto PG, Cristofani F et al. Anomalous surface distribution of glycosyl phosphatidyl inositol-anchored proteins in neurons lacking acid sphingomyelinase. Mol Biol Cell 2008; 19(2):509-522.
61. Yu ZF, Nikolova-Karakashian M, Zhou D et al. Pivotal role for acidic sphingomyelinase in cerebral ischemia-induced ceramide and cytokine production and neuronal apoptosis. J Mol Neurosci 2000; 15(2):85-97.
62. Petrache I, Natarajan V, Zhen L et al. Ceramide upregulation causes pulmonary cell apoptosis and emphysema-like disease in mice. Nat Med 2005; 11(5):491-498.
63. Teichgraber V, Ulrich M, Endlich N et al. Ceramide accumulation mediates inflammation, cell death and infection susceptibility in cystic fibrosis. Nat Med 2008; 14(4):382-391.
64. Grassme H, Jendrossek V, Riehle A et al. Host defense against Pseudomonas aeruginosa requires ceramide-rich membrane rafts. Nat Med 2003; 9(3):322-330.
65. Paris F, Fuks Z, Kang A et al. Endothelial apoptosis as the primary lesion initiating intestinal radiation damage in mice. Science 2001; 293(5528):293-297.
66. Tomiuk S, Hofmann K, Nix M et al. Cloned mammalian neutral sphingomyelinase: functions in sphingolipid signaling? Proc Natl Acad Sci USA1998; 95(7):3638-3643.
67. Hofmann K, Tomiuk S, Wolff G et al. Cloning and characterization of the mammalian brain-specific, Mg2+-dependent neutral sphingomyelinase. Proc Natl Acad Sci USA 2000; 97(11):5895-5900.
68. Zumbansen M, Stoffel W. Neutral sphingomyelinase 1 deficiency in the mouse causes no lipid storage disease. Mol Cell Biol 2002; 22(11):3633-3638.
69. Guenet JL, Stanescu R, Maroteaux P et al. Fragilitas ossium: a new autosomal recessive mutation in the mouse. J Hered 1981; 72(6):440-441.
70. Sillence DO, Ritchie HE, Dibbayawan T et al. Fragilitas ossium (fro/fro) in the mouse: a model for a recessively inherited type of osteogenesis imperfecta. Am J Med Genet 1993; 45(2):276-283.
71. Aubin I, Adams CP, Opsahl S et al. A deletion in the gene encoding sphingomyelin phosphodiesterase 3 (Smpd3) results in osteogenesis and dentinogenesis imperfecta in the mouse. Nat Genet 2005; 37(8):803-805.
72. Stoffel W, Jenke B, Block B et al. Neutral sphingomyelinase 2 (smpd3) in the control of postnatal growth and development. Proc Natl Acad Sci USA2005; 102(12):4554-4559.
73. Huitema K, van den Dikkenberg J, Brouwers JF et al. Identification of a family of animal sphingomyelin synthases. EMBO 2004; 23(1):33-44.
74. Yamaoka S, Miyaji M, Kitano T et al. Expression cloning of a human cDNA restoring sphingomyelin synthesis and cell growth in sphingomyelin synthase-defective lymphoid cells. J Biol Chem 2004; 279(18):18688-18693.
75. Miyaji M, Jin ZX, Yamaoka S et al. Role of membrane sphingomyelin and ceramide in platform formation for Fas-mediated apoptosis. J Exp Med 2005; 202(2):249-259.
76. Van der Luit AH, Budde M, Zerp S et al. Resistance to alkyl-lysophospholipid-induced apoptosis due to downregulated sphingomyelin synthase 1 expression with consequent sphingomyelin- and cholesterol-deficiency in lipid rafts. Biochem J 2007; 401(2):541-549.
77. Hailemariam TK, Huan C, Liu J et al. Sphingomyelin synthase 2 deficiency attenuates NFkappaB activation. Arterioscler Thromb Vasc Biol 2008; 28(8):1519-1526.

78. Van Veldhoven PP, Gijsbers S, Mannaerts GP et al. Human sphingosine-1-phosphate lyase: cDNA cloning, functional expression studies and mapping to chromosome 10q22(1). Biochimica et Biophysica Acta 2000; 1487(2-3):128-134.
79. Van Veldhoven PP. Sphingosine-1-phosphate lyase. Methods in Enzymology 2000; 311:244-254.
80. Genter MB, Van Veldhoven PP, Jegga AG et al. Microarray-based discovery of highly expressed olfactory mucosal genes: potential roles in the various functions of the olfactory system. Physiol Genomics 2003; 16(1):67-81.
81. Schwab SR, Pereira JP, Matloubian M et al. Lymphocyte sequestration through S1P lyase inhibition and disruption of S1P gradients. Science 2005; 309(5741):1735-1739.
82. Reiss U, Oskouian B, Zhou J et al. Sphingosine-phosphate lyase enhances stress-induced ceramide generation and apoptosis. J Biol Chem 2004; 279(2):1281-1290.
83. Oskouian B, Sooriyakumaran P, Borowsky AD et al. Sphingosine-1-phosphate lyase potentiates apoptosis via p53- and p38-dependent pathways and is down-regulated in colon cancer. Proc Natl Acad Sci USA 2006; 103(46):17384-17389.
84. Schmahl J, Raymond CS, Soriano P. PDGF signaling specificity is mediated through multiple immediate early genes. Nat Genet 2007; 39(1):52-60.
85. Holland WL, Brozinick JT, Wang LP et al. Inhibition of ceramide synthesis ameliorates glucocorticoid-, saturated-fat- and obesity-induced insulin resistance. Cell Metab 2007; 5(3):167-179.
86. Coetzee T, Fujita N, Dupree J et al. Myelination in the absence of galactocerebroside and sulfatide: normal structure with abnormal function and regional instability. Cell 1996; 86(2):209-219.
87. Graf C, Zemann B, Rovina P et al. Neutropenia with impaired immune response to Streptococcus pneumoniae in ceramide kinase-deficient mice. J Immunol 2008; 180(5):3457-3466.
88. Jeckel D, Karrenbauer A, Burger KN et al. Glucosylceramide is synthesized at the cytosolic surface of various Golgi subfractions. J Cell Biol 1992; 117(2):259-267.
89. Yamashita T, Wada R, Sasaki T et al. A vital role for glycosphingolipid synthesis during development and differentiation. Proc Natl Acad Sci USA 1999; 96(16):9142-9147.

Chapter 8

Ceramide-1-Phosphate in Cell Survival and Inflammatory Signaling

Antonio Gómez-Muñoz,* Patricia Gangoiti, María H. Granado, Lide Arana and Alberto Ouro

Abstract

An important metabolite of ceramide is ceramide-1-phosphate (C1P). This lipid second messenger was first demonstrated to be mitogenic for fibroblasts and macrophages and later shown to have antiapoptotic properties. C1P is also an important mediator of the inflammatory response, by stimulating the release of arachidonic acid through activation of group IVA cytosolic phospholipase A_2, the initial rate-limiting step of eicosanoid biosynthesis. C1P is formed from ceramide by the action of a specific ceramide kinase (CerK), which is distinct from the sphingosine kinases that synthesize sphingosine-1-phosphate. CerK is specific for natural ceramides with the erythro configuration in the base component and esterified to long-chain fatty acids. CerK can be activated by different agonists, including interleukin 1-beta, macrophage colony stimulating factor, or calcium ions. Most of the effects of C1P so far described seem to take place in intracellular compartments; however, the recent observation that C1P stimulates cell migration implicates a specific plasma membrane receptor that is coupled to a G_i protein. Therefore, C1P has a dual regulatory capacity acting as an intracellular second messenger to regulate cell survival, or as extracellular receptor ligand to stimulate chemotaxis.

Introduction

Normal development of an organism requires the intervention of complex biological processes that are strictly regulated to maintain cell and tissue homeostasis. These include systems to control cell growth and survival, as well as mechanisms to prevent disease. Alteration of any of these processes can lead to metabolic dysfunction or cause illnesses such as autoimmune diseases, chronic inflammation, neural degeneration, cardiovascular disorders, or cancer.[1-3]

Many sphingolipids are crucial metabolites to control cell activation. Some of them have been described as key regulators of signal transduction processes that are essential for normal development. In particular, ceramides inhibit cell growth and are potent inducers of apoptosis, a form of programmed cell death.[4-17] In neurons however, the situation is controversial as ceramides have been shown to promote either apoptosis or cell survival.[18-22] Also, ceramides play important roles in the regulation of cell differentiation, survival and inflammation[9,13,15,23-28] and are key mediators of radiation and chemotherapy effects on tumors, bacterial and viral infections, heat or UVA injury and ischemia-reperfusion injury (Reviewed by Gulbins and Kolesnick[29]). In addition, ceramides have been associated with insulin resistance through activation of protein phosphatase 2A and the subsequent dephosphorylation and inactivation of protein kinase B (PKB).[30-32] By contrast,

*Corresponding Author: Antonio Gómez-Muñoz—Department of Biochemistry and Molecular Biology. Faculty of Science and Technology. University of the Basque Country, P.O. Box 644, 48080 Bilbao, Spain. Email: antonio.gomez@ehu.es

Sphingolipids as Signaling and Regulatory Molecules, edited by Charles Chalfant and Maurizio Del Poeta.

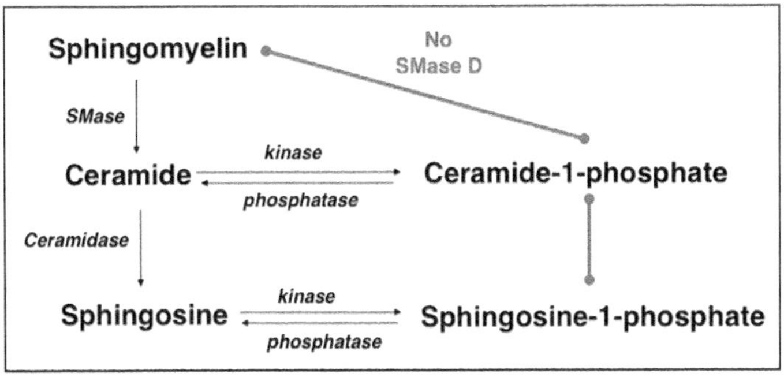

Figure 1. Biosynthesis of sphingosine 1-phosphate and ceramide-1-phosphate.

sphingosylphosphorylcholine,[33] sphingosine-1-phosphate (S1P)[8,34-37] and ceramide-1-phosphate (C1P)[24,38-40] are potent stimulators of cell proliferation. As mentioned in previous chapters of this book, ceramides are generated by de novo synthesis, or can be produced by the action of different sphingomyelinases (SMases). Details on SMase activities, enzymology and compartmentalization are reviewed in several chapters included in this book. Natural ceramides typically have long *N*-acyl chains ranging from 16 to 26 carbons in length[7,43,44] and some times longer in tissues such as skin. Many studies have used a short-chain analog (N-acetylsphingosine, or C_2-ceramide) in experiments with cells in culture because it is more water soluble than long-chain ceramides and it has been presumed that this compound did not occur in vivo. However, recent studies demonstrated that C_2-ceramide does exist in mammalian tissues. In particular, C_2-ceramide was found in rat liver cells[45,46] and brain tissue.[46]

Formation of ceramide is also relevant because it is the precursor of important bioactive sphingolipids that can also regulate cellular functions. For instance, stimulation of ceramidases results in generation of sphingosine (Fig. 1), which is a physiological inhibitor of protein kinase C (PKC).[14] There are numerous reports showing that PKC is inhibited by exogenous sphingosine and Merrill and coworkers demonstrated that addition of the ceramide synthase inhibitor fumonisin B1 to J774 macrophages to increase the levels of endogenous sphingoid bases, also inhibited protein kinase C.[47] Sphingosine can control the activity of other key enzymes involved in the regulation of metabolic or cell signaling pathways such as the Mg^{2+} dependent form of phosphatidate phosphohydrolase,[48,49] phospholipase D (PLD),[50] or diacylglycerol kinase (DAGK)[51,52] in different cell types. Sphingosine, in turn, can be phosphorylated by the action of sphingosine kinases to generate S1P, which is a potent mitogenic agent and can also inhibit apoptosis in many cell types.[8,34,35,37,53,54] More recently, it was demonstrated that S1P stimulates cortisol[55] and aldosterone secretion[56] in cells of the zona fasciculata and zona glomerulosa, respectively, implicating S1P in the regulation of steroidogenesis.

A major metabolite of ceramide is ceramide-1-phosphate (C1P), which is generated through direct phosphorylation of ceramide by ceramide kinase (CerK) (Fig. 1). There is increasing evidence suggesting that C1P can regulate cell proliferation and apoptosis (Reviewed in refs. 24, 38) and the Chalfant laboratory group demonstrated that C1P is a key factor in inflammatory responses (Reviewed in refs. 57, 58). In addition, C1P plays a key role in phagocytosis (please see chapter by Hinkovska-Galcheva et al).[59,60] The aim of the present chapter is to review recent progress related to the control of cell survival and the inflammatory response by C1P.

Ceramide-1-Phosphate Synthesis and Degradation

The only enzyme so far identified to produce C1P in mammalian cells is ceramide kinase (CerK). CerK was first observed in brain synaptic vesicles[61] and later found in human leukemia HL-60 cells.[11] This activity was first reported to be confined to the microsomal membrane fraction, but has also been reported to be mainly located in the citosol.[62] These discrepancies might be due to different degrees of enzyme expression in different cell types and it may also be possible that subcellular localization of this enzyme changes depending on the metabolic status of cells. In fact, Van Veldhoven's group found that tagged forms of human CerK (FLAG-HsCerK and EGFP-HsCerK fusions), upon expression in Chinese Hamster Ovary (CHO) cells, were mainly localized to the plasma membrane, whereas no evidence for an endoplasmic reticulum (ER) association was found.[63] These findings agree with those of Boath et al[64] who have recently reported that ceramides are not phosphorylated at the ER but must be transported to the Golgi apparatus for phosphorylation by CerK. Once generated, C1P traffics from the Golgi network along the secretory pathway to the plasma membrane, where it can be back-exchanged into the extracellular milieu and then bind to acceptor proteins such as albumin or lipoproteins.[64] These observations are consistent with those of Chalfant laboratory;[65] using mass spectrometry and confocal microscopy, Lamour and coworkers demonstrated that CerK utilizes ceramide transported to the trans-Golgi apparatus by ceramide transport protein (CERT). Downregulation of CERT by RNA interference resulted in strong inhibition of newly synthesized C1P, suggesting that CERT plays a critical role in C1P formation. However, this observation has been recently challenged by Boat et al[64] who reported that the transport of ceramides to the vicinity of CerK is not dependent on CERT. The reason for such discrepancy is unknown at present. However, whereas Lamour and coworkers used siRNA technology in their studies to inhibit CERT,[65] Boat et al utilized pharmacological inhibitors.[64] Also, it might be possible that different cell types might have different subcellular distribution of CerK and that expression of this enzyme may not be equal in all cell types. Concerning regulation of enzyme activity, besides its ability to move intracellularly from one compartment to another and its dependency on cations (mainly Ca^{2+} ions) for activity, CerK was proposed to be regulated by phosphorylation/dephosphorylation processes.[66] In addition, CerK is myristoylated, a feature that is related to targeting proteins to membranes. However, elimination of myristoylation did not affect the intracellular localization of the enzyme. Interestingly, CerK location and activity seem to require the integrity of its PH domain, which includes the myristoylation site.[66]

Another important observations was that bone marrow-derived macrophages (BMDM) from CerK-null mice (CerK-/-) still had significant levels of C1P, suggesting the existence of a metabolic pathway, other than ceramide/CerK, for generation of C1P.[64] In particular, formation of C16-C1P, which is a major species of C1P in cells, was not abolished in BMDM. Two alternative pathways for generation of C1P in cells might be acylation of S1P by a putative acyl transferase, or cleavage of sphingomyelin (SM) by activation of SMase D. However, work from our own lab[39] and that of others[64] demonstrated that acylation of S1P to form C1P does not occur in mammalian cells. In addition, C1P could potentially be formed by the action of SMase D, which is a major component of the venom of a variety of arthropodes including spiders of the gender *Loxosceles* (the brown recluse spider *L. reclusa, L. amazonica, L. arizonica, L. intermedia*, or *L. laeta*) and also in the toxins of some bacteria such as *Corynebacterium pseudotuberculosis*, or *Vibrio damsela*.[67] Although we found no evidence for an analogous activity of SMase D when using rat fibroblasts,[39] this possibility has not been exhaustively explored.

Human CERK was recently reported to be highly dependent on Mg^{2+} ions and less dependent on Ca^{2+}.[63] This enzyme was cloned by Sugiura and coworkers.[69] The protein sequence has 537 amino acids with two protein sequence motifs, an N-terminus that encompasses a sequence motif known as a pleckstrin homology (PH) domain (amino acids 32-121) and a C-terminal region containing a Ca^{2+}/calmodulin binding domain (amino acids 124-433). Using site-directed mutagenesis, it was found that leucine 10 in the PH domain is essential for its catalytic activity.[70] Also, it was shown that interaction between the PH domain of CERK and phosphatidylinositol 4,5-bisphosphate regulates the plasma membrane targeting and C1P levels.[71]

With regards to substrate specificity, it was reported that phosphorylation of ceramide by CERK is stereospecific.[72] A minimum of a 12-carbon acyl chain was required for normal CERK activity, whereas the short-chain ceramide analogues C_8-ceramide, C_4-ceramide, or C_2-ceramide were poor substrates for CERK. It was concluded that CERK phosphorylates only the naturally occurring D-erythro-ceramides.[72] However, Van Overloops and coworkers[46] reported that

C_2-ceramide is a good substrate for CerK, when albumin is used as a carrier and that C_2-ceramide can be converted to C_2-C1P within cells. This raises the possibility that C_2-C1P is also a natural sphingolipid, capable of eliciting important biologic effects, as previously demonstrated (i.e., stimulation of cell proliferation[39]). These observations suggest that substrate presentation is an important factor when assaying CerK activity. The importance of CERK in cell signaling was highlighted using specific RNAi to downregulate this enzyme activity. This treatment inhibited arachidonic acid (AA) release and PGE_2 production in response to ATP, the calcium ionophore A23187 and interleukin 1-β.[57,73] The relevance of this enzyme in cell biology was also highlighted in studies using CerK−/− mice; specifically, Bornancin's group found a potent reduction in the amount of neutrophils in blood and spleen of these mice, whereas de amount of leukocytes, other than neutrophils, was increased in these animals. These observations pointed to an important role of CerK in neutrophil homeostasis.[74]

Recently, a human ceramide kinase-like (CERKL) enzyme was identified in retina[75] and subsequently cloned.[76] However, this enzyme was unable to phosphorylate ceramide, or other related lipids, under conditions commonly used to measure CERK activity and therefore its role in cell biology is unclear.

In a previous report, Hinkovska-Galcheva et al[60] observed that endogenous C1P can be generated during the phagocytosis of antibody-coated erythrocytes in human neutrophils that were primed with formylmethionylleucylphenylalanine and more recently, the same group established that C1P is a mediator of phagocytosis.[59] It was also reported that C1P can be formed in neutrophils upon addition of exogenous cell-permeable (^{3}H)*N*-hexanoylsphingosine (C_6-ceramide) to cells[77] and Riboni and coworkers[78] demonstrated that C1P can be generated in cerebellar granule cells both from SM-derived ceramide and through the recycling of sphingosine produced by ganglioside catabolism. C1P can be also generated by the action of interleukin 1-β on A549 lung adenocarcinoma cells,[73] or by M-CSF on bone marrow-derived macrophages[79] and plays an important role in inflammation.[57,73,80-83] We found that C1P is present in normal bone marrow-derived macrophages isolated from healthy mice[84] and that C1P levels were substantially decreased in apoptotic macrophages, suggesting that C1P plays an important role in cell survival.[38,84]

The identification of C1P phosphatase in rat brain[85] and hepatocytes,[86] together with the existence of CERK suggested that ceramide and C1P are interconvertible in cells. C1P phosphatase is enriched in brain synaptosomes and liver plasma membrane fractions and appeared to be distinct from the phosphatase that hydrolyzes phosphatidic acid (PA), PA phosphohydrolase. Nonetheless, C1P can also be converted to ceramide by the action of a PA phosphohydrolase that is specifically located in the plasma membrane of cells.[87] The latter enzyme belongs to a family of at least three mammalian lipid phosphate phosphatases (LPPs).[88] LPPs have recently been shown to regulate cell survival by controlling the levels of intracellular PA and S1P pools[89] and also to regulate leukocyte infiltration and airway inflammation.[90] Dephosphorylation of C1P might be a way of terminating its regulatory effects, although the resulting formation of ceramide could potentially be detrimental for cells. Controlling the levels of ceramide and C1P by the coordinated action of CERK and C1P phosphatases, may be of crucial importance for the metabolic or signaling pathways that are regulated by these two sphingolipids. Another possibility for degradation of C1P might be its deacylation to S1P, which can then be cleaved to render a fatty aldehyde and ethanolamine phosphate by lyase activity,[4] or to sphingosine by the action of S1P phosphatases (Fig. 1). However, no C1P deacylase or lyase has so far been identified in mammalian cells.

Ceramide-1-Phosphate: A Key Regulator of Cell Growth and Survival

C1P was first found to have mitogenic properties in rat or mouse fibroblasts.[39,40] Using primary macrophages, Gangoiti and coworkers found that like for most growth factors, the mechanisms whereby C1P exerts its mitogenic effects implicate stimulation of the mitogen-activated protein kinase kinase (MEK)/Extracellularly regulated kinases 1-2 (ERK1-2), phosphatidylinositol 3-kinase (PI3-K)/protein kinase B (PKB, also known as Akt) and c-Jun terminal kinase (JNK) pathways.[79] In addition, C1P caused stimulation of the DNA binding activity of the transcription factor NF-κB.

Another major target of PKB is glycogen synthase kinase-3β (GSK-3 β), which expression was also increased by C1P. This led to up-regulation of cyclin D1 and c-Myc, two important markers of cell proliferation that are targets of GSK-3β. Other effectors usually involved in the regulation of cell growth, such as diacylglycerol and subsequent activation of protein kinase C (PKC), phospholipase D (PLD), intracellular calcium levels, or cAMP were not affected by C1P.[39,40] However, the short-chain analogues, C_2-C1P- or C_8-C1P, induced Ca^{2+} mobilization in calf pulmonary artery endothelial (CAPE) cells,[91] thyroid FRTL-5,[92] or Jurkat T-cells.[93] By contrast, the short-chain C1Ps did not induce Ca^{2+} mobilization in fibroblasts[39,40] or neutrophils[77] and natural C_{16}-C1P failed to alter intracellular Ca^{2+} concentrations in A549 cells.[80] Whether or not natural C1P is able to affect Ca^{2+} homeostasis in any cell type still remains to be determined.

We previously demonstrated that natural C1P blocked cell death in bone marrow-derived macrophages that were incubated in the absence of macrophage-colony stimulating factor (M-CSF),[84] a condition known to induce apoptosis in these cells.[94,95] We found that C1P blocked both DNA fragmentation and the stimulation of the caspase-9/caspase-3 pathway, thereby suggesting that the prosurvival effect of C1P was due to inhibition of apoptosis.[84] Consistent with these observations, Mitra and coworkers[96] found that down-regulation of CerK in mammalian cells reduced growth, promoted apoptosis and blocked epithelial growth factor-induced cell proliferation. In addition, C1P was reported to induce the synthesis of S1P,[92] an important pro-mitogenic and anti-apoptotic sphingolipid metabolite. However, contrary to these observations, Graf and coworkers showed that addition of the cell permeable C_2-ceramide to cells overexpressing CerK led to C_2-C1P formation and apoptosis.[97] Indeed, overexpression of CerK would potently increase formation of intracellular C1P, particularly if cells were supplied with high concentration of exogenous ceramide and this would result toxic for cells. In this context, we reported previously that relatively high concentrations of C_2-C1P are less effective at stimulating cell division[39] or inhibiting apoptosis[84] than low concentrations and that C_2-C1P or natural C1P are toxic for cells at high concentrations.[84]

Another relevant finding was that apoptotic bone marrow-derived macrophages posses high acid SMase activity and high levels of ceramides compared to healthy cells.[95,98] Investigation into the mechanism whereby C1P exerts its anti-apoptotic effects demonstrated complete inhibition of acid SMase and ceramide accumulation by C1P in intact macrophages.[84] C1P also blocked the activity of acid SMase in cell homogenates suggesting that inhibition of this enzyme occurs by direct physical interaction with C1P. The concentrations of C1P needed to fully block SMase activity in intact cells were in the micromolar range (30 μM);[84] however, only 4% of C1P was incorporated by cells.[40] It was concluded that C1P is a natural inhibitor of acid SMase and that inhibition of this enzyme activity is a major mechanism whereby C1P promotes cell survival.[84] Also, this observation suggests that inhibition of acid SMase by C1P is not mediated through receptor interaction. Acid SMase was also inhibited by S1P in intact macrophages,[98] but unlike C1P the inhibitory effect of S1P did not involve direct interaction with the enzyme. Also of interest, activation of acid SMase plays an important role in pulmonary infections as it facilitates internalization of bacteria into lung epithelial cells.[29] Therefore, inhibition of acid SMase by C1P could be important to reduce or prevent infection in the lung. Recent work by Granado and coworkers (unpublished work) showed that ceramide levels are also increased in apoptotic alveolar NR8383 macrophages. There was only marginal activation of neutral and acidic SMases, suggesting a different source for ceramide formation in these cells. Investigation into the mechanism whereby ceramide levels increased in alveolar macrophages revealed that activation of serine palmitoyltransferase (SPT), the key regulatory enzyme of the de novo pathway of ceramide synthesis, was a major factor in this process. Interestingly, inhibition of SPT activation by treatment with C1P prevented the macrophages from entering apoptosis. It can be concluded that C1P promotes macrophage survival by blocking ceramide accumulation through inhibition of either SMases, or SPT, depending on cell type.

The physiological relevance of the prosurvival effect of C1P is underscored by the demonstration that intracellular levels of C1P are substantially decreased in apoptotic macrophages. It was hypothesized that the decrease in C1P concentration could result in the release of acid SMase from inhibition, thereby triggering ceramide generation an apoptotic cell death.[84]

A major mechanism whereby growth factors promote cell survival is activation of phosphatidylinositol 3-kinase (PI3-K), which can lead to stimulation of the transcription factor NF-κB and expression of antiapoptotic genes. Using two different experimental approaches, it was demonstrated that PI3-K was a target of C1P in bone marrow-derived macrophages.[99] PI3-K activation was demonstrated by immunoprecipitation of the enzyme from whole cell lysates and assayed in vitro using ^{32}P-phosphatidylinositol. In addition, an in vivo approach provided evidence of phosphatidylinositol (3,4,5)-trisphosphate (PIP3) formation in intact cells that were prelabeled with ^{32}P-orthophosphate.[99] PIP3 is a major product of PI3-K activity and was shown to directly inhibit acid SMase.[100] Therefore, PI3-K activation may potentiate the inhibitory effect of C1P on acid SMase through generation of PIP3. Whether C1P and PIP3 bind to the same or different domains of acid SMase remains to be elucidated. C1P stimulated the phosphorylation of protein kinase B (PKB), which is a target of kinases from different signaling pathways including PI3-K,[101,102] cAMP or cAMP-dependent protein kinase (PKA)[103,104] and PKC-ζ.[105] C1P-induced phosphorylation of PKB was sensitive to inhibition by wortmannin or LY294002, which are inhibitors of PI3-K activity. These two inhibitors also blocked the prosurvival effect of C1P indicating that PKB is downstream of PI3-K in macrophages and important for the antiapoptotic effect of C1P.[99] Another relevant finding was that C1P caused IκB phosphorylation and stimulation of the DNA binding activity of NF-κB in primary cultures of mouse macrophages.[99] Of note, C1P up-regulated the expression of anti-apoptotic Bcl-X_L, which is a downstream target of NF-κB. The latter results provided the first evidence for a novel biological role of natural C1P in the regulation of cell survival by the PI3-K/PKB/NF-κB pathway in mammalian cells.[99]

As mentioned above, C1P can be metabolized to ceramide by phosphatase activity and then further converted to sphingosine and S1P by ceramidases and sphingosine kinases. Therefore, it could be speculated that the effects of C1P might be mediated through C1P-derived metabolites. However, ceramides and C1P are antagonistic signals and C1P is unable to mimic many of the effects of sphingosine or S1P (i.e., PLD activation, adenylyl cyclase inhibition, or Ca^{2+} mobilization).[24,39,40,106,107] Also, ceramides can decrease the expression of Bcl-X_L,[57] whereas C1P causes its up-regulation.[99] Finally, no ceramidases capable of converting C1P to S1P have so far been identified and S1P and C1P inhibit acid SMase by different mechanisms.[84,98] Therefore, it can be concluded that C1P acts on its own right to regulate cell functions.

It is obvious from the above observations that the activity of the enzymes involved in ceramide and C1P metabolism must be strictly regulated so that cells can maintain appropriate levels of pro- versus anti-apoptotic metabolites. Any alteration in the balance between ceramides and C1P could potentially result in disease, or be fatal for cells. Detailed investigation into the mechanisms controlling ceramide and C1P levels may facilitate the development of new molecular strategies for counteracting metabolic disorders, or for prevention and treatment of disease.

Ceramide-1-Phosphate and the Control of Inflammatory Responses

Initially, inflammation is beneficial for protecting the organism against infection or injury, but it can be detrimental when it becomes out of control. It was proposed that inflammation evolved as an adaptive response for restoring homeostasis. In general, the acute inflammatory response triggered by infection or tissue injury involves the coordinated delivery of blood components (plasma and leukocytes) to the site of injury or infection.[108] Inflammatory mediators include chemokines, cytokines, vasoactive amines, products of proteolytic cascades, phospholipases, or lipids such as eicosanoids and sphingolipids. Concerning phospholipases, a major mediator of inflammatory responses is PLA_2. In particular, group IV cytosolic PLA_2 exhibits properties of a receptor regulated enzyme and has been involved in receptor-dependent and independent eicosanoid production. With regards to lipid metabolites, some sphingolipids have been described as key mediators of inflammatory responses. This is the case of ceramide, which was initially described as pro-inflammatory for different cell types.[109-112] More recently a role for ceramide in the development of allergic asthmatic responses and airway inflammation was established[113] and exogenous addition of C_2-ceramide to cultured astrocytes induced 12-lipoxigenase leading to generation of reactive oxygen species (ROS) and inflammation.[114] Also, acid sphingomyelinase-derived ceramide was involved in PAF-mediated pulmonary edema.[115] Subsequently, it was proposed that at least some of the pro-inflammatory effects of ceramides may actually be mediated by its conversion to C1P. The first report on the regulation of arachidonic acid (AA) release and the production of prostaglandins by C1P was by the Chalfant laboratory.[73] These authors demonstrated that C1P potently and specifically stimulated AA release and prostanoid synthesis in A549 lung adenocarcinoma cells. In the same report, the authors showed that C1P could be generated intracellularly through stimulation of CerK by the action of interleukin 1-β. In a later report, the same group demonstrated that the mechanism whereby C1P stimulates AA release occurs through direct activation of $cPLA_2$.[80] Subsequently, Subramanian and coworkers[83] found that C1P is a positive allosteric activator of group IV $cPLA_2$ and that it enhances the interaction of the enzyme with phosphatidylcholine. The authors concluded that C1P may function to recruit $cPLA_2\alpha$ to intracellular membranes and that it allosterically increases the catalytic ability of the membrane-associated enzyme.[83] Recent studies by Chalfant and coworkers also demonstrated that activation of group IV $cPLA_2$ by C1P is chain length-specific; in particular, C1P bearing acyl chains equal or higher than 6 carbons were able to efficiently activate $cPLA_2\alpha$ in vitro, whereas shorter acyl chains (in particular C_2-C1P) were unable to activate the enzyme. It was concluded that the biological activity of C_2-C1P does not occur via eicosanoid synthesis.[116] Also, C1P was shown to act in coordination with S1P to ensure maximal production of prostaglandins. Specifically, S1P induces cyclooxigenase-2 (COX-2) activity, which then uses $cPLA_2$-derived AA as substrate to synthesize prostaglandins.[81] For details on the role of C1P in inflammatory response the reader is referred to elegant reviews by Lamour and Chalfant;[65] Wijesinghe et al[117] and Chalfant and Spiegel.[57] Lastly, it should be pointed out that C1P is also involved in other inflammation processes including stimulation of phagocytosis in neutrophils[59,60] and activation of degranulation in mast cells.[62]

Ceramide-1-Phosphate Mediates Macrophage Migration

Macrophages are an important component of both innate and adaptive immunity. They are also involved in a number of chronic diseases characterized by unregulated chronic inflammation, such as autoimmune diseases, atherosclerosis,[118] or multiple sclerosis[119] and in tumor progression and metastasis.[120] Macrophage populations in tissues are determined by the rates of recruitment of monocytes from the bloodstream into the tissue, the rates of macrophage proliferation and apoptosis and the rate of macrophage migration or efflux. Recently, our group demonstrated that exogenous addition of C1P to cultured Raw 264.7 macrophages induced cell migration.[121] This action could only be observed when C1P was applied exogenously and not by increasing the intracellular levels of C1P (i.e., through agonist stimulation of CerK). This fact led us to identify a specific receptor through which C1P stimulates chemotaxis. This is a low affinity receptor with a K_d of approximately 7.8 μM. Although relatively high concentrations (10-20 μM) of C1P are needed for optimal activation of the receptor, it should be borne in mind that C1P was added forming vesicles (sonicated in water) and therefore the concentration that was actually available to the cells was much lower than was added. In addition, C1P tightly binds to serum proteins and the bovine serum albumin that are present in the culture medium (at 0.2 and 0.1 %, respectively) making C1P even less available to the cells. The C1P receptor is coupled to G_i proteins and causes phosphorylation of ERK1-2 and PKB upon ligation with C1P. It was found that inhibition of either of these pathways completely abolished C1P-stimulated macrophage migration. In addition, C1P stimulated the DNA binding activity of NF-κB and blockade of this transcription factor resulted in complete inhibition of macrophage migration. These observations suggested that MEK/ERK1, PI3-K/PKB (Akt) and NF-κB are crucial components of the cascade of events leading to stimulation of cell migration by C1P. It can be concluded that this newly identified receptor could be an important drug target for treatment of illnesses that are associated to inflammatory processes, or to diseases in which cell migration is a major cause of pathology, as it occurs in atherosclerosis or in metastatic tumors. Recently, a C1P analogue named phosphoceramide analogue-1 (PCERA-1) did not block activation of NF-κB in Raw 264.7 macrophages and was suggested to have

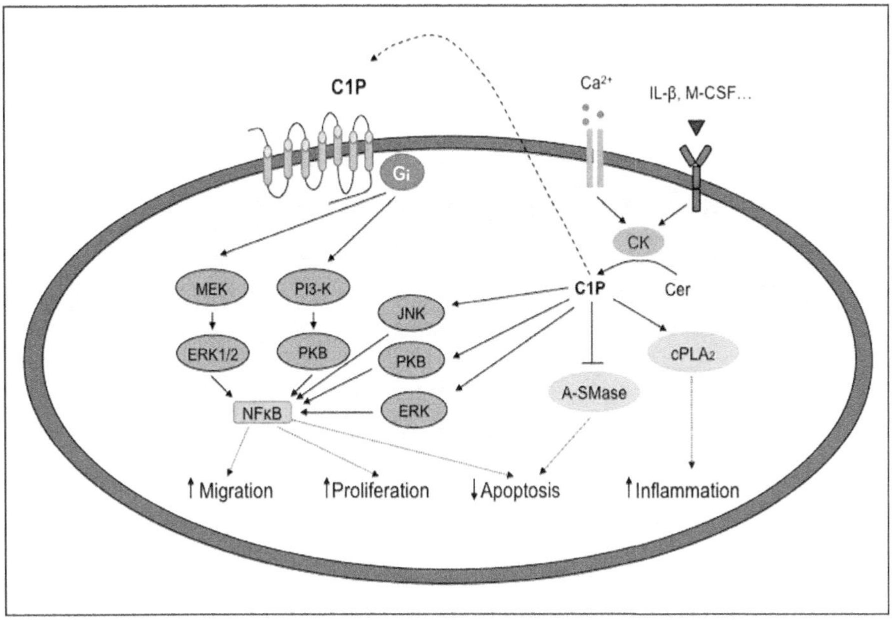

Figure 2. Working model for the induction of cell survival and inflammatory responses by ceramide-1-phosphate.

anti-inflammatory properties. The anti-inflammatory activity of PCERA-1 seems to be mediated by a cell membrane receptor that seems to be distinct to the C1P receptor. This compound, or some of its derivatives, may result in promising tools for blocking inflammation.[122]

Conclusion

In light of the pro-survival and pro-inflammatory actions of C1P, the enzyme responsible for C1P formation, CerK, might be an important target for development of novel therapeutic compounds for treatment of inflammatory illnesses, including atherosclerosis or tumorigenesis. Overexpression of $cPLA_2$, a major target of C1P, has been observed in several human cancers and downregulation of this enzyme causes reduction in the size of tumors. Blockade of $cPLA_2$ through inhibition of C1P formation might have similar effects as knocking down $cPLA_2$, thereby causing a reduction of AA release and prostanoid formation. Further characterization and cloning of the newly identified C1P receptor may provide the means for development of new pharmacological tools for treatment of these illnesses. A working model for the action of C1P is highlighted in Figure 2.

Acknowledgements

The authors wish to acknowledge the support from Ministerio de Educación y Ciencia (Madrid, Spain), Departamento de Educación, Universidades e Investigación del Gobierno Vasco (Gazteiz-Vitoria, Basque Country), Departamento de Industria, Comercio y Turismo del Gobierno Vasco (Gazteiz-Vitoria, Basque Country), Universidad del País Vasco UPV/EHU (Bilbao, Basque Country) and Fundación Ramón Areces (Madrid, Spain).

References

1. Zhang J, Xu M. Apoptotic DNA fragmentation and tissue homeostasis. Trends Cell Biol 2002; 12:84-9.
2. Vaux DL, Korsmeyer SJ. Cell death in development. Cell 1999; 96:245-54.
3. Thompson CB. Apoptosis in the pathogenesis and treatment of disease. Science 1995; 267:1456-62.
4. Merrill AH Jr, Jones DD. An update of the enzymology and regulation of sphingomyelin metabolism. Biochim Biophys Acta 1990; 1044:1-12.
5. Merrill AH Jr, Schmelz EM, Dillehay DL et al. Sphingolipids—the enigmatic lipid class: biochemistry, physiology and pathophysiology. Toxicol Appl Pharmacol 1997; 142:208-25.
6. Merrill AH Jr. Cell regulation by sphingosine and more complex sphingolipids. J Bioenerg Biomembr 1991; 23:83-104.
7. Merrill AH Jr. De novo sphingolipid biosynthesis: a necessary, but dangerous, pathway. J Biol Chem 2002; 277:25843-6.
8. Spiegel S, Merrill AH Jr. Sphingolipid metabolism and cell growth regulation. Faseb J 1996; 10:1388-97.
9. Kolesnick R, Golde DW. The sphingomyelin pathway in tumor necrosis factor and interleukin-1 signaling. Cell 1994; 77:325-8.
10. Kolesnick RN. 1,2-Diacylglycerols but not phorbol esters stimulate sphingomyelin hydrolysis in GH3 pituitary cells. J Biol Chem 1987; 262:16759-62.
11. Kolesnick RN, Hemer MR. Characterization of a ceramide kinase activity from human leukemia (HL-60) cells. Separation from diacylglycerol kinase activity. J Biol Chem 1990; 265:18803-8.
12. Kolesnick RN, Goni FM, Alonso A. Compartmentalization of ceramide signaling: physical foundations and biological effects. J Cell Physiol 2000; 184:285-300.
13. Hannun YA. The sphingomyelin cycle and the second messenger function of ceramide. J Biol Chem 1994; 269:3125-8.
14. Hannun YA, Loomis CR, Merrill AH Jr et al. Sphingosine inhibition of protein kinase C activity and of phorbol dibutyrate binding in vitro and in human platelets. J Biol Chem 1986; 261:12604-9.
15. Hannun YA, Obeid LM. Ceramide: an intracellular signal for apoptosis. Trends Biochem Sci 1995; 20:73-7.
16. Hannun YA. Functions of ceramide in coordinating cellular responses to stress. Science 1996; 274:1855-9.
17. Hannun YA, Obeid LM. The Ceramide-centric universe of lipid-mediated cell regulation: stress encounters of the lipid kind. J Biol Chem 2002; 277:25847-50.
18. Brann AB, Scott R, Neuberger Y et al. Ceramide signaling downstream of the p75 neurotrophin receptor mediates the effects of nerve growth factor on outgrowth of cultured hippocampal neurons. J Neurosci 1999; 19:8199-206.
19. Goodman Y, Mattson MP. Ceramide protects hippocampal neurons against excitotoxic and oxidative insults and amyloid beta-peptide toxicity. J Neurochem 1996; 66:869-72.
20. Ping SE, Barrett GL. Ceramide can induce cell death in sensory neurons, whereas ceramide analogues and sphingosine promote survival. J Neurosci Res 1998; 54:206-13.
21. Plummer G, Perreault KR, Holmes CF et al. Activation of serine/threonine protein phosphatase-1 is required for ceramide-induced survival of sympathetic neurons. Biochem J 2005; 385:685-93.
22. Song MS, Posse de Chaves EI. Inhibition of rat sympathetic neuron apoptosis by ceramide. Role of p75NTR in ceramide generation. Neuropharmacology 2003; 45:1130-50.
23. Dressler KA, Mathias S, Kolesnick RN. Tumor necrosis factor-alpha activates the sphingomyelin signal transduction pathway in a cell-free system. Science 1992; 255:1715-8.
24. Gómez-Muñoz A. Modulation of cell signalling by ceramides. Biochim Biophys Acta 1998; 1391:92-109.
25. Mathias S, Dressler KA, Kolesnick RN. Characterization of a ceramide-activated protein kinase: stimulation by tumor necrosis factor alpha. Proc Natl Acad Sci USA 1991; 88:10009-13.
26. Mathias S, Kolesnick R. Ceramide: a novel second messenger. Adv Lipid Res 1993; 25:65-90.
27. Okazaki T, Bielawska A, Bell RM et al. Role of ceramide as a lipid mediator of 1 alpha,25-dihydroxyvitamin D3-induced HL-60 cell differentiation. J Biol Chem 1990; 265:15823-31.
28. Menaldino DS, Bushnev A, Sun A et al. Sphingoid bases and de novo ceramide synthesis: enzymes involved, pharmacology and mechanisms of action. Pharmacol Res 2003; 47:373-81.
29. Gulbins E, Kolesnick R. Raft ceramide in molecular medicine. Oncogene 2003; 22:7070-7.
30. Adams JM 2nd, Pratipanawatr T, Berria R et al. Ceramide content is increased in skeletal muscle from obese insulin-resistant humans. Diabetes 2004; 53:25-31.
31. Schmitz-Peiffer C. Protein kinase C and lipid-induced insulin resistance in skeletal muscle. Ann N Y Acad Sci 2002; 967:146-57.

32. Stratford S, Hoehn KL, Liu F et al. Regulation of insulin action by ceramide: dual mechanisms linking ceramide accumulation to the inhibition of Akt/protein kinase B. J Biol Chem 2004; 279:36608-15.
33. Sun L, Xu L, Henry FA et al. A new wound healing agent—sphingosylphosphorylcholine. J Invest Dermatol 1996; 106:232-7.
34. Spiegel S, Foster D, Kolesnick R. Signal transduction through lipid second messengers. Curr Opin Cell Biol 1996; 8:159-67.
35. Spiegel S, Milstien S. Sphingosine 1-phosphate, a key cell signaling molecule. J Biol Chem 2002; 277:25851-4.
36. Spiegel S, Olivera A, Carlson RO. The role of sphingosine in cell growth regulation and transmembrane signaling. Adv Lipid Res 1993; 25:105-29.
37. Spiegel S, Milstien S. Sphingosine-1-phosphate: an enigmatic signalling lipid. Nat Rev Mol Cell Biol 2003; 4:397-407.
38. Gómez-Muñoz A. Ceramide-1-phosphate: a novel regulator of cell activation. FEBS Lett 2004; 562:5-10.
39. Gómez-Muñoz A, Duffy PA, Martin A et al. Short-chain ceramide-1-phosphates are novel stimulators of DNA synthesis and cell division: antagonism by cell-permeable ceramides. Mol Pharmacol 1995; 47:833-9.
40. Gómez-Muñoz A, Frago LM, Alvarez L et al. Stimulation of DNA synthesis by natural ceramide 1-phosphate. Biochem J 1997; 325 (Pt 2):435-40.
41. Goni FM, Alonso A. Sphingomyelinases: enzymology and membrane activity. FEBS Lett 2002; 531:38-46.
42. Cremesti AE, Goni FM, Kolesnick R. Role of sphingomyelinase and ceramide in modulating rafts: do biophysical properties determine biologic outcome? FEBS Lett 2002; 531:47-53.
43. Merrill AH Jr, Sullards MC, Allegood JC et al. Sphingolipidomics: high-throughput, structure-specific and quantitative analysis of sphingolipids by liquid chromatography tandem mass spectrometry. Methods 2005; 36:207-24.
44. Pettus BJ, Bielawska A, Kroesen BJ et al. Observation of different ceramide species from crude cellular extracts by normal-phase high-performance liquid chromatography coupled to atmospheric pressure chemical ionization mass spectrometry. Rapid Commun Mass Spectrom 2003; 17:1203-11.
45. Merrill AH Jr, Sullards MC, Wang E et al. Sphingolipid metabolism: roles in signal transduction and disruption by fumonisins. Environ Health Perspect 2001; 109(Suppl 2):283-9.
46. Van Overloop H, Denizot Y, Baes M et al. On the presence of C_2-ceramide in mammalian tissues: possible relationship to etherphospholipids and phosphorylation by ceramide kinase. Biol Chem 2007; 388:315-24.
47. Smith ER, Jones PL, Boss JM et al. Changing J774A.1 cells to new medium perturbs multiple signaling pathways, including the modulation of protein kinase C by endogenous sphingoid bases. J Biol Chem 1997; 272:5640-6.
48. Gómez-Muñoz A, Hamza EH, Brindley DN. Effects of sphingosine, albumin and unsaturated fatty acids on the activation and translocation of phosphatidate phosphohydrolases in rat hepatocytes. Biochim Biophys Acta 1992; 1127:49-56.
49. Jamal Z, Martin A, Gómez-Muñoz A et al. Plasma membrane fractions from rat liver contain a phosphatidate phosphohydrolase distinct from that in the endoplasmic reticulum and cytosol. J Biol Chem 1991; 266:2988-96.
50. Natarajan V, Jayaram HN, Scribner WM et al. Activation of endothelial cell phospholipase D by sphingosine and sphingosine-1-phosphate. Am J Respir Cell Mol Biol 1994; 11:221-9.
51. Sakane F, Yamada K, Kanoh H. Different effects of sphingosine, R59022 and anionic amphiphiles on two diacylglycerol kinase isozymes purified from porcine thymus cytosol. FEBS Lett 1989; 255:409-13.
52. Yamada K, Sakane F, Imai S et al. Sphingosine activates cellular diacylglycerol kinase in intact Jurkat cells, a human T-cell line. Biochim Biophys Acta 1993; 1169:217-24.
53. Wu J, Spiegel S, Sturgill TW. Sphingosine 1-phosphate rapidly activates the mitogen-activated protein kinase pathway by a G protein-dependent mechanism. J Biol Chem 1995; 270:11484-8.
54. Olivera A, Spiegel S. Sphingosine-1-phosphate as second messenger in cell proliferation induced by PDGF and FCS mitogens. Nature 1993; 365:557-60.
55. Rabano M, Pena A, Brizuela L et al. Sphingosine-1-phosphate stimulates cortisol secretion. FEBS Lett 2003; 535:101-5.
56. Brizuela L, Rabano M, Pena A et al. Sphingosine 1-phosphate: a novel stimulator of aldosterone secretion. J Lipid Res 2006; 47:1238-49.
57. Chalfant CE, Spiegel S. Sphingosine 1-phosphate and ceramide 1-phosphate: expanding roles in cell signaling. J Cell Sci 2005; 118:4605-12.
58. Lamour NF, Chalfant CE. Ceramide-1-phosphate: the "missing" link in eicosanoid biosynthesis and inflammation. Mol Interv 2005; 5:358-67.

59. Hinkovska-Galcheva V, Boxer LA, Kindzelskii A et al. Ceramide 1-phosphate, a mediator of phagocytosis. J Biol Chem 2005; 280:26612-21.
60. Hinkovska-Galcheva VT, Boxer LA, Mansfield PJ et al. The formation of ceramide-1-phosphate during neutrophil phagocytosis and its role in liposome fusion. J Biol Chem 1998; 273:33203-9.
61. Bajjalieh SM, Martin TF, Floor E. Synaptic vesicle ceramide kinase. A calcium-stimulated lipid kinase that copurifies with brain synaptic vesicles. J Biol Chem 1989; 264:14354-60.
62. Mitsutake S, Kim TJ, Inagaki Y et al. Ceramide kinase is a mediator of calcium-dependent degranulation in mast cells. J Biol Chem 2004; 279:17570-7.
63. Van Overloop H, Gijsbers S, Van Veldhoven PP. Further characterization of mammalian ceramide kinase: substrate delivery and (stereo)specificity, tissue distribution and subcellular localization studies. J Lipid Res 2006; 47:268-83.
64. Boath A, Graf C, Lidome E et al. Regulation and traffic of ceramide 1-phosphate produced by ceramide kinase: comparative analysis to glucosylceramide and sphingomyelin. J Biol Chem 2008; 283:8517-26.
65. Lamour NF, Stahelin RV, Wijesinghe DS et al. Ceramide kinase uses ceramide provided by ceramide transport protein: localization to organelles of eicosanoid synthesis. J Lipid Res 2007; 48:1293-304.
66. Baumruker T, Bornancin F, Billich A. The role of sphingosine and ceramide kinases in inflammatory responses. Immunol Lett 2005; 96:175-85.
67. Truett AP 3rd, King LE Jr. Sphingomyelinase D: a pathogenic agent produced by bacteria and arthropods. Adv Lipid Res 1993; 26:275-91.
68. Lee S, Lynch KR. Brown recluse spider (Loxosceles reclusa) venom phospholipase D (PLD) generates lysophosphatidic acid (LPA). Biochem J 2005; 391:317-23.
69. Sugiura M, Kono K, Liu H et al. Ceramide kinase, a novel lipid kinase. Molecular cloning and functional characterization. J Biol Chem 2002; 277:23294-300.
70. Kim TJ, Mitsutake S, Kato M et al. The leucine 10 residue in the pleckstrin homology domain of ceramide kinase is crucial for its catalytic activity. FEBS Lett 2005; 579:4383-8.
71. Kim TJ, Mitsutake S, Igarashi Y. The interaction between the pleckstrin homology domain of ceramide kinase and phosphatidylinositol 4,5-bisphosphate regulates the plasma membrane targeting and ceramide 1-phosphate levels. Biochem Biophys Res Commun 2006; 342:611-7.
72. Wijesinghe DS, Massiello A, Subramanian P et al. Substrate specificity of human ceramide kinase. J Lipid Res 2005; 46:2706-16.
73. Pettus BJ, Bielawska A, Spiegel S et al. Ceramide kinase mediates cytokine- and calcium ionophore-induced arachidonic acid release. J Biol Chem 2003; 278:38206-13.
74. Graf C, Zemann B, Rovina P et al. Neutropenia with Impaired Immune Response to Streptococcus pneumoniae in Ceramide Kinase-Deficient Mice. J Immunol 2008; 180:3457-66.
75. Tuson M, Marfany G, Gonzalez-Duarte R. Mutation of CERKL, a novel human ceramide kinase gene, causes autosomal recessive retinitis pigmentosa (RP26). Am J Hum Genet 2004; 74:128-38.
76. Bornancin F, Mechtcheriakova D, Stora S et al. Characterization of a ceramide kinase-like protein. Biochim Biophys Acta 2005; 1687:31-43.
77. Rile G, Yatomi Y, Takafuta T et al. Ceramide 1-phosphate formation in neutrophils. Acta Haematol 2003; 109:76-83.
78. Riboni L, Bassi R, Anelli V et al. Metabolic formation of ceramide-1-phosphate in cerebellar granule cells: evidence for the phosphorylation of ceramide by different metabolic pathways. Neurochem Res 2002; 27:711-6.
79. Gangoiti P, Granado MH, Wang SW et al. Ceramide 1-phosphate stimulates macrophage proliferation through activation of the PI3-kinase/PKB, JNK and ERK1/2 pathways. Cell Signal 2008; 20:726-36.
80. Pettus BJ, Bielawska A, Subramanian P et al. Ceramide 1-phosphate is a direct activator of cytosolic phospholipase A2. J Biol Chem 2004; 279:11320-6.
81. Pettus BJ, Kitatani K, Chalfant CE et al. The coordination of prostaglandin E2 production by sphingosine-1-phosphate and ceramide-1-phosphate. Mol Pharmacol 2005; 68:330-5.
82. Nakamura H, Hirabayashi T, Shimizu M et al. Ceramide-1-phosphate activates cytosolic phospholipase A2alpha directly and by PKC pathway. Biochem Pharmacol 2006; 71:850-7.
83. Subramanian P, Stahelin RV, Szulc Z et al. Ceramide 1-phosphate acts as a positive allosteric activator of group IVA cytosolic phospholipase A2 alpha and enhances the interaction of the enzyme with phosphatidylcholine. J Biol Chem 2005; 280:17601-7.
84. Gómez-Muñoz A, Kong JY, Salh B et al. Ceramide-1-phosphate blocks apoptosis through inhibition of acid sphingomyelinase in macrophages. J Lipid Res 2004; 45:99-105.
85. Shinghal R, Scheller RH, Bajjalieh SM. Ceramide 1-phosphate phosphatase activity in brain. J Neurochem 1993; 61:2279-85.
86. Boudker O, Futerman AH. Detection and characterization of ceramide-1-phosphate phosphatase activity in rat liver plasma membrane. J Biol Chem 1993; 268:22150-5.

87. Waggoner DW, Gómez-Muñoz A, Dewald J et al. Phosphatidate phosphohydrolase catalyzes the hydrolysis of ceramide 1-phosphate, lysophosphatidate and sphingosine 1-phosphate. J Biol Chem 1996; 271:16506-9.
88. Brindley DN, Waggoner DW. Mammalian lipid phosphate phosphohydrolases. J Biol Chem 1998; 273:24281-4.
89. Long J, Darroch P, Wan KF et al. Regulation of cell survival by lipid phosphate phosphatases involves the modulation of intracellular phosphatidic acid and sphingosine 1-phosphate pools. Biochem J 2005; 391:25-32.
90. Zhao Y, Usatyuk PV, Cummings R et al. Lipid phosphate phosphatase-1 regulates lysophosphatidic acid-induced calcium release, NF-kappaB activation and interleukin-8 secretion in human bronchial epithelial cells. Biochem J 2005; 385:493-502.
91. Gijsbers S, Mannaerts GP, Himpens B et al. N-acetyl-sphingenine-1-phosphate is a potent calcium mobilizing agent. FEBS Lett 1999; 453:269-72.
92. Hogback S, Leppimaki P, Rudnas B et al. Ceramide 1-phosphate increases intracellular free calcium concentrations in thyroid FRTL-5 cells: evidence for an effect mediated by inositol 1,4,5-trisphosphate and intracellular sphingosine 1-phosphate. Biochem J 2003; 370:111-9.
93. Colina C, Flores A, Castillo C et al. Ceramide-1-P induces Ca2+ mobilization in Jurkat T-cells by elevation of Ins(1,4,5)-P3 and activation of a store-operated calcium channel. Biochem Biophys Res Commun 2005; 336:54-60.
94. Jaworowski A, Wilson NJ, Christy E et al. Roles of the mitogen-activated protein kinase family in macrophage responses to colony stimulating factor-1 addition and withdrawal. J Biol Chem 1999; 274:15127-33.
95. Hundal RS, Gómez-Muñoz A, Kong JY et al. Oxidized low density lipoprotein inhibits macrophage apoptosis by blocking ceramide generation, thereby maintaining protein kinase B activation and Bcl-XL levels. J Biol Chem 2003; 278:24399-408.
96. Mitra P, Maceyka M, Payne SG et al. Ceramide kinase regulates growth and survival of A549 human lung adenocarcinoma cells. FEBS Lett 2007; 581:735-40.
97. Graf C, Rovina P, Tauzin L et al. Enhanced ceramide-induced apoptosis in ceramide kinase overexpressing cells. Biochem Biophys Res Commun 2007; 354:309-14.
98. Gómez-Muñoz A, Kong J, Salh B et al. Sphingosine-1-phosphate inhibits acid sphingomyelinase and blocks apoptosis in macrophages. FEBS Lett 2003; 539:56-60.
99. Gómez-Muñoz A, Kong JY, Parhar K et al. Ceramide-1-phosphate promotes cell survival through activation of the phosphatidylinositol 3-kinase/protein kinase B pathway. FEBS Lett 2005; 579:3744-50.
100. Testai FD, Landek MA, Goswami R et al. Acid sphingomyelinase and inhibition by phosphate ion: role of inhibition by phosphatidyl-myo-inositol 3,4,5-triphosphate in oligodendrocyte cell signaling. J Neurochem 2004; 89:636-44.
101. Scheid MP, Huber M, Damen JE et al. Phosphatidylinositol (3,4,5)P3 is essential but not sufficient for protein kinase B (PKB) activation; phosphatidylinositol (3,4)P2 is required for PKB phosphorylation at Ser-473: studies using cells from SH2-containing inositol-5-phosphatase knockout mice. J Biol Chem 2002; 277:9027-35.
102. Scheid MP, Woodgett JR. Unravelling the activation mechanisms of protein kinase B/Akt. FEBS Lett 2003; 546:108-12.
103. Filippa N, Sable CL, Filloux C et al. Mechanism of protein kinase B activation by cyclic AMP-dependent protein kinase. Mol Cell Biol 1999; 19:4989-5000.
104. Sable CL, Filippa N, Hemmings B et al. cAMP stimulates protein kinase B in a Wortmannin-insensitive manner. FEBS Lett 1997; 409:253-7.
105. Van Kolen K, Gilany K, Moens L et al. P2Y12 receptor signalling towards PKB proceeds through IGF-I receptor cross-talk and requires activation of Src, Pyk2 and Rap1. Cell Signal 2006; 18:1169-81.
106. Gómez-Muñoz A, Waggoner DW, O'Brien L et al. Interaction of ceramides, sphingosine and sphingosine 1-phosphate in regulating DNA synthesis and phospholipase D activity. J Biol Chem 1995; 270:26318-25.
107. Gómez-Muñoz A, Aboulsalham A, Kikuchi Y et al. Role of Sphingolipids in Regulating the Phospholipase D Pathway and Cell Division in Sphingolipid-mediated Signal Transduction (ed. Hannun YA) (New York, RG Landes Co 1997; 103-120).
108. Medzhitov R. Origin and physiological roles of inflammation. Nature 2008; 454:428-35.
109. Serhan CN, Haeggstrom JZ, Leslie CC. Lipid mediator networks in cell signaling: update and impact of cytokines. Faseb J 1996; 10:1147-58.
110. Manna SK, Aggarwal BB. IL-13 suppresses TNF-induced activation of nuclear factor-kappa B, activation protein-1 and apoptosis. J Immunol 1998; 161:2863-72.
111. Newton R, Hart L, Chung KF et al. Ceramide induction of COX-2 and PGE(2) in pulmonary A549 cells does not involve activation of NF-kappaB. Biochem Biophys Res Commun 2000; 277:675-9.

112. Hayakawa M, Jayadev S, Tsujimoto M et al. Role of ceramide in stimulation of the transcription of cytosolic phospholipase A2 and cyclooxygenase 2. Biochem Biophys Res Commun 1996; 220:681-6.
113. Masini E, Giannini L, Nistri S et al. Ceramide: a key signaling molecule in a Guinea pig model of allergic asthmatic response and airway inflammation. J Pharmacol Exp Ther 2008; 324:548-57.
114. Prasad VV, Nithipatikom K, Harder DR. Ceramide elevates 12-hydroxyeicosatetraenoic acid levels and upregulates 12-lipoxygenase in rat primary hippocampal cell cultures containing predominantly astrocytes. Neurochem Int 2008; 53:220-9.
115. Goggel R, Winoto-Morbach S, Vielhaber G et al. PAF-mediated pulmonary edema: a new role for acid sphingomyelinase and ceramide. Nat Med 2004; 10:155-60.
116. Wijesinghe DS, Subramanian P, Lamour NF et al. The chain length specificity for the activation of group IV cytosolic phospholipase A2 by ceramide-1-phosphate. Use of the dodecane delivery system for determining lipid-specific effects. J Lipid Res 2008; In press.
117. Wijesinghe DS, Lamour NF, Gómez-Muñoz A et al. Ceramide kinase and ceramide-1-phosphate. Methods Enzymol 2007; 434:265-92.
118. Rader DJ, Daugherty A. Translating molecular discoveries into new therapies for atherosclerosis. Nature 2008; 451:904-13.
119. Hendriks JJ, Teunissen CE, de Vries HE et al. Macrophages and neurodegeneration. Brain Res Rev 2005; 48:185-95.
120. Condeelis J, Pollard JW. Macrophages: obligate partners for tumor cell migration, invasion and metastasis. Cell 2006; 124:263-6.
121. Granado MH, Gangoiti P, Ouro A et al. Ceramide 1-phosphate (C1P) promotes cell migration Involvement of a specific C1P receptor. Cell Signal 2008; 21:405-12.
122. Goldsmith M, Avni D, Levy-Rimler G et al. A ceramide-1-phosphate analogue, PCERA-1, simultaneously suppresses tumour necrosis factor-alpha and induces interleukin-10 production in activated macrophages. Immunology 2008.

CHAPTER 9

Ceramide-1-Phosphate in Phagocytosis and Calcium Homeostasis

Vania Hinkovska-Galcheva* and James A. Shayman

Abstract

Sphingolipids are well established sources of important signaling molecules. For example, ceramide (Cer) has been described as a potent inhibitor of cell growth and inducer of apoptosis. In contrast, ceramide-1-phosphate (C1P) has been reported to have mitogenic properties and to inhibit apoptosis. Our understanding of the distinct biological roles of C1P in the regulation of DNA synthesis, inflammation, membrane fusion and intracellular Ca^{2+} increase has rapidly expanded. C1P is a bioactive sphingolipid formed by the phosphorylation of ceramide catalyzed by ceramide kinase (CERK). This chapter specifically focuses on the role of C1P in phagocytosis and Ca^{2+} homeostasis. Studies of the metabolism of C1P during phagocytosis, may lead to a better understanding of its role in signaling. Potentially, the inhibition of CERK and C1P formation may be a therapeutic target for inflammation.

Ceramide-1-Phosphate in Phagocytosis

The clearance of pathogens by the phagocytosis of opsonized, infectious agents is a vital biological process that is part of the innate immune system.[1] Phagocytosis is usually triggered by the interaction of target-bound opsonins with specific receptors on the surface of phagocytes. These receptors include the Fc receptors (FcRs), which bind to the Fc portion of immunoglobins[2] and the complement receptors,[3] which bind to the complement deposited on targets. FcRs recognize the Fc portion of immunoglobins and are expressed differentially on many cell types of the immune system.[1] Receptors for IgG (FcγR), IgE (FcεR) and IgA (FγA) have been characterized.[1] There are three classes of FcγRs: FcγRI, FcγRII and FcγRIII. Each class consists of several receptor isoforms that are the product of different genes and splicing variants.[2] The interaction of FcRs with their immunoglobulin ligands triggers a series of leukocyte responses that include phagocytosis, the respiratory burst, antibody-dependent cell mediated cytotoxicity, the release of pro-inflammatory mediators and the production of cytokines.[1,4] The activation of these receptors leads also to a reorganization of the plasma membrane that profoundly affects the function of phagocytes. The plasma membrane forms pseudopods that extend around an extracellular particle followed by fusion to form a membrane-bounded intracellular vesicle, termed the phagosome. As the process of phagocytosis proceeds, cytoplasmic granules fuse with the phagosome membrane to deliver hydrolytic and antibacterial enzymes to the phagosome.[5] Phagolysosome formation requires an increase in intracellular Ca^{2+} and in addition to the recruitment of a complex containing docking and fusion proteins.[6,7] In contrast to apoptotic cells, Fc receptor-mediated phagocytosis of micro-organisms is often associated with a robust inflammatory response. Recently it has been shown that in immune cells, sphingolipid metabolism

*Corresponding Author: Vania Hinkovska-Galcheva—109 Zina Pitcher Place, 4460 BSRB, Ann Arbor, Michigan 48109-2200, USA. Email: tzgalch@umich.edu

Sphingolipids as Signaling and Regulatory Molecules, edited by Charles Chalfant and Maurizio Del Poeta.

triggered by phagocytosis results in the formation of several lipid second messengers, including ceramide (Cer), sphingosine, C1P and sphingosine-1-phosphate (S1P).[8] It has been observed that the sphingolipid Cer is generated coincident with the termination of the respiratory burst and phagocytosis. Furthermore, the addition of cell-permeable ceramide blocks oxidant release and Fc-mediated phagocytosis.[9] In related work, ceramide kinase (CERK) has been identified as a central enzyme that regulates the levels of Cer via its phosphorylation to the bioactive sphingolipid metabolite, C1P.[10]

Ceramide kinase is a highly conserved lipid kinase, present in animals and plants,[11,12] brain synaptic vesicles,[13] human leukemia (HL-60) cells[14] and primary neutrophils.[10] The cDNA sequence for CERK was cloned by Sugiura and colleagues in 2002.[11] hCERK encodes a protein of 537 amino acids that has a catalytic region with a high degree of similarity to the glycerol kinase catalytic domain. hCERK also has a putative N-myristoylation site on its NH2 terminus followed by a pleckstrin homology domain (PH). The PH domain in its N-terminus is known to bind the β/γ subunit of heterotrimeric G-proteins,[15] phosphoinositol-4,5-bisphosphate[16] and phosphorylated tyrosine residues.[17,18] Several studies have demonstrated that the PH domain may be an important regulatory site for CERK and is required for the proper localization of the enzyme in cells. CERK also contains a Ca^{2+}/calmodulin (Ca^{2+}/CaM) binding motif.[19] Recently, Igarashi and colleagues demonstrated that the activation of CERK and the formation of its product, C1P, in response to an increased intracellular concentration of Ca^{2+} were dependent on CaM.[20] Using calcium chelator BAPTA Boath demonstrated the dependence of CERK on Ca^{2+} ions.[21]

The cloning of CERK also afforded the opportunity to examine, in addition to the structure, the function of the enzyme and its product.[11] C1P has been reported to have mitogenic effects[22] and to mediate arachidonic acid release.[23] In addition to cell growth, C1P has been found to mediate various inflammatory responses, such as the translocation of cytosolic phospholipase $A_2\alpha$ ($cPLA_2\alpha$) to the Golgi apparatus and directly interacts with $cPLA_2\alpha$ in vitro.[24] Other studies have further documented a role of C1P as a mediator of Ca^{2+}-dependent degranulation in mast cells[25] and its important role in phagolysosome formation in polymorphonuclear leukocytes (PMN) and Ca^{2+} signaling.[10,26]

The signaling pathways involved in phagocytosis are determined by the activation state of the PMN, as well as by the type of agonist used for activation. Membrane fusion plays an important role in the degranulation process by creating a pathway by which granule contents have access to phagosomes or to the extracellular milieu. Using a cell-free fusion assay, Ca^{2+} alone cannot promote fusion between neutrophil granules and plasma membrane fractions.[27] Therefore, other components must be required to bring about phagolysosomal formation, factors such as annexin, VAMP-2 and lipids.[28] The existence of CERK activity in PMNs has been previously established.[10] Calcium-dependent CERK is localized to both the PMN plasma membrane and secretory vesicles based on colocalization with the plasma membrane marker HLA and with the secretory vesicle marker latent alkaline phosphatase.[10]

The presence of CERK activity in brain synaptic vesicles and the plasma membrane of PMNs led to the hypothesis that C1P may attenuate membrane charge, regulate vesicle transport, or play a role in regulating the secretion of neurotransmitters by promoting the fusion of vesicle membranes.[13] Phospholipid composition is known to play a significant role in membrane fusogenicity. Because phosphorylation of ceramide would produce an acidic phospholipid similar to phosphatidic acid (PA), a lipid shown to be highly fusogenic,[29] the role of C1P in phospholipid-dependent vesicle fusion was examined in several studies. The addition of exogenous C1P was shown to promote liposome fusion in a cell-free system.[10] In mast cells, C1P formation is associated with Ca^{2+}-dependent degranulation and that C1P formation is enhanced during activation induced by IgE-antigen complex or by the Ca^{2+}-ionophore A23187.[25] Exogenous introduction of CERK into permeabilized RBL-2H3 cells is also sufficient to cause degranulation.[25] Additionally a newly developed synthetic inhibitor of CERK efficiently inhibits mast cell degranulation.[30]

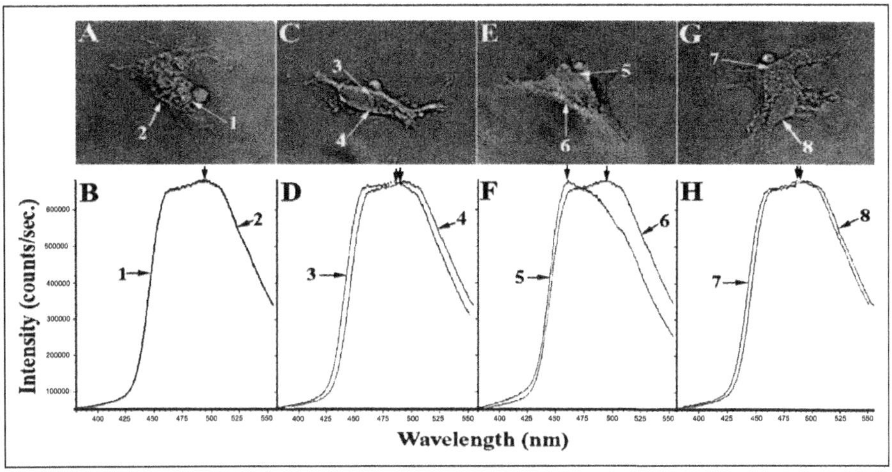

Figure 1. Lipid-ordered membrane domains in association with sites of phagocytosis in COS-1 cells expressing FcγRIIA, FcγRIIA/vector, FcγRIIA/hCERK and FcγRIIA/G198DhCERK. COS-1 cells were labeled with Laurdan. Regions 1, 3, 5 and 7 (panels A,C,E and G) correspond to membrane areas participating in phagocytosis. The Laurdan emission spectra for those regions are denoted with the same numbers as shown below each micrograph. FcγRIIA cells demonstrated a 0-nm shift (panel B) and the cells treated with FcγRIIA/vector demonstrated a ~3-nm shift (panel D). The FcγRIIA/hCERK transfectants exhibited a shift in emission maximum of ~37 nm (panel F) and FcγRIIA/G198DhCERK-transfected cells (panel G) exhibited a shift in the emission of 2.2+0.6-nm shift (panel H). Quantitatively, the calculated general polarization of membrane regions 1, 2, 3, 6 and 7 were (–0.030), region 8 was (–0.024), region 4 was (–0.016) and membrane regions 5 was (+0.088). Fifteen separate spectra were averaged to generate the spectra shown in panels B, D, F and H. The difference in both emission maximum and general polarization was highly significant.

Both phagocytosis and degranulation require a membrane fusion step. The fusion of opposed membranes requires the destabilization of the membranes to render them susceptible to fusion. This destabilization may result from the Ca^{2+}-induced phase separation of rigid (more ordered L_o) crystalline domains of acidic phospholipids (e.g., phosphatidic acid) within mixed lipid membranes.[31] Fusion can be initiated between closely opposed membranes at the boundaries between crystalline and the surrounding noncrystalline domains. Such boundaries represent structurally unstable points and thus offer focal points for the mixing of molecules from opposed membranes. To establish whether C1P might contribute to such destabilization, COS-1 cells expressing FcγRIIA were employed. The overexpression of CERK in these cells enhanced C1P generation and phagocytosis during activation with EIgG.[32] Labeling these cells with Laurdan demonstrated a dramatic shift in emission during phagocytosis, corresponding to a distinct change in lipid-ordered structure (Fig. 1). The change in Laurdan emission provides strong evidence of a lipid raft-like L_o domain in COS-1 cells at the site of phagosome formation.[a]

[a] Lo phase lipids have been observed in reconstructed lipid rafts using Laurdan.[33] Dietrich C, Bagatolli L, Volovyk ZN, Thompson NL, Levi M, Jacobson K, Gratton E. Lipid rafts reconstituted in model membranes. Biophys J 2001; 80:1417-1428. The demonstration of Lo phase lipids in nontransformed cells has been limited to neutrophils at the site of the lamellipodium[34] Kindzelski A, Sitrin R, Petty H. Cutting edge: Optical microspectrophotometry supports the existence of gel phase lipid rafts at the lamellipodium of neutrophils: apparent role in calcium signaling. J Immunol 2004; 172:4681-4685.

Ceramide-1-Phosphate as a Regulator of Calcium Homeostasis

Calcium is a ubiquitous intracellular messenger, controlling a diverse range of cellular processes, such as gene transcription, muscle contraction, cell proliferation and apoptosis.[35-37] Of importance to immunity, Ca^{2+} waves have been observed in migrating polymorphonuclear leukocytes (PMN), fibroblasts and tumor cells.[38,39] The level of intracellular Ca^{2+} is determined by a balance between the "on" reactions that introduce Ca^{2+} into the cytoplasm and the "off" reactions through which this signal is removed by the combined action of buffers, pumps and exchangers. Each cell type expresses a unique set of components for Ca^{2+} signaling which create Ca^{2+}-signaling systems with different spatial and temporal properties.[39] Calcium influx is mediated through both Ca^{2+} release from intracellular stores and Ca^{2+} entry from the extracellular environment. In the case of the latter, there are many different plasma membrane channels that control Ca^{2+} entry from the external medium in response to stimuli that include membrane depolarization, stretch, noxious stimuli, extracellular agonists, intracellular messengers and the depletion of intracellular stores.[39] The plasma membrane Ca^{2+} channels can be divided into several different types. Voltage operated channels (VOC) are employed largely by excitable cell types such as muscle and neuronal cells where they are activated by plasma membrane depolarization. Receptor operated channels are structurally and functionally diverse channels found on secretory cells and nerve terminals. Small molecule operated channels are activated by a number of small messenger molecules, such as diacylglycerol[40] and arachidonic acid.[41] In addition to these more clearly defined channel-opening mechanisms, store operated channels are sensitive to a diverse array of stimuli. Many of these channels belong to the large transient receptor protein (TRP) ion-channel family, which are encoded by up to 29 different genes.[42] The mammalian TRP superfamily of ion channels consists of voltage-independent, nonselective cation channels that are expressed in excitable and nonexcitable cells. The biologic roles of TRP channels are diverse and include vascular tone, thermo sensation, irritant stimuli sensing and flow sensing in the kidney. Growing evidence supports the notion that most cells possess different TRP channels (TRPCs) that are located in association with Ca^{2+} stores where they are capable of functioning as Ca^{2+}-release channels.[40,43,44]

It been established that members of a subgroup of closely related TRP channels (TRPC3/6/7) can be activated by diacylglycerol, a product of PLC activation.[40,45,46] However another subgroup of TRP channels (TRPC1/4/5), although dependent on receptor-induced PLC activation, are completely unresponsive to DAG,[47] suggesting that different TRPC proteins may have different mechanisms of activation. Notably, recent data have shown that TRPC1 and TRPC5 can be activated by S1P.[48] Sphingolipids, including sphingosine, S1P and sphingosylphosphorylcholine, have diverse effects on the regulation of intracellular free Ca^{2+} concentration in nonexcitable and excitable cells.[49-51] C1P has emerged as a putative modulator of cellular functions that are in part regulated by Ca^{2+} signaling.[52] Studies in the role of C1P in modulating Ca^{2+} flux have produced somewhat controversial results. In some reports C1P did not modulate $[Ca^{2+}]_i$ nor did it affect Ca^{2+} mobilization in mouse fibroblasts;[22,53-55] however, others have clearly shown that C1P enhanced store-operated Ca^{2+} entry into thyroid cells.[26,56] The precise role of C1P in Ca^{2+} signaling is therefore not yet well established and is discussed in more detail below.

Gijsbers et al reported that C1P exogenously added in calf pulmonary artery endothelial cells is more potent than S1P for causing a fast and transient intracellular rise in Ca^{2+}.[57] Colina et al showed that C1P increased intracellular Ca^{2+} in Jurkat T-cells. In this study C1P elevated the concentration of $InsP_3$, inducing the liberation of Ca^{2+} from the endoplasmic reticulum, which in turn provoked the opening of a store operated Ca^{2+} channel at the plasma membrane.[58] Hogback et al[59] reported that C1P evoked a concentration-dependent increase in $[Ca]_i$, both in calcium-containing and calcium-free buffer in FRTL-5 cells. In this report, the effect of C1P was mediated, at least in part, by a pertussis toxin–sensitive G protein. The phospholipase C inhibitor U73122 attenuated the effect of C1P. C1P invoked a small, but significant increase in inositol $InsP_3$. However, the effect of C1P on Ca^{2+} was not inhibited by Xestospongin C, 2-aminoethoxydiphenylborate, or neomycin indicating independent activation of IP_3R. The effect of C1P on Ca^{2+} was potently attenuated by dihydrosphingosine and dimethylsphingosine, two inhibitors of sphingosine kinase. This attenuation may be the result of the C1P evoked increase in the production of intracellular S1P.[59] C1P also induced Ca^{2+} mobilization in GH4C1 rat pituitary cells, but indirectly, through voltage-operated Ca^{2+} channels.[57]

Most studies to date have examined the mechanism of C1P by its exogenous addition to cells. The cloning of CERK provided a new tool to study the role of C1P in Ca^{2+} signaling. Using COS-1 cells stably transfected with FcγRIIA and hCERK, our laboratory previously showed that the activation of CERK with the concomitant accumulation of C1P altered Ca^{2+} signaling near the phagosome and significantly promoted phagocytosis and phagolysosomal formation.[26] EIgG-mediated ligation of FcγRIIA leads to the accumulation of CERK and TRPC-1 within lipid rafts, which are key sites associated with signal transduction. High-speed microscopy was used to study the contribution of CERK to phagosomal Ca^{2+} signaling. The high-speed imaging indicated that Ca^{2+} signaling in this model system occurs at "hot spots" that represent brief, but intense, Ca^{2+} release events. These sites of enhanced Ca^{2+} signaling are not random but often repeat at the same spatial location. These Ca^{2+} signaling "hot spots" exist for a period of time (<100 ms) that is consistent with channel-gating times. The quantal Ca^{2+} burst seen in these studies at the pseudopods and phagosome membranes may represent Ca^{2+} release at local regions of signaling (Fig. 2).[26] It was shown previously in neutrophils that C1P is formed by a calcium-dependent CERK located in the plasma membrane during IgG-dependent phagocytosis.[10] Although CERK may participate in Ca^{2+} signaling, the kinase does not directly mediate Ca^{2+} movement across cellular membranes. These studies suggest that TRPC-1 is a leading candidate for a potential signaling partner of CERK/C1P. The store operated Ca^{2+} channel blockers $CdCl_2$, CAI and SKF93365 significantly decreased the ability of COS-1 cells transfected with hCERK to undergo Fc-mediated phagocytosis and phagolysosomal fusion.[26] These findings are consistent with the idea that store operated calcium channels, possibly mediated by TRP channels, are participants in phagocyte function. These findings are also consistent with the previous findings that exogenously added C1P enhances store operated Ca^{2+} entry[56] and that store operated calcium channels participate in phagocyte function in vivo and in vitro.[25,60] To date the mechanism of CERK/C1P modulation of Ca^{2+} signaling has not been determined.

A recent paper by Beech provides a brief and focused review of their latest findings that show that TRPC5 is a sensor of important signaling phospholipids including S1P.[61] When Cav1.2 channels are lost, there is no concomitant loss of Ca^{2+} entry. Ca^{2+} entry is instead enabled by other ion channels (TRPC), which are often resistant to therapeutic concentrations of Ca^{2+} antagonists and permeable to Na^+ and K^+ as well as Ca^{2+}. In a screen of potential lipid regulators of TRPCs, Xu et al identified S1P as an activator of TRPC5. Ion permeation involving TRPC5 is crucial because S1P-evoked motility is also suppressed by the channel blocker 2-aminoethoxydiphenyl borate or a TRPC5 ion-pore mutant.[48] Although TRP channels are structurally related to voltage-gated ion channels, they do not require depolarization in order to be active; instead they are activated by several different endogenous chemical substances.

As regulation of entry of Ca^{2+} and other cations through store operated Ca^{2+} channels occurs in a number of cells, identifying TRP channels and their function has broad significance in a variety of states of cell activation. The regulation of TRPC1/TRPC5 may depend on different lipids. S1P, which is suggested to have pivotal roles in mural cell recruitment during both vascular development[62] and atherosclerosis[63] is a novel bipolar activator of the TRPC1/5 heteromultimeric channel. Blocking these channels could potentially be used for drug targeting.

At present, the manner in which C1P regulates Ca^{2+} signaling is not firmly established. One hypothesis is that the calmodulin binding motif could be responsible for the increase in Ca^{2+} signaling during activation of CERK. However, it has been shown that there is residual Ca^{2+} signaling after the deletion of this motif.[20]

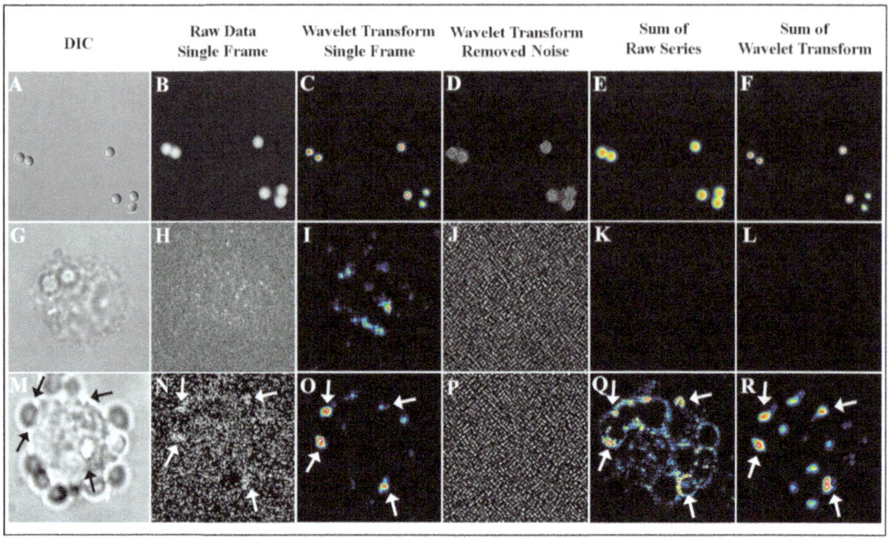

Figure 2. Analyses of high speed images. Differential interference contrast (DIC; column 1(A, G,M)) and fluorescence images (columns 2 through 6) are shown. A. Panels A-F denote a series of experiments using blue fluorescent beads. A single exposure of the flash-lamp yields a high quality image of the beads (B). A-F shows data which were obtained using the same excitation/emission filters and excitation flash as those used for the cells. When the beads were exposed to a single 6 μsec. flash, the fluorescence image of Figure 2B was obtained. This image, suggests that the illumination field provided by the flash lamp was acceptable. Wavelet software is an important emerging technology and has many advantages including the ability to remove uncorrelated noise.[64,65] Because the cell data utilize wavelet filtration for noise removal, the image of fluorescence beads was processed using this software for comparison. Wavelet transformation of the single frame in Figure 2B returned a very similar but somewhat sharper image (Fig. 2C). As the image was very clean, very little noise was removed (Fig. 2D). If image stacks of 100 frames are summed of the raw or wavelet filtered images, essentially the same images are again returned (Figs. 2E and F, respectively). B. Panels G-L and M-R are comparable images for two different transfectants (G-L for FcγRIIA transfected cells and M-R for FcγRIIA/hCERK transfected cells). The result using a wavelet coefficient of 1 is shown in panel D. When 100 frames of a high speed series are summed, the raw data and the wavelet filtered data yield matching results. Transfectants expressing only FcγRIIA did not yield a definitive fluorescence signal (G-L). However, transfectants expressing FcγRIIA/hCERK displayed a signal near pseudopods and phagosomes. As indicated in the raw data of Figure 2N, two peri-phagosomal calcium signals were noted in this frame. After wavelet filtration of the data was performed to minimize the contribution of noise to this image, these same two regions were quite apparent in the filtered image (Fig. 2O). A substantial component of the uncorrelated noise removed from the original raw image is shown in Figure 2P. To confirm that the wavelet-filtered single frame of Figure 2O was contained in the summed (long time scale) image, the 500 raw images were collapsed into a single frame (Fig. 2Q) then compared with the filtered single frame (Fig. 2O). When the 500 wavelet-filtered images are collapsed onto a single frame, all of the significant signaling areas represented in the summed raw data (Fig. 2Q) are represented in the summed wavelet-filtered image (Fig. 2R). Regions with noted repetitions throughout the sequence of 500 images are indicated with arrows. Wavelet filtration indicated two particularly strong signals near the same target (wavelet coefficient = 16). Much of the noise removed from the image is shown in panel 2P, which employed a wavelet coefficient of 2. These data indicate that CERK transfection leads to punctate Ca^{2+} signaling events regions of signaling near targets at phagosomes or phagocytic cups.

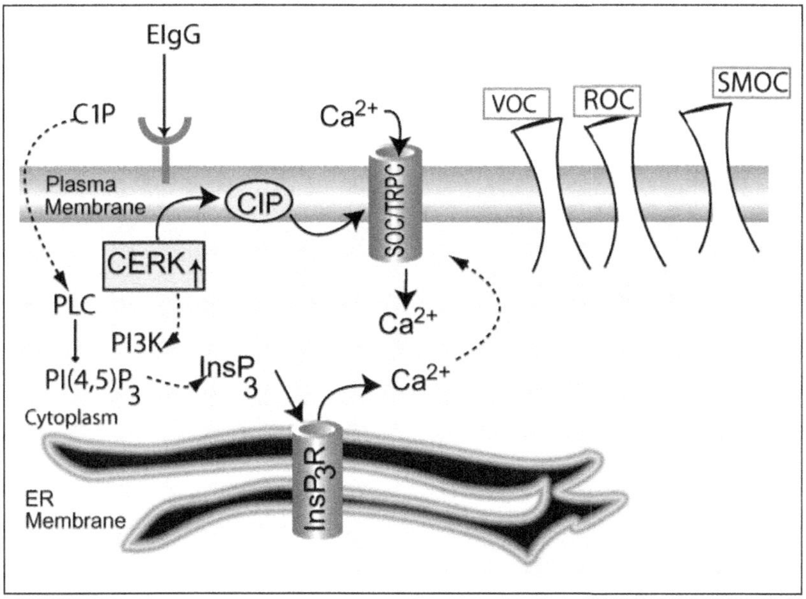

Figure 3. Proposed mechanisms of Ca^{2+} entry in hCERK transfected COS-1 Cells. The roles of ceramide kinase (CERK) and ceramide-1-phosphate (C1P) in modulating Ca^{2+} flux are controversial. C1P may act through a store operated (SOC). Alternatively, C1P may mediate voltage operated channels (VOC) or exert its effects through an increase in inositol-1,4,5-trisphosphate (InsP3) levels. In the latter case InsP3 in concert with phosphatidylinositol-3-kinase (PI3K) activation and a subsequent activation of InsP3 receptor (InsP3R)-operated Ca^{2+} channels from the endoplasmic reticulum (ER). Overexpression of hCERK results in the elevation of C1P during activation by opsonized erythrocytes (EIgG) and stimulates Ca^{2+} entry through SOC/TRPCs. PI(4,5)P2 Phosphatidylinositol-4,5-bisphosphate, PLC phospholipase C, TRPC, transient potential channels.

Conclusion

Sphingolipid-metabolizing enzymes control the dynamic balance of the cellular levels of the bioactive lipid ceramide. C1P is a bioactive lipid which is extensively studied in inflammation. Collectively, studies addressing the role of C1P in Ca^{2+} signaling are limited and often their findings conflict. Recent findings from studies of the role of C1P in Ca^{2+} signaling emphasize the importance of discerning the mechanism of Ca^{2+} signaling under different physiological and pathological conditions. Data obtained from our model have provided the basis of our hypothesis of phagocytosis-triggered cellular signaling, wherein EIgG-mediated ligation of FcγRIIA leads to the activation of CERK and the subsequent accumulation of C1P. This leads to punctuate distribution of Ca^{2+} release at pseudopods and the periphagosomal vicinity. The higher Ca^{2+} signal observed in hCERK transfected cells as well as the fact that CERK colocalized with EIgG during phagocytosis support our hypothesis that Ca^{2+} signaling is an important factor for increasing phagocytosis and is regulated by CERK in a manner that likely involves TRPCs (Fig. 3). Future studies will be required to focus on understanding the activation of different channels involved in these processes with the hope of elucidating the role of TRPC as lipid responsive ionotropic receptors.

Acknowledgements

We thank Robin G. Kunkel Department of Pathology for helping with the figures in the chapter. We want to thank AHA for the support (0930039N).

References

1. Jones S, Lindberg FP, Brown EJ. Phagocytosis. In: Paul WE, ed. Fundamental Immunology. Philadelphia:Lippincott-Raven, 1999:997-1020.
2. Ravetch J, Bolland S. IgG Fc receptors. Annu Rev Immunol 2001; 19:275-290.
3. Brown E. Complement receptors, adhesions and phagocytosis. Infect Agents Dis 1992; 1:63-70.
4. Sanchez T, Hla T. Structural and functional characteristics of S1P receptors. J Cell Biochem 2004; 92:913.
5. Beron W, Alvarez-Dominquez C, Mayorga L et al. Memebrane traficing along the phagocytic pathway. Trends Cell Biol 1995; 5:100-104.
6. Burgoyne R, Geisow MJ. The annexin family of calcium-binding proteins. Cell Calcium 1989; 10:1-10.
7. Borregard N, Boxer LA. Williams Hematology, in Williams Hematology, 7th ed. In: Lichtman M, Beutler E, Kipps TJ, Seligsohn V, Kaushansky K, Prchal JT, ed. New York: McGraw-Hill Publishers, 2005;921-957.
8. Hannun Y, Bell RM. Functions of sphingolipids and sphingolipid breakdown products in cellular regulation. Science 1989; 243:500-506.
9. Hinkovska-Galcheva V, Kjeldsen L, Mansfield J et al. Activation of a plasma membrane associated neutral sphingomyelinase and concomitant ceramide accumulation during IgG dependent phagocytosis in human polymorphonuclear leukocytes. Blood 1998; 91:4761-4769.
10. Hinkovska-Galcheva V, Boxer L, Mansfield P et al. The formation of ceramide-1-phosphate during neutrophil phagocytosis and its role in liposome fusion. J Biol Chem 1998; 273:33203-33209.
11. Sugiura M, Kono K, Liu H et al. Ceramide Kinase, a novel lipid kinase. Molecular cloning and functional characterization. J Biol Chem 2002; 277:23294-23300.
12. Liang H, Yao N, Song JT et al. Ceramides modulate programed cell death in plants. Genes Dev 2003; 17:2636-2641.
13. Bajjalieh S, Martin T, Floor E. Synaptic vesicle ceramide kinase. A calcium-stimulated lipid kinase that copurifies with brain synaptic vesicles. J Biol Chem 1989; 264:14354-14360.
14. Kolesnick R, Hemer M. Characterization of a ceramide kinase activity from human leukemia (HL-60) cells. Separation from diacylglycerol kinase activity. J Biol Chem 1990; 265:18803-18808.
15. Lemmon M, Ferguson KM, Schlessinger J. PH domains: diverse sequences with a common fold recruit signaling molecules to the cell surface. Cell 1996; 85:621-624.
16. Harlan J, Hajduc PJ, Yoon HS et al. Pleckstrin homology domains bind to phosphatidylinositol-4,5-bisphosphate. Nature 1994; 371:168-170.
17. Gibson T, Hyvonen M, Musacchio A et al. PH domain: the first anniversary. Trends Biochem Sci 1994; 19:349-353.
18. Kavanough W, Turck CW, Williams LT. PTB domain binding to signaling proteins through a sequence motif containing phosphotyrosine. Science 1995; 268:1177-1179.
19. Rhoads A, Friedberg F. Sequence motifs for calmudolin recognition. FASEB J 1997; 11:331-340.
20. Mitsutake S, Igarashi Y. Calmodulin is involved in the Ca^{2+}—dependent activation of ceramide kinase as a calcium sensor. J Biol Chem 2005; 280:40436-44044.
21. Boath A, Gray C, Lidome E et al. Regulation and traffic of ceramide-1-phosphate produced by ceramide kinase, J Biol Chem 2008; 283:8517-8526.
22. Gomez-Munoz A, Frago LM, Alvarez L et al. Stimulation of DNA synthesis by natural ceramide 1-phosphate. Biochemical Journal 1997; 325:435-440.
23. Pethus B, Bielawski A, Spiegel S et al. Ceramide kinase mediates cytokine- and calcium ionophore-induced arachidonic and release. Biol Chem 2003; 278:38206-38213.
24. Stahelin R, Subramanian P, Vora M et al. Ceramide-1-phosphate binds group IVA cytosolic phospholipase A_2 via a Novel Site in the C2 Domain. J Biol Chem 2007;20467-20474.
25. Mitsutake S, Kim T-J, Inagaki Y et al. Ceramide kinase is a mediator of calcium-dependent degranulation in mast cells. J Biol Chem 2004; 279:17570-17577.
26. Hinkovska-Galcheva V, Clark A, VanWay S et al. Ceramide kinase promotes Ca^{2+} signaling near IgG-opsonized targets and enhances phagolysosomal fusion in COS-1 cells. J Lipid Res 2008; 49:531-542.
27. Francis J, Smolen W, Balazovich KJ et al. Calcium-dependent fusion on the plasma membrane fraction from human neutrophils with liposomes. Biochim Biophys Acta 1990; 1025:1-9.
28. Nakamura T, Abe A, Balazovich K et al. Ceramide regulates oxidant release in adherent human neutrophils. J Biol Chem 1994; 269:18384-18389.
29. Blackwood R, Transue T, Harsh M et al. PLA_2 promotes fusion between PMN-specific granules and complex liposomes. J Leuk Biol 1996; 59:663-667.
30. Kim J, Inagaki Y, Mitsutake S et al. Suppression of mast cell degranulation by a novel ceramide kinase inhibitor, the F-12509A olefin isomer K1. Biochim Biophys Acta 2005; 1738:82-90.

31. Papahadjopoulos D, Vail W, Newton C et al. Studies on membrane fusion. III The role of calcium -induced phase changes. Biochim Biophys Acta 1977; 465:579-598.
32. Hinkovska-Galcheva V, Boxer L, Kindzelcki A et al. Ceramide-1-phosphate: A mediator of phagocytosis. J Biol Chem 2005; 280:26612-26621.
33. Dietrich C, Bagatolli L, Volovyk ZN et al. Lipid rafts reconstituted in model membranes. Biophys J 2001; 80:1417-1428.
34. Kindzelski A, Sitrin R, Petty H. Cutting edge:Optical microspectrophotometry supports the existence of gel phase lipid rafts at the lamellipodium of neutrophils: apparent role in calcium signalling. J Immunol 2004; 172:4681-4685.
35. Pollock J, McFarlane SM, Connell MC et al. TNF-α receptors simultaneously activate Ca^{2+} mobilisation and stress kinases in cultured sensory neurons. Neuropharmacology 2002; 42:95-106.
36. Lennartz M. Phospholipases and phagocytosis: the role of phospholipid-derived second messengers in phagocytosis. Int J Biochem 1999; 31:415-430.
37. Tas P, Koschel K. Sphingosine-1-phosphate induces a Ca^{2+} signal in primary astrocytes and a Ca^{2+} signal and shape change in C6 rat glioma cells. J Neurosci Res 1998; 52:427-434.
38. Mandeville J, Ghosh R, Maxfield E. Intracellular calcium levels correlate with speed and persistent forward motion in migrating neutrophils. Biophys J 1995; 68:1207-1217.
39. Berridge M, Bootman MD, Roderick HL. Calcium signalling: Dynamics, homeostasis and remodeling. Nature Reviews/Molecular and Cellular Biology 2003; 4:517-529.
40. Hofmann T, Obukhov G, Schaefer M et al. Direct activation of TRPC-6 and TRPC-3 chanels by diacylglycerol. Nature 1999; 397:259-263.
41. Mignen O, Shuttleworth T. I_{ARC}, a novel arachidonate-regulated, noncapacitative Ca^{2+} entry channel. J Biol Chem 2000; 275:9114-9119.
42. Clapham D, Runnels LW, Stubing C. The TRP ion channel family. Nat Rev Neurosci 2001; 2:387-396.
43. Hofmann T, Schaeffer M, Schultz G et al. Subunit composition of mammalian receptor potential channels in living cells. Proc Natl Acad Sci USA 2002; 99:7461-7466.
44. Turner H, Fleig A, Stokes A et al. Discrimination of intracellular calcium store subcompartments using TRPV1 (transient receptor potential channel, vanilloid subfamily member 1) release activity. Biochem J 2003; 371:341-350.
45. Okada T, Inoue R, Yamazaki K et al. Molecular and Functional Characterization of a Novel Mouse Transient Receptor Potential Protein Homologue TRP7. Ca^{2+}—permeable cation channel that is constitutively activated and enhanced by stimulation of G-protein-coupled receptor. J Biol Chem 1999; 274:27359-27370.
46. Okamoto T, Schlegel A, Scherer P et al. Caveolins, a family of scaffolding proteins for organizing "Preassembled signaling complexes" at the plasma membrane. J Biol Chem 1998; 273:5419-5422.
47. Schaefer M, Plant TD, Obukhow AG et al. Receptor-mediated Regulation of the Nonselective Cation Channels TRPC4 and TRPC5. J Biol Chem 2000; 275:17517-17526.
48. Xu S-Z, Muraki K, Zeng F et al. A sphingosine-1-phosphate-activated calcium channel controlling vascular smooth muscle cell motility, Circulation Res 2006; 98:1381.
49. Spiegel S, Milstein S. Sphingosine-1-phosphate, a key signaling molecule. J Biol Chem 2002; 277:25851-25854.
50. Mayer zu Heringdort D, van Koppen CJ, Jacobs KH. Molecular diversity of sphingolipid signaling. FEBS Lett 1997; 410:34-38.
51. Hinkovska-Galcheva V, VanWay SM, Shanley T et al. The role of sphingosine-1-phosphate and ceramide-1-phosphate in calcium homeostasis. Curent Opinion in Investigational Drugs 2008; 9:1192-1205.
52. Gomez-Munoz A. Ceramide-1phosphate: a novel regulator of cell activation, FEBS Letters 2004; 562:5-10.
53. Pettus B, Bielawska A, Subramanian P et al. Ceramide 1-phosphate is a direct activator of cytosolic phospholipase A_2. J Biol Chem 2004; 279:11320-11326.
54. Gomez-Munoz A, Duffy PA, Martin A et al. Short-chain ceramide-1-phosphates are novel stimulators of DNA synthesis and cell division: antagonism by cellpermeable ceramides. Molecular Pharmacology 1995; 47:833-899.
55. Rile G, Yatomi Y, Takafuta T et al. Ceramide 1-phosphate formation in neutrophils. Acta Haematologica 2003; 109:76-83.
56. Törnquist K, Ramström C, Rudnäs B et al. Ceramide 1-(2-cyanoethyl) phosphate enhances store-operated Ca^{2+} entry in thyroid FRTL-5 cells. Eur J Biochem 2002; 253:1-11.
57. Gijsbers S, Mannaerts GP, Himpens B et al. N-acetyl-sphingenine-1-phosphate is a potent calcium mobilizing agent. FEBS Letters 1999; 453:269-272.

58. Colina C, Flores A, Castillo C et al. Ceramide-1-phosphate induces Ca^{2+} mobiliztion in Jurkat T-cell by elevation of Ins (1,4,5) P_3 and activation of a store-operated channel. Biochem Biophys Res Commun 2005; 336:54-60.
59. Hogback S, Leppimaki P, Rudnas B et al. Ceramide-1-phosphate increses intracellular free calcium in thyroid FRTL-5 cells: evidence for an effect mediated by inositol 1,4,5-trisphosphate and intracellular sphingosine-1-phosphate. Biochem J 2003; 370:111-119.
60. Itagaki K, Kannan KB, Livingston DH et al. Store-operated calcium entry in human neutrophils reflects multiple contributions from independently regulated pathways. J Immunol 2002; 168:4063-4069.
61. Beech D. Ion channel switching and activation in smoot-muscle cells of occlusive vascular diseases. Biochemical Society Transactions 2007; 5:890-894.
62. Jain R. Molecular regulation of vessel maturation, Nat Med 2003; 9:685-693.
63. Siess W. Athero-and thrombogenic actions of lysophosphatidic acid and sphingosine-1-phosphate. Biochim Biophys Acta 2002; 1582:204-215.
64. Carafoli E. Calcium Signaling: a tale for all seasons. Proc Natl Acad Sci USA 2002; 99:115-1122.
65. Unser M, Albroubi A. A review of wavelets in biochemical applications. Proc IEEE 1996; 84:626-638.

Chapter 10

Extracellular and Intracellular Actions of Sphingosine-1-Phosphate

Graham M. Strub, Michael Maceyka, Nitai C. Hait, Sheldon Milstien and Sarah Spiegel*

Abstract

Sphingosine-1-phosphate (S1P) is a bioactive lipid mediator with crucial roles in a wide variety of cellular functions across a broad range of organisms. Though a simple molecule in structure, S1P functions are complex. The formation of S1P is catalyzed by one of two sphingosine kinases that have differential cellular distributions as well as both overlapping and opposing functions and which are activated by many different stimuli. S1P can act on a family of G protein-coupled receptors (S1PRs) that are also differentially expressed in different cell types, which influences the cellular responses to S1P. In addition to acting on receptors located on the plasma membrane, S1P can also function inside the cell, independently of S1PRs. It also appears that both the intracellular location and the isotype of sphingosine kinase involved are major determinants of inside-out signaling of S1P in response to many extracellular stimuli. This chapter is focused on the current literature on extracellular and intracellular actions of S1P.

Introduction

In just over a decade, the sphingolipid metabolite, sphingosine-1-phosphate (S1P), has emerged as a key regulator of numerous physiological functions, including cell growth and survival, angiogenesis, cell motility and migration and lymphocyte trafficking.[1] S1P promotes cell growth and inhibits apoptosis, while its precursors, ceramide and sphingosine, typically inhibit cell growth and induce apoptosis.[1] Therefore, the cellular balance of these three sphingolipid metabolites, the "sphingolipid rheostat", is of crucial importance in regulating cell fate.[1] Moreover, sphingolipid metabolism has been found to be dysregulated in many human diseases, including cancer, inflammation, atherosclerosis and asthma.[2] S1P is produced intracellularly by two sphingosine kinases (SphK1 and SphK2) and is the ligand for a family of five G protein-coupled receptors, termed $S1P_{1-5}$. However, there are some actions of S1P that appear to be independent of the known S1PRs. This review will focus on the emerging evidence for S1P as a second messenger.

Sphingolipid Metabolism

Sphingolipids are ubiquitous components of the lipid bilayer of eukaryotic cells. Like glycerolipids, numerous agonists regulate sphingolipid metabolism to generate signaling molecules, including ceramide (N-acyl sphingosine), sphingosine and sphingosine-1-phosphate (S1P).[1,3,4] Ceramide, the backbone of all sphingolipids, is produced both by de novo synthesis and by turnover of sphingolipids, such as hydrolysis of sphingomyelin by sphingomyelinases. It is important to

*Corresponding Author: Sarah Spiegel—Department of Biochemistry and Molecular Biology, VCU School of Medicine, 1101 E. Marshall Street, 2-011 Sanger Hall, Richmond, Virginia 23298 USA. Email: sspiegel@vcu.edu

Sphingolipids as Signaling and Regulatory Molecules, edited by Charles Chalfant and Maurizio Del Poeta.

note that sphingosine is not produced de novo and is only formed by catabolism of sphingolipids. The 3-ketosphinganine formed is rapidly reduced to dihydrosphingosine, which is subsequently N-acylated by one of a family of six (dihydro)ceramide synthases (CerS, formerly referred to as LASS) to form dihydroceramide, with the CerSs having differing but overlapping preferences for acyl chains from 16 to 26 carbons long. Finally, a 4-5 trans double bond is introduced in the sphingoid base to produce ceramide. Ceramide is then trafficked from the ER to the Golgi, where a variety of head groups are added, forming sphingolipids. After removal of these head groups during catabolism, deacylation of ceramide by ceramidases yields sphingosine, the most common sphingoid base in mammals. It is important to note that sphingosine and dihydrosphingosine (sphinganine) are not produced de novo and are only formed by catabolism of sphingolipids. These sphingoid bases can be re-utilized for complex sphingolipid biosynthesis or phosphorylated by two sphingosine kinases (SphK1 and SphK2) to form S1P. S1P can be degraded either by reversible dephosphorylation to sphingosine by phosphatases, including lysosomal phosphatases, members of the LPP family of lipid-specific phosphatases and two S1P-specific phosphatases, termed SPP1 and SPP2 (reviewed in 5), or degraded by irreversible cleavage to ethanolamine phosphate and hexadecenal by S1P lyase (SPL).[6]

Ceramide and sphingosine, the precursors of S1P, are important regulatory components of stress responses, typically inducing growth arrest and apoptosis.[3,4] In contrast, S1P has been implicated in motility and cytoskeletal rearrangements, formation of adherens junctions, proliferation, survival, angiogenesis and the trafficking of immune cells.[1,7-9] Thus, the dynamic balance between S1P and ceramide and the consequent regulation of opposing signaling pathways, is an important factor that determines cell fate.[10] S1P produces many of its effects by acting as a ligand for one or more of its five receptors, discussed below. However, although intracellular targets for both ceramide and sphingosine have been identified, indicating they are both bona fide second messengers, intracellular targets for S1P have remained elusive.

Sphingosine Kinases

Oddly, like mammalian SphKs, two SphK isoenzymes are also found in organisms as diverse as yeast, slime molds, worms, flies and mammals.[7] In mammals, both SphKs have a broad and overlapping tissue distribution, with SphK1 predominating in lung and spleen and SphK2 predominating in the heart, brain and liver.[11,12] Both SphKs are members of the diacylglycerol kinase family, containing five conserved domains responsible for substrate binding and catalytic activity.[13] Differential activation of SphKs can be determined by in vitro assays, as detergents stimulate SphK1 and inhibit SphK2.[12] Conversely, SphK2 is stimulated by high salt, whereas SphK1 is inhibited. SphK1 and SphK2 have similar endogenous substrate specificities, with both being able to phosphorylate d-*erythro*-dihydrosphingosine and d-*erythro*-sphingosine, the two mammalian sphingoid bases.[11,13] Cells contain both S1P and dihydro-S1P and both activate cell surface S1PRs with essentially the same affinity (see below). S1P predominates both in cells and plasma and thus for the purposes of this review "S1P" will refer to all phosphorylated sphingoid bases, including phyto-S1P found in yeast and plants, unless otherwise indicated. SphK1 and SphK2 are primarily cytosolic, although their distributions are altered in different cell types and by various signals (see below). Homozygous single knockouts of either isoenzyme are viable and there are also no obvious phenotypes in yeast[14] or mice,[15,16] suggesting that they may have redundant, overlapping, or compensatory functions. Intriguingly, there is no functional redundancy in *Drosophila*, as deletion of Sk2 but not Sk1 results in flight defects and reduced fecundity.[17] In yeast, even the double knockouts are viable.[14] However, when *Sphk1*$^{-/-}$ and *Sphk2*$^{-/-}$ mice were crossed, mice lacking 1 to 3 of the *Sphk* alleles appeared normal, but mice lacking all four alleles died in utero due to defective brain and cardiovascular system development.[16] This suggests that SphK1 and SphK2 are redundant in mammals, at least for viability. However, subtle yet significant phenotypic differences in *Sphk1*$^{-/-}$ and *Sphk2*$^{-/-}$ mice are emerging and the apparent functional redundancy may not apply to a variety of pathophysiological conditions, suggesting that isozyme-specific targeting of SphKs may be an effective means of disease control or prevention.

SphK1

SphK1a is a 48 kDa splice variant that was cloned utilizing peptide sequences from SphK1 purified to homogeneity from rat kidneys.[11] Most investigations of the role of SphK1 have focused on SphK1a, the major splice form, although two N-terminal extension splice variants have been described in humans, SphK1b (+14 aa) and SphK1c (+86 aa), all with similar enzymatic properties.[18] SphK1 isoforms are predominantly cytosolic but have slightly different subcellular distributions, with SphK1b and SphK1c having greater plasma membrane localization.[19] Interestingly, extracellular secretion of SphK1a, by unknown mechanisms through a Golgi-independent route has been described,[19] although the biological significance of this is not yet clear. SphK1 was initially purified in part based on its ability to bind to a calcium-calmodulin column.[20] Subsequently, it was shown that SphK1 has a calmodulin binding site between residues 191-206, the mutation of which blocks agonist-induced SphK1 translocation to the plasma membrane, but not its activation.[21] An important advance in understanding regulation of SphK1 was the demonstration that SphK1 is phosphorylated on Ser225 by ERK1/2,[22] which was necessary but not sufficient for its targeting to the plasma membrane.[23] SphK1 has been shown to physically interact with a number of proteins that may affect its cellular localization, including RPK118,[24] PECAM-1,[25] Acy1[26] and δ-catenin/NPRAP,[27] among others. SphK1 is activated by diverse stimuli, including hormones, growth factors, immunoglobulin receptor crosslinking, cytokines, chemokines and lysolipids, including S1P.[1] Functionally, numerous studies have demonstrated a role for S1P produced by SphK1 in protection of cells from apoptosis, promotion of cell growth, stimulation of motility and tumorigenesis and as an essential component of many signaling pathways, activating kinases, phospholipases and inducing calcium release (reviewed in refs. 1, 8, 28, 29). Indeed, many of the pro-growth and anti-apoptotic effects observed by exogenous addition of S1P can be reproduced by overexpression of SphK1.

SphK2

Much less is known about SphK2, which was cloned in mammalian cells based on its homology to SphK1.[12] SphK2 has two splice variants. The smaller, originally cloned SphK2-S, contains 618 amino acids and the N-terminally extended SphK2-L, consists of 654 amino acids.[30] SphK2-L is expressed in human cells but not in mice and is reported to be the predominant SphK2 mRNA in most cell lines and tissues except brain and kidney.[30] SphK1 and SphK2 have similar abilities to phosphorylate endogenous substrates, but SphK2 is mainly responsible for phosphorylation of the immunosuppressive pro-drug, FTY720.[15,31,32] SphK2 was found to be a two hybrid interactor with the cytoplasmic domain of the IL-12 receptor β1 and SphK2 overexpression enhanced IL-12-mediated signaling.[33] SphK2 also binds calmodulin at a site that is conserved between the SphKs, although unlike SphK1, interaction with calmodulin does not alter SphK2 localization.[21] SphK2 contains a putative BH3 domain that has been shown to contribute to the ability of overexpressed SphK2 to induce apoptosis in a variety of cell types.[34] In contrast to SphK1, only a few agonists have been reported to activate SphK2, including EGF,[35] IgE receptor crosslinking[36] and IL-1β and TNFα.[37] Of note, EGF activates SphK2 and induces phosphorylation on Ser351 and Thr578, both dependent on ERK1.[38] Moreover, this phosphorylation was required for SphK2-enhanced cell migration towards EGF.[38] Intriguingly, although SphK2 expression typically inhibits growth, promotes apoptosis and chemosensitizes several cell types,[30,34,39,40] SphK2 protects MCF-7 breast cancer cells and HCT116 colon cancer cells from doxorubicin-induced apoptosis by a pathway mediated by p53-independent upregulation of p21.[41]

SphK1 vs. SphK2

Even though mice with single knockouts of *Sphk1* and *Sphk2* are viable with few obvious phenotypes, suggesting that the SphKs have redundant functions, the kinases exhibit many differences in a variety of experimental systems. Indeed, several studies have shown that the two SphKs in yeast, Lcb3p and Lcb4p, do not complement one another (e.g., ref. 42), although this may be due to the much higher activity of Lcb4p. Moreover, both SphKs, in yeast as well as mammals, have overlapping but distinct subcellular localizations, suggesting that they may interact with different proteins and/or lipids and utilize different sphingoid base substrate pools. The first clues that eukaryotic SphK1 and SphK2 may differ functionally came from the observations that, in contrast to many reports demonstrating a pro-growth, anti-apoptotic role for SphK1, SphK2 overexpression induced growth arrest and cell death.[30,34,39] Consistent with these results, SphK1 decreased, while SphK2 increased, the sensitivity of several different cell lines to a variety of chemotherapeutic drugs.[34,40] However, other studies have revealed that SphK2 knockdown reduced proliferation of glioblastoma cell lines.[43] The cell culture results demonstrating a role for SphK1 in cell growth and apoptosis are likely pathophysiologically relevant, as SphK1 is overexpressed in a number of solid tumors, including breast, ovary, kidney, brain, stomach and kidney.[43,44] Perhaps because of its role as a promoter of apoptosis, SphK2 but not SphK1 was reported to be responsible for the secretion of S1P during apoptosis of Jurkat T-cells induced by staurosporin.[45] Conversely, in hematopoietic cancer cells, it was recently shown that SphK1 produced the S1P that was secreted in response to doxorubicin-induced apoptosis.[46]

The differential effects of the two SphKs on cell fate are due in part to their different roles in regulating ceramide levels. SphK1 expression decreases ceramide levels, likely by inhibiting ceramide synthases.[47,48] Conversely, SphK2 expression increases ceramide levels by increasing the salvage of sphingoid bases.[47] In a pathway that is conserved in yeast,[42] exogenous sphingoid bases must first be phosphorylated by a specific SphK (Lcb4p or SphK2), then dephosphorylated by a specific SPP (Lcb3p or SPP1).[49,50] This cycle enables cells to re-acylate sphingoid bases to ceramides and likely represents a control point that cells use to regulate the amount of ceramide and sphingolipids at the levels of de novo synthesis (SphK1) and salvage of sphingoid bases (SphK2). SphK1 and SphK2 also appear to have different roles in the uptake of S1P, with SphK1 but not SphK2 cooperating in another phosphorylation/dephosphorylation cycle that utilizes cell surface LPP-1 to promote sphingoid base uptake.[51]

Differential effects of SphK1 and SphK2 are also apparent in transduction of signals from cell surface receptors. Perhaps because it was cloned first and thus is more well studied, SphK1-dependent formation of S1P has been found to be an important component of numerous receptor signaling pathways, but even in cases where participation of both SphK1 and SphK2 have been examined, it is still SphK1 that is typically implicated. To cite a few examples, estradiol activates SphK1, but not SphK2, ultimately leading to EGFR activation in MCF-7 cells[52]; similarly, VEGF-induced activation of ERK1/2 in T24 bladder cancer cells requires SphK1 but not SphK2[53]; moreover, targeting SphK1 but not SphK2 with siRNA also blocks TNFα-induced COX-2 induction and PGE_2 secretion.[54]

Many studies have demonstrated a role for S1P and S1PRs in cell motility (reviewed in ref. 55) and SphK1, but not SphK2, has often been reported to be the source of S1P. For example, EGF stimulates the activity and translocation of SphK1 to lamellipodia in MCF-7 cells and overexpression of SphK1 enhances migration towards EGF.[56] EGF also has been shown to stimulate both SphK1 and SphK2 in HEK 293 cells, but only SphK1 was required for EGF-induced motility.[35] The lack of requirement for SphK2 in these cells is the more remarkable because SphK2 is already localized at the plasma membrane and suggests that a SphK may need to be targeted to a specific sub-compartment of the plasma membrane, perhaps in close proximity to its substrate and specific S1PRs and/or perhaps S1P transporter(s). Intriguingly, EGF also stimulates both SphK1 and SphK2 in MDA-MB-453 cells, but in these breast cancer cells, both SphKs are required for EGF-induced motility.[35]

The roles of S1P in the immune system are complex and deciphering which SphK isozyme is involved is a challenging task. It has been shown that SphK1, but not SphK2, is required for degranulation of rat RBL-2H3 mast cells in response to IgE receptor crosslinking.[57] In contrast, in mouse bone marrow-derived mast cells, both SphK1 and SphK2 are activated by IgE receptor crosslinking in a Fyn-dependent manner.[36] Interestingly, SphK1 and SphK2 have different requirements for effectors downstream of Fyn, suggesting other levels of regulation of SphK activation. These authors also showed that exogenous S1P could only partially restore degranulation to *Fyn*$^{-/-}$ mice, hinting at an intracellular role for S1P (discussed below). This group later took advantage of

Sphk1$^{-/-}$ and *Sphk2*$^{-/-}$ double knockout mice to demonstrate that in fetal liver-derived and bone marrow-derived mast cells, only SphK2 was responsible for IgE receptor triggered degranulation and cytokine release.[58] However, in a passive systemic anaphylaxis model, they found that *Sphk2*$^{-/-}$ mice fared as well or slightly worse than wild-type mice in terms of increased plasma histamine levels. Conversely, *Sphk1*$^{-/-}$ mice had reduced plasma histamine levels. *Sphk1*$^{-/-}$ mice also had reduced plasma S1P levels compared to wild type and *Sphk2*$^{-/-}$ mice and the intensity of the histamine released positively correlated with circulating S1P levels. The triple allele knockout *Sphk1*$^{+/-}$*Sphk2*$^{-/-}$ mice had the lowest histamine responses and had plasma S1P levels as low as the *Sphk1*$^{-/-}$ mice. Thus, mast cell function in mice is determined both by SphK2 in mast cells (intrinsic S1P) and circulating S1P levels determined by nonmast cell SphK1 (extrinsic S1P). In contrast, in both human LAD2 mast cells and human umbilical cord blood-derived mast cells, knockdown of SphK1 expression decreased degranulation, cytokine release and motility in response to IgE/antigen.[59] Conversely, SphK2 was dispensable for antigen-induced degranulation, motility, or release of most cytokines.[59] S1P likely also plays important roles in other types of immune cells, although neutrophil function in cells isolated from *Sphk1*$^{-/-}$ or *Sphk2*$^{-/-}$ mice, or even in the whole animals themselves, showed little observable differences between the knockouts and wild type. The *Sphk2*$^{-/-}$ mice did have increased disease progression in a lung infection model.[60] Differences between mouse and human immune systems remain to be resolved, but will likely require SphK isozyme-specific inhibitors to elucidate the roles for the SphK1 and SphK2 in mast cell functions and development.

S1P Receptors

S1P is a ligand for five specific GPCRs, $S1P_{1\text{-}5}$, formerly called endothelial differentiation gene (EDG) receptors, which are differentially expressed in different tissues. The cell type specific expression of S1PRs, as well as their differential coupling to different G proteins, explains the diverse signaling of S1P.[61] As mentioned above, many stimuli, including hormones, immunoglobulin receptor ligation, growth factors and cytokines, activate cytosolic SphKs and the production of S1P that is required for the full activity of these agonists. In many cases, the S1P produced activates cell surface S1PRs in a paracrine and/or autocrine manner (reviewed in ref. 62). Indeed, many of the downstream effects of these stimuli require transactivation of one or more S1PRs, also called "inside-out" signaling. For example, in MCF-7 cells, estradiol stimulates ERK1/2 though a mechanism that requires at least two autocrine signaling loops.[52] In the first loop, estradiol stimulates SphK1 and formation of S1P leading to activation of $S1P_3$. In the second loop, $S1P_3$ activates the metalloproteinase MMP-9, which in turn releases EGF from the EGF-heparin binding protein and activates EGFR, finally leading to ERK1/2 phosphorylation.[52] How transactivation of S1PRs is accomplished is an intriguing puzzle as the SphKs and S1P production are both on the cytosolic side of the plasma membrane and S1PRs bind S1P on the exoplasmic side. While it has been suggested that SphK proteins themselves may be secreted and produce S1P extracellularly,[19] it has been convincingly shown that the ABCC1 transporter mediates secretion of intracellularly produced S1P from mast cells.[63] Moreover, S1PR transactivation is not only regulated by S1P secretion, cellular levels of S1P are also important, as decreasing levels by overexpression of SPP1 inhibits transactivation,[64] while decreasing S1P degradation by inhibition of SPL promotes it.[65] Such inside-out transactivation loops may be a general phenomenon, as chemotactic signals for neutrophils acting through their receptors induce secretion of ATP that then locally activates cell surface nucleotide receptors to coordinate directed cell migration.[66]

$S1P_1$

$S1P_1$ is ubiquitously expressed, with high levels in brain, lung, spleen, cardiovascular system and kidney. It was originally identified as an orphan GPCR involved in differentiation of endothelial cells.[67] Since its discovery, many of the important physiological functions of S1P have been attributed to ligation of this receptor. It is now known that $S1P_1$ plays a key role in angiogenesis, because its deletion in mice is embryonic lethal due to hemorrhage resulting from incomplete vascular maturation as smooth muscle cells and pericytes fail to migrate and envelop nascent endothelial tubes.[68] Endothelial cell conditional $S1P_1$ knockout mice have been generated using the Cre/Lox system and these mice display the same vascular deficiencies, suggesting that $S1P_1$ receptors on endothelial cells are also responsible for vessel coverage by smooth muscle cells.[69] $S1P_1$ also plays an important role in maintenance of endothelial and epithelial barrier integrity by functioning in conjunction with $S1P_2$ and $S1P_3$ to increase vascular integrity.[70-72] Disruption of endothelial barriers leads to increased vascular permeability, often found in tumors and in inflammation. Silencing of $S1P_1$ expression with siRNA blocks barrier enhancement, determined by transendothelial monolayer electrical resistance, while silencing of $S1P_3$ inhibits vascular disruption.[71] These experiments also identified the downstream signaling molecules Akt and Rac as effectors of $S1P_1$ actions on vascular integrity.[71] A role for $S1P_1$ in vascular integrity has also been demonstrated in vivo[73-76] and experiments using a $S1P_1$ selective antagonist demonstrated that $S1P_1$ is crucial in maintaining vascular tone.[77] Of particular interest are the conditional SphK1/SphK2 double knockout mice, which have normal vascular integrity despite having over 100-fold lower plasma S1P levels than wild type mice,[78] suggesting that even a very low level of S1P is sufficient as long as $S1P_1$ expression is normal.

$S1P_1$ is also intimately involved in immune cell function. In particular, its expression is required for lymphocyte egress from lymph nodes. Resting T- and B-cells express primarily $S1P_1$ and its downregulation or deletion results in lymphopenia due to the inability of lymphocytes to exit from the lymph nodes.[15,79] Additionally, transplantation of $S1P_1$ deficient thymocytes and lymphocytes into normal mice results in their sequestration in lymph nodes and Peyer's patches. Moreover, T-cells overexpressing $S1P_1$ preferentially distribute into blood.[80] These results confirm that $S1P_1$ controls lymphocyte recirculation.

$S1P_2$

$S1P_2$ is also widely expressed in a variety of different cell types. Unlike $S1P_1$ knockout mice, newborn $S1P_2$ deficient mice do not demonstrate any striking abnormalities, although they have been reported to develop sporadic seizures between 3-7 weeks of age.[81] Neocortical pyramidal cells from these mice also display an increase in excitability.[81] In addition, $S1P_2^{-/-}$ mice are deaf indicating that $S1P_2$ is required for proper development of the auditory and vestibular systems.[82,83] $S1P_2$ is also required for proper degranulation of mast cells.[57] $S1P_2$ is generally considered to be a repellant receptor as its activation inhibits cell migration and appears to work in opposition to $S1P_1$ and $S1P_3$, which both enhance cell migration.[84] Similarly, activation of $S1P_2$ activates ROCK/Rho and leads to increases in vascular permeability.[72] Finally, although the visceral organs of $S1P_2^{-/-}$ mice develop normally, expression of $S1P_2$ promotes liver tissue remodeling in response to acute injury.[85]

$S1P_3$

Much like the lack of phenotypic effects of deletion of $S1P_2$, deletion of $S1P_3$ in mice does not generate any obvious phenotype. $S1P_3$ is expressed in the cardiovascular system, lungs, kidney, intestines, spleen and cartilage.[86] Knockouts of both $S1P_2$ and $S1P_3$ increases perinatal lethality, but not to a great extent.[87] However, the triple knockout of $S1P_{1-3}$ leads to embryonic lethality due to massive vascular deficiencies perhaps even worse than those resulting from knockout of $S1P_1$ alone.[88] $S1P_3$ is also an important regulator of vascular permeability signaling through the downstream effectors ROCK and Rho.[72] A clear role for $S1P_3$ has also been demonstrated in the regulation of heart rate,[89] as $S1P_3$ expression is localized to myocytes and perivascular smooth muscle cells and its activation results in bradycardia and hypertension.

$S1P_4$ and $S1P_5$

$S1P_{4-5}$ have much narrower patterns of expression than the dominant $S1P_{1-3}$ receptors, localizing in human leukocytes, NK cells, airway smooth muscle cells and white matter of CNS tracts.[90-93] $S1P_4$ is primarily expressed in lymphoid tissues, including the thymus, spleen, bone marrow, appendix and peripheral leukocytes.[94] $S1P_4$ directly couples to $G\alpha_i$ and $G\alpha_{12/13}$ subunits of trimeric G proteins and Jurkat cells overexpressing $S1P_4$ display enhanced pertussis toxin-sensitive cell motility in the absence of S1P.[95] $S1P_4$ stimulation also activates the mitogen activated-protein kinases ERK1/2, activates phospholipase C and modulates the opening of intracellular calcium stores.[96,97] Stimulation of $S1P_4$ ectopically expressed on CHO-K1 cells induced cytoskeletal rearrangements and cell rounding, as well as its internalization following S1P stimulation.[95] Whether or not $S1P_4$ has a role in cell motility remains unclear. In D10G4.1 mouse Th2 cells and EL4.IL-2 mouse T-cells lacking endogenous S1P receptors but transfected with $S1P_4$, its activation failed to transduce chemotactic responses.[98] These cells also displayed enhanced secretion of IL-10 and decreased proliferation in response to S1P.[98]

$S1P_5$ is highly expressed in oligodendrocytes;[92] however, silencing of $S1P_5$ expression does not inhibit myelination or produce any other obvious phenotype in these cells.[99] Binding of S1P to $S1P_5$ induces phosphatase-dependent inhibition of ERK1/2, resulting in an antiproliferative phenotype.[91,100,101] In addition, stimulation of rat oligodendrocytes with PDGF increases $S1P_1$ expression with a concomitant downregulation of $S1P_5$, resulting in an amplified mitogenic response.[102] A recent report demonstrated that $S1P_5$ is present in natural killer cells (NK). Mice deficient in $S1P_5$ display aberrant NK cell homing and mobilization of NK cells to inflamed organs.[93] Finally, despite a multitude of studies focusing on physiological functions of S1PRs, some actions of S1P resulting from activation of SphKs are independent of S1PRs.

Evidence for Intracellular Targets of S1P

As discussed above, cellular levels of S1P are controlled both by its synthesis and by its degradation. S1P can be degraded either by dephosphorylation back to sphingosine or irreversibly degraded by SPL to ethanolamine phosphate and fatty aldehyde. In fact, the latter is the only pathway in eukaryotic cells for degradation of sphingoid bases. Thus, S1P formation and subsequent degradation by SPL is one means for decreasing sphingolipid levels within the cell. Intriguingly, S1P is also an intermediate in the formation of sphingolipids from salvaged sphingoid bases. Though these pathways regulating sphingolipid levels demonstrate a central role for S1P, emerging evidence from yeast, plants and mammals points to S1P as a classical, intracellular second messenger.

S1P in Saccharomyces cerevisiae

The yeast genome does not encode a recognizable cell surface receptor for S1P and exogenous S1P does not affect yeast growth.[103] Thus, any role for S1P in yeast physiology must therefore be intracellular. While no direct target for S1P has been found, S1P does indeed have intracellular functions in yeast. First, it has been demonstrated that yeast cells deleted of S1P phosphatase (*lcb3/ysr2/lbp1*) and SPL (*dpl1*) accumulate large amounts of S1P and are nonviable or very slow growing. This growth arrest can be can be rescued if the major yeast SphK (*lcb4*) is deleted, indicating that intracellular S1P suppresses yeast cell growth.[104,105] Indeed, sphingosine itself induces growth arrest in yeast and this effect can be blocked by mutational inactivation of SphK.[103] The growth inhibition may be due to elevated levels cytosolic calcium, as S1P has been shown to increase intracellular calcium levels.[106] In contrast, while heat shock-induced cell cycle arrest is mediated by the sphingoid base itself[107] heat shock also increases SphK activity[103] and S1P accumulation,[108] suggesting a functional role for S1P in heat shock responses. Preventing S1P metabolism by deletion of either the S1P phosphatase Lcb3p[109,110] or SPL[108] led to enhanced levels of S1P and increased heat shock tolerance. Moreover, mutational inactivation of SphK reversed the protective effect of SPL deletion.[103] That deletion of either an S1P phosphatase or SPL promotes heat tolerance indicates that S1P has direct actions, rather than merely functioning as a metabolic intermediate. Identification of molecular targets of S1P in yeast would aid this quest in mammalian cells.

S1P in Arabidopsis thaliana

Abscisic acid is a plant hormone responsible for mediating responses to drought conditions such as closure of stomata to prevent water loss. It was recently demonstrated that drought increased S1P levels in plants and that abscisic acid-induced stomatal closure was reduced by SphK inhibitors.[111] Subsequently, it was shown that abscisic acid activated SphK in *Arabidopsis*.[112] Intriguingly, these authors showed that the effects of S1P on stomatal openings were dependent on the single canonical G_α protein in plants, GPA1. This finding suggests that S1P might be acting through plant GPCRs. However, *Arabidopsis* has only one GPCR-like protein, designated GCR1, which is not homologous to the known S1PRs, does not bind phosphorylated sphingoid bases and GCR1 mutants are hypersensitive to S1P-induced stomatal closure.[113] Thus, S1P likely regulates stomatal apertures and drought responses intracellularly in plants.

S1P in Mammalian Cells

The observation that expression of SphK1, but not SphK2, decreases ceramide levels and increases dihydrosphingosine levels[47] suggests that S1P produced by SphK1 may negatively regulate one or more of the six (dihydro)ceramide synthases (CerS), leading to accumulation of its substrate, dihydrosphingosine. Subsequently, using lysates from cells over-expressing individual CerSs, S1P was shown to be a noncompetitive in vitro inhibitor of CerS2, but not of other CerSs.[48] In silico analysis identified two domains in CerS2 with predicted homology to S1PRs and mutation of one arginine to alanine in each of these two domains removed S1P inhibition without altering CerS activity.[48] The authors concluded that S1P directly binds to and inhibits, CerS2. However, it is also possible that S1P acts through intermediate proteins in cell lysates to inhibit CerS2. Moreover, CerS2 does not use C16 or C18 acyl CoAs as substrates, yet it is C16 and C18 ceramides that are most affected by expression of SphK1.[47] Additionally, S1P produced by SphK1 reverses CerS1-induced chemosensitivity, suggesting that S1P directly affects CerS1.[40] Finally, it is not yet clear whether inhibition of CerS2 by S1P is a signal-mediated effect or whether it is simply feedback product inhibition. Further work is needed to confirm that S1P binds to and inhibits CerS2.

As mentioned previously, autocrine and/or paracrine transactivation of S1PRs is typically accompanied by translocation of SphK1 to the plasma membrane. However, both SphK1 and SphK2 have been shown to translocate to other cellular compartments, suggesting that S1P may be produced locally in these compartments to act on specific intracellular targets. For example, SphK1 is targeted to internal membranes through interaction with RPK118[24] and Acy1.[26] SphK1 translocated to nascent phagosomes promotes maturation into mature phagolysosomes.[114] SphK1 has also been shown to translocate to the nuclear envelope during S-phase[115] and this translocation may play a role in the ability of SphK1 overexpression to promote the G_1/S transition.[116] Conversely, SphK2 has been shown to reside in the nucleoplasm in certain cells, where it functions to arrest cells in G_1.[30,39] Moreover, SphK2 has a nuclear export signal sequence that is activated by phorbol ester-induced phosphorylation, likely through protein kinase D.[117] Whether the nuclear export signal serves to promote SphK2/S1P signaling in the cytosol or to decrease SphK2/S1P signaling in the nucleus is unclear. In some cells, SphK2 has been shown to translocate to the ER under stress conditions and promote apoptosis.[47] Indeed, targeting SphK1 to the ER induced apoptosis,[34] suggesting that S1P produced at the ER has specific targets.

A direct target for S1P in the ER has not conclusively been identified, although strong evidence indicates that S1P can activate thapsigargin-sensitive calcium channels, likely in the ER. S1P induced inositol trisphosphate receptor-independent release of calcium from permeablized cells[118] and from cell fractions rich in rough, but not smooth, ER.[119] Consistent with these results, cells overexpressing SphK2, which localizes in part to the ER, were also shown to have elevated intracellular calcium.[47] Conversely, fetal liver derived-mast cells from *Sphk2*$^{-/-}$ mice have a defect in calcium mobilization in response to IgE receptor crosslinking that cannot be restored with exogenous S1P,[58] again suggesting an internal S1P target. Exogenous S1P increases calcium in HEK293 cells, a response that can be inhibited with pertussis toxin. However, microinjection of S1P increases calcium bypassing a pertussis toxin block, again supporting a role for intracellular S1P in calcium release.[120] Similarly, caged S1P can elicit calcium mobilization in cells that do not respond to exogenous S1P.[121] Likewise, UTP stimulates calcium mobilization in a SphK-dependent manner, but this effect is not mimicked by exogenous S1P.[122] Finally, exogenous S1P itself can stimulate SphK and SphK inhibitors reduced calcium release induced by S1P.[123] Intracellular calcium release by S1P may be an evolutionarily ancient pathway, as it has been shown that S1P can increase calcium in yeast.[106] In sum, these results strongly suggest that S1P is a second messenger that can activate calcium channels.

Finally, there are other effects of S1P that cannot be explained by activation of the known S1PRs. For example, overexpression of SphK1 promotes survival of endothelial cells in part through PECAM-1 expression and activation of Akt, an effect that is not reproduced by exogenous S1P.[124] Dihydro-S1P and S1P bind and activate S1PRs with similar affinities, but dihydro-S1P cannot recapitulate all of the effects of S1P.[125-128] For example, S1P, but not dihydro-S1P, protects male germ cells from apoptosis, an effect that was linked to inhibition of NF-κB and activation of Akt.[128] A similar protection pattern was observed for HL-60 and PC-12 cells.[125] Conversely, S1P-phosphonate, which does not bind to S1PRs, also protects these cells from apoptosis,[125] suggesting an intracellular action. Other studies using embryonic fibroblasts from S1PR knockout mice ruled out S1PR involvement, demonstrating that SphK1 overexpression stimulated growth and survival in wild type and in S1PR negative cells.[129] In the $APC^{min/+}$ model of intestinal tumorigenesis, it was shown that $Sphk1^{-/-}$ mice, but not $S1p_1^{-/+}$, $S1p_2^{-/-}$, or $S1p_3^{-/-}$ mice, had reduced tumor progression and size,[130] suggesting an intracellular role for S1P. Moreover, SphK1 null mice had elevated levels of sphingosine but not S1P, so it is possible that reduced tumor progression and size was due to inhibitory effects of sphingosine rather than to the absence of S1P stimulation. S1PR knockouts were similarly used to demonstrate that SphK2-induced apoptosis was S1PR independent.[34] Consistent with a role for SphK2 in apoptosis, the FTY720 analog AAL(R), which is phosphorylated by SphK2 but only poorly by SphK1, induces apoptosis in Jurkat cells and primary splenocytes.[131] The authors showed that AAL(R) had to be phosphorylated to affect cells as the nonphosphorylated AAL(R) isomer did not induce apoptosis in cells from $Sphk2^{-/-}$ mice. However, exogenous addition of phosphorylated AAL(R) did not induce apoptosis, suggesting both that phospho-AAL(R) must be generated at or near its site of action and that S1PRs are not involved in SphK2-induced apoptosis.

Intracellular S1P has also been linked to regulation of inflammatory responses. $CD4^+$ T-cells from SphK2 knockout mice displayed a hyperactivated phenotype, increased proliferation and enhanced secretion of cytokines and STAT5 activation in response to IL-2.[132] This phenotype was physiologically relevant, as T-cells from SphK2 knockout mice induced a much more rapid response than T-cells from wild type littermates in an adoptive transfer model of inflammatory bowel disease. The hyperresponsiveness to IL-2 could not be reversed with exogenous S1P, indicating that internal S1P normally suppresses IL-2-induced inflammatory responses. Interestingly, activation of the T-cell receptor in Th1 and Th2 cells leads to increased expression of SphK1, but not SphK2 and SphK1 negatively regulates chemokine expression, although the authors did not examine whether exogenous S1P reproduced chemokine suppression.[133] Similarly, it has been shown that in primary umbilical vein endothelial cells, the pro-inflammatory cytokine TNFα greatly increased expression of the S1P phosphatase SPP2.[134] Induction of SPP2 was required for the TNFα-induced production of IL-1β and IL-8. SPP2 is an integral membrane protein of the ER and its requirement for induction of inflammatory responses suggests that it acts at the ER to remove an inhibitory S1P signal. Moreover, downregulation of SPP1 demonstrated that it played no role in TNFα-induced cytokine secretion. As SPP1 is also an ER resident protein, these results suggest a role for S1P at a specific subcompartment of the ER. Moreover, parallel findings have been reported in yeast, where the S1P-specific phosphatases, YSR2 and YSR3, are both localized to the ER and have overlapping but distinct functions.[5]

Many studies have utilized nonisozyme specific SphK inhibitors to implicate S1P in various signaling pathways (e.g., refs. 135,136), leading to the conclusion that S1P was acting intracellularly in these systems, but S1P release was not detected. Still, care should be used in interpreting these results as S1P release may have been below the limits of detection. Moreover, exogenous S1P can activate all surface S1PRs, perhaps resulting in net opposing effects. It is also possible that signaling events localize secretion of S1P to regions of the plasma membrane where specific receptors to be activated are localized. Thus, lack of detectable S1P secretion does not definitively show that S1PRs are not involved in a particular signaling pathway.

Furthermore, studies in endothelial cells using a pan sphingosine kinase inhibitor demonstrated that endogenously generated S1P functions as a positive modulator of calcium entry via store operated channels (SOC), whereas exogenously administered S1P initiated calcium release from the ER similar to histamine and decreased endothelial cell permeability.[137] These results suggest that the production of intracellular S1P and not the secretion of S1P to act on other immunoregulatory cells, is the primary determinant of the inflammatory response.

Recently, a new study demonstrated that the vascular permeability inhibitor Ang-1 stimulated SphK1 and led to increases in intracellular S1P and decreases in vascular permeability. This function remained undisturbed even when expression of $S1P_{1\text{-}3}$ was downregulated.[138] Exogenous S1P also decreased vascular permeability, but not in cells where S1PRs were downregulated.[138] These results suggest that Ang-1/SphK1 actions on vascular permeability are mediated via an intracellular mechanism.

Implications, Future Directions, and Conclusion

It is now clear that the bioactive lipid mediator S1P exerts effects both intracellularly and extracellularly. In addition, SphK1 and SphK2 are not only distributed differently throughout tissues and within cells, S1P produced by these kinases can have different, sometimes opposite, downstream effects. How can one simple molecule like S1P have such a wide range of effects? It seems that in the case of S1P, the location of its production may be a major determinant of the resulting phenotype. The key to unlocking this riddle will be the inevitable discovery of intracellular S1P binding partners, but to date, none have been unequivocally identified. Extensive studies of proteins in the nucleus, particularly those involved in the transcriptional machinery and the regulation of the cell cycle, are clearly indicated to elucidate the mechanism by which S1P exerts its effects there. Additionally, the consequences of activation or inhibition SphK1 vs. SphK2 must be studied with more scrutiny, as it is apparent that despite catalyzing the same reaction, they have different functions. In addition, because the intracellular actions of S1P involve many functions related to cancer and other diseases, a full understanding of both the extracellular and intracellular actions of S1P will be needed to best design clinical therapies targeting SphKs and S1P production as well as its receptors.

Acknowledgements

Supported by National Institutes of Health Grants R37GM043880, RO1CA61774, RO1AI050094 and U19AI077435-018690 (SS), NIH National Research Service Award F30NS058008 (GS) and the NIMH Intramural Research Program (SM).

References

1. Spiegel S, Milstien S. Sphingosine-1-phosphate: an enigmatic signalling lipid. Nat Rev Mol Cell Biol 2003; 4:397-407.
2. Takabe K, Paugh SW, Milstien S et al. "Inside-out" signaling of sphingosine-1-phosphate: therapeutic targets. Pharmacol Rev 2008; 60:181-195.
3. Ogretmen B, Hannun YA. Biologically active sphingolipids in cancer pathogenesis and treatment. Nature Rev Cancer 2004; 4:604-616.
4. Reynolds CP, Maurer BJ, Kolesnick RN. Ceramide synthesis and metabolism as a target for cancer therapy. Cancer Lett 2004; 206:169-180.
5. Maceyka M, Milstien S, Spiegel S. Measurement of mammalian sphingosine-1-phosphate phosphohydrolase activity in vitro and in vivo. Methods Enzymol 2007; 434:243-256.
6. Bandhuvula P, Saba JD. Sphingosine-1-phosphate lyase in immunity and cancer: silencing the siren. Trends Mol Med 2007; 13:210-217.
7. Maceyka M, Payne SG, Milstien S et al. Sphingosine kinase, sphingosine-1-phosphate and apoptosis. Biochim Biophys Acta 2002; 1585:193-201.
8. Saba JD, Hla T. Point-counterpoint of sphingosine 1-phosphate metabolism. Circ Res 2004; 94:724-734.
9. Cyster JG. Chemokines, sphingosine-1-phosphate and cell migration in secondary lymphoid organs. Annu Rev Immunol 2005; 23:127-159.
10. Cuvillier O, Pirianov G, Kleuser B et al. Suppression of ceramide-mediated programmed cell death by sphingosine-1-phosphate. Nature 1996; 381:800-803.
11. Kohama T, Olivera A, Edsall L et al. Molecular cloning and functional characterization of murine sphingosine kinase. J Biol Chem 1998; 273:23722-23728.

12. Liu H, Sugiura M, Nava VE et al. Molecular cloning and functional characterization of a novel mammalian sphingosine kinase type 2 isoform. J Biol Chem 2000; 275:19513-19520.
13. Liu H, Chakravarty D, Maceyka M et al. Sphingosine kinases: a novel family of lipid kinases. Prog Nucl Acid Res 2002; 71:493-511.
14. Nagiec MM, Skrzypek M, Nagiec EE et al. The LCB4 (YOR171c) and LCB5 (YLR260w) genes of Saccharomyces encode long chain base kinases. J Biol Chem 1998; 273:19437-19442.
15. Allende ML, Sasaki T, Kawai H et al. Mice deficient in sphingosine kinase 1 are rendered lymphopenic by FTY720. J Biol Chem 2004; 279:52487-52492.
16. Mizugishi K, Yamashita T, Olivera A et al. Essential role for sphingosine kinases in neural and vascular development. Mol Cell Biol 2005; 25:11113-11121.
17. Herr DR, Fyrst H, Creason MB et al. Characterization of the Drosophila sphingosine kinases and requirement for Sk2 in normal reproductive function. J Biol Chem 2004; 279:12685-12694.
18. Billich A, Bornancin F, Devay P et al. Phosphorylation of the immunomodulatory drug FTY720 by sphingosine kinases. J Biol Chem 2003; 278:47408-47415.
19. Venkataraman K, Thangada S, Michaud J et al. Extracellular export of sphingosine kinase-1a contributes to the vascular S1P gradient. Biochem J 2006; 397:461-471.
20. Olivera A, Kohama T, Tu Z et al. Purification and characterization of rat kidney sphingosine kinase. J Biol Chem 1998; 273:12576-12583.
21. Sutherland CM, Moretti PA, Hewitt NM et al. The calmodulin-binding site of sphingosine kinase and its role in agonist-dependent translocation of sphingosine kinase 1 to the plasma membrane. J Biol Chem 2006; 281:11693-11701.
22. Pitson SM, Moretti PA, Zebol JR et al. Activation of sphingosine kinase 1 by ERK1/2-mediated phosphorylation. EMBO J 2003; 22:5491-5500.
23. Pitson SM, Xia P, Leclercq TM et al. Phosphorylation-dependent translocation of sphingosine kinase to the plasma membrane drives its oncogenic signalling. J Exp Med 2005; 201:49-54.
24. Hayashi S, Okada T, Igarashi N et al. Identification and characterization of RPK118, a novel sphingosine kinase-1-binding protein. J Biol Chem 2002; 277:33319-33324.
25. Fukuda Y, Aoyama Y, Wada A et al. Identification of PECAM-1 association with sphingosine kinase 1 and its regulation by agonist-induced phosphorylation. Biochim Biophys Acta 2004; 1636:12-21.
26. Maceyka M, Nava VE, Milstien S et al. Aminoacylase 1 is a sphingosine kinase 1-interacting protein. FEBS Lett 2004; 568:30-34.
27. Fujita T, Okada T, Hayashi S et al. Delta-catenin/NPRAP (neural plakophilin-related armadillo repeat protein) interacts with and activates sphingosine kinase 1. Biochem J 2004; 382:717-723.
28. Hait NC, Oskeritzian CA, Paugh SW et al. Sphingosine kinases, sphingosine 1-phosphate, apoptosis and diseases. Biochim Biophys Acta 2006; 1758:2016-2026.
29. Hannun YA, Obeid LM. Principles of bioactive lipid signalling: lessons from sphingolipids. Nat Rev Mol Cell Biol 2008; 9:139-150.
30. Okada T, Ding G, Sonoda H et al. Involvement of N-terminal-extended form of sphingosine kinase 2 in serum-dependent regulation of cell proliferation and apoptosis. J Biol Chem 2005; 280:36318-36325.
31. Paugh SW, Payne SG, Barbour SE et al. The immunosuppressant FTY720 is phosphorylated by sphingosine kinase type 2. FEBS Lett 2003; 554:189-193.
32. Sanchez T, Estrada-Hernandez T, Paik JH et al. Phosphorylation and action of the immunomodulator FTY720 inhibits VEGF-induced vascular permeability. J Biol Chem 2003; 278:47281-47290.
33. Yoshimoto T, Furuhata M, Kamiya S et al. Positive modulation of IL-12 signaling by sphingosine kinase 2 associating with the IL-12 receptor beta1 cytoplasmic region. J Immunol 2003; 171:1352-1359.
34. Liu H, Toman RE, Goparaju S et al. Sphingosine kinase type 2 is a putative BH3-only protein that induces apoptosis. J Biol Chem 2003; 278:40330-40336.
35. Hait NC, Sarkar S, Le Stunff H et al. Role of sphingosine kinase 2 in cell migration towards epidermal growth factor. J Biol Chem 2005; 280:29462-29469.
36. Olivera A, Urtz N, Mizugishi K et al. IgE-dependent activation of sphingosine kinases 1 and 2 and secretion of sphingosine 1-phosphate requires Fyn kinase and contributes to mast cell responses. J Biol Chem 2006; 281:2515-2525.
37. Mastrandrea LD, Sessanna SM, Laychock SG. Sphingosine kinase activity and sphingosine-1 phosphate production in rat pancreatic islets and INS-1 cells: response to cytokines. Diabetes 2005; 54:1429-1436.
38. Hait NC, Bellamy A, Milstien S et al. Sphingosine kinase type 2 activation by ERK-mediated phosphorylation. J Biol Chem 2007; 282:12058-12065.
39. Igarashi N, Okada T, Hayashi S et al. Sphingosine kinase 2 is a nuclear protein and inhibits DNA synthesis. J Biol Chem 2003; 278:46832-46839.

40. Min J, Mesika A, Sivaguru M et al. (Dihydro)ceramide synthase 1 regulated sensitivity to cisplatin is associated with the activation of p38 mitogen-activated protein kinase and is abrogated by sphingosine kinase 1. Mol Cancer Res 2007; 5:801-812.
41. Sankala HM, Hait NC, Paugh SW et al. Involvement of sphingosine kinase 2 in p53-independent induction of p21 by the chemotherapeutic drug doxorubicin. Cancer Res 2007; 67:10466-10474.
42. Funato K, Lombardi R, Vallée B et al. Lcb4p is a key regulator of ceramide synthesis from exogenous long chain sphingoid base in Saccharomyces cerevisiae. J Biol Chem 2003; 278:7325-7334.
43. Van Brocklyn JR, Jackson CA, Pearl DK et al. Sphingosine kinase-1 expression correlates with poor survival of patients with glioblastoma multiforme: roles of sphingosine kinase isoforms in growth of glioblastoma cell lines. J Neuropathol Exp Neurol 2005; 64:695-705.
44. French KJ, Schrecengost RS, Lee BD et al. Discovery and evaluation of inhibitors of human sphingosine kinase. Cancer Res 2003; 63:5962-5969.
45. Weigert A, Johann AM, von Knethen A et al. Apoptotic cells promote macrophage survival by releasing the antiapoptotic mediator sphingosine-1-phosphate. Blood 2006; 108:1635-1642.
46. Gude DR, Alvarez SE, Paugh SW et al. Apoptosis induces expression of sphingosine kinase 1 to release sphingosine-1-phosphate as a "come-and-get-me" signal. FASEB J 2008; 22:2629-2638.
47. Maceyka M, Sankala H, Hait NC et al. Sphk1 and Sphk2: Sphingosine kinase isoenzymes with opposing functions in sphingolipid metabolism. J Biol Chem 2005; 280:37118-37129.
48. Laviad EL, Albee L, Pankova-Kholmyansky I et al. Characterization of ceramide synthase 2: tissue distribution, substrate specificity and inhibition by sphingosine 1-phosphate. J Biol Chem 2008; 283:5677-5684.
49. Le Stunff H, Galve-Roperh I, Peterson C et al. Sphingosine-1-phosphate phosphohydrolase in regulation of sphingolipid metabolism and apoptosis. J Cell Biol 2002; 158:1039-1049.
50. Le Stunff H, Giussani P, Maceyka M et al. Recycling of sphingosine is regulated by the concerted actions of sphingosine-1-phosphate phosphohydrolase 1 and sphingosine kinase 2. J Biol Chem 2007; 282:3472-3480.
51. Zhao Y, Kalari SK, Usatyuk PV et al. Intracellular generation of sphingosine 1-phosphate in human lung endothelial cells: Role of lipid phosphate phosphatase-1 and sphingosine kinase 1. J Biol Chem 2007; 282:14165-14177.
52. Sukocheva O, Wadham C, Holmes A et al. Estrogen transactivates EGFR via the sphingosine 1-phosphate receptor Edg-3: the role of sphingosine kinase-1. J Cell Biol 2006; 173:301-310.
53. Shu X, Wu W, Mosteller RD et al. Sphingosine kinase mediates vascular endothelial growth factor-induced activation of ras and mitogen-activated protein kinases. Mol Cell Biol 2002; 22:7758-7768.
54. Pettus BJ, Bielawski J, Porcelli AM et al. The sphingosine kinase 1/sphingosine-1-phosphate pathway mediates COX-2 induction and PGE2 production in response to TNF-alpha. FASEB J 2003; 17:1411-1421.
55. Donati C, Bruni P. Sphingosine 1-phosphate regulates cytoskeleton dynamics: Implications in its biological response. Biochim Biophys Acta 2006; 1758:2037-2048.
56. Sarkar S, Maceyka M, Hait NC et al. Sphingosine kinase 1 is required for migration, proliferation and survival of MCF-7 human breast cancer cells. FEBS Lett 2005; 579:5313-5317.
57. Jolly PS, Bektas M, Olivera A et al. Transactivation of sphingosine-1-phosphate receptors by FcepsilonRI triggering is required for normal mast cell degranulation and chemotaxis. J Exp Med 2004; 199:959-970.
58. Olivera A, Mizugishi K, Tikhonova A et al. The sphingosine kinase-sphingosine-1-phosphate axis is a determinant of mast cell function and anaphylaxis. Immunity 2007; 26:287-297.
59. Oskeritzian CA, Alvarez SE, Hait NC et al. Distinct roles of sphingosine kinases 1 and 2 in human mast-cell functions. Blood 2008; 111:4193-4200.
60. Zemann B, Urtz N, Reuschel R et al. Normal neutrophil functions in sphingosine kinase type 1 and 2 knockout mice. Immunol Lett 2007; 109:56-63.
61. Brinkmann V. Sphingosine 1-phosphate receptors in health and disease: mechanistic insights from gene deletion studies and reverse pharmacology. Pharmacol Ther 2007; 115:84-105.
62. Alvarez SE, Milstien S, Spiegel S. Autocrine and paracrine roles of sphingosine-1-phosphate. Trends Endocrinol Metab 2007; 18:300-307.
63. Mitra P, Oskeritzian CA, Payne SG et al. Role of ABCC1 in export of sphingosine-1-phosphate from mast cells. Proc Natl Acad Sci USA 2006; 103:16394-16399.
64. Le Stunff H, Mikami A, Giussani P et al. Role of sphingosine-1-phosphate phosphatase 1 in epidermal growth factor-induced chemotaxis. J Biol Chem 2004; 279:34290-34297.
65. Schwab SR, Pereira JP, Matloubian M et al. Lymphocyte sequestration through S1P lyase inhibition and disruption of S1P gradients. Science 2005; 309:1735-1739.
66. Chen Y, Corriden R, Inoue Y et al. ATP release guides neutrophil chemotaxis via P2Y2 and A3 receptors. Science 2006; 314:1792-1795.

67. Hla T, Maciag T. An abundant transcript induced in differentiating human endothelial cells encodes a polypeptide with structural similarities to G-protein coupled receptors. J Biol Chem 1990; 265:9308-9313.
68. Liu Y, Wada R, Yamashita T et al. Edg-1, the G protein-coupled receptor for sphingosine-1-phosphate, is essential for vascular maturation. J Clin Invest 2000; 106:951-961.
69. Allende ML, Yamashita T, Proia RL. G-protein coupled receptor S1P1 acts within endothelial cells to regulate vascular maturation. Blood 2003; 102:3665-3667.
70. McVerry BJ, Garcia JG. Endothelial cell barrier regulation by sphingosine 1-phosphate. J Cell Biochem 2004; 92:1075-1085.
71. Singleton PA, Dudek SM, Ma SF et al. Transactivation of sphingosine 1-phosphate receptors is essential for vascular barrier regulation. Novel role for hyaluronan and CD44 receptor family. J Biol Chem 2006; 281:34381-34393.
72. Sanchez T, Skoura A, Wu MT et al. Induction of vascular permeability by the sphingosine-1-phosphate receptor-2 (S1P2R) and its downstream effectors ROCK and PTEN. Arterioscler Thromb Vasc Biol 2007; 27:1312-1318.
73. Singleton PA, Dudek SM, Chiang ET et al. Regulation of sphingosine 1-phosphate-induced endothelial cytoskeletal rearrangement and barrier enhancement by S1P1 receptor, PI3 kinase, Tiam1/Rac1 and alpha-actinin. FASEB J 2005; 19:1646-1656.
74. Finigan JH, Dudek SM, Singleton PA et al. Activated protein C mediates novel lung endothelial barrier enhancement: Role of sphingosine 1-phosphate receptor transactivation. J Biol Chem 2005; 280:17286-17293.
75. McVerry BJ, Garcia JG. In vitro and in vivo modulation of vascular barrier integrity by sphingosine 1-phosphate: mechanistic insights. Cell Signal 2005; 17:131-139.
76. Lee JF, Zeng Q, Ozaki H et al. Dual roles of tight junction associated protein, zonula occludens-1, in sphingosine-1-phosphate mediated endothelial chemotaxis and barrier integrity. J Biol Chem 2006; 281:29190-29200.
77. Sanna MG, Wang SK, Gonzalez-Cabrera PJ et al. Enhancement of capillary leakage and restoration of lymphocyte egress by a chiral S1P(1) antagonist in vivo. Nat Chem Biol 2006; 2:434-441.
78. Pappu R, Schwab SR, Cornelissen I et al. Promotion of lymphocyte egress into blood and lymph by distinct sources of sphingosine-1-phosphate. Science 2007; 316:295-298.
79. Matloubian M, Lo CG, Cinamon G et al. Lymphocyte egress from thymus and peripheral lymphoid organs is dependent on S1P receptor 1. Nature 2004; 427:355-360.
80. Chi H, Flavell RA. Cutting edge: regulation of T-cell trafficking and primary immune responses by sphingosine 1-phosphate receptor 1. J Immunol 2005; 174:2485-2488.
81. MacLennan AJ, Carney PR, Zhu WJ et al. An essential role for the H218/AGR16/Edg-5/LP(B2) sphingosine 1-phosphate receptor in neuronal excitability. Eur J Neurosci 2001; 14:203-209.
82. Kono M, Belyantseva IA, Skoura A et al. Deafness and stria vascularis defects in $S1P_2$ receptor null mice. J Biol Chem 2007; 282:10690-10696.
83. Herr DR, Chun J. Effects of LPA and S1P on the nervous system and implications for their involvement in disease. Curr Drug Targets 2007; 8:155-167.
84. Lepley D, Paik JH, Hla T et al. The G protein-coupled receptor S1P2 regulates Rho/Rho kinase pathway to inhibit tumor cell migration. Cancer Res 2005; 65:3788-3795.
85. Serriere-Lanneau V, Teixeira-Clerc F, Li L et al. The sphingosine 1-phosphate receptor S1P2 triggers hepatic wound healing. FASEB J 2007; 21:2005-2013.
86. Ishii I, Friedman B, Ye X et al. Selective loss of sphingosine 1-phosphate signaling with no obvious phenotypic abnormality in mice lacking its G protein-coupled receptor, LP(B3)/EDG-3. J Biol Chem 2001; 276:33697-33704.
87. Ishii I, Ye X, Friedman B et al. Marked perinatal lethality and cellular signaling deficits in mice Null for the two sphingosine 1-phosphate receptors, S1P2/LPB2/EDG-5 and S1P3/LPB3/EDG-3. J Biol Chem 2002; 277:25152-25159.
88. Kono M, Mi Y, Liu Y et al. The sphingosine-1-phosphate receptors S1P1, S1P2 and S1P3 function coordinately during embryonic angiogenesis. J Biol Chem 2004; 279:29367-29373.
89. Forrest M, Sun SY, Hajdu R et al. Immune cell regulation and cardiovascular efects of sphingosine 1-phosphate receptoragonists in rodents are mediated via distinct receptor sub-types. J Pharmacol Exp Ther 2004; 309:758-768.
90. Gräler MH, Bernhardt G, Lipp M. EDG6, a novel G-protein-coupled receptor related to receptors for bioactive lysophospholipids, is specifically expressed in lymphoid tissue. Genomics 1998; 53:164-169.
91. Im DS, Heise CE, Ancellin N et al. Characterization of a novel sphingosine 1-phosphate receptor, Edg-8. J Biol Chem 2000; 275:14281-14286.
92. Terai K, Soga T, Takahashi M et al. Edg-8 receptors are preferentially expressed in oligodendrocyte lineage cells of the rat CNS. Neuroscience 2003; 116:1053-1062.

93. Walzer T, Chiossone L, Chaix J et al. Natural killer cell trafficking in vivo requires a dedicated sphingosine 1-phosphate receptor. Nat Immunol 2007; 8:1337-1344.
94. Kluk MJ, Hla T. Signaling of sphingosine-1-phosphate via the S1P/EDG-family of G-protein-coupled receptors. Biochim Biophys Acta 2002; 1582:72-80.
95. Graler MH, Grosse R, Kusch A et al. The sphingosine 1-phosphate receptor S1P4 regulates cell shape and motility via coupling to Gi and G12/13. J Cell Biochem 2003; 89:507-519.
96. Van Brocklyn JR, Graler MH, Bernhardt G et al. Sphingosine-1-phosphate is a ligand for the G protein-coupled receptor EDG-6. Blood 2000; 95:2624-2629.
97. Yamazaki Y, Kon J, Sato K et al. Edg-6 as a putative sphingosine 1-phosphate receptor coupling to Ca(2+) signaling pathway. Biochem Biophys Res Commun 2000; 268:583-589.
98. Wang W, Graeler MH, Goetzl EJ. Type 4 sphingosine 1-phosphate G protein-coupled receptor (S1P4) transduces S1P effects on T-cell proliferation and cytokine secretion without signaling migration. FASEB J 2005; 19:1731-1733.
99. Jaillard C, Harrison S, Stankoff B et al. Edg8/S1P5: an oligodendroglial receptor with dual function on process retraction and cell survival. J Neurosci 2005; 25:1459-1469.
100. Gonda K, Okamoto H, Takuwa N et al. The novel sphingosine 1-phosphate receptor AGR16 is coupled via pertussis toxin-sensitive and -insensitive G-proteins to multiple signalling pathways. Biochem J 1999; 337:67-75.
101. Malek RL, Toman RE, Edsall LC et al. Nrg-1 belongs to the endothelial differentiation gene family of G protein-coupled sphingosine-1-phosphate receptors. J Biol Chem 2001; 276:5692-5699.
102. Jung CG, Kim HJ, Miron VE et al. Functional consequences of S1P receptor modulation in rat oligodendroglial lineage cells. Glia 2007; 55:1656-1667.
103. Lanterman MM, Saba JD. Characterization of sphingosine kinase (SK) activity in Saccharomyces cerevisiae and isolation of SK-deficient mutants. Biochem J 1998; 332:525-531.
104. Kim S, Fyrst H, Saba J. Accumulation of phosphorylated sphingoid long chain bases results in cell growth inhibition in Saccharomyces cerevisiae. Genetics 2000; 156:1519-1529.
105. Zhang X, Skrzypek MS, Lester RL et al. Elevation of endogenous sphingolipid long-chain base phosphates kills Saccharomyces cerevisiae cells. Curr Genet 2001; 40:221-233.
106. Birchwood CJ, Saba JD, Dickson RC et al. Calcium influx and signaling in yeast stimulated by intracellular sphingosine 1-phosphate accumulation. J Biol Chem 2001; 276:11712-11718.
107. Jenkins GM, Hannun YA. Role for de novo sphingoid base biosynthesis in the heat-induced transient cell cycle arrest of Saccharomyces cerevisiae. J Biol Chem 2001; 276:8574-8581.
108. Skrzypek MS, Nagiec MM, Lester RL et al. Analysis of phosphorylated sphingolipid long-chain bases reveals potential roles in heat stress and growth control in Saccharomyces. J Bacteriol 1999; 181:1134-1140.
109. Mao C, Saba JD, Obeid LM. The dihydrosphingosine-1-phosphate phosphatases of Saccharomyces cerevisiae are important regulators of cell proliferation and heat stress responses. Biochem J 1999; 342:667-675.
110. Mandala SM, Thornton R, Galve-Roperh I et al. Molecular cloning and characterization of a lipid phosphohydrolase that degrades sphingosine-1-phosphate and induces cell death. Proc Natl Acad Sci USA 2000; 97:7859-7864.
111. Ng CK, Carr K, McAinsh MR et al. Drought-induced guard cell signal transduction involves sphingosine-1-phosphate. Nature 2001; 410:596-599.
112. Coursol S, Fan LM, Le Stunff H et al. Sphingolipid signalling in Arabidopsis guard cells involves heterotrimeric G proteins. Nature 2003; 423:651-654.
113. Pandey S, Assmann SM. The Arabidopsis putative G protein-coupled receptor GCR1 interacts with the G protein alpha subunit GPA1 and regulates abscisic acid signaling. Plant Cell 2004; 16:1616-1632.
114. Thompson CR, Iyer SS, Melrose N et al. Sphingosine kinase 1 (SK1) is recruited to nascent phagosomes in human macrophages: inhibition of SK1 translocation by Mycobacterium tuberculosis. J Immunol 2005; 174:3551-3561.
115. Kleuser B, Maceyka M, Milstien S et al. Stimulation of nuclear sphingosine kinase activity by platelet-derived growth factor. FEBS Lett 2001; 503:85-90.
116. Olivera A, Kohama T, Edsall LC et al. Sphingosine kinase expression increases intracellular sphingosine-1-phosphate and promotes cell growth and survival. J Cell Biol 1999; 147:545-558.
117. Ding G, Sonoda H, Yu H et al. Protein kinase D-mediated phosphorylation and nuclear export of sphingosine kinase 2. J Biol Chem 2007; 282:27493-27502.
118. Mattie M, Brooker G, Spiegel S. Sphingosine-1-phosphate, a putative second messenger, mobilizes calcium from internal stores via an inositol trisphosphate-independent pathway. J Biol Chem 1994; 269:3181-3188.
119. Ghosh TK, Bian J, Gill DL. Sphingosine 1-phosphate generated in the endoplasmic reticulum membrane activates release of stored calcium. J Biol Chem 1994; 269:22628-22635.

120. Meyer zu Heringdorf D, Lass H, Alemany R et al. Sphingosine kinase-mediated Ca^{2+} signalling by G-protein-coupled receptors. EMBO J 1998; 17:2830-2837.
121. Meyer zu Heringdorf D, Liliom K, Schaefer M et al. Photolysis of intracellular caged sphingosine--1-phosphate causes Ca2+ mobilization independently of G-protein-coupled receptors. FEBS Lett 2003; 554:443-449.
122. Alemany R, Sichelschmidt B, zu Heringdorf DM et al. Stimulation of sphingosine-1-phosphate formation by the P2Y(2) receptor in HL-60 cells: Ca(2+) requirement and implication in receptor-mediated Ca(2+) mobilization, but not MAP kinase activation. Mol Pharmacol 2000; 58:491-497.
123. Blom T, Slotte JP, Pitson SM et al. Enhancement of intracellular sphingosine-1-phosphate production by inositol 1,4,5-trisphosphate-evoked calcium mobilisation in HEK-293 cells: endogenous sphingosine-1-phosphate as a modulator of the calcium response. Cell Signal 2005; 17:827-836.
124. Limaye VS, Li X, Hahn C et al. Sphingosine kinase-1 enhances endothelial cell survival through a PECAM-1-dependent activation of PI-3K/Akt and regulation of Bcl-2 family members. Blood 2005; 105:3169-3177.
125. Van Brocklyn JR, Lee MJ, Menzeleev R et al. Dual actions of sphingosine-1-phosphate: extracellular through the G_i-coupled orphan receptor edg-1 and intracellular to regulate proliferation and survival. J Cell Biol 1998; 142:229-240.
126. Xia P, Gamble JR, Rye KA et al. Tumor necrosis factor-a induces adhesion molecule expression through the sphingosine kinase pathway. Proc Natl Acad Sci USA 1998; 95:14196-14201.
127. Morita Y, Perez GI, Paris F et al. Oocyte apoptosis is suppressed by disruption of the acid sphingomyelinase gene or by sphingosine-1-phosphate therapy. Nature Med 2000; 6:1109-1114.
128. Suomalainen L, Pentikainen V, Dunkel L. Sphingosine-1-phosphate inhibits nuclear factor kappaB activation and germ cell apoptosis in the human testis independently of its receptors. Am J Pathol 2005; 166:773-781.
129. Olivera A, Rosenfeldt HM, Bektas M et al. Sphingosine kinase type 1 Induces G12/13-mediated stress fiber formation yet promotes growth and survival independent of G protein coupled receptors. J Biol Chem 2003; 278:46452-46460.
130. Kohno M, Momoi M, Oo ML et al. Intracellular role for sphingosine kinase 1 in intestinal adenoma cell proliferation. Mol Cell Biol 2006; 26:7211-7223.
131. Don AS, Martinez-Lamenca C, Webb WR et al. Essential requirement for sphingosine kinase 2 in a sphingolipid apoptosis pathway activated by FTY720 analogs. J Biol Chem 2007; 282:15833-15842.
132. Samy ET, Meyer CA, Caplazi P et al. Cutting Edge: Modulation of Intestinal Autoimmunity and IL-2 Signaling by Sphingosine Kinase 2 Independent of Sphingosine 1-Phosphate. J Immunol 2007; 179:5644-5648.
133. Yang J, Castle BE, Hanidu A et al. Sphingosine kinase 1 Is a negative regulator of CD4+ Th1 cells. J Immunol 2005; 175:6580-6588.
134. Mechtcheriakova D, Wlachos A, Sobanov J et al. Sphingosine 1-phosphate phosphatase 2 is induced during inflammatory responses. Cell Signal 2007; 19:748-760.
135. Wang F, Van Brocklyn JR, Edsall L et al. Sphingosine-1-phosphate inhibits motility of human breast cancer cells independently of cell surface receptors. Cancer Res 1999; 59:6185-6191.
136. Hammad SM, Crellin HG, Wu BX et al. Dual and distinct roles for sphingosine kinase 1 and sphingosine 1 phosphate in the response to inflammatory stimuli in RAW macrophages. Prostaglandins Other Lipid Mediat 2007; 85:107-114.
137. Itagaki K, Yun JK, Hengst JA et al. Sphingosine 1-phosphate has dual functions in the regulation of endothelial cell permeability and Ca2+ metabolism. J Pharmacol Exp Ther 2007; 323:186-191.
138. Li X, Stankovic M, Bonder CS et al. Basal and angiopoietin-1-mediated endothelial permeability is regulated by sphingosine kinase-1. Blood 2008; 111:3489-3497.

CHAPTER 11

Glucosylceramide in Humans

Maria C. Messner and Myles C. Cabot*

Abstract

Glucosylceramide has a unique and often ambiguous role in mammalian cells. Activation of glucosylceramide synthase, the enzyme that places a glucosyl moiety onto ceramide, is the first pathway-committed step to the production of more complex glycosphingolipids such as lactosylceramide and gangliosides. Alterations in the level of glucosylceramide are noted in cells and tissues in response to cardiovascular disease, diabetes, skin disorders and cancer. Overall, upregulation of glucosylceramide offers cellular protection and primes certain cells for proliferation. However, prolonged overabundance of glucosylceramide is detrimental, as seen in Gaucher disease in humans.

Introduction

Glucosylceramide (GlcCer) is common to plants, fungi and animals and it is essential for viability of multicellular organisms. Cells in culture can survive a deletion of glucosylceramide synthase (GCS), the gene responsible for the glucosylation of ceramide, whereas embryos cannot.[1] GlcCer is the precursor of lactosylceramide (LacCer) and the rest of the neutral and acidic oligo glycosphingolipids. LacCer and several gangliosides have also been shown to function as signaling molecules, however this chapter will focus only on the properties of GlcCer and its regulation.

Glucosylceramide Synthesis and Degradation

Ceramide made in the endoplasmic reticulum (ER) is transported by ceramide transport protein CERT to the cytosolic side of the trans-Golgi where it is converted to GlcCer by GCS[2] (Fig. 1). GlcCer must cross into the lumen of the Golgi in order to be converted to LacCer and higher order glycosphingolipids. It has been shown that short-chain GlcCer can be pumped into the Golgi by ABC transporters, however it is unclear whether natural GlcCer flips into the Golgi lumen, is pumped into the Golgi by ABC transport proteins, or is shuttled back to the ER by the transport protein FAPP2.[3] GlcCer shuttled to the ER has been shown to be transported or flipped into the ER lumen where it enters the trans-Golgi network to undergo transformation to LacCer and GM3.[4] GlcCer from the cytosolic leaf of the Golgi may also be pumped or flipped into the lumen.[5] The GlcCer that makes it into the Golgi lumen is converted to LacCer by lactosylceramide synthase (LCS).[6] Most of the GlcCer that is synthesized is not converted to LacCer, suggesting that GCS and LCS are not functionally coupled. There appears to be a distinct pool of GlcCer that is used for LacCer and downstream oligo glycosphingolipid synthesis, which raises the possibility that the cell utilizes GlcCer for different functions.[7,8]

The glucose moiety on GlcCer is removed by lysosomal GlcCer-β-glucosidase and by the ER-localized saposin C enzyme to yield ceramide. Deficiencies in these enzymes in humans lead to Gaucher disease. In Gaucher disease, cells cannot degrade GlcCer efficiently and the excess is stored throughout the cell,[9] leading to overall increased GlcCer levels. The most pronounced

*Corresponding Author: Myles C. Cabot—John Wayne Cancer Institute, 2200 Santa Monica Blvd, Santa Monica, California 90404, USA. Email: cabotm@jwci.org

Sphingolipids as Signaling and Regulatory Molecules, edited by Charles Chalfant and Maurizio Del Poeta.

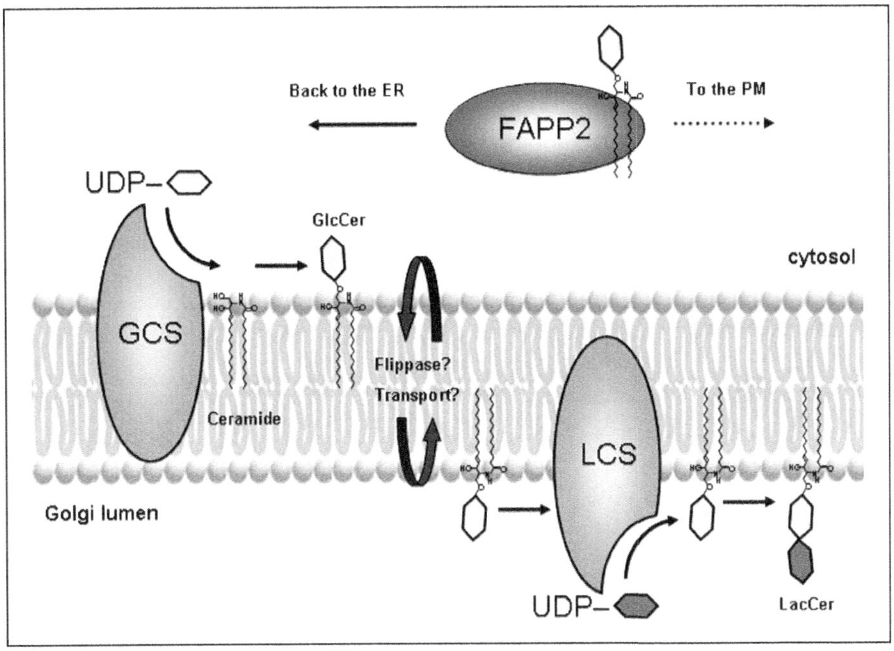

Figure 1. Synthesis of glucosylceramide. Ceramide transported from the ER by CERT to the trans-Golgi is glycosylated on the cytosolic leaf by glucosylceramide synthase (GCS). Most of the glucosylceramide (GlcCer) is shuttled back to the endoplasmic reticulum (ER) by transport protein FAPP2 where it will re-enter the Golgi network on the lumenal side. Some GlcCer is transported to the plasma membrane (PM) or endosomes. Alternatively, GlcCer is either flipped or pumped into the Golgi lumen to undergo further modification to lactosylceramide (LacCer) by lactosylceramide synthase (LCS).

effect occurs in phagocytotic macrophages that participate in ingesting blood cells and apoptotic lymphocytes. These macrophages become engorged with GlcCer and other lipids and release cytokines due to chronic activation.[10] Similar to many other diseases, there are differences in severity and clinical manifestations.

Regulation of the rate-limiting enzyme of glycosphingolipid synthesis, GCS, is not well-understood. The protein c-Fos is an established transcription factor, but has also been shown to bind to the ER and activate the biosynthesis of phospholipids.[11] Recently, c-Fos was shown to activate GCS, but does not activate GlcCer galactosyltransferase 1 and LacCer sialyltransferase 1. Additionally, c-Fos co-immunoprecipitated with GCS in neuronal PC12 cells.[12] GCS transcription may also be regulated in part through Sp1.[13,14] Overexpression of a reticulon family member, RTN-1C, sensitized neuroepithelial cells to fenretinide, a synthetic retinoid derivative that induces apoptosis. Di Sano et al[15] also showed that RTN-1C interacts with GCS and modulates its activity. Their results indicate that GCS/RTN-1C interaction may mediate signals between Golgi and ER compartments, including the cellular response to apoptotic stimuli. This may occur because RTN-1C also interacts with the proapoptotic protein Bcl-xL.[16] GCS may also be regulated by PI3K in natural killer cells.[17] The GCS antagonist D-*threo*-1-phenyl-2-palmitoylamino-3-pyrrolidino-1-propanol (P4),[18] used to decrease GlcCer levels, induced tyrosine phosphorylation of PLCγ and increased inositol 1,4,5 trisphosphate (a PLCγ product).[17] In support of this, phosphorylation of PLCγ was abolished by the exogenous addition of GlcCer.[19]

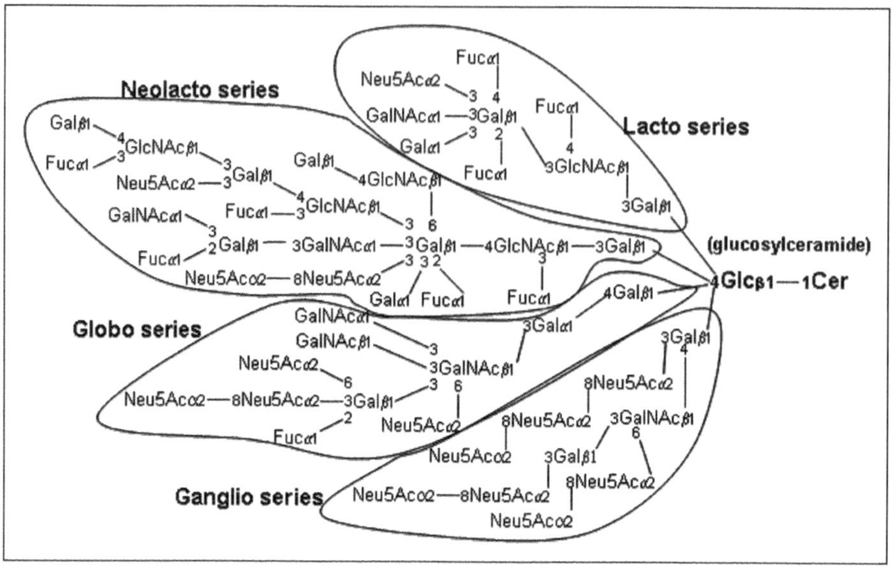

Figure 2. Glucosylceramide is the major precursor of complex glycosphingolipids. Glucosylceramide (GlcCer) is the backbone upon which hundreds of neutral and acidic oligoglycosphingolipids are formed. Lactosylceramide is Galβ1-4 Glucβ1-1 Ceramide. Adapted from http://www.genome.ad.jp/kegg/pathway.html.

Multiple Functions of Glucosylceramide

Template for Higher Order Glycosphingolipids

One of the most important functions of GlcCer is to serve as a template for production of LacCer and complex glycosphingolipids (Fig. 2). Many of the higher order oligo glycosphingolipids have distinct signaling functions that are beyond the scope of this chapter, but that are required for proper development in humans.

Membrane and Lipid Raft Constituent

GlcCer and other glycosphingolipids have higher affinity for cholesterol compared to the major membrane constituent phosphatidylcholine and they are known to induce formation of liquid-ordered phases.[20] Recent analysis of detergent-resistant membrane fractions reveal that in addition to sphingmyelin and cholesterol, glycolipids are also present.[21] Furthermore, GlcCer is elevated in the detergent-resistant membranes from Gaucher cells.[21] It has been shown that GlcCer, unlike galactosylceramide, is located preferentially on the cytoplasmic surface of cell membranes and therefore likely utilized in lipid rafts at the ER and Golgi.[4] Studies have implicated a role for GlcCer in vesicular and glycolipid sorting in the endocytic pathway via lipid rafts. GlcCer on the cytosolic surface of the Golgi appears to aid in sorting membrane proteins to melanosomes and it is believed to be essential for the activity of tyrosinase, a key enzyme in melanin biosynthesis.[22] Glycosylphosphatidylinositol-anchored proteins prefer glycosphingolipid-enriched domains and recently it was demonstrated that PKA-RIIa docks to glycosphingolipids in HepG2 cells and affects apical targeting of P-glycoprotein (P-gp).[23,24] Treatment with D-*threo*-1-phenyl-2-decanoylamino-3-morpholino-1-propanol (PDMP),[18] an inhibitor of GCS and LCS, delays the appearance of P-gp, suggesting that GlcCer synthesis is important for Golgi-to-apical transport of P-gp.[24] However, conflicting data suggest that the role of glycosphingolipids in P-gp transport may be cell-specific.[25] Although the fate of specific proteins may differ, there is a higher concentration of GlcCer in the apical domain of polarized cells such as intestine and kidney epithelial cells.[26,27] The directed sorting of GlcCer in brain capillary endothelial cells, which also have continuous tight junctions and polar distributed proteins, suggests that GlcCer aids in the proper function of polarized cell types.[28]

Cellular Protection in the Skin

The outer layer of our skin is composed of a system of lamellar sheets of lipids in the intracellular space of the stratum corneum. GlcCer constitutes approximately 4% of the total epidermal lipid mass, but is one of the main components of lamellar bodies.[29,30] Lamellar bodies are lipid-storage organelles that are exocytosed from keratinocytes. The skin uses this storage function to respond to the environment quickly by altering the level of ceramide, which is an essential lipid for maintaining proper water and electrolyte levels and for skin barrier repair.[31] By developing an epidermis-specific GlcCer-β-glucosidase knock-out mouse, Jenneman et al[32] showed that lack of GlcCer in the epidermis led to an irregular arrangement of the lamellar bodies and of extracellular lipids in the stratum corneum. In the absence of GlcCer-β-glucosidase, sphingomyelin became the default reservoir for ceramide. It was shown, however, certain subtypes of ceramides necessary to maintain barrier function were lacking. These data provide strong evidence that GlcCer is absolutely essential to provide the correct ceramides needed for barrier function.

Another function of GlcCer in skin is to signal proliferation in keratinocytes. In a study by Uchida et al,[33] human keratinocytes treated with exogenous sphingomyelinase initially increased ceramide and, as a consequence, proliferation was stalled. By 24 h, cells began to proliferate parallel to an increase in GlcCer and GCS expression. Cells cotreated with exogenous GlcCer further augmented proliferation, while GCS inhibitor treatment enhanced the antiproliferative effects of sphingomyelinase. Additionally, HaCaT cells engineered to overexpress GCS were resistant to the effects of sphingomyelinase.[33] These data suggest that a balance between ceramide and GlcCer aids in cell differentiation.

Skin diseases like atopic dermatitis and psoriasis exhibit hyperproliferation, inflammation and impaired barrier function.[34] The mRNA and protein expression of GlcCer-β-glucosidase was shown to be lower in nonlesional skin from patients with psoriatic epidermis compared to control skin, yet was higher in lesional compared with nonlesional skin from psoratic patients.[35] These data suggest that flare-ups of certain skin conditions likely occur due to the misregulation of GlcCer. In a clinical study, patients with atopic eczema that ingested GlcCer saw a reduction of transepidermal water loss.[36] These results were reiterated in a chronic skin perturbation mouse model that exhibited enhanced skin improvement following dietary supplementation of plant-based GlcCer.[37] Although seemingly contradictory, these results are likely due to a systemic anti-inflammatory effect of GlcCer, especially plant-derived GlcCer (see section entitled "Cellular protection in carcinomas").

Cellular Protection in the Cardiovasculature

GlcCer may offer protective effects in the cardiovasculature by serving as an anticoagulant. An analysis of plasma from venous thrombosis patients showed lower GlcCer levels than in healthy blood donors.[38] In the same study, plasma from healthy individuals treated with GlcCer β-glucosidase to decrease GlcCer, resulted in a reduction of plasma sensitivity to activated protein C (APC)/protein S. Exogenously added GlcCer dose-dependently prolonged clotting times in the presence of APC/protein S.[38] Furthermore, lipid vesicles containing GlcCer enhanced binding of APC and enhanced factor Va inactivation by protein S-dependent APC.[38,39] These results suggest that GlcCer facilitates anticoagulation activity.

Analysis of the glycosphingolipid composition in artery layers from normal and atherosclerotic human aorta show striking differences. The intima layer, which interacts with blood, showed an accumulation of GlcCer, LacCer and ganglioside GM2 in fatty streaks and atherosclerotic plaques, but not in normal intima.[40] Oxidized LDL, which occurs during the progression of atherosclerosis, harbors GlcCer and complex oligo glycosphingolipids. However the oligo glycosphingolipids do not accumulate as GlcCer does, suggesting that GlcCer is synthesized in the artery wall.[40] Heart attacks often occur following clot formation from the rupture of an unstable plaque and one of the body's protective mechanisms is to induce proliferation of smooth muscle cells to form a cap over the plaque. In another study it was demonstrated that LacCer, but not GlcCer derived from plaque intima, exerted an increase in proliferation of cultured human aortic smooth muscle cells.[41] Thus, the upregulation of GlcCer and LacCer is possibly preventing the development of an unstable plaque by the proliferative effect of LacCer and by the anticoagulative effect of GlcCer.

Cellular Protection in the Brain

Cerebrospinal fluid from Alzheimer's disease patients and in affected brain regions show decreased GCS protein and increased levels of ceramide.[42,43] In cell culture experiments, cerebellar granule cells exposed to the GCS inhibitor P4 responded by increasing ceramide and subsequently, viability decreased.[42] Experiments in rodents to induce stroke suggest that glycosphingolipids play a role in decreasing the severity of damage. Following middle cerebral artery occlusion in rats, brain ceramide increased with concomitant decrease in GCS activity.[44] The stereoisomer of PDMP, L-*threo*-PDMP, has been used previously to activate GCS in neuronal cells.[45] Following permanent occlusion of the left middle cerebral artery, resulting in chronic ischemia, intraperitoneal injection of L-*threo*-PDMP induced an increase in GlcCer in the ischemic cortex, without changes in sphingomyelin or ceramide. Importantly, the rats showed improved learning and memory compared to artery-occluded rats that did not receive the GCS activator.[46] In the case of Alzheimer's disease, it is unclear whether GlcCer or a downstream glycosphingolipid is the molecule affording protection against cell death. Although it is possible that GlcCer is the molecule protecting ischemic brain areas because it is an anticoagulant,[38,39] more evidence would be needed to draw this conclusion.

Cellular Protection in the Immune System

The role GlcCer plays in the immune system is not well understood. Decreased GlcCer in macrophages and T-cells appears to shift the cell's response from "attack" to "self-protect". Isolated peritoneal macrophages from mice treated with N-alkylnojirimycin,[47] a GCS inhibitor, showed a decreased capacity to release TNFα or IL-6 in response to whole killed *E. coli*.[48] Renal cell carcinoma cells induce apoptosis in T-cells via RelA degradation as shown in coculture experiments.[49] Pretreatment with the GCS inhibitor P4 protected T-cells from tumor-induced RelA degradation and subsequent apoptosis.[49] Systemic increase in GlcCer appeared to facilitate disease recovery. Diabetic and insulin-resistant rodent models given daily intraperitoneal injections of GlcCer or given inhibitors of GCS responded with increased insulin sensitivity and decreased hyperglycemia.[50-52] In Ob/Ob mice, GlcCer injections also led to an increase in IL-10 and peripheral/intrahepatic natural killer T-cell ratio and a corresponding decrease in serum INF-γ.[53] Natural killer T-cells are a subset of lymphocytes that lead to the release of either pro-inflammatory or anti-inflammatory chemokines when activated. In one study, mice induced to develop hepatic sclerosis were given intraperitoneal injections of GlcCer. In the experimental arm, GlcCer ameliorated liver damage, measured by decreased transaminase serum levels. Furthermore, there was a decrease in intrahepatic natural killer T and CD8 lymphocytes and a decrease in T(h)1 and T(h)2 cytokines.[54] An increase in GlcCer may also prevent certain pathogens from binding to cells, which is the first step of invasion. It was demonstrated that bacteria isolated from patients with cystic fibrosis and pneumonia do not bind to GlcCer, LacCer and several gangliosides, but they do bind to asialo-GM1, asialo-GM2 and fucosylasialo-GM1.[55]

Ceramide is doubled in human skeletal muscle from patients with insulin-resistance, and patients with Gaucher disease often develop peripheral insulin resistance, possibly through the influence of glycosphingolipids on insulin receptor functioning.[56,57] The insulin resistance that develops may occur following the conversion of GlcCer to higher order gangliosides. It has been shown that excess GlcCer is metabolized to neutral glycosphingolipids; and gangliosides such as GM3 were shown to participate in the pathogenesis of insulin resistance.[58,59]

Cellular Protection in Carcinomas

Cancer cells are more sensitive to the killing effects of increased ceramide. Accordingly, many chemotherapeutic drugs have been developed that work, in part, by increasing ceramide. Some cancer cells become multidrug resistant by increasing the metabolism of ceramide into GlcCer.[60] Several studies have shown that decreasing GCS expression using GCS inhibitors in multidrug resistant carcinoma cells enhances the effects of chemotherapeutic drugs.[61,62] Blocking GCS in multidrug resistant cancer cells (NCI/ADR-RES) by siRNA also reversed drug resistance, shown by enhanced drug uptake and increased ceramide-induced apoptosis.[63,64] Not surprising, the reverse is seen following overexpression of GCS; drug resistance increased [62,65] as did resistance to TNFα-induced apoptosis.[65,66] Many researchers in the cancer field are interested in P-gp, an ABC transport protein shown to flip short-chain GlcCer into the Golgi.[2] P-gp expression is also elevated in multidrug resistant cancer cells.[67] Gouaze et al[68] showed that P-gp expression decreased following RNA interference or chemical blockade of GCS; however, overexpression of GCS did not induce P-gp expression.[65] In another study, a short-chain analog of ceramide increased P-gp expression, and this effect was due to the cellular conversion of the analog to GlcCer.[69] These data suggest a relationship between GicCer, P-gp, and drug resistance in cancer cells. GlcCer has been show to increase proliferation of normal smooth muscle cells and keratinocytes,[33,41] and Gaucher disease patients have and increased risk for malignant B and other lymphoid cancers.[70,71] It is not a far stretch of the imagination to think that increased GlcCer not only protects cancer cells from chemotherapy, but may actually enhance tumor growth and malignancy. Thus, blocking the conversion of ceramide to GlcCer may be a promising approach to reverse multidrug resistance in cancer cells.

Although upregulation of endogenous GlcCer in cancer cells may protect them from death, ingestion of soy GlcCer appears to have anticancer effects, demonstrated by the suppression of colon cancer growth in mice.[72] The apoptotic effects of soy GlcCer may be due to the high amount of δ8-double bond in the backbone as opposed to mammalian cells that often contain a δ4-double bond. Soy GlcCer also has high amounts of α-hydroxy fatty acids compared to mammalian cells.[73] In addition to soy GlcCer, GlcCer extracted from malt feed (derived from beer brewing waste) also suppressed growth of skin, lung and liver cancer cells.[74] This surprising twist may also be an anticancer avenue to explore.

Conclusion

Because GlcCer is the precursor to hundreds of downstream glycosphingolipids that have unique functions, understanding GlcCer function fully has proved difficult. It is clear, however, that the presence of GlcCer is essential for normal cellular function and viability of multicellular organisms. Additionally, transient increases in GlcCer bestow protection to individual cells and to whole-organisms. For example, GlcCer has been shown to anchor certain proteins to membranes, serve as an anticoagulant in cardiovascular disease and possibly stroke and is essential for the barrier function of our skin. GlcCer is impartial in providing protection, however, because cancer cells also utilize GlcCer to evade our body's immune response and the effects of chemotherapeutic drugs. In comparison to the beneficial effects of temporary GlcCer elevation, sustained high levels of GlcCer lead to adverse effects, as noted in patients with Gaucher disease who have a litany of symptoms depending on severity of the disease. Some groups are developing tissue-specific knock-downs of GlcCer β-glucosidase in an effort to tease apart the complex pathology of Gaucher disease.[75] Further advancement in the understanding of glycosphingolipid regulation will be of great benefit. Future studies aimed at increasing GlcCer in diseased tissues or decreasing GlcCer in cancer will likely improve the human condition.

Acknowledgements

This work was supported in part by the USPHS grant GM77391, the Association for Breast and Prostate Cancer Studies (Santa Monica, CA) and the Fashion Footwear Association of New York Charitable Foundation (New York, NY).

References

1. Yamashita T, Wada R, Sasaki T et al. A vital role for glycosphingolipid synthesis during development and differentiation. Proc Natl Acad Sci USA 1999; 96:9142-9147.
2. Hanada K, Kumagai K, Yasuda S et al. Molecular machinery for nonvesicular trafficking of ceramide. Nature 2003; 426:803-809.
3. D'Angelo G, Polishchuk E, Di Tullio G et al. Glycosphingolipid synthesis requires FAPP2 transfer of glucosylceramide. Nature 2007; 449:62-67.
4. Halter D, Neumann S, van Dijk SM et al. Pre- and postGolgi translocation of glucosylceramide in glycosphingolipid synthesis. J Cell Biol 2007; 179:101-115.
5. Lavie Y, Cao H, Volner A et al. Agents that reverse multidrug resistance, tamoxifen, verapamil and cyclosporin A, block glycosphingolipid metabolism by inhibiting ceramide glycosylation in human cancer cells. J Biol Chem 1997; 272:1682-1687.
6. Lannert H, Gorgas K, Meissner I et al. Functional organization of the Golgi apparatus in glycosphingolipid biosynthesis. Lactosylceramide and subsequent glycosphingolipids are formed in the lumen of the late Golgi. J Biol Chem 1998; 273:2939-2946.
7. Maxzud MK, Maccioni HJ. Glucosylceramide synthesized in vitro from endogenous ceramide is uncoupled from synthesis of lactosylceramide in Golgi membranes from chicken embryo neural retina cells. Neurochem Res 2000; 25:145-152.
8. Veldman RJ, Klappe K, Hinrichs J et al. Altered sphingolipid metabolism in multidrug-resistant ovarian cancer cells is due to uncoupling of glycolipid biosynthesis in the Golgi apparatus. Faseb J 2002; 16:1111-1113.
9. Fuller M, Rozaklis T, Lovejoy M et al. Glucosylceramide accumulation is not confined to the lysosome in fibroblasts from patients with Gaucher disease. Mol Genet Metab 2008; 93:437-443.
10. Cox TM. Gaucher disease: understanding the molecular pathogenesis of sphingolipidoses. J Inherit Metab Dis 2001; 24 Suppl 2:106-121; discussion 187-108.
11. Gil GA, Bussolino DF, Portal MM et al. c-Fos activated phospholipid synthesis is required for neurite elongation in differentiating PC12 cells. Mol Biol Cell 2004; 15:1881-1894.
12. Crespo PM, Silvestre DC, Gil GA et al. c-Fos activates glucosylceramide synthase and glycolipid synthesis in PC12 cells. J Biol Chem 2008; 283:31163-31171.
13. Liu YY, Yu JY, Yin D et al. A role for ceramide in driving cancer cell resistance to doxorubicin. Faseb J 2008; 22:2541-2551.
14. Uchida Y, Itoh M, Taguchi Y et al. Ceramide reduction and transcriptional up-regulation of glucosylceramide synthase through doxorubicin-activated Sp1 in drug-resistant HL-60/ADR cells. Cancer Res 2004; 64:6271-6279.
15. Di Sano F, Fazi B, Citro G et al. Glucosylceramide synthase and its functional interaction with RTN-1C regulate chemotherapeutic-induced apoptosis in neuroepithelioma cells. Cancer Res 2003; 63:3860-3865.
16. Tagami S, Eguchi Y, Kinoshita M et al. A novel protein, RTN-XS, interacts with both Bcl-XL and Bcl-2 on endoplasmic reticulum and reduces their anti-apoptotic activity. Oncogene 2000; 19:5736-5746.
17. Taguchi Y, Kondo T, Watanabe M et al. Interleukin-2-induced survival of natural killer (NK) cells involving phosphatidylinositol-3 kinase-dependent reduction of ceramide through acid sphingomyelinase, sphingomyelin synthase and glucosylceramide synthase. Blood 2004; 104:3285-3293.
18. Abe A, Wild SR, Lee WL et al. Agents for the treatment of glycosphingolipid storage disorders. Curr Drug Metab 2001; 2:331-338.
19. Shu L, Lee L, Shayman JA. Regulation of phospholipase C-gamma activity by glycosphingolipids. J Biol Chem 2002; 277:18447-18453.
20. Ahmed SN, Brown DA, London E. On the origin of sphingolipid/cholesterol-rich detergent-insoluble cell membranes: physiological concentrations of cholesterol and sphingolipid induce formation of a detergent-insoluble, liquid-ordered lipid phase in model membranes. Biochemistry 1997; 36:10944-10953.
21. Hein LK, Meikle PJ, Hopwood JJ et al. Secondary sphingolipid accumulation in a macrophage model of Gaucher disease. Mol Genet Metab 2007; 92:336-345.
22. Sprong H, Degroote S, Claessens T et al. Glycosphingolipids are required for sorting melanosomal proteins in the Golgi complex. J Cell Biol 2001; 155:369-380.
23. Sheets ED, Lee GM, Simson R et al. Transient confinement of a glycosylphosphatidylinositol-anchored protein in the plasma membrane. Biochemistry 1997; 36:12449-12458.
24. Wojtal KA, de Vries E, Hoekstra D et al. Efficient trafficking of MDR1/P-glycoprotein to apical canalicular plasma membranes in HepG2 cells requires PKA-RIIalpha anchoring and glucosylceramide. Mol Biol Cell 2006; 17:3638-3650.
25. Dijkhuis AJ, Klappe K, Kamps W et al. Gangliosides do not affect ABC transporter function in human neuroblastoma cells. J Lipid Res 2006; 47:1187-1195.

26. van ISC, Hoekstra D. (Glyco)sphingolipids are sorted in sub-apical compartments in HepG2 cells: a role for nonGolgi-related intracellular sites in the polarized distribution of (glyco)sphingolipids. J Cell Biol 1998; 142:683-696.
27. van der Bijl P, Lopes-Cardozo M, van Meer G. Sorting of newly synthesized galactosphingolipids to the two surface domains of epithelial cells. J Cell Biol 1996; 132:813-821.
28. Tewes BJ, Galla HJ. Lipid polarity in brain capillary endothelial cells. Endothelium 2001; 8:207-220.
29. Grayson S, Johnson-Winegar AG, Wintroub BU et al. Lamellar body-enriched fractions from neonatal mice: preparative techniques and partial characterization. J Invest Dermatol 1985; 85:289-294.
30. Hamanaka S, Nakazawa S, Yamanaka M et al. Glucosylceramide accumulates preferentially in lamellar bodies in differentiated keratinocytes. Br J Dermatol 2005; 152:426-434.
31. Hamanaka S, Hara M, Nishio H et al. Human epidermal glucosylceramides are major precursors of stratum corneum ceramides. J Invest Dermatol 2002; 119:416-423.
32. Jennemann R, Sandhoff R, Langbein L et al. Integrity and barrier function of the epidermis critically depend on glucosylceramide synthesis. J Biol Chem 2007; 282:3083-3094.
33. Uchida Y, Murata S, Schmuth M et al. Glucosylceramide synthesis and synthase expression protect against ceramide-induced stress. J Lipid Res 2002; 43:1293-1302.
34. Segre JA. Epidermal barrier formation and recovery in skin disorders. J Clin Invest 2006; 116:1150-1158.
35. Alessandrini F, Pfister S, Kremmer E et al. Alterations of glucosylceramide-beta-glucosidase levels in the skin of patients with psoriasis vulgaris. J Invest Dermatol 2004; 123:1030-1036.
36. Miyanishi K, Shiono N, Shirai H et al. Reduction of transepidermal water loss by oral intake of glucosylceramides in patients with atopic eczema. Allergy 2005; 60:1454-1455.
37. Tsuji K, Mitsutake S, Ishikawa J et al. Dietary glucosylceramide improves skin barrier function in hairless mice. J Dermatol Sci 2006; 44:101-107.
38. Deguchi H, Fernandez JA, Pabinger I et al. Plasma glucosylceramide deficiency as potential risk factor for venous thrombosis and modulator of anticoagulant protein C pathway. Blood 2001; 97:1907-1914.
39. Yegneswaran S, Deguchi H, Griffin JH. Glucosylceramide, a neutral glycosphingolipid anticoagulant cofactor, enhances the interaction of human- and bovine-activated protein C with negatively charged phospholipid vesicles. J Biol Chem 2003; 278:14614-14621.
40. Mukhin DN, Chao FF, Kruth HS. Glycosphingolipid accumulation in the aortic wall is another feature of human atherosclerosis. Arterioscler Thromb Vasc Biol 1995; 15:1607-1615.
41. Chatterjee SB, Dey S, Shi WY et al. Accumulation of glycosphingolipids in human atherosclerotic plaque and unaffected aorta tissues. Glycobiology 1997; 7:57-65.
42. Marks N, Berg MJ, Saito M et al. Glucosylceramide synthase decrease in frontal cortex of Alzheimer brain correlates with abnormal increase in endogenous ceramides: consequences to morphology and viability on enzyme suppression in cultured primary neurons. Brain Res 2008; 1191:136-147.
43. Satoi H, Tomimoto H, Ohtani R et al. Astroglial expression of ceramide in Alzheimer's disease brains: a role during neuronal apoptosis. Neuroscience 2005; 130:657-666.
44. Takahashi K, Ginis I, Nishioka R et al. Glucosylceramide synthase activity and ceramide levels are modulated during cerebral ischemia after ischemic preconditioning. J Cereb Blood Flow Metab 2004; 24:623-627.
45. Inokuchi J, Kuroda Y, Kosaka S et al. L-threo-1-phenyl-2-decanoylamino-3-morpholino-1-propanol stimulates ganglioside biosynthesis, neurite outgrowth and synapse formation in cultured cortical neurons and ameliorates memory deficits in ischemic rats. Acta Biochim Pol 1998; 45:479-492.
46. Hisaki H, Okazaki T, Kubota M et al. L-PDMP improves glucosylceramide synthesis and behavior in rats with focal ischemia. Neurol Res 2008; 30:979-984.
47. Butters TD, Dwek RA, Platt FM. Imino sugar inhibitors for treating the lysosomal glycosphingolipidoses. Glycobiology 2005; 15:43R-52R.
48. Yohe HC, Kitzmiller TJ, Bement WJ et al. Reduced inflammatory potential of peritoneal macrophages recruited in mice pretreated with a glycolipid synthesis inhibitor. Inflammopharmacology 2006; 14:170-175.
49. Thornton MV, Kudo D, Rayman P et al. Degradation of NF-kappa B in T-cells by gangliosides expressed on renal cell carcinomas. J Immunol 2004; 172:3480-3490.
50. Aerts JM, Ottenhoff R, Powlson AS et al. Pharmacological inhibition of glucosylceramide synthase enhances insulin sensitivity. Diabetes 2007; 56:1341-1349.
51. Fox TE, Han X, Kelly S et al. Diabetes alters sphingolipid metabolism in the retina: a potential mechanism of cell death in diabetic retinopathy. Diabetes 2006; 55:3573-3580.
52. Zigmond E, Zangen S, Pappo O et al. Beta-glycosphingolipids improved glucose intolerance and hepatic steatosis of the Cohen diabetic rat. Am J Physiol Endocrinol Metab 2008.
53. Margalit M, Shalev Z, Pappo O et al. Glucocerebroside ameliorates the metabolic syndrome in OB/OB mice. J Pharmacol Exp Ther 2006; 319:105-110.

54. Safadi R, Zigmond E, Pappo O et al. Amelioration of hepatic fibrosis via beta-glucosylceramide-mediated immune modulation is associated with altered CD8 and NKT-lymphocyte distribution. Int Immunol 2007; 19:1021-1029.
55. Krivan HC, Roberts DD, Ginsburg V. Many pulmonary pathogenic bacteria bind specifically to the carbohydrate sequence GalNAc beta 1-4Gal found in some glycolipids. Proc Natl Acad Sci USA 1988; 85:6157-6161.
56. Adams JM 2nd, Pratipanawatr T, Berria R et al. Ceramide content is increased in skeletal muscle from obese insulin-resistant humans. Diabetes 2004; 53:25-31.
57. Langeveld M, Ghauharali KJ, Sauerwein HP et al. Type I Gaucher disease, a glycosphingolipid storage disorder, is associated with insulin resistance. J Clin Endocrinol Metab 2008; 93:845-851.
58. Saito M, Rosenberg A. The fate of glucosylceramide (glucocerebroside) in genetically impaired (lysosomal beta-glucosidase deficient) Gaucher disease diploid human fibroblasts. J Biol Chem 1985; 260:2295-2300.
59. Tagami S, Inokuchi Ji J, Kabayama K et al. Ganglioside GM3 participates in the pathological conditions of insulin resistance. J Biol Chem 2002; 277:3085-3092.
60. Lavie Y, Cao H, Bursten SL et al. Accumulation of glucosylceramides in multidrug-resistant cancer cells. J Biol Chem 1996; 271:19530-19536.
61. Rath G, Schneider C, Langlois B et al. De novo ceramide synthesis is responsible for the anti-tumor properties of camptothecin and doxorubicin in follicular thyroid carcinoma. Int J Biochem Cell Biol 2008.
62. Xie P, Shen YF, Shi YP et al. Overexpression of glucosylceramide synthase in associated with multidrug resistance of leukemia cells. Leuk Res 2008; 32:475-480.
63. Liu YY, Han TY, Yu JY et al. Oligonucleotides blocking glucosylceramide synthase expression selectively reverse drug resistance in cancer cells. J Lipid Res 2004; 45:933-940.
64. Sun YL, Zhou GY, Li KN et al. Suppression of glucosylceramide synthase by RNA interference reverses multidrug resistance in human breast cancer cells. Neoplasma 2006; 53:1-8.
65. Liu YY, Han TY, Giuliano AE et al. Glycosylation of ceramide potentiates cellular resistance to tumor necrosis factor-alpha-induced apoptosis. Exp Cell Res 1999; 252:464-470.
66. Liu YY, Cabot MC. Development of a mammalian Tet-on expression cell line: glucosylceramide synthase regulates TNF-alpha-induced apoptosis. Methods Mol Biol 2004; 249:177-192.
67. Gottesman MM, Fojo T, Bates SE. Multidrug resistance in cancer: role of ATP-dependent transporters. Nat Rev Cancer 2002; 2:48-58.
68. Gouaze V, Liu YY, Prickett CS et al. Glucosylceramide synthase blockade down-regulates P-glycoprotein and resensitizes multidrug-resistant breast cancer cells to anticancer drugs. Cancer Res 2005; 65:3861-3867.
69. Gouaze-Andersson V, Yu JY, Kreitenberg AJ et al. Ceramide and glucosylceramide upregulate expression of the multidrug resistance gene MDR1 in cancer cells. Biochim Biophys Acta 2007; 1771:1407-1417.
70. Burstein Y, Zakuth V, Rechavi G et al. Abnormalities of cellular immunity and natural killer cells in Gaucher's disease. J Clin Lab Immunol 1987; 23:149-151.
71. Lalazar G, Preston S, Zigmond E et al. Glycolipids as immune modulatory tools. Mini Rev Med Chem 2006; 6:1249-1253.
72. Symolon H, Schmelz EM, Dillehay DL et al. Dietary soy sphingolipids suppress tumorigenesis and gene expression in 1,2-dimethylhydrazine-treated CF1 mice and ApcMin/+ mice. J Nutr 2004; 134:1157-1161.
73. Sullards MC, Lynch DV, Merrill AH Jr et al. Structure determination of soybean and wheat glucosylceramides by tandem mass spectrometry. J Mass Spectrom 2000; 35:347-353.
74. Oku H, Wongtangtintharn S, Iwasaki H et al. Tumor specific cytotoxicity of glucosylceramide. Cancer Chemother Pharmacol 2007; 60:767-775.
75. Sinclair GB, Jevon G, Colobong KE et al. Generation of a conditional knockout of murine glucocerebrosidase: utility for the study of Gaucher disease. Mol Genet Metab 2007; 90:148-156.

Chapter 12

Gangliosides as Regulators of Cell Membrane Organization and Functions

Sandro Sonnino and Alessandro Prinetti*

Abstract

Gangliosides, characteristic complex lipids present in the external layer of plasma membranes, deeply influence the organization of the membrane as a whole and the function of specific membrane associated proteins due to lipid-lipid and lipid-protein lateral interaction. Here we discuss the basis for the membrane-organizing potential of gangliosides, examples of ganglioside-regulated membrane protein complexes and the mechanisms for the regulation of ganglioside membrane composition.

Introduction

Glycosphingolipids (GSL) are components of all animal cell membranes and, among these, gangliosides, sialic acid containing compounds, are abundant in the plasma membranes of neurons. Ceramide,[1] the hydrophobic backbone of all sphingolipids, is constituted by a long chain amino alcohol, 2*S*,3*R* 2-amino-1,3-dihydroxy-octadec-4-ene (sphingosine), linked to a fatty acid by an amide bond. The hydrophilic head group of glycosphingolipids is an oligosaccharide chain, highly variable due to the sugar structure, content, sequence and connections. Sialic acid[2] is the name that identifies all the derivatives of 5-amino-3,5-dideoxy-D-*glycero*-D-*galacto*-non-2--ulopyranosonic acid, or neuraminic acid. Three main sialic acids are known: the 5-*N*-acetyl-, the 5-*N*-acetyl-9-*O*-acetyl- and the 5-*N*-glycolyl-derivative. Healthy humans have only the first two,[3-6] the 5-*N*-acetyl derivative being 85-90% of the total. The high Ka and the resulting negative charge confer them a strong amphiphilc character.

Due to the high heterogeneity of the both the hydrophibic and hydrophilic portion, glycosphingolipids are a very large family of natural compounds as shown in Table 1, were the main gangliosides from nervous system are reported.

Gangliosides (ganglioside nomenclature is in accordance with the IUPAC-IUBMB recommendations[7]) belong to external leaflet of the plasma membrane[8] and are highly enriched in the pre and postsynaptic membranes of the synaptic terminals. With their hydrophilic oligosaccharide chains, they face the extracellular environment, providing ideal sites of interaction with extracellular molecules, including toxins, matrix components, adhesion molecules and specific receptors and enzymes on the surface of adjacent cells (trans interactions). Moreover, they have proven to be able to laterally interact with proteins belonging to the same membrane, modulating their biological functions (cis interactions).[9-29]

In the last 10 years it has emerged the concept that GSL possess a high potential for the creation of lateral order in biological membranes, contributing to the creation and/or stabilization of membrane

*Corresponding Author: Alessandro Prinetti—Dipartimento di Chimica, Biochimica e Biotecnogie per la Medicina, Via Fratelli Cervi 93, 20090 Segrate (Mi), Italy.
Email: alessandro.prinetti@unimi.it

Sphingolipids as Signaling and Regulatory Molecules, edited by Charles Chalfant and Maurizio Del Poeta.

Table 1. Structures of the main Neu5Ac containing gangliosides in vertebrates

Series	Series Abbreviation	Series Structure	Ganglioside Abbreviation	Ganglioside Structure
Gal	GalCer	Gal-(1-1)-Cer	GM4	Neu5AcGalCer
Lac	LacCer	b-Gal-(1-4)-b-Glc-(1-1)-Cer	GM3	II3Neu5AcLacCer
			GD3	II3(Neu5Ac)2LacCer
			O-acetyl-GD3	II3[Neu5,9Ac2-(2-8)-Neu5Ac]LacCer
Ganglio-3	Gg3Cer	b-GalNAc-(1-4)-b-Gal-(1-4)-b-Glc-(1-1)-Cer	GM2	II3Neu5AcGg3Cer
			GD2	II3(Neu5Ac)2Gg3Cer
Ganglio-4	Gg4Cer	b-Gal-(1-3)-b-GalNAc-(1-4)-b-Gal-(1-4)-b-Glc-(1-1)-Cer	GM1	II3Neu5AcGg4Cer
			GM1b	IV3Neu5AcGg4Cer
			GD1a	IV3Neu5AcII3Neu5AcGg4Cer
			GD1b	II3(Neu5Ac)2Gg4Cer
			GD1b-lactone	II3[Neu5Ac-(2-8,1-9)-Neu5Ac]Gg4Cer
			GT1a	IV3(Neu5Ac)2II3Neu5AcGg4Cer
			GT1b	IV3Neu5AcII3(Neu5Ac)2Gg4Cer
			O-Acetyl-GT1b	IV3Neu5AcII3[Neu5,9Ac2-(2-8)-Neu5Ac]Gg4Cer
			GT1c	II3(Neu5Ac)3Gg4Cer
			GQ1b	IV3(Neu5Ac)2II3(Neu5Ac)2Gg4Cer
			GQ1c	IV3Neu5AcII3(Neu5Ac)3Gg4Cer
			GP1c	IV3(Neu5Ac)2II3(Neu5Ac)3Gg4Cer
			Fuc-GM1	IV2αFucII3Neu5AcGg4Cer

continued on following page

Table 1. Continued

Series	Series Abbreviation	Series Structure	Ganglioside Abbreviation	Ganglioside Structure
			Fuc-GD1b	IV2αFucII3Neu5Ac2Gg4Cer
			GD1α	IV3Neu5AcIII6Neu5AcGg4Cer
			Chol-1α-a	IV3Neu5AcIII6Neu5AcII3Neu5AcGg4Cer
			Chol-1β	III6Neu5AcII3(Neu5Ac)2Gg4Cer
			GT1α	IV3Neu5AcIII6(Neu5Ac)2Gg4Cer
			GQ1α	IV3(Neu5Ac)2III6(Neu5Ac)2Gg4Cer
			Chol-1α-b	IV3Neu5AcIII6Neu5AcII3(Neu5Ac)2Gg4Cer
Ganglio-5	Gg5Cer	b-GalNAc-(1-4)-b-Gal-(1-3)-b-GalNAc-(1-4)-b-Gal-(1-4)-b-Glc-(1-1)-Cer	GalNAc-GM1	II3Neu5AcGg5Cer
			GalNAc-GD1a	IV3Neu5AcII3Neu5AcGg5Cer
Neolacto-4	nLc4Cer	b-Gal-(1-4)-b-GlcNAc-(1-3)- b-Gal-(1-4)-b-Glc-(1-1)-Cer	3′-LM1	IV3nLc4Cer

macro- and microdomains.[30] Lipid membrane microdomains, or lipid rafts, areas in the membrane characterized by a lateral organization dictated by the properties of their lipid components, have been involved in several biological events.[31-39] The coexistence of lipids in different physical phases within the same model membrane was probably the first evidence leading to the concept of lipid rafts. Lipid bilayers at physiological temperature usually exist in a liquid-disordered (*ld*) phase characterized by high fluidity, in which the lipid acyl chains are disordered and highly mobile. Lowering the temperature below the melting point freezes the lipid acyl chains in an ordered gel phase with very limited freedom of movement. Membrane lipids can also exist in a third physical phase, the liquid-ordered (*lo*) phase, in which the acyl chains of lipids are extended and ordered, as in the gel phase, but have higher lateral mobility in the bilayer. The lo phase is stabilized by the presence of cholesterol, that fill the hydrophobic gaps between the phospholipid or glycolipid acyl chains.[40,41] Lipid rafts and GSL-rich microdomains are more ordered than the *ld* phase, being in this regard similar to a *lo* or a metastable gel phase.

Lateral phase separation of complex lipids in phospholipid bilayers can be observed in binary mixtures of diacyl lecithins differing in chain length and/or saturation[42-44] and in ternary mixtures of palmitoyloleyl PC, dioleyl PC and cholesterol. However, GSL strongly differ from glycerolipids for their molecular structure and conformational properties, thus leading to a strong tendency to segregate within phospholipid bilayers (reviewed in ref. 41). Gangliosides display the ability to associate with a pre-existing ordered domain or to segregate in their own domains, that can be distinct from the cholesterol-enriched phase. The segregation of one ganglioside component with respect to the other has been observed in mixed micelles of two gangliosides with the same acyl chain but different hydrophilic headgroups.[45,46] Lateral separation of gangliosides also occurs in one-component as well as in two-component, two-phase phosphatidylcholine bilayers and in phospholipid bilayers in the presence of cholesterol and/or sphingomyelin.[47-55] The formation of separate GM1–enriched and cholesterol-enriched liquid-ordered phases was observed[56] in ternary sphingomyelin/GM1/cholesterol vesicles and in lipid monolayers.[57]

Studies conducted on model membranes led to the following statements. (1) The presence of an amide nitrogen, of a carbonyl oxygen and of a hydroxyl group at the water/lipid interface in the common hydrophobic ceramide backbone confers to sphingolipids to act both as donors and acceptors for the formation of hydrogen bonds[58]; this feature is unique for sphingolipids among complex membrane lipids and the formation of a hydrogen bond network at the water/lipid interface strongly stabilizes the segregation of a rigid segregated phase enriched in sphingolipids. (2) In the case of GSL, the presence of the bulky oligosaccharide hydrophilic headgroup termodinamically favours their segregation within a biological membrane. Phase separation with clustering of GSL in a phospholipid bilayer is a spontaneous process driven by the minimization of the interfacial free energy. The interfacial area increases with the size of the oligosaccharide chain, with a corresponding more pronounced segregation.[59-71] These predictions based on the geometrical parameters of the GSL molecules are experimentally confirmed by the observation that phase separation is present in mixed micelles of two different gangliosides[45,46] with the same fatty acid composition and that the extent of ganglioside phase separation in glycerophospholipid bilayers depend on the surface area occupied by the GSL oligosaccharide chain, that is usually increasing with the number of sugar residues.[50-52] (3) Membrane complex lipids, in particular glycerophospholipids, are highly heterogeneous in their fatty acid composition. The hydrophobic portion of complex lipids is very heterogeneous, but in sphingomyelin and gangliosides (at least in the nervous system), saturated acyl chains like palmitic and stearic acid are the main components[72] (Table 2). The presence of saturated acyl chains (that can be tightly packed with a high degree of order in the hydrophobic core of a bilayer) favours the phase separation of a rigid, liquid-ordered phase. In the case of GSL, for GM1 it has been shown that distribution in the fluid phase of a phospholipid bilayer[54] is inversely correlated with the acyl chain length and directly correlated with the degree of unsaturation.

These data clearly demonstrate that sphingolipids form laterally separated phases characterized by reduced fluidity and hydrocarbon chain mobility. An hypothetical model for sphingolipid

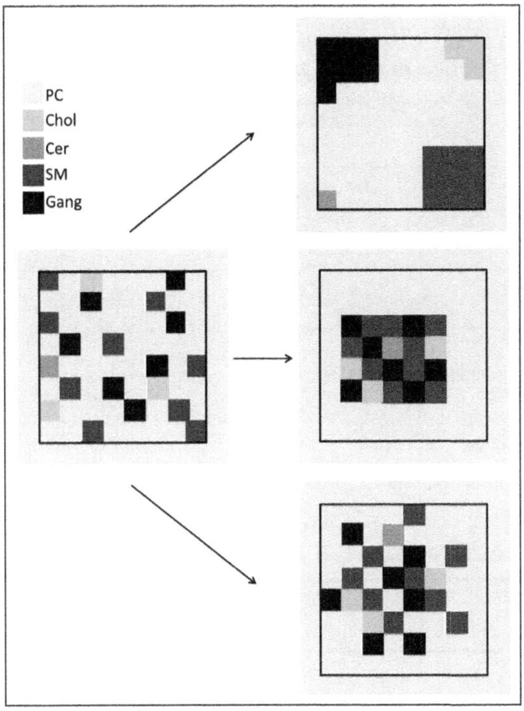

Figure 1. The scheme shows the different metabolic pathways that determine the membrane ganglioside content and pattern. Neo-biosynthesis occurs through the ER and Golgi apparatus the, but shedding of vesicles and monomers, together with a sialidase-sialyltransferase cycle participate to determine the finale ganglioside content and composition of the membrane.

segregation in cell membranes is presented in Figure 1. Segregation of membrane sphingolipids is responsible for the creation of less fluid membrane regions, where membrane-associated proteins can be confined, favouring lateral interactions between sphingolipids and/or proteins that are segregated in the same lipid domain or preventing interactions between proteins that are associated with different domains.

Segregation of Membrane Lipids and Detergent-Resistant Membrane Domains

The interest for lipid membrane domains became very strong when many proteins deputed to cell signaling were found to be preferentially associated with an environment of lipids highly enriched in sphingolipids and cholesterol. The possible role of lipid membrane domains in the transport of GPI-anchored proteins from the Golgi apparatus to the apical plasma membrane of polarized cells originated the now widely used term "lipid rafts".[73]

Originally, the existence of a membrane fraction characterized by a peculiar lipid composition leading to a liquid-ordered or highly organized phase was operationally defined on the basis of the insolubility in aqueous nonionic detergents.[74] Most components of the cell membrane are solubilized by detergents.[75] In contrast, many cellular components are insoluble in nonionic (Triton X-100) under certain experimental conditions. After detergent treatment, the detergent insoluble membrane domain can be separated from the rest of the cell thanks to its relative light density (due to its richness in lipids),[74] using continuous or discontinuous density gradients. Low-density, detergent-insoluble fractions were isolated from a wide variety of cultured cells, including almost all

Table 2. Ratio between the components of sphingolipid-cholesterol enriched membrane fraction in rat cerebellar granule cells differentiated in culture

	Molar Ratio, Gangliosides as 1	Molar Ratio,%	Asymmetry, Layer
Proteins	0.04	0.26	Both, largely inner
Ceramide	0.20	1.32	Outer
Sphingomyelin	1.40	9.20	Outer
Gangliosides	1.00	6.63	Outer
Phosphatidylcholine	7.00	46.45	Outer
Plasmalophosphatidylcholine	0.01	0.07	Outer?
Phosphatidylserine	0.40	2.65	Inner
Phosphatidylethanolamine	0.70	4.64	Largely Inner
Plasmalophosphatidylethanolamine	0.10	0.66	Inner?
Phosphatidylinoisitol	0.10	0.66	Largely Inner
Phosphatidylinoisitolmonophosphate	0.01	0.07	Inner
Phosphatidylinoisitoldiphosphate	0.01	0.07	Inner
Cholesterol	4.10	27.21	Both, largely inner

mammalian cell types[76-91] and tissues[92-98] investigated so far, yeasts[99] and protozoans.[100] In following time, a wide range of different detergents[95,96,101,102] was used and several "detergent-free" procedures for the separation of low-density membrane fractions corresponding to lipid membrane domains were developed.[103,104] When comparatively analyzed, low-density membrane fractions obtained after cell lysis under the dramatically different experimental conditions described above, are very similar but not identical.[77,90,91,105-116] Low-density membrane fractions always contain a highly resistant supramolecular structure possible corresponding to the native core of lipid membrane domains suggesting that the low-density membrane fraction composition correspond to that of physiological lipid membrane domains and that it is not determined by a random rearrangement of cell components induced by the experimental conditions used. Nevertheless, it cannot be excluded that the differences observed by some Authors might be simply due to contingent situations, since the standardization of the experimental procedures is sometime difficult and the overall composition of DRM fractions or the association of specific molecules with it seem to be affected by even tiny modifications of several conditions, including agents used for membrane disruption,[77,93,95,98,101,104,110] mechanical procedures used to obtain or aid membrane solubilization,[110] temperature[74,101,116,117] and ratio between detergent and biological material.[98,118] However, several studies indicate that at least in some cases the differences observed in the composition of low density DRM fractions isolated by different methods might reflect the existence of different levels of order within lipid membrane domains and/or of biochemically distinct lipid membrane domains within the plasma membrane of the same cell. This is particularly clear if the results obtained by the use of different detergents are compared.[74,101,119-122] Thus, differential detergent solubilization might prove to be a powerful tool to study different lipid membrane domain subpopulations.

The existence of lipid membrane domains in natural cell membranes was suggested by the observation that glycosphingolipids at the cell surface form clusters, which have been visualized by immuno-electron microscopy using antiglycosphingolipid antibodies.[79,123-126] Several approaches, relying on more advanced technologies including single-particle tracking or single fluorophore tracking microscopy,[127-132] fluorescence recovery after photobleaching,[133] fluorescence resonance energy transfer[134,135] and atom force microscopy[136] are now available allowing the detection and the study of lipid membrane domains in intact cell membranes.[137,138] However, these techniques are very heterogeneous and and data obtained with different approaches are sometime conflicting. As example, there is no agreement on their average size that ranges from 26 nm to about 2 μm.[129,133,135,136,139-144]

Lipid Membrane Domain Functions

Gangliosides and Lipid Membrane Domains in the Nervous System

Neuronal and glial lipid membrane domains, prepared from cultures of neural cells or nervous tissues[118,145,146] are rich in gangliosides, sphingomyelin, cholesterol and proteins involved in mechanisms of signal transduction that are relevant for neuronal functions.[90,91,94,102,106,108,117,147-168] Lipid membrane domains in the nervous system cells has been involved in neurotrophic factor signaling,[149-152] cell adhesion and migration,[150,158,162] axon guidance, synaptic transmission,[150,157] neuron-glia interactions[163,164] and myelin genesis.[165] The involvement of lipid membrane domains in neuronal and glial signal transduction includes several different ways: 1) receptors and effector proteins permanently resident in lipid membrane domains can be activated, giving rise to signal propagation that involves other components intrinsically present in the lipid membrane domain. Examples are neurotrophin receptors of the trk family, EGFR, PDGFR, p75NTR, GFRα[149-152] and the neural cell adhesion molecule TAG-1.[106,166,169] Src family tyrosine kinases are among the effector signaling proteins that are most commonly engaged in these cases; 2) the activation of membrane receptors is followed by the recruitment to lipid membrane domains of receptors themselves or effector signaling proteins that are not located in lipid membrane domains under basal conditions, or, the activation of receptors that are associated with lipid membrane domains under resting conditions determines their translocation outside lipid membrane domains. Examples of the former are the receptor tyrosine kinase c-Ret, recruited into lipid membrane domains by its GPI-anchored coreceptor GFRα[149,150,152] and the neuronal adhesion receptor NCAM, recruited into lipid membrane domains by cis- or trans- interaction with its membrane bound, GPI-anchored ligand, prion protein.[162]

Both modes implies changes in the reciprocal interactions of lipid membrane domain components. Changes in the lipid and/or protein composition of lipid membrane domains and in the interactions of the lipid (in particular glycosphingolipid) components with specific proteins of functional relevance could thus be very relevant during the process of neuronal adhesion, survival, migration, differentiation and senescence.

Properties of lipid membrane domains from rat cerebellar granule cells at different stages of development in culture are available.[72] The surface occupied by these structures increased during development, with the maximum ganglioside density in fully differentiated neurons. On the other hand, a high content of ceramide was found in the domains of aging neurons. The sphingolipid/glycerophospholipid molar ratio was more than doubled during the initial stage of development, corresponding to axonal sprouting and neurite extension, whereas the cholesterol/glycerophospholipid molar ratio gradually decreased during in vitro differentiation. Phosphorylated phosphoinositides were very scant in the domains of undifferentiated cells and dramatically increased during differentiation and aging.

Src family protein tyrosine kinases, (c-Src, Lyn and Fyn) known to participate to the process of neuronal differentiation, were associated with the lipid membrane domains in a way specific for the type of kinase and for the developmental stage of the cell.[72] Within the lipid membrane domains, ganglioside GM3 has been found closely associated with c-Src and Csk in neuroblastoma Neuro2a cells[90] and GD3 associated with Src-family kinase Lyn and the neural cell adhesion molecule TAG-1 in rat brain[94,169] and cerebellar granule cells. In these cells, a complex lipid environment characterized by the presence of many ganglioside species and other membrane lipids (mainly cholesterol and dipalmitoylphosphatidylcholine) is essential for the interaction with the domain of c-Src, Lyn, Fyn, TAG-1 and prion protein.[72,102,117,148] The presence of Src family nonreceptor tyrosine kinases in lipid membrane domains of neurons is particularly interesting, because many facts indicate that c-Src and other kinases of this family are important in the process of neuronal differentiation and in neuronal function.[170-177] As mentioned above, in neuroblastoma Neuro2a cells, c-Src and Csk, are associated with GM3 ganglioside within lipid membrane domain and neuritogenic concentration of gangliosides are able to induce c-Src activation followed by mitogen-activated protein kinases activation.[90] In these cells, anti-GM3 antibody is also able to induce differentiation.[174] In rat cerebellum and cerebellar neurons, GD3 ganglioside is associated with Lyn and the neural cell adhesion molecule TAG-1 and antibody-mediated cross-linking of TAG-1 or GD3 induce Lyn activation.[94,106,169] Glycosphingolipids were essential for TAG-1-dependent signaling via Lyn and for the maintenance of the differentiated neuronal phenotype, since incubation of cerebellar neurons with the glycosphingolipid-degrading enzyme endoglycoceramidase in the presence of its activator protein reduced the levels of cell surface glycosphingolipids, caused the redistribution of TAG-1 from nondomain membranes to the lipid membrane domain fraction, abolished TAG-mediated Lyn activation and consequent phosphorylation of p80 and induced neurite retraction.[169]

A possible role of lipid membrane domains in the pathogenesis of spontaneous and transmissible neurodegenerative diseases was recently highlighted by the discovery that a number of molecules causally connected to such diseases are associated with these domains. The most prominent examples are represented by the amyloid precursor protein (APP) in Alzheimer's disease and by the prion protein. In both cases, the generation of the aberrant forms of these proteins, which are responsible for the onset of the disease, seems to be localized in the lipid membrane domains and/or dependent from the structure of the domain itself.[178,179]

The Glycosynapse

Altered GM3 ganglioside expression plays a multiple role in the control of tumor cell motility, invasiveness and survival. Adhesion of B16 melanoma cells (expressing high levels of GM3) to endothelial cells, that express LacCer and Gg3, is mediated by GM3-LacCer or GM3-Gg3 interaction and leads to enhanced B16 cell motility and thereby initiates metastasis.[105,180] GM3 is highly expressed in noninvasive, superficial bladder tumors compared with invasive bladder tumors, where the activities of glycosyltransferases responsible for GM3 synthesis were consistently upregulated.[181,182] Enhanced GM3 expression achieved by pharmacological treatment with brefeldin A,[181,183] or the exogenous administration of GM3[182] suppressed the tumorigenic activity and/or the invasive potential of human colonic and bladder tumor cell lines and the stable overexpression of GM3 synthase in a mouse bladder carcinoma cell line reduced cell proliferation, motility and invasion with concomitant increase in the number of apoptotic cells.[184] High expression levels of GM3 with concomitant expression of the tetraspanin CD9 in colorectal[185,186] and bladder[181] cancer cell lines inhibited Matrigel and laminin-5-dependent cell motility.

At the molecular level, GM3 control on the properties of tumor cells requires a complex supermolecular membrane organization that defines highly specialized detergent-insoluble lipid membrane domains. The term "glycosynapse" has been proposed by S. Hakomori[187,188] to generally describe a membrane microdomain involved in carbohydrate-dependent adhesion. Carbohydrate-dependent adhesion in glycosynapse, occurring through GSL-GSL interactions or through GSL-dependent modulation of adhesion protein receptors (such as integrins) leads to signal transduction events reflecting in deep changes in the motility and invasiveness of tumor cells. In the case of GM3-dependent adhesion of melanoma cells, it has been shown that GM3 is closely associated with c-Src, Rho and Ras within glycosphingolipid-enriched membrane domains and binding with Gg3 or anti-GM3 antibody stimulates focal adhesion kinase phosphorylation and c-Src activity.[105] This molecular assembly defines a classically Triton X-100 insoluble GSL-enriched microdomain ("glycosynapse 1"), that can be isolated and separated from a caveolin-containing low-density membrane fraction in B16 cells.[189] A similar association between a sialoglycolipid and c-Src and other related signaling molecules was observed for GM3 also in neuroblastoma cell,[90] for disialylgalactosylgloboside in renal carcinoma cells[190] and for monosialyl-Gb5 in breast carcinoma cells.[191] Tetraspanin CD9 and integrin $\alpha 3$ or $\alpha 5$ also are colocalized within a distinct low-density, Brij 98-insoluble glycolipid-enriched domain ("glycosynapse 3"). The presence of GM3 positively modulated CD9/integrin association. In fact, association between CD9 and integrin in the Chinese hamster ovary mutant cell line ldlD14 (deficient in UDP-Gal-4-epimerase) has been shown by co-immunoprecipitation experiments when cells were grown in the presence of galactose, allowing GM3 synthesis, or supplementing cells with exogenous GM3 Colocalization of CD9, α3 and GM3 in cells was observed in intact ldlD14 cells in the presence but not in the absence of galactose.[192] The formation of α3/CD9/GM3 complexes strongly inhibited the laminin-5-dependent motility in ldlD14 cells. On the other hand, it has been shown that CD9/GM3 complexes are essential for the regulation of integrin-mediated cell adhesion and signal transduction in oncogenic transformation, suggesting a crucial role for GM3 complexed with CD9 and integrin α3β1 or α5β1 in the control of tumor cell motility and invasiveness. v-Jun-transformed mouse and chicken embryo fibroblasts were characterized by lower GM3 levels and down—regulated GM3 synthase mRNA levels respect to the nontransformed counterparts.[193] Reversion of oncogenic phenotype of v-Jun-transformed cells to normal could be achieved by enhanced GM3 synthesis through its gene transfection. When v-Jun-transformed cells were transfected with GM3 synthase expression plasmid, leading to increased GM3 synthase activity and GM3 cellular levels, their ability anchorage-independent growth in agar was strongly reduced. During phenotypic reversion induced by GM3 synthase transfection, the association of CD9/α5β1 complex (shown by co-immunoprecipitation and confocal microscopy experiments) was increased. Remarkably, the N-glycosylated form of β1 integrin was preferentially associated with the complex in GM3 synthase gene transfectants.[193] GM3 levels were 4-5 times higher in the noninvasive KK47 cell line (originated from superficial human bladder cancer) than in the invasive YTS1 human bladder cancer cell line. Knock down of CD9 or pharmacologically achieved GM3 depletion in KK47 cells induces the phenotypic conversion to invasive variants. On the other hand, exogenous GM3 addition induces the phenotypic reversion of the highly invasive and metastatic cell lines YTS1 to low motility variants. The changes in cell motility were strictly correlated with the association of CD9 with α3 integrin. This interaction was higher in nonivasive than in highly invasive cells and was modulated by the cellular levels of GM3: CD9/α3 integrin association was reduced by GM3 depletion in KK47 and conversely enhanced by exogenous GM3 addition in YTS1 cells. GM3 levels in glycosynapse controls not only CD9/α3 integrin association, but also the activation state of c-Src. c-Src is present in higher amount in the glycosynapse fraction in YTS1 cells and it is activated in cells with low GM3 levels and high invasive potential (YTS1 or GM3-depleted KK47). On the other hand, exogenous addition of GM3 to YST1 cells caused Csk traslocation to the detergent-insoluble fraction and consequent inactivation of c-Src, influencing cell motility.[194]

GM3 and EGF Receptor

The function of growth factor receptors can be modulated by gangliosides[13] and lon time ago epidermal growth factor receptor (EGFR) was identified as the target of the inhibitory action of GM3.[195] GM3 inhibited EGFR autophosphorylation without competing with EGF for receptor binding[13,196,197] and without affecting receptor dimerization.[198] The sialyllactose oligosaccharide is essential for ganglioside-receptor interaction and that the substitution with any other sugar negatively affects the binding.[199] However, the molecular basis of this interaction has been only recently fully elucidated, emphasizing the importance of side-by-side carbohydrate-carbohydrate interactions between GM3 oligosaccharide and a N-linked glycan bearing multiple GlcNAc terminal residues on the receptor.[200,201] GM3/EGFR interaction is facilitated by the enrichment of EGFR in classical ganglioside-enriched, cholesterol-sensitive, Triton X-100 insoluble membrane domains.[202,203] However, other GSL- and lipid raft-dependent factors can affect EGFR function. Caveolae and caveolin-1 are involved in the modulation of EGFR signaling[204,205] and EGFR is localized within a caveolin-rich fraction in A431 cells. However, EGFR-containing membrane fragments can be separated from caveolae.[110,206] In a keratinocyte-derived cell line, GM3 overexpression induced a shift of caveolin-1 to EGFR-rich membrane regions, allowing its functional interaction with the EGFR receptor, that caused inhibition of EGFR tyrosine phosphorylation and dimerization.[207] Thus, GM3 influences EGFR signaling by a second distinct molecular mechanism modulating EGFR/caveolin-1 association. Moreover, GM3 negatively regulates as well the ligand-independent cross-talk of EGFR with integrin receptor signaling, disrupting when accumulated in cultured cells, the interaction of integrin β1 subunit with EGFR.[208]

GM3, Caveolae and the Regulation of Insulin Receptor and PDGF Receptor

Insulin receptors (IR) are present in detergent-resistant membranes from normal adipocytes[209] and localized in caveolae in intact cells,[210] where the β-subunit of IR interacts with caveolin-1 through a binding motif recognizing the scaffold domain of caveolin-1.[211] IR can form distinct complexes with caveolin-1 and GM3 within lipid membrane domains.[212] The interaction between GM3 and IR is direct and specific and was abolished in IR mutants where the lysine residue at 944 was replaced with arginine, valine, serine, or glutamine, suggesting that an electrostatic interaction between the negatively charged sialyllactose chain of GM3 and the positively charged amino group of lysine 944, located in close proximity to the transmembrane domain sequence of IR, is essential for the formation of the GM3/IR complex. In 3T3-L1 adipocytes, the induction of insulin resistance by treatment with TNFα was accompanied by the upregulation of GM3 synthase, leading to an increase of cellular GM3,[210,213] that accumulated in detergent-resistant membranes. In insulin resistance, the association of IR with GM3 was increased, while its association with caveolin-1 was decreased, indicating that the excess amount of GM3 in lipid membrane domains leads to the displacement of IR from the complex with caveolin-1, thus suggesting that the regulation of IR/caveolin-1 by GM3 could be responsible for the changes in insulin response in adipocytes.[212]

A similar regulatory mechanism has been recently observed for PDGFR.[214] Overexpression of the N-terminal domain of PAG, caused the accumulation of GM1 at the cell surface, with the consequent displacement of PDGFR from caveolin-rich fractions and caveolae, without altering the caveolar distribution of caveolin-1. The same redistribution of PDGFR has been observed after incubation of cells with exogenous GM1. Increased GM1 cellular levels lead to the displacement of another growth factor receptor, PDGF, from caveolae,[214] negatively regulating Src mitogenic signaling. However, in this case it is not known whether the formation of a PDGFR-GM1 complex is required for its uncoupling from caveolae. Since caveolin-1 can direct bind sphingolipids, including GM1, in this case it cannot be excluded that GM1 forms a complex with caveolin-1, or that an enrichment in GM1 inside the caveola induces a deep reorganization of caveolar membrane, thus excluding PDGFR from caveolae.

The Regulation of Glycosphingolipid Composition of the Plasma Membranes

Changes in the glycosphingolipid composition of the plasma membrane would predictably lead to very important biological consequences, thus all mechanisms possibly contributing to these changes have a high functional significance. Changes in the activities of enzymes of the biosynthetic pathway residing in Golgi apparatus have been associated with the changes in GSL expression that are associated with neoplastic transformation or neuronal differentiation. However, both catabolic and biosynthetic enzymes for glycosphingolipids have been found associated with the plasma membranes, thus local mechanisms for the regulation of the glycosphingolipid composition of plasma membranes or restricted plasma membrane areas could be very important. The membrane-bound sialidase Neu3 has been identified and cloned[215-217] and its *trans* activity in modifying the structure of gangliosides of adjacent cells has been proven.[218-220] Moreover, the presence of sialyltransferase activities at the cell surface has been also reported.[221-225] Thus, sialylation/desialylation cycles might be very important mechanisms responsible for rapid and possibly transient changes of the plasma membrane ganglioside content and pattern. The presence of other active glycosylhydrolases, β-glucosidase, β-galactosidase and β-hexosaminidase,[220,226] in the plasma membrane has been demonstrated, implying that local hydrolysis of glycosphingolipids at the cell surface might represent a general mechanism for the control of glycosphingolipid composition. Moreover, gangliosides, as well as the other

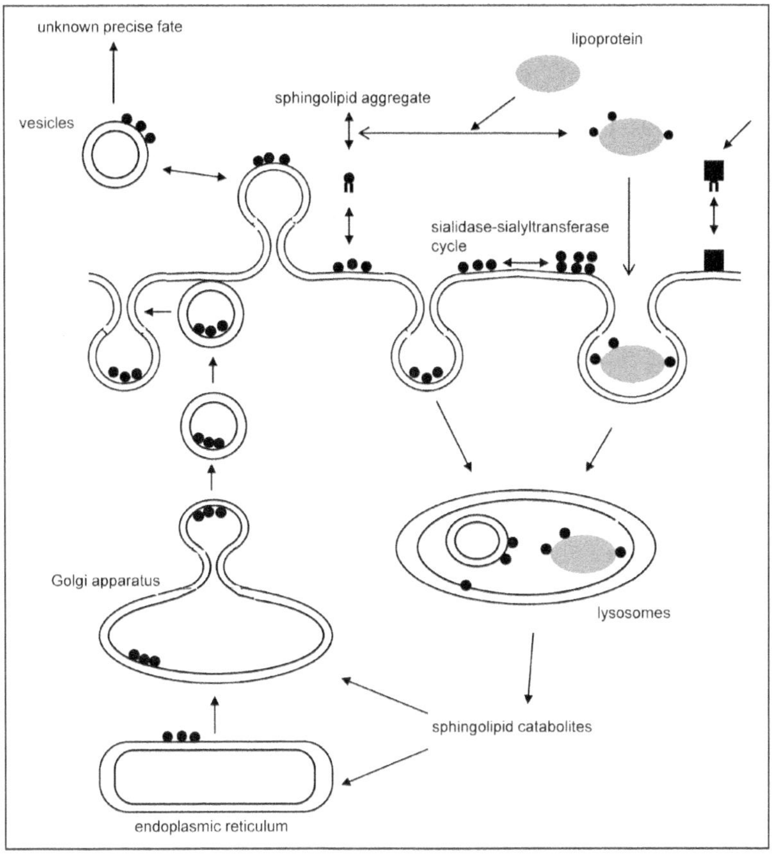

Figure 2. The figure shows the possible distribution of lipids belonging to the outer layer of a lipid raft according to the pattern reported in Table 2. According to the well known concept that the outer layer requires a larger surface with respect to the inner layer, that sphingolipids and phosphatidylcholine are components of the outer layer and all together cover about 60% of the total lipid raft content, the larger portion of cholesterol must be located in the inner layer. This is in agreement with published data.[230,231] Protein are not shown, but the larger portion should be associated to the inner layer, the main proteins recognised in lipid rafts, at least from a quantitative point of view, being those of cytoskeleton.[166,232,233] In the panels only the elements inserted in the square belong to the lipid rafts. The left panel shows a random distribution of phosphatidylcholine, sphingomyelin, ceramide, gangliosides and a few molecules of cholesterol. These molecule must be organised as separated phases. We present three different possibility in the right portion of figure. The upper panel shows five different area each one covered by the same species. The central panel shows all the sphingolipids in a single phase. The lower panel shows a phase where some phosphatidylcholine is distributed within sphingolipids. This lipid distribution has been experimentally proved.[52] Specific sphingolipid-protein interactions can modify this phase separation promoting subdomain organization.

sphingolipids, can be released from the cell surface in different forms, including shedding vesicles,[227-229] whose controlled release from specific sphingo(glyco)lipid-enriched membrane areas, could represent a further way to modify the lipid membrane domain composition and organization. The mechanisms responsible for he control of plasma membrane glycosphingolipid composition are depicted in Figure 2.

Conclusion

In this chapter, we describe the properties of gangliosides that underlie their roles as organizers of membrane organization and function. The potential for gangliosides in determining the properties of specific membrane areas characterized by a selective enrichment in these lipids as well as their ability to engage specific lateral interactions with membrane proteins have been discussed with the aid of examples covering the recent literature. In addition, we discussed the possibility that several metabolic pathways could affect the local ganglioside composition of specific membrane districts, thus regulating signaling pathways originated within these membrane regions.

References

1. Karlsson KA. On the chemistry and occurrence of sphingolipid long-chain bases. Chem Phys Lipids 1970; 5(1):6-43.
2. Schauer R. Chemistry, metabolism and biological functions of sialic acids. Adv Carbohydr Chem Biochem 1982; 40:131-234.
3. Yu RK, Ledeen RW. Gangliosides of human, bovine and rabbit plasma. J Lipid Res 1972; 13(5):680-6.
4. Kamerling JP, Vliegenthart JF. Identification of O-cetylated N-acylneuraminic acids by mass spectrometry. Carbohydr Res 1975; 41:7-17.
5. Ghidoni R, Sonnino S, Tettamanti G et al. Isolation and characterization of a trisialoganglioside from mouse brain, containing 9-O-acetyl-N-acetylneuraminic acid. J Biol Chem 1980; 255(14):6990-5.
6. Sonnino S, Ghidoni R, Chigorno V et al. Chemistry of gangliosides carrying o-acetylated sialic acid. Adv Exp Med Biol 1982; 152:55-69.
7. IUPAC-IUBMB JCoBN. Nomenclature of glycolipids. Carbohydr Res 1998; 312:167-175.
8. Feizi T. Demonstration by monoclonal antibodies that carbohydrate structures of glycoproteins and glycolipids are onco-developmental antigens. Nature 1985; 314(6006):53-7.
9. Caputto R, Maccioni AH, Caputto BL. Activation of deoxycholate solubilized adenosine triphosphatase by ganglioside and asialoganglioside preparations. Biochem Biophys Res Commun 1977; 74(3):1046-52.
10. Partington CR, Daly JW. Effect of gangliosides on adenylate cyclase activity in rat cerebral cortical membranes. Mol Pharmacol 1979; 15(3):484-91.
11. Davis CW, Daly JW. Activation of rat cerebral cortical 3',5'-cyclic nucleotide phosphodiesterase activity by gangliosides. Mol Pharmacol 1980; 17(2):206-11.
12. Leon A, Facci L, Toffano G et al. Activation of (Na+, K+)-ATPase by nanomolar concentrations of GM1 ganglioside. J Neurochem 1981; 37(2):350-7.
13. Bremer EG, Hakomori S, Bowen-Pope DF et al. Ganglioside-mediated modulation of cell growth, growth factor binding and receptor phosphorylation. J Biol Chem 1984; 259(11):6818-25.
14. Goldenring JR, Otis LC, Yu RK et al. Calcium/ganglioside-dependent protein kinase activity in rat brain membrane. J Neurochem 1985; 44(4):1229-34.
15. Kim JY, Goldenring JR, DeLorenzo RJ et al. Gangliosides inhibit phospholipid-sensitive Ca^{2+}-dependent kinase phosphorylation of rat myelin basic proteins. J Neurosci Res 1986; 15(2):159-66.
16. Kreutter D, Kim JY, Goldenring JR et al. Regulation of protein kinase C activity by gangliosides. J Biol Chem 1987; 262(4):1633-7.
17. Chan KF. Ganglioside-modulated protein phosphorylation. Partial purification and characterization of a ganglioside-inhibited protein kinase in brain. J Biol Chem 1988; 263(1):568-74.
18. Chan KF. Ganglioside-modulated protein phosphorylation in muscle. Activation of phosphorylase b kinase by gangliosides. J Biol Chem 1989; 264(31):18632-7.
19. Yates AJ, Walters JD, Wood CL et al. Ganglioside modulation of cyclic AMP-dependent protein kinase and cyclic nucleotide phosphodiesterase in vitro. J Neurochem 1989; 53(1):162-7.
20. Yates AJ, Rampersaud A. Sphingolipids as receptor modulators. An overview. Ann NY Acad Sci 1998; 845:57-71.
21. Bassi R, Chigorno V, Fiorilli A et al. Exogenous gangliosides GD1b and GD1b-lactone, stably associated to rat brain P2 subcellular fraction, modulate differently the process of protein phosphorylation. J Neurochem 1991; 57(4):1207-11.
22. Rybak S, Ginzburg I, Yavin E. Gangliosides stimulate neurite outgrowth and induce tubulin mRNA accumulation in neural cells. Biochem Biophys Res Commun 1983; 116(3):974-80.
23. Morgan JI, Seifert W. Growth factors and gangliosides: a possible new perspective in neuronal growth control. J Supramol Struct 1979; 10(2):111-24.

24. Roisen FJ, Bartfeld H, Nagele R et al. Ganglioside stimulation of axonal sprouting in vitro. Science 1981; 214(4520):577-8.
25. Tsuji S, Arita M, Nagai Y. GQ1b, a bioactive ganglioside that exhibits novel nerve growth factor (NGF)-like activities in the two neuroblastoma cell lines. J Biochem (Tokyo) 1983; 94(1):303-6.
26. Facci L, Leon A, Toffano G et al. Promotion of neuritogenesis in mouse neuroblastoma cells by exogenous gangliosides. Relationship between the effect and the cell association of ganglioside GM1. J Neurochem 1984; 42(2):299-305.
27. Glebov OO, Nichols BJ. Distribution of lipid raft markers in live cells. Biochem Soc Trans 2004; 32(Pt 5):673-5.
28. Glebov OO, Nichols BJ. Lipid raft proteins have a random distribution during localized activation of the T-cell receptor. Nat Cell Biol 2004; 6(3):238-43.
29. Lin J, Shaw AS. Getting downstream without a Raft. Cell 2005; 121(6):815-6.
30. Sonnino S, Mauri L, Chigorno V et al. Gangliosides as components of lipid membrane domains. Glycobiology 2006.
31. Lajoie P, Partridge EA, Guay G et al. Plasma membrane domain organization regulates EGFR signaling in tumor cells. J Cell Biol 2007; 179(2):341-56.
32. Guirland C, Zheng JQ. Membrane lipid rafts and their role in axon guidance. Adv Exp Med Biol 2007; 621:144-55.
33. Benarroch EE. Lipid rafts, protein scaffolds and neurologic disease. Neurology 2007; 69(16):1635-9.
34. Hanzal-Bayer MF, Hancock JF. Lipid rafts and membrane traffic. FEBS Lett 2007; 581(11):2098-104.
35. Riethmuller J, Riehle A, Grassme H et al. Membrane rafts in host-pathogen interactions. Biochim Biophys Acta 2006; 1758(12):2139-47.
36. Debruin LS, Harauz G. White matter rafting—membrane microdomains in myelin. Neurochem Res 2007; 32(2):213-28.
37. Delacour D, Jacob R. Apical protein transport. Cell Mol Life Sci 2006; 63(21):2491-505.
38. Taylor DR, Hooper NM. The prion protein and lipid rafts. Mol Membr Biol 2006; 23(1):89-99.
39. Manes S, Viola A. Lipid rafts in lymphocyte activation and migration. Mol Membr Biol 2006; 23(1):59-69.
40. Brown DA, London E. Structure and function of sphingolipid- and cholesterol-rich membrane rafts. J Biol Chem 2000; 275(23):17221-4.
41. Sonnino S, Prinetti A, Mauri L et al. Dynamic and structural properties of sphingolipids as driving forces for the formation of membrane domains. Chem Rev 2006; 106(6):2111-25.
42. Rock P, Allietta M, Young WW Jr. et al. Ganglioside GM1 and asialo-GM1 at low concentration are preferentially incorporated into the gel phase in two-component, two-phase phosphatidylcholine bilayers. Biochemistry 1991; 30(1):19-25.
43. Grant CW, Wu SH, McConnell HM. Lateral phase separations in binary lipid mixtures: correlation between spin label and freeze-fracture electron microscopic studies. Biochim Biophys Acta 1974; 363(2):151-8.
44. Knoll W, Schmidt G, Rotzer H et al. Lateral order in binary lipid alloys and its coupling to membrane functions. Chem Phys Lipids 1991; 57(2-3):363-74.
45. Cantu L, Corti M, Sonnino S et al. Evidence for spontaneous segregation phenomena in mixed micelles of gangliosides. Chem Phys Lipids 1990; 55(3):223-9.
46. Cantù L, Corti M, Casellato R et al. Aggregation properties of GD1b, II3Neu5Ac2GgOse4Cer and of GD1b-lactone, II3(alpha-Neu5Ac-(2——8, 1——9)-alpha-Neu5Ac)GgOse4Cer, in aqueous solution. Chem Phys Lipids 1991; 60:111-118.
47. Sharom FJ, Grant CW. A model for ganglioside behaviour in cell membranes. Biochim Biophys Acta 1978; 507(2):280-93.
48. Bertoli E, Masserini M, Sonnino S et al. Electron paramagnetic resonance studies on the fluidity and surface dynamics of egg phosphatidylcholine vesicles containing gangliosides. Biochim Biophys Acta 1981; 647(2):196-202.
49. Rock P, Allietta M, Young WWJ et al. Ganglioside GM1 and asialo-GM1 at low concentration are preferentially incorporated into the gel phase in two-component, two-phase phosphatidylcholine bilayers. Biochemistry 1990; 29:8484-8490.
50. Goins B, Masserini M, Barisas BG et al. Lateral diffusion of ganglioside GM1 in phospholipid bilayer membranes. Biophys J 1986; 49(4):849-56.
51. Masserini M, Palestini P, Venerando B et al. Interactions of proteins with ganglioside-enriched microdomains on the membrane: the lateral phase separation of molecular species of GD1a ganglioside, having homogeneous long-chain base composition, is recognized by Vibrio cholerae sialidase. Biochemistry 1988; 27(20):7973-8.

52. Masserini M, Palestini P, Freire E. Influence of glycolipid oligosaccharide and long-chain base composition on the thermotropic properties of dipalmitoylphosphatidylcholine large unilamellar vesicles containing gangliosides. Biochemistry 1989; 28(12):5029-34.
53. Terzaghi A, Tettamanti G, Masserini M. Interaction of glycosphingolipids and glycoproteins: thermotropic properties of model membranes containing GM1 ganglioside and glycophorin. Biochemistry 1993; 32(37):9722-5.
54. Palestini P, Allietta M, Sonnino S et al. Gel phase preference of ganglioside GM1 at low concentration in two-component, two-phase phosphatidylcholine bilayers depends upon the ceramide moiety. Biochim Biophys Acta 1995; 1235(2):221-30.
55. Palestini P, Masserini M, Tettamanti G. Exposure to galactose oxidase of GM1 ganglioside molecular species embedded into phospholipid vesicles. FEBS Lett 1994; 350(2-3):219-22.
56. Ferraretto A, Pitto M, Palestini P et al. Lipid domains in the membrane: thermotropic properties of sphingomyelin vesicles containing GM1 ganglioside and cholesterol. Biochemistry 1997; 36(30):9232-6.
57. Dietrich C, Volovyk ZN, Levi M et al. Partitioning of Thy-1, GM1 and cross-linked phospholipid analogs into lipid rafts reconstituted in supported model membrane monolayers. Proc Natl Acad Sci USA 2001; 98(19):10642-7.
58. Pascher I. Molecular arrangements in sphingolipids. Conformation and hydrogen bonding of ceramide and their implication on membrane stability and permeability. Biochim Biophys Acta 1976; 455(2):433-51.
59. Sonnino S, Cantu L, Corti M et al. Aggregative properties of gangliosides in solution. Chem Phys Lipids 1994; 71(1):21-45.
60. Acquotti D, Poppe L, Dabrowski J et al. Three-dimensional structure of the oligosaccaride chain of GM1 ganglioside revealed by a distance-mapping procedure: A rotating and laboratory frame nuclear overhauser enhancement investigation of native glycolipid in dimethyl sulfoxide and in water-dodecylphosphocholine solutions. J Am Chem Soc 1990; 112:7772-7778.
61. Brocca P, Cantu L, Sonnino S. Aggregation properties of semisynthetic GD1a ganglioside (IV3Neu5AcII3Neu5AcGgOse4Cer) containing an acetyl group as acyl moiety. Chem Phys Lipids 1995; 77(1):41-9.
62. Sonnino S, Cantu L, Corti M et al. Aggregation properties of semisynthetic GM1 ganglioside (II3Neu5AcGgOse4Cer) containing an acetyl group as acyl moiety. Chem Phys Lipids 1990; 56(1):49-57.
63. Sonnino S, Cantu L, Acquotti D et al. Aggregation properties of GM3 ganglioside (II3Neu5AcLacCer) in aqueous solutions. Chem Phys Lipids 1990; 52(3-4):231-41.
64. Poppe L, van Halbeek H, Acquotti D et al. Carbohydrate dynamics at a micellar surface: GD1a headgroup transformations revealed by NMR spectroscopy. Biophys J 1994; 66(5):1642-52.
65. Brocca P, Berthault P, Sonnino S. Conformation of the oligosaccharide chain of G(M1) ganglioside in a carbohydrate-enriched surface. Biophys J 1998; 74(1):309-18.
66. Siebert HC, Reuter G, Schauer R et al. Solution conformations of GM3 gangliosides containing different sialic acid residues as revealed by NOE-based distance mapping, molecular mechanics and molecular dynamics calculations. Biochemistry 1992; 31(30):6962-71.
67. Brocca P, Acquotti D, Sonnino S. Nuclear Overhauser effect investigation on GM1 ganglioside containing N-glycolyl-neuraminic acid (II3Neu5GcGgOse4Cer). Glycoconj J 1996; 13(1):57-62.
68. Scarsdale JN, Prestegard JH, Yu RK. NMR and computational studies of interactions between remote residues in gangliosides. Biochemistry 1990; 29(42):9843-55.
69. Levery SB. 1H-NMR study of GM2 ganglioside: evidence that an interresidue amide-carboxyl hydrogen bond contributes to stabilization of a preferred conformation. Glycoconj J 1991; 8(6):484-92.
70. Acquotti D, Fronza G, Ragg E et al. Three dimensional structure of GD1b and GD1b-monolactone gangliosides in dimethylsulphoxide: a nuclear Overhauser effect investigation supported by molecular dynamics calculations. Chem Phys Lipids 1991; 59(2):107-25.
71. Acquotti D, Cantu L, Ragg E et al. Geometrical and conformational properties of ganglioside GalNAc-GD1a, IV4GalNAcIV3Neu5AcII3Neu5AcGgOse4Cer. Eur J Biochem 1994; 225(1):271-88.
72. Prinetti A, Chigorno V, Prioni S et al. Changes in the lipid turnover, composition and organization, as sphingolipid-enriched membrane domains, in rat cerebellar granule cells developing in vitro. J Biol Chem 2001; 276(24):21136-45.
73. Simons K, Ikonen E. Functional rafts in cell membranes. Nature 1997; 387(6633):569-72.
74. Brown DA, Rose JK. Sorting of GPI-anchored proteins to glycolipid-enriched membrane subdomains during transport to the apical cell surface. Cell 1992; 68(3):533-44.
75. Marchesi VT, Andrews EP. Glycoproteins: isolation from cellmembranes with lithium diiodosalicylate. Science 1971; 174(15):1247-8.

76. Chigorno V, Palestini P, Sciannamblo M et al. Evidence that ganglioside enriched domains are distinct from caveolae in MDCK II and human fibroblast cells in culture. Eur J Biochem 2000; 267(13):4187-97.
77. Sargiacomo M, Sudol M, Tang Z et al. Signal transducing molecules and glycosyl-phosphatidylinositol-linked proteins form a caveolin-rich insoluble complex in MDCK cells. J Cell Biol 1993; 122(4):789-807.
78. Zurzolo C, van't Hof W, van Meer G et al. VIP21/caveolin, glycosphingolipid clusters and the sorting of glycosylphosphatidylinositol-anchored proteins in epithelial cells. Embo J 1994; 13(1):42-53.
79. Sorice M, Parolini I, Sansolini T et al. Evidence for the existence of ganglioside-enriched plasma membrane domains in human peripheral lymphocytes. J Lipid Res 1997; 38(5):969-80.
80. Fra AM, Williamson E, Simons K et al. Detergent-insoluble glycolipid microdomains in lymphocytes in the absence of caveolae. J Biol Chem 1994; 269(49):30745-8.
81. Iwabuchi K, Nagaoka I. Lactosylceramide-enriched glycosphingolipid signaling domain mediates superoxide generation from human neutrophils. Blood 2002; 100(4):1454-64.
82. Waheed AA, Shimada Y, Heijnen HF et al. Selective binding of perfringolysin O derivative to cholesterol-rich membrane microdomains (rafts). Proc Natl Acad Sci USA 2001; 98(9):4926-31.
83. Samuel BU, Mohandas N, Harrison T et al. The role of cholesterol and glycosylphosphatidylinositol-anchored proteins of erythrocyte rafts in regulating raft protein content and malarial infection. J Biol Chem 2001; 276(31):29319-29.
84. Mendez AJ, Lin G, Wade DP et al. Membrane lipid domains distinct from cholesterol/sphingomyelin-rich rafts are involved in the ABCA1-mediated lipid secretory pathway. J Biol Chem 2001; 276(5):3158-66.
85. Ledesma MD, Simons K, Dotti CG. Neuronal polarity: essential role of protein-lipid complexes in axonal sorting. Proc Natl Acad Sci USA 1998; 95(7):3966-71.
86. Kasahara K, Sanai Y. Functional roles of glycosphingolipids in signal transduction via lipid rafts. Glycoconj J 2000; 17(3-4):153-62.
87. Shogomori H, Futerman AH. Cholera toxin is found in detergent-insoluble rafts/domains at the cell surface of hippocampal neurons but is internalized via a raft-independent mechanism. J Biol Chem 2001; 276(12):9182-8.
88. Vey M, Pilkuhn S, Wille H et al. Subcellular colocalization of the cellular and scrapie prion proteins in caveolae-like membranous domains. Proc Natl Acad Sci 1996; 93(25):14945-14949.
89. Naslavsky N, Stein R, Yanai A et al. Characterization of detergent-insoluble complexes containing the cellular prion protein and its scrapie isoform. J Biol Chem 1997; 272(10):6324-31.
90. Prinetti A, Iwabuchi K, Hakomori S. Glycosphingolipid-enriched signaling domain in mouse neuroblastoma Neuro2a cells. Mechanism of ganglioside-dependent neuritogenesis. J Biol Chem 1999; 274(30):20916-24.
91. Prinetti A, Chigorno V, Tettamanti G et al. Sphingolipid-enriched membrane domains from rat cerebellar granule cells differentiated in culture. A compositional study. J Biol Chem 2000; 275(16):11658-65.
92. Lisanti MP, Scherer PE, Vidugiriene J et al. Characterization of caveolin-rich membrane domains isolated from an endothelial-rich source: implications for human disease. J Cell Biol 1994; 126(1):111-26.
93. Chang WJ, Ying YS, Rothberg KG et al. Purification and characterization of smooth muscle cell caveolae. J Cell Biol 1994; 126(1):127-38.
94. Kasahara K, Watanabe Y, Yamamoto T et al. Association of Src family tyrosine kinase Lyn with ganglioside GD3 in rat brain. Possible regulation of Lyn by glycosphingolipid in caveolae-like domains. J Biol Chem 1997; 272(47):29947-53.
95. Madore N, Smith KL, Graham CH et al. Functionally different GPI proteins are organized in different domains on the neuronal surface. EMBO J 1999; 18(24):6917-6926.
96. Parkin ET, Turner AJ, Hooper NM. Amyloid precursor protein, although partially detergent-insoluble in mouse cerebral cortex, behaves as an atypical lipid raft protein. Biochem J 1999; 344(Pt 1):23-30.
97. Kawabuchi M, Satomi Y, Takao T et al. Transmembrane phosphoprotein Cbp regulates the activities of Src-family tyrosine kinases. Nature 2000; 404(6781):999-1003.
98. Parkin ET, Turner AJ, Hooper NM. Differential effects of glycosphingolipids on the detergent-insolubility of the glycosylphosphatidylinositol-anchored membrane dipeptidase. Biochem J 2001; 358(Pt 1):209-16.
99. Kubler E, Dohlman HG, Lisanti MP. Identification of Triton X-100 insoluble membrane domains in the yeast Saccharomyces cerevisiae. Lipid requirements for targeting of heterotrimeric G-protein subunits. J Biol Chem 1996; 271(51):32975-80.
100. Zhang X, Thompson GA, Jr. An apparent association between glycosylphosphatidylinositol-anchored proteins and a sphingolipid in Tetrahymena mimbres. Biochem J 1997; 323(Pt 1):197-206.
101. Taylor CM, Coetzee T, Pfeiffer SE. Detergent-insoluble glycosphingolipid/cholesterol microdomains of the myelin membrane. J Neurochem 2002; 81(5):993-1004.

102. Loberto N, Prioni S, Bettiga A et al. The membrane environment of endogenous cellular prion protein in primary rat cerebellar neurons. J Neurochem 2005; 95(3):771-83.
103. Song KS, Li S, Okamoto T et al. Copurification and direct interaction of Ras with caveolin, an integral membrane protein of caveolae microdomains. Detergent-free purification of caveolae microdomains. J Biol Chem 1996; 271(16):9690-7.
104. Smart EJ, Ying YS, Mineo C et al. A detergent-free method for purifying caveolae membrane from tissue culture cells. Proc Natl Acad Sci USA 1995; 92(22):10104-8.
105. Iwabuchi K, Yamamura S, Prinetti A et al. GM3-enriched microdomain involved in cell adhesion and signal transduction through carbohydrate-carbohydrate interaction in mouse melanoma B16 cells. J Biol Chem 1998; 273(15):9130-8.
106. Kasahara K, Watanabe K, Takeuchi K et al. Involvement of gangliosides in glycosylphosphatidylinositol-anchored neuronal cell adhesion molecule TAG-1 signaling in lipid rafts. J Biol Chem 2000; 275(44):34701-9.
107. Liu P, Ying Y, Anderson RG. Platelet-derived growth factor activates mitogen-activated protein kinase in isolated caveolae. Proc Natl Acad Sci USA 1997; 94(25):13666-70.
108. Wu C, Butz S, Ying Y et al. Tyrosine kinase receptors concentrated in caveolae-like domains from neuronal plasma membrane. J Biol Chem 1997; 272(6):3554-9.
109. Couet J, Sargiacomo M, Lisanti MP. Interaction of a receptor tyrosine kinase, EGF-R, with caveolins. Caveolin binding negatively regulates tyrosine and serine/threonine kinase activities. J Biol Chem 1997; 272(48):30429-38.
110. Waugh MG, Lawson D, Hsuan JJ. Epidermal growth factor receptor activation is localized within low-buoyant density, noncaveolar membrane domains. Biochem J 1999; 337(Pt 3):591-7.
111. Bilderback TR, Gazula VR, Lisanti MP et al. Caveolin interacts with Trk A and p75(NTR) and regulates neurotrophin signaling pathways. J Biol Chem 1999; 274(1):257-63.
112. Silva WI, Maldonado HM, Lisanti MP et al. Identification of caveolae and caveolin in C6 glioma cells. Int J Dev Neurosci 1999; 17(7):705-14.
113. Bravo-Zehnder M, Orio P, Norambuena A et al. Apical sorting of a voltage- and Ca^{2+}-activated K+ channel alpha-subunit in Madin-Darby canine kidney cells is independent of N-glycosylation. Proc Natl Acad Sci USA 2000; 97(24):13114-9.
114. Waugh MG, Lawson D, Tan SK et al. Phosphatidylinositol 4-phosphate synthesis in immunoisolated caveolae-like vesicles and low buoyant density noncaveolar membranes. J Biol Chem 1998; 273(27):17115-21.
115. Eckert GP, Igbavboa U, Muller WE et al. Lipid rafts of purified mouse brain synaptosomes prepared with or without detergent reveal different lipid and protein domains. Brain Res 2003; 962(1-2):144-50.
116. Kim KB, Kim SI, Choo HJ et al. Two-dimensional electrophoretic analysis reveals that lipid rafts are intact at physiological temperature. Proteomics 2004; 4(11):3527-35.
117. Prinetti A, Prioni S, Chigorno V et al. Immunoseparation of sphingolipid-enriched membrane domains enriched in Src family protein tyrosine kinases and in the neuronal adhesion molecule TAG-1 by anti-GD3 ganglioside monoclonal antibody. J Neurochem 2001; 78(5):1162-7.
118. Scandroglio F, Venkata JK, Loberto N et al. Lipid content of brain, brain membrane lipid domains and neurons from acid sphingomyelinase deficient mice. J Neurochem 2008; 107(2):329-38.
119. Fiedler K, Kobayashi T, Kurzchalia TV et al. Glycosphingolipid-enriched, detergent-insoluble complexes in protein sorting in epithelial cells. Biochemistry 1993; 32(25):6365-73.
120. Kim T, Pfeiffer SE. Myelin glycosphingolipid/cholesterol-enriched microdomains selectively sequester the noncompact myelin proteins CNP and MOG. J Neurocytol 1999; 28(4-5):281-93.
121. Simons M, Kramer EM, Thiele C et al. Assembly of myelin by association of proteolipid protein with cholesterol- and galactosylceramide-rich membrane domains. J Cell Biol 2000; 151(1):143-54.
122. Roper K, Corbeil D, Huttner WB. Retention of prominin in microvilli reveals distinct cholesterol-based lipid micro-domains in the apical plasma membrane. Nat Cell Biol 2000; 2(9):582-92.
123. Tillack TW, Allietta M, Moran RE et al. Localization of globoside and Forssman glycolipids on erythrocyte membranes. Biochim Biophys Acta 1983; 733(1):15-24.
124. Rahmann H, Rosner H, Kortje KH et al. Ca(2+)-ganglioside-interaction in neuronal differentiation and development. Prog Brain Res 1994; 101:127-45.
125. Hakomori S, Handa K, Iwabuchi K et al. New insights in glycosphingolipid function: "glycosignaling domain," a cell surface assembly of glycosphingolipids with signal transducer molecules, involved in cell adhesion coupled with signaling. Glycobiology 1998; 8(10):xi-xix.
126. Iwabuchi K, Prinetti A, Sonnino S et al. Involvement of very long fatty acid-containing lactosylceramide in lactosylceramide-mediated superoxide generation and migration in neutrophils. Glycoconj J 2008; 25(4):357-74.
127. Saxton MJ, Jacobson K. Single-particle tracking: applications to membrane dynamics. Annu Rev Biophys Biomol Struct 1997; 26:373-99.

128. Jacobson K, Sheets ED, Simson R. Revisiting the fluid mosaic model of membranes. Science 1995; 268(5216):1441-2.
129. Sheets ED, Lee GM, Simson R et al. Transient confinement of a glycosylphosphatidylinositol-anchored protein in the plasma membrane. Biochemistry 1997; 36(41):12449-58.
130. Chen Y, Lagerholm BC, Yang B et al. Methods to measure the lateral diffusion of membrane lipids and proteins. Methods 2006; 39(2):147-53.
131. Jacobson K, Mouritsen OG, Anderson RG. Lipid rafts: at a crossroad between cell biology and physics. Nat Cell Biol 2007; 9(1):7-14.
132. Jacobson K, Dietrich C. Looking at lipid rafts? Trends Cell Biol 1999; 9(3):87-91.
133. Varma R, Mayor S. GPI-anchored proteins are organized in submicron domains at the cell surface. Nature 1998; 394(6695):798-801.
134. Rao M, Mayor S. Use of Forster's resonance energy transfer microscopy to study lipid rafts. Biochim Biophys Acta 2005; 1746(3):221-33.
135. Pralle A, Keller P, Florin EL et al. Sphingolipid-cholesterol rafts diffuse as small entities in the plasma membrane of mammalian cells. J Cell Biol 2000; 148(5):997-1008.
136. Poole K, Meder D, Simons K et al. The effect of raft lipid depletion on microvilli formation in MDCK cells, visualized by atomic force microscopy. FEBS Lett 2004; 565(1-3):53-8.
137. Ishitsuka R, Sato SB, Kobayashi T. Imaging lipid rafts. J Biochem (Tokyo) 2005; 137(3):249-54.
138. Lagerholm BC, Weinreb GE, Jacobson K et al. Detecting microdomains in intact cell membranes. Annu Rev Phys Chem 2005; 56:309-36.
139. Schutz GJ, Kada G, Pastushenko VP et al. Properties of lipid microdomains in a muscle cell membrane visualized by single molecule microscopy. Embo J 2000; 19(5):892-901.
140. Wilson BS, Pfeiffer JR, Oliver JM. Observing FcepsilonRI signaling from the inside of the mast cell membrane. J Cell Biol 2000; 149(5):1131-42.
141. Brugger B, Graham C, Leibrecht I et al. The membrane domains occupied by glycosylphosphatidylinositol-anchored prion protein and Thy-1 differ in lipid composition. J Biol Chem 2004; 279(9):7530-6.
142. Sharma DK, Brown JC, Choudhury A et al. Selective stimulation of caveolar endocytosis by glycosphingolipids and cholesterol 10.1091/mbc.E04-03-0189. Mol Biol Cell 2004; 15(7):3114-3122.
143. Douglass AD, Vale RD. Single-molecule microscopy reveals plasma membrane microdomains created by protein-protein networks that exclude or trap signaling molecules in T-cells. Cell 2005; 121(6):937-50.
144. Lommerse PH, Spaink HP, Schmidt T. In vivo plasma membrane organization: results of biophysical approaches. Biochim Biophys Acta 2004; 1664:119-131.
145. Maekawa S, Iino S, Miyata S. Molecular characterization of the detergent-insoluble cholesterol-rich membrane microdomain (raft) of the central nervous system. Biochim Biophys Acta 2003; 1610(2):261-70.
146. Suzuki T. Lipid rafts at postsynaptic sites: distribution, function and linkage to postsynaptic density. Neurosci Res 2002; 44(1):1-9.
147. Chini B, Parenti M. G-protein coupled receptors in lipid rafts and caveolae: how, when and why do they go there? J Mol Endocrinol 2004; 32(2):325-38.
148. Prinetti A, Marano N, Prioni S et al. Association of Src-family protein tyrosine kinases with sphingolipids in rat cerebellar granule cells differentiated in culture. Glycoconj J 2000; 17(3-4):223-32.
149. Paratcha G, Ibanez CF. Lipid rafts and the control of neurotrophic factor signaling in the nervous system: variations on a theme. Curr Opin Neurobiol 2002; 12(5):542-9.
150. Tsui-Pierchala BA, Encinas M, Milbrandt J et al. Lipid rafts in neuronal signaling and function. Trends Neurosci 2002; 25(8):412-7.
151. Nagappan G, Lu B. Activity-dependent modulation of the BDNF receptor TrkB: mechanisms and implications. Trends Neurosci 2005; 28(9):464-71.
152. Saarma M. GDNF recruits the signaling crew into lipid rafts. Trends Neurosci 2001; 24(8):427-9.
153. Yang XL, Xiong WC, Mei L. Lipid rafts in neuregulin signaling at synapses. Life Sci 2004; 75(21):2495-504.
154. Gauthier LR, Robbins SM. Ephrin signaling: One raft to rule them all? One raft to sort them? One raft to spread their call and in signaling bind them? Life Sci 2003; 74(2-3):207-16.
155. Becher A, McIlhinney RA. Consequences of lipid raft association on G-protein-coupled receptor function. Biochem Soc Symp 2005; 72:151-64.
156. Barnett-Norris J, Lynch D, Reggio PH. Lipids, lipid rafts and caveolae: Their importance for GPCR signaling and their centrality to the endocannabinoid system. Life Sciences Targeted Lipidomics: Endocannabinoids and other Endolipid Modulators 2005; 77(14):1625-1639.
157. Tooze SA, Martens GJ, Huttner WB. Secretory granule biogenesis: rafting to the SNARE. Trends Cell Biol 2001; 11(3):116-22.

158. Decker L, Baron W, Ffrench-Constant C. Lipid rafts: microenvironments for integrin-growth factor interactions in neural development. Biochem Soc Trans 2004; 32(Pt 3):426-30.
159. Mineo C, James GL, Smart EJ et al. Localization of epidermal growth factor-stimulated Ras/Raf-1 interaction to caveolae membrane. J Biol Chem 1996; 271(20):11930-11935.
160. Bilderback TR, Grigsby RJ, Dobrowsky RT. Association of p75(NTR) with caveolin and localization of neurotrophin-induced sphingomyelin hydrolysis to caveolae. J Biol Chem 1997; 272(16):10922-10927.
161. Masserini M, Palestini P, Pitto M. Glycolipid-enriched caveolae and caveolae-like domains in the nervous system. J Neurochem 1999; 73(1):1-11.
162. Santuccione A, Sytnyk V, Leshchyns'ka I et al. Prion protein recruits its neuronal receptor NCAM to lipid rafts to activate $p59^{fyn}$ and to enhance neurite outgrowth. J Cell Biol 2005; 169(2):341-354.
163. McKerracher L. Ganglioside rafts as MAG receptors that mediate blockade of axon growth. Proc Natl Acad Sci USA 2002; 99(12):7811-3.
164. Vyas AA, Patel HV, Fromholt SE et al. Gangliosides are functional nerve cell ligands for myelin-associated glycoprotein (MAG), an inhibitor of nerve regeneration. Proc Natl Acad Sci USA 2002; 99(12):8412-7.
165. Boggs JM, Wang H, Gao W et al. A glycosynapse in myelin? Glycoconj J 2004; 21(3-4):97-110.
166. Loberto N, Prioni S, Prinetti A et al. The adhesion protein TAG-1 has a ganglioside environment in the sphingolipid-enriched membrane domains of neuronal cells in culture. J Neurochem 2003; 85(1):224-33.
167. Rivaroli A, Prioni S, Loberto N et al. Reorganization of prion protein membrane environment during low potassium-induced apoptosis in primary rat cerebellar neurons. J Neurochem 2007; 103(5):1954-67.
168. Pierchala BA, Milbrandt J, Johnson EM, Jr. Glial cell line-derived neurotrophic factor-dependent recruitment of Ret into lipid rafts enhances signaling by partitioning Ret from proteasome-dependent degradation. J Neurosci 2006; 26(10):2777-87.
169. Kasahara K, Watanabe K, Kozutsumi Y et al. Association of GPI-anchored protein TAG-1 with src-family kinase Lyn in lipid rafts of cerebellar granule cells. Neurochem Res 2002; 27(7-8):823-9.
170. Lynch SA, Brugge JS, Levine JM. Induction of altered c-src product during neural differentiation of embryonal carcinoma cells. Science 1986; 234(4778):873-6.
171. Mellstrom K, Bjelfman C, Hammerling U et al. Expression of c-src in cultured human neuroblastoma and small-cell lung carcinoma cell lines correlates with neurocrine differentiation. Mol Cell Biol 1987; 7(12):4178-84.
172. Bjelfman C, Meyerson G, Cartwright CA et al. Early activation of endogenous pp60src kinase activity during neuronal differentiation of cultured human neuroblastoma cells. Mol Cell Biol 1990; 10(1):361-70.
173. den Hertog J, Pals CE, Peppelenbosch MP et al. Receptor protein tyrosine phosphatase alpha activates pp60c-src and is involved in neuronal differentiation. Embo J 1993; 12(10):3789-98.
174. Chakraborty M, Anderson GM, Chakraborty A et al. Accumulation of high level of pp60c-srcN is an early event during GM3-antibody mediated differentiation of neuro-2a neuroblastoma cells. Brain Res 1993; 625(2):197-202.
175. Cartwright CA, Simantov R, Kaplan PL et al. Alterations in pp60c-src accompany differentiation of neurons from rat embryo striatum. Mol Cell Biol 1987; 7(5):1830-40.
176. Yu XM, Salter MW. Src, a molecular switch governing gain control of synaptic transmission mediated by N-methyl-D-aspartate receptors. Proc Natl Acad Sci USA 1999; 96(14):7697-704.
177. Foster-Barber A, Bishop JM. Src interacts with dynamin and synapsin in neuronal cells. Proc Natl Acad Sci USA 1998; 95(8):4673-7.
178. Hooper N. Roles of proteolysis and lipid rafts in the processing of the amyloid precursor protein and prion protein. Biochem Soc Trans 2005; 33:335-8.
179. Kazlauskaite J, Pinheiro TJ. Aggregation and fibrillization of prions in lipid membranes. Biochem Soc Symp 2005; 72:211-22.
180. Kojima N, Shiota M, Sadahira Y et al. Cell adhesion in a dynamic flow system as compared to static system. Glycosphingolipid-glycosphingolipid interaction in the dynamic system predominates over lectin- or integrin-based mechanisms in adhesion of B16 melanoma cells to non-activated endothelial cells. J Biol Chem 1992; 267(24):17264-17270.
181. Satoh M, Ito A, Nojiri H et al. Enhanced GM3 expression, associated with decreased invasiveness, is induced by brefeldin A in bladder cancer cells. Int J Oncol 2001; 19(4):723-731.
182. Kawamura S, Ohyama C, Watanabe R et al. Glycolipid composition in bladder tumor: A crucial role of GM3 ganglioside in tumor invasion. International Journal of Cancer 2001; 94(3):343-347.
183. Nojiri H, Yamana H, Shirouzu G et al. Glycotherapy for cancer: remodeling of ganglioside pattern as an effective approach for cancer therapy. Cancer Detec Preven 2002; 26(2):114-120.

184. Watanabe R, Ohyama C, Aoki H et al. Ganglioside GM3 overexpression induces apoptosis and reduces malignant potential in murine bladder cancer. Cancer Res 2002; 62(13):3850-3854.
185. Ono M, Handa K, Sonnino S et al. GM3 ganglioside inhibits CD9-facilitated haptotactic cell motility: coexpression of GM3 and CD9 is essential in the downregulation of tumor cell motility and malignancy. Biochemistry 2001; 40(21):6414-6421.
186. Ono M, Handa K, Withers DA et al. Motility inhibition and apoptosis are induced by metastasis-suppressing gene product CD82 and its analogue CD9, with concurrent glycosylation. Cancer Res 1999; 59(10):2335-2339.
187. Hakomori S. Glycosylation defining cancer malignancy: new wine in an old bottle. Proc Natl Acad Sci USA 2002; 99(16):10231-3.
188. Hakomori S. Glycosynapses: microdomains controlling carbohydrate-dependent cell adhesion and signaling. An Acad Bras Cienc 2004; 76(3):553-572.
189. Iwabuchi K, Handa K, Hakomori S. Separation of "glycosphingolipid signaling domain" from caveolin-containing membrane fraction in mouse melanoma B16 cells and its role in cell adhesion coupled with signaling. J Biol Chem 1998; 273(50):33766-73.
190. Satoh M, Nejad FM, Ohtani H et al. Association of renal cell carcinoma antigen, disialylgalactosylgloboside, with c-Src and Rho A in clustered domains at the surface membrane. Int J Oncol 2000; 16(3):529-536.
191. Steelant WF, Kawakami Y, Ito A et al. Monosialyl-Gb5 organized with cSrc and FAK in GEM of human breast carcinoma MCF-7 cells defines their invasive properties. FEBS Letters 2002; 531(1):93-98.
192. Kawakami Y, Kawakami K, Steelant WF et al. Tetraspanin CD9 is a "proteolipid," and its interaction with alpha 3 integrin in microdomain is promoted by GM3 ganglioside, leading to inhibition of laminin-5-dependent cell motility. J Biol Chem 2002; 277(37):34349-58.
193. Miura Y, Kainuma M, Jiang H et al. Reversion of the Jun-induced oncogenic phenotype by enhanced synthesis of sialosyllactosylceramide (GM3 ganglioside). Proc Natl Acad Sci USA 2004; 101(46):16204-16209.
194. Mitsuzuka K, Handa K, Satoh M et al. A specific microdomain ("glycosynapse 3") controls phenotypic conversion and reversion of bladder cancer cells through GM3-mediated interaction of alpha3beta1 integrin with CD9. J Biol Chem 2005; 280(42):35545-53.
195. Bremer EG, Schlessinger J, Hakomori S. Ganglioside-mediated modulation of cell growth. Specific effects of GM3 on tyrosine phosphorylation of the epidermal growth factor receptor. J Biol Chem 1986; 261(5):2434-40.
196. Hanai N, Nores GA, MacLeod C et al. Ganglioside-mediated modulation of cell growth. Specific effects of GM3 and lyso-GM3 in tyrosine phosphorylation of the epidermal growth factor receptor. J Biol Chem 1988; 263(22):10915-21.
197. Song WX, Vacca MF, Welti R et al. Effects of gangliosides GM3 and De-N-acetyl GM3 on epidermal growth factor receptor kinase activity and cell growth. J Biol Chem 1991; 266(16):10174-81.
198. Zhou Q, Hakomori S, Kitamura K et al. GM3 directly inhibits tyrosine phosphorylation and de-N-acetyl-GM3 directly enhances serine phosphorylation of epidermal growth factor receptor, independently of receptor-receptor interaction. J Biol Chem 1994; 269(3):1959-65.
199. Miljan EA, Meuillet EJ, Mania-Farnell B et al. Interaction of the extracellular domain of the epidermal growth factor receptor with gangliosides. J Biol Chem 2002; 277(12):10108-13.
200. Yoon SJ, Nakayama K, Hikita T et al. Epidermal growth factor receptor tyrosine kinase is modulated by GM3 interaction with N-linked GlcNAc termini of the receptor. Proc Natl Acad Sci USA 2006; 103(50):18987-91.
201. Yoon SJ, Nakayama K, Takahashi N et al. Interaction of N-linked glycans, having multivalent GlcNAc termini, with GM3 ganglioside. Glycoconj J 2006; 23(9):639-49.
202. Ringerike T, Blystad FD, Levy FO et al. Cholesterol is important in control of EGF receptor kinase activity but EGF receptors are not concentrated in caveolae. J Cell Sci 2002; 115(6):1331-1340.
203. Roepstorff K, Thomsen P, Sandvig K et al. Sequestration of epidermal growth factor receptors in non-caveolar lipid rafts inhibits ligand binding. J Biol Chem 2002; 277(21):18954-60.
204. Pike LJ. Growth factor receptors, lipid rafts and caveolae: an evolving story. Biochim Biophys Acta 2005; 1746(3):260-73.
205. Pike LJ, Han X, Gross RW. Epidermal growth factor receptors are localized to lipid rafts that contain a balance of inner and outer leaflet lipids: a shotgun lipidomics study. J Biol Chem 2005; 280(29):26796-804.
206. Waugh MG, Minogue S, Anderson JS et al. Signalling and noncaveolar rafts. Biochem Soc Trans 2001; 29(Pt 4):509-511.
207. Wang X-Q, Sun P, Paller AS. Ganglioside induces caveolin-1 redistribution and interaction with the epidermal growth factor receptor. J Biol Chem 2002; 277(49):47028-47034.

208. Wang XQ, Sun P, Paller AS. Ganglioside GM3 blocks the activation of epidermal growth factor receptor induced by integrin at specific tyrosine sites. J Biol Chem 2003; 278(49):48770-8.
209. Gustavsson J, Parpal S, Karlsson M et al. Localization of the insulin receptor in caveolae of adipocyte plasma membrane. Faseb J 1999; 13(14):1961-71.
210. Kabayama K, Sato T, Kitamura F et al. TNFalpha-induced insulin resistance in adipocytes as a membrane microdomain disorder: involvement of ganglioside GM3. Glycobiology 2005; 15(1):21-9.
211. Couet J, Li S, Okamoto T et al. Identification of peptide and protein ligands for the caveolin-scaffolding domain. Implications for the interaction of caveolin with caveolae-associated proteins. J Biol Chem 1997; 272(10):6525-33.
212. Kabayama K, Sato T, Saito K et al. Dissociation of the insulin receptor and caveolin-1 complex by ganglioside GM3 in the state of insulin resistance. Proc Natl Acad Sci USA 2007; 104(34):13678-83.
213. Tagami S, Inokuchi Ji J, Kabayama K et al. Ganglioside GM3 participates in the pathological conditions of insulin resistance. J Biol Chem 2002; 277(5):3085-92.
214. Veracini L, Simon V, Richard V et al. The Csk-binding protein PAG regulates PDGF-induced Src mitogenic signaling via GM1. J Cell Biol 2008; 182(3):603-14.
215. Wada T, Yoshikawa Y, Tokuyama S et al. Cloning, expression and chromosomal mapping of a human ganglioside sialidase. Biochem Biophys Res Commun 1999; 261(1):21-7.
216. Miyagi T, Wada T, Iwamatsu A et al. Molecular cloning and characterization of a plasma membrane-associated sialidase specific for gangliosides. J Biol Chem 1999; 274(8):5004-11.
217. Hasegawa T, Yamaguchi K, Wada T et al. Molecular cloning of mouse ganglioside sialidase and its increased expression in neuro2a cell differentiation. J Biol Chem 2000; 275(19):14778.
218. Kalka D, von Reitzenstein C, Kopitz J et al. The plasma membrane ganglioside sialidase cofractionates with markers of lipid rafts. Biochem Biophys Res Commun 2001; 283(4):989-93.
219. Papini N, Anastasia L, Tringali C et al. The plasma membrane-associated sialidase MmNEU3 modifies the ganglioside pattern of adjacent cells supporting its involvement in cell-to-cell interactions. J Biol Chem 2004; 279(17):16989-95.
220. Valaperta R, Chigorno V, Basso L et al. Plasma membrane production of ceramide from ganglioside GM3 in human fibroblasts. Faseb J 2006; 20(8):1227-9.
221. Preti A, Fiorilli A, Lombardo A et al. Occurrence of sialyltransferase activity in the synaptosomal membranes prepared from calf brain cortex. J Neurochem 1980; 35(2):281-96.
222. Matsui Y, Lombard D, Massarelli R et al. Surface glycosyltransferase activities during development of neuronal cell cultures. J Neurochem 1986; 46(1):144-50.
223. Durrie R, Saito M, Rosenberg A. Endogenous glycosphingolipid acceptor specificity of sialosyltransferase systems in intact Golgi membranes, synaptosomes and synaptic plasma membranes from rat brain. Biochemistry 1988; 27(10):3759-64.
224. Durrie R, Rosenberg A. Anabolic sialosylation of gangliosides in situ in rat brain cortical slices. J Lipid Res 1989; 30(8):1259-66.
225. Iwamori M, Iwamori Y. Changes in the glycolipid composition and characteristic activation of GM3 synthase in the thymus of mouse after administration of dexamethasone. Glycoconj J 2005; 22(3):119-26.
226. Mencarelli S, Cavalieri C, Magini A et al. Identification of plasma membrane associated mature beta-hexosaminidase A, active towards GM2 ganglioside, in human fibroblasts. FEBS Lett 2005; 579(25):5501-6.
227. Kong Y, Li R, Ladisch S. Natural forms of shed tumor gangliosides. Biochim Biophys Acta 1998; 1394(1):43-56.
228. Deng W, Li R, Ladisch S. Influence of cellular ganglioside depletion on tumor formation. J Natl Cancer Inst 2000; 92(11):912-7.
229. Chigorno V, Giannotta C, Ottico E et al. Sphingolipid uptake by cultured cells: complex aggregates of cell sphingolipids with serum proteins and lipoproteins are rapidly catabolized. J Biol Chem 2005; 280(4):2668-75.
230. Schroeder F, Nemecz G, Wood WG et al. Transmembrane distribution of sterol in the human erythrocyte. Biochim Biophys Acta 1991; 1066(2):183-92.
231. Igbavboa U, Avdulov NA, Chochina SV et al. Transbilayer distribution of cholesterol is modified in brain synaptic plasma membranes of knockout mice deficient in the low-density lipoprotein receptor, apolipoprotein E, or both proteins. J Neurochem 1997; 69(4):1661-7.
232. Palestini P, Pitto M, Tedeschi G et al. Tubulin anchoring to glycolipid-enriched, detergent-resistant domains of the neuronal plasma membrane. J Biol Chem 2000; 275(14):9978-85.
233. Viola A, Gupta N. Tether and trap: regulation of membrane-raft dynamics by actin-binding proteins. Nat Rev Immunol 2007; 7(11):889-96.

CHAPTER 13

Cancer Treatment Strategies Targeting Sphingolipid Metabolism

Babak Oskouian and Julie D. Saba*

Abstract

Ceramide and sphingosine-1-phosphate are related sphingolipid metabolites that can be generated through a de novo biosynthetic route or derived from the recycling of membrane sphingomyelin. Both these lipids regulate cellular responses to stress, with generally opposing effects. Sphingosine-1-phosphate functions as a growth and survival factor, acting as a ligand for a family of G protein-coupled receptors, whereas ceramide activates intrinsic and extrinsic apoptotic pathways through receptor-independent mechanisms. A growing body of evidence has implicated ceramide, sphingosine-1-phosphate and the genes involved in their synthesis, catabolism and signaling in various aspects of oncogenesis, cancer progression and drug- and radiation resistance. This may be explained in part by the finding that both lipids impinge upon the PI3K/AKT pathway, which represses apoptosis and autophagy. In addition, sphingolipids influence cell cycle progression, telomerase function, cell migration and stem cell biology. Considering the central role of ceramide in mediating physiological as well as pharmacologically stimulated apoptosis, ceramide can be considered a tumor-suppressor lipid. In contrast, sphingosine-1-phosphate can be considered a tumor-promoting lipid, and the enzyme responsible for its synthesis functions as an oncogene. Not surprisingly, genetic mutations that result in reduced ceramide generation, increased sphingosine-1-phosphate synthesis or which reduce steady state ceramide levels and increase sphingosine-1-phosphate levels have been identified as mechanisms of tumor progression and drug resistance in cancer cells. Pharmacological tools for modulating sphingolipid pathways are being developed and represent novel therapeutic strategies for the treatment of cancer.

Introduction

Over the past twenty years, lipid metabolites have become recognized for their participation in membrane functions and signaling events that control a wide array of cellular activities. Two major sphingolipid metabolites, ceramide and sphingosine-1-phosphate (S1P), are involved in signaling pathways that regulate cell proliferation, apoptosis, motility, differentiation, stress responses, protein synthesis, carbohydrate metabolism, innate and adaptive immunity and angiogenesis.[1] Ceramide and S1P exert opposing effects on cell survival, ceramide being pro-apoptotic and S1P generally promoting cell survival. In relation to their influence over cell fate, these two lipids and the proteins that mediate their signaling functions and metabolism have been implicated in various aspects of tumor biology and chemo- and radio-resistance. This review will summarize the main steps of sphingolipid metabolism and our current understanding of the interactions between ceramide, S1P and the cellular pathways that control cell fate. Recent in vitro and in vivo evidence supporting the roles these lipids play in oncogenesis, tumor progression and cancer therapeutics

*Corresponding Author: Julie D. Saba—Children's Hospital Oakland Research Institute, 5700 Martin Luther King Jr. Way, Oakland, California, 94609-1673, USA. Email: jsaba@chori.org

Sphingolipids as Signaling and Regulatory Molecules, edited by Charles Chalfant and Maurizio Del Poeta.

will be discussed. Readers interested in other aspects of sphingolipid metabolism and signaling are referred to several excellent recent review articles.[2-7]

Sphingolipid Metabolism

Ceramides are a class of hydrophobic molecules that contain a fatty acid moiety linked to sphingosine or a related long chain base.[8] Ceramides are the basic constituents of higher order sphingolipids and also serve as bioactive intermediates that promote cell death. Ceramide is generated via a de novo biosynthetic pathway that is initiated by a condensation reaction between serine and palmitoyl CoA catalyzed by the essential enzyme serine palmitoyltransferase (SPT), resulting in the formation of 3-oxo-sphinganine (Fig. 1). The NADPH-dependent enzyme 3-oxo-sphinganine reductase converts 3-oxo-sphinganine to dihydrosphingosine. The latter is converted into dihydroceramide via an acylation reaction catalyzed by dihydroceramide synthase, encoded by the CerS genes, formerly known as LASS genes. Oxidation of dihydroceramide by dihydroceramide desaturase generates ceramide. Ceramide may be converted to sphingomyelin by sphingomyelin synthase, which catalyzes the transfer of phosphocholine from phosphatidylcholine to ceramide.[9] Alternatively, ceramide may be modified via the addition of carbohydrate groups at the carbon 1 position by glucosylceramide synthase or galactosyl ceramide synthase and subsequent biochemical steps to generate glycosphingolipids. Ceramide can also be generated through a recycling pathway known as the "sphingomyelin cycle".[10] This process involves the hydrolysis of plasma membrane sphingomyelin, yielding ceramide and

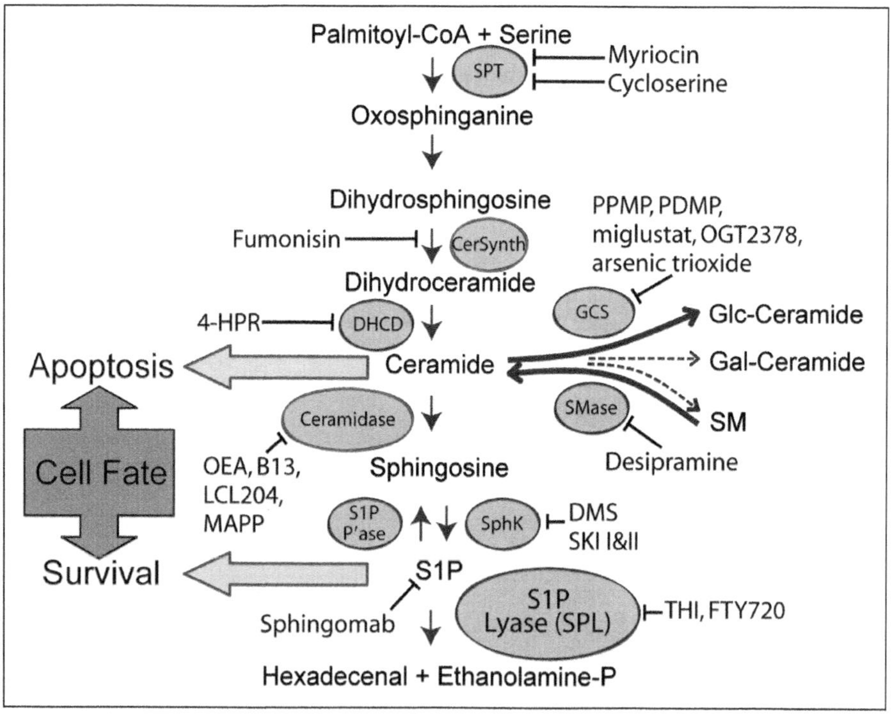

Figure 1. Sphingolipid metabolism. The de novo pathway for generation and degradation of sphingolipids is depicted along with the enzymes that catalyze each step and their inhibitors. SPT, serine-palmitoyl transferase; DHCD, dihydroceramide desaturase; SphK, sphigosine kinase; S1P P'ase, S1P phosphatase; GCS, glucosyl ceramide synthase; PPMP, 1-phenyl-2-palmitoylamino-3-morpholino-1-propanol; PDMP, l-threo-1-phenyl-2-decanoylamino-3-morpholino-1-propanol; DMS, dimethyl sphingosine; 4-HPR, N-(4-hydroxyphenyl) retinamide; OEA, oleylethanolamide; MAPP, 2-(N-myristoylamino)-1-phenyl-1-propanol.

phosphorylcholine. Sphingomyelinases are specific phospholipases that carry out this reaction.[11] Three different categories of sphingomyelinases have been characterized, based on their distinct pH optima. Similar hydrolytic reactions may generate ceramide from other higher order sphingolipids.[12] Ceramide generated through de novo biosynthesis or sphingolipid recycling can be deacylated by ceramidases, giving rise to the long chain base sphingosine. Sphingosine is a bioactive lipid which inhibits cell cycle progression and/or promotes apoptosis, depending upon the cell type.[13] Sphingosine is phosphorylated through the actions of sphingosine kinases (SK), encoded by *SK1* and *SK2*, producing the mitogenic signaling lipid, S1P.[14] S1P can be recycled back to sphingosine by lipid phosphatases.[15,16] Alternatively, it can be irreversibly degraded by the enzyme S1P lyase (SPL), producing ethanolamine phosphate and hexadecenal.[17]

Ceramide Generated via Different Biochemical Routes Can Induce Apoptosis

Neutral sphingomyelinase (N-SMase) is found in the plasma membrane, cytoplasm and the cell nucleus.[18,19] N-SMase has been implicated in cell death mediated by many stimuli including serum starvation,[20] hypoxia,[21] ethanol,[22] nitric oxide (NO),[23] reactive oxygen species (ROS) and chemotherapeutic agents such as daunorubicin[24] and etoposide.[25] In glioma cells treated with etoposide, activation of N-SMase was shown to occur downstream of p53-mediated ROS formation.[26] N-SMase is activated by arachidonate,[27] IL-1β and TNFαR1,[28] iPLA2β,[29] radiation[30,31] and other cellular stimuli but is inhibited by glutathione, which is cytoprotective.[32] These studies establish N-SMase as a central responder to cytokine- and stress-induced cell death.[33]

Acidic sphingomyelinase (A-SMase) is a lysosomal enzyme. An inherited deficiency of A-SMase activity is responsible for the lipid storage disorder Niemann–Pick disease (types A and B).[34] It was shown as early as 1996 that lymphoblasts from Niemann–Pick patients are resistant to apoptosis induced by ionizing radiation.[35,36] This observation served as a prelude to subsequent knockout mouse experiments revealing the significant role that A-SMase plays in mediating radiation-induced apoptosis through the generation of intracellular ceramide. A-SMase is upregulated by signaling events linked to induction of apoptosis including TNFα receptor ligation,[37] UV-irradiation,[38] and ischemia.[39] A-SMase can also be secreted to the extracellular space.[40] In a small study of patients receiving spatially fractionated high dose radiation, secretory SMase and ceramide increased in serum in response to radiation treatment and elevation of these markers correlated with clinical response.[41] Although there have been no reports of A-SMase mutations in human cancers, an A-SMase-like protein was found to be elevated in bladder cancer.[42]

Alkaline sphingomyelinase activity is localized in the epithelial cells of the small intestines.[43] The enzyme hydrolyses sphingomyelin in both intestinal lumen and the mucosal membrane in a bile salt-dependent manner.[44] There is currently no direct evidence demonstrating that alkaline sphingomyelinase promotes cell death. However, the enzyme was recently found to be downregulated in colon cancer and aberrant forms of the enzyme with reduced activity have been associated with liver tumors, raising the possibility that ceramide generated in the enteric system by this enzyme could have an influence on gastrointestinal tumor biology.[45]

The identification of the sphingomyelin cycle represented the first causal connection between ceramide and cell fate. Later studies demonstrated that ceramide generated through the de novo pathway can also promote apoptosis. A recent study by Bose and colleagues showed that the anticancer drug daunorubicin induces apoptosis in P388 and U937 leukemia cells by increasing ceramide levels.[46] The elevation in ceramide and apoptosis could be abrogated by addition of fumonisin B1, an inhibitor of ceramide synthase, indicating that de novo ceramide synthesis and not sphingomyelin hydrolysis, was required for daunorubicin-mediated cell death. Another study demonstrated that treatment of HL60 leukemia cells with H_2O_2 increased the intracellular levels of ceramide and sphinganine.[47] Neither the levels of sphingomyelin nor sphingomyelinase activity were affected by H_2O_2 treatment, which indirectly suggested that H_2O_2 treatment may be activating the de novo pathway. Numerous additional examples of de novo ceramide synthesis contributing to apoptosis have been reported.[48-52]

Ceramide as a Mediator of Cell Death by Chemopreventive Agents

In addition to its role in promoting radiation- and chemotherapy-induced apoptosis, ceramide has been implicated in mediating the effects of chemopreventive agents including curcumin, resveratrol and nonsteroidal anti-inflammatory drugs. A number of reports have ascribed an anticarcinogenic effect and pro-apoptotic functions to curcumin, a polyphenol found in the spice turmeric. Moussavi et al[53] recently demonstrated that curcumin induces ceramide generation and apoptosis in colon cancer cells. As apoptosis could be partially inhibited by myriocin, an inhibitor of SPT, it was concluded that ceramide generation via the de novo pathway was responsible.[53] Resveratrol is a natural antioxidant found in grapes and wine that may potentially act as an anticancer agent.[54] Resveratrol affects key targets involved in the regulation of cell cycle progression and apoptosis. For example, DU145 cells resistant to radiation became sensitive upon pretreatment with resveratrol. The sensitization effect of resveratrol was due to the induction of de novo synthesis of ceramide, since myriosin blocked the cell death caused by resveratrol/radiation.[55] Nonsteroidal anti-inflammatory drugs (NSAIDs) comprise an important class of chemopreventive agents currently in use to protect patients prone to colon cancer. Studies indicate that NSAIDs stimulate ceramide production, which may contribute to their anticancer effects.[56]

Ceramide Influences Both the Intrinsic and Extrinsic Apoptotic Pathways

Apoptosis can be elicited in cells through two major routes, an extrinsic route initiated through activation of cell surface receptors such as TNFαR1 and Fas and an intrinsic route that is activated by permeabilization of the outer mitochondrial membrane via pro-apoptotic members of the Bcl-2 protein family (Fig. 2). Both pathways converge on caspases, which are responsible for executing subsequent steps in the apoptotic program.

Numerous studies have shown that ceramide promotes apoptosis through the mitochondrial pathway, in part due to its effects on Bcl-2 family proteins.[57-62] The Bcl-2 family can be subdivided into anti-apoptotic members such as Bcl-2 and Bcl-x(L) and pro-apoptotic family members such as Bcl-x(s), Bax, Bak, Bok, Bad, Bik, Bid, Puma, Bim, Bmf and Noxa. The posttranscriptional processing of the Bcl-x gene determines whether anti-apoptotic Bcl-x(L) or pro-apoptotic Bcl-x(s) will be expressed.[63] The Bcl-x splice variant, Bcl-x(s), is produced by activation of an upstream 5′ splice site within Bcl-x exon 2. Importantly, ceramide was shown to regulate the 5′ splice site selection within the Bcl-x exon 2.[64] Treatment of A549 lung adenocarcinoma cells with cell-permeable ceramide and/or agents that induce the synthesis of de novo ceramide downregulated Bcl-x(L) mRNA and protein levels and concomitantly increased Bcl-x(s) mRNA and protein.[64] This effect correlated with increased sensitivity of A549 cells to daunorubicin. Furthermore, A549 cells resistant to chemotherapeutic agents and cell-permeable ceramides demonstrated increased Bcl-x(L) levels. Others have reported that UV light-induced Bax activation and ensuing cytochrome C release and apoptosis, require the actions of A-SMase.[38] Thus, in HeLa cells treated with siRNA against A-SMase or in A-SMase$^{-/-}$ cells from NPD patients, UV light induction of Bax conformation change was drastically reduced. Further, restoration of A-SMase or addition of exogenous ceramide to A-SMase-deficient cells restored the UV pro-apoptotic response. These findings suggest that ceramide activates the intrinsic apoptotic pathway through its effects on Bcl-2 family proteins.[38,64]

Apoptosis can also be activated through the extrinsic, or death receptor pathway (Fig. 2). TNF receptor 1 and other members of the TNF family initiate this process when activated by ligand. Once activated, these receptors interact with an adaptor protein called FADD, leading to assembly of a protein complex that activates caspase-8, which in turn cleaves and activates the Bcl-2 family member, Bid. Bid then translocates to the mitochondrial outer membrane, initiating the intrinsic apoptotic pathway. Interestingly, Bid can also be cleaved by the lysosomal aspartate protease, capthepsin D. It was recently established that activation of cathepsin D by TNFα requires A-SMase activity.[65,66] Further, ceramide was shown to bind directly to cathepsin D, causing autocatalytic proteolysis of the pre-pro-cathepsin D to form the enzymatically active isoforms of the enzyme, thereby implicating ceramide in regulation of Bid processing.

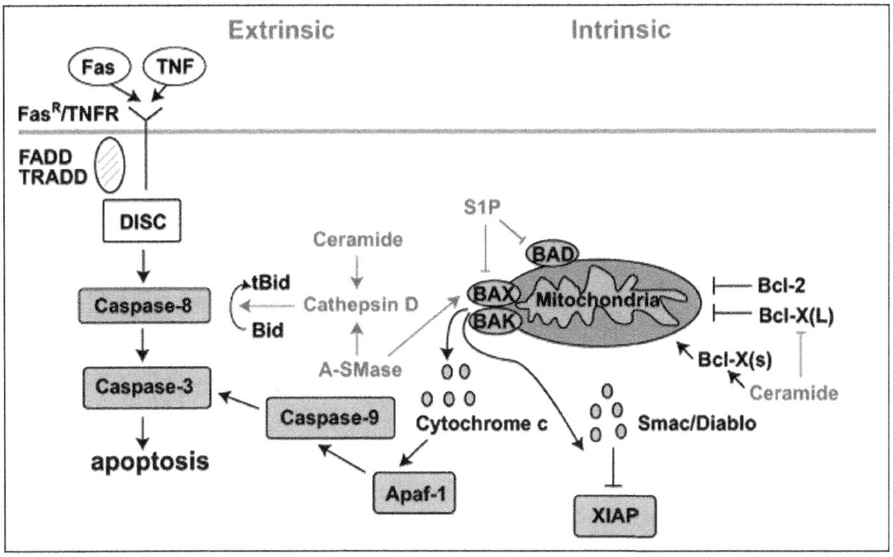

Figure 2. Sphingolipid effects on intrinsic and extrinsic apoptotic pathways. The two pathways to apoptosis are shown. The extrinsic pathway begins by ligand binding to cell surface receptors such as FasR or TNFR, followed by recruitment of Death Domain containing protein adaptors (e.g., TRADD) resulting in the formation of Death-Inducing Signaling Complex (DISC). This complex then activates the caspase cascade, culminating in cell death. The intrinsic or mitochondrial pathway begins when pro-apoptotic members of the Bcl-2 family cause mitochondrial release of cytochrome c, which binds to and activates Apaf-1 (apoptotic protease-activating factor), resulting in subsequent activation of the caspase cascade. Smac/DIABLO (Second mitochondria-derived activator of caspase/direct inhibitor of apoptosis-binding protein with low pI) is also released, inhibiting XIAP. XIAP (X-linked Inhibitor of Apoptosis Protein) functions as an inhibitor of caspases 3, 7 and 9. Shown are also the influence of ceramide, A-SMase and S1P on different components of the pathway.

Lastly, some of ceramide's effects may be due to its biophysical properties and influence on membrane organization. Endogenous ceramide aggregates into membrane microdomains, known as lipid rafts.[67,68] These ceramide-enriched domains serve to cluster activated receptor molecules such as the Fas receptor[67] and may thereby facilitate receptor-mediated (extrinsic pathway) apoptotic signaling events. Whether S1P exerts additional or opposing effects on membrane organization remains to be determined.

Sphingosine-1-Phosphate as a Counterbalance to Ceramide

Sphingosine 1-phosphate (S1P) is the final common product of sphingolipid catabolism, produced via phosphorylation of sphingosine by SK. S1P signaling inhibits apoptotic pathways, stimulates the proliferation, differentiation and migration of various cell types and promotes the development and maturation of nascent blood vessels, including tumor vessels.[22,69-72] These effects are due to the ability of S1P to promote homotypic and heterotypic cell-cell interactions. S1P signaling is also required for lymphocyte egress from peripheral lymphoid organs and inhibition of this pathway using the pharmacological agent FTY720 is being tested as an immunomodulatory intervention in solid organ transplantation and autoimmune disease.[73]

A majority of S1P's effects appear to be mediated through its ability to ligate and activate cell surface $S1P_{1-5}$ receptors.[74-76] These proteins represent a family of G protein-coupled receptors (GPCRs), first identified as endothelial differentiation genes. Due to its derivation from ceramide, its stimulatory effect on cell growth and proliferation and its ability to block apoptotic pathways,

S1P has been considered a counterbalance to ceramide in the regulation of cell fate.[76] Further, the anti-apoptotic and pro-angiogenic effects observed in response to activation of S1P receptors have led to the hypothesis that these signaling pathways may contribute to tumorigenesis and tumor progression. This notion is supported by the observation that SK functions as an oncogene in cell models of cancer and that blocking antibodies against S1P receptors inhibit xenograft growth in nude mice.[77,78]

Inhibitory Effects of S1P on Apoptotic Pathways

In contrast to the effects of ceramide, S1P signaling has been shown to block apoptotic pathways, primarily by inhibiting changes in mitochondrial membrane potential and preventing cytochrome c release from mitochondria.[79] S1P stimulation of cells leads to downregulation of Bax expression and protection against Fas-induced apoptosis.[80] In Jurkat T lymphoblasts, caspase-3 and -7 activation resulting from Fas ligation could be prevented by S1P addition or PKC activation via phorbol esters.[81] S1P was also shown to inactivate Bad and prevent Bax translocation to mitochondria in response to Fas ligation.[82] Interestingly, cathepsin B is responsible for proteolysis of SK1 during the onset of apoptosis.[83] In light of the effects of ceramide on cathepsin D described above, it is intriguing to note the prominent role that lysosomes play in mediating effects of sphingolipids on cell fate decisions.

Interestingly, S1P may have different effects on cell fate depending upon its subcellular location. Spiegel and colleagues showed that the enzyme SK2 functions as a BH3-domain protein that counteracts the effects of SK1 and induces apoptosis.[84] Experiments targeting SK1 expression to the ER, where SK2 normally resides, showed that SK1 can also promote apoptosis, indicating that S1P production at this site could have effects on local structures and also influences the biochemical conversion of sphingosine into ceramide, thereby perturbing the balance of sphingolipids that regulate cell fate.[85,86]

Sphingolipids Regulate Key Signaling Pathways That Control Cell Fate

Another mechanism by which sphingolipids influence cell fate involves their ability to modulate key signaling pathways that control cell metabolism and survival in response to stress. One such pathway is the PI_3 kinase/AKT pathway, which has emerged as a central signaling node responsible for adjusting the metabolic and biosynthetic operations of the cell in response to changing nutrient and environmental (stress) conditions (Fig. 3).[87,88]

Activation of the AKT pathway facilitates cell cycle progression, inhibits apoptotic responses, suppresses autophagy (see below) and enhances nutrient uptake, metabolism, ribosome biogenesis and protein synthesis.[88] The cumulative effects of PI_3K/AKT signaling lead to increased cell growth (cell mass) and cell survival. This critical function is coordinated through a system of cell surface receptors, typified by the insulin receptor which, when stimulated by ligand, leads to activation of the class I phosphoinositol-3 kinase (PI_3K). Once activated, PI_3K phosphorylates phosphatidylinositol-4,5-bisphosphate (PIP_2), generating phosphatidylinositol-3,4,5-trisphosphate (PIP_3). Generation of PIP_3 at the plasma membrane leads to membrane recruitment of the serine/threonine protein kinase AKT, along with PDK1, each by virtue of their lipophilic pleckstrin homology domains (PHD). PDK1 then phosphorylates AKT on Thre308 of its activation loop. A second AKT residue (Ser473) is phosphorylated by mTORC2. Activated AKT subsequently phosphorylates downstream targets involved in protein synthesis, cell cycle progression and DNA repair, growth, survival and metabolism, angiogenesis, carbohydrate uptake, migration and proteins involved in feedback regulation of the AKT pathway itself. Antagonists to AKT signaling include the lipid phosphatase PTEN, which catalyzes the conversion of PIP_3 to PIP_2 and PHLPP, which dephosphorylates AKT at Ser 473. Importantly, activating mutations in the PI_3K/AKT pathway have been identified in many types of human cancers.[89-91] Considering the broad effects of PI_3K/AKT signaling on cell growth and survival, it is not surprising that oncogenic mutations in the AKT pathway have been implicated in carcinogenesis, cytotoxic therapy, epithelial-mesenchymal transition and tumor progression within an inhospitable tumor environment.

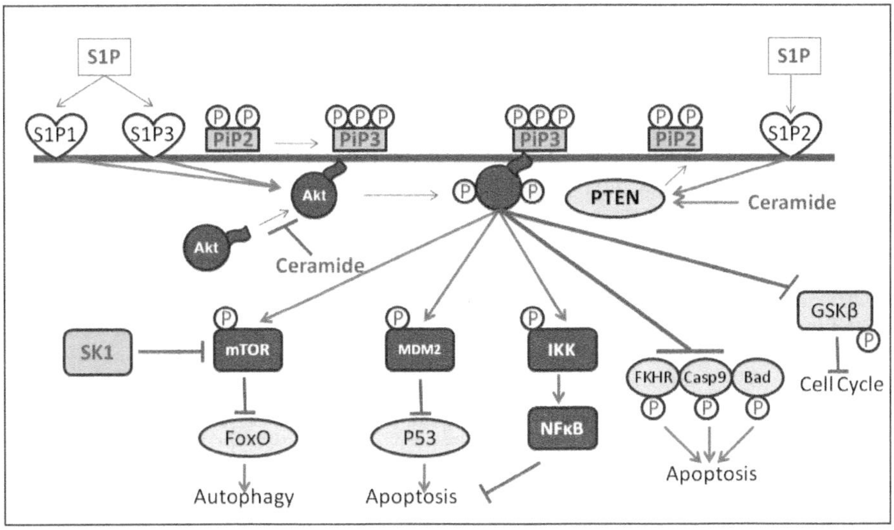

Figure 3. Effect of sphingolipids on the PI_3K/AKT pathway. Upon phosphorylation of PIP_2 to PIP_3 in response to growth stimulatory signals, Akt and PDK1 (not shown) are recruited to the plasma membrane by virtue of their pleckstrin homology domains. Once there, PDK phosphorylates and activates Akt, which in turn either activates or inhibits a large number of substrates by phosphorylating them, thus exerting influence over various cell functions. The schematic focuses on the role of Akt in inhibition of apoptosis. The Akt substrates that have an inhibitory role on apoptosis are depicted in dark blue rectangles and those that exert a positive influence (which phosphorylation by Akt abrogates) are shown in light blue ovals. The figure also illustrates how different sphingolipids might influence the pathway both negatively (Ceramide) or positively (S1P). A color verison of this figure is available on www.landesbioscience.com/curie.

Ceramide acts as a negative regulator of the PI_3K/AKT pathway.[58,92-94] In vascular smooth muscle cells, ceramide was shown to block PDGF-induced AKT activation in a PKCζ-dependent manner.[92] Ceramide disrupts interactions between 14-3-3 scaffolding proteins and PKCζ, thereby recruiting the latter to lipid rafts. Additional studies in 3T3 L1 pre-adipocytes have shown that ceramide blocks AKT translocation from cytosol to plasma membrane via its PHD.[95] In addition to its effect on AKT transloction, ceramide also appears to block AKT signaling through dephosphorylation of the active AKT protein by PP2A, a direct target of ceramide.[95-97]

These findings indicate several independent effects of ceramide on AKT signaling. They are consistent with the frequent observations correlating ceramide accumulation in skeletal muscle and liver with insulin resistance and decreased insulin-dependent glycogen synthesis and glucose uptake, which are AKT-dependent activities. Similar effects of ceramide have been observed in parasite-infected macrophages, neuronal cells, myotubes and intestinal epithelial cells.[58,98-100] Whereas most reports have described the effects of ceramide to be independent of phosphatidylinositol levels (i.e., downstream of PI_3K), one study in PC12 pheochromocytoma cells found that ceramide recruited PTEN to lipid rafts and that ceramide-dependent cell death was dependent on PTEN.[101] Interestingly, another lipid mediator, ceramide-1-phosphate, was found to activate AKT, likely through indirect effects on sphingomyelinase resulting in reduced ceramide generation.[102]

Whereas the effects of ceramide are inhibitory to AKT signaling, S1P activates this pathway. In endothelial cells and cardiac myocytes, S1P activation of AKT led to eNOS activation, hypertrophy, stress fiber formation and resistance to ischemia/reperfusion injury.[103,104] S1P receptors are required for each of these effects, as demonstrated by studies employing chemical inhibitors, siRNA and murine knockout models. In endothelial cells, the effect of S1P on AKT appears to require S1P receptor-mediated activation of AMPK, followed by activation of the small GTPase, Rac.[105] VEGF treatment of endothelial cells was found to stimulate expression of $S1P_3$, which stabilized Akt3 mRNA and increased Akt3 protein expression.[106] These results suggest that S1P may exert short-term, long-term and Akt isoform-specific effects on the PI_3K/AKT system in a variety of cell types. Transactivation of the VEGF receptor Flk-1/KDR by ligation of S1P receptors resulted in AKT activation and eNOS phosphorylation.[107] Further, overexpression of SK1 was found to induce AKT signaling, leading to enhancement of endothelial cell-cell interactions.[108] In a CHO cell expression system, S1P activated AKT signaling in a PLD- and S1P3-dependent manner, leading to membrane ruffling.[109] S1P also appears to activate PI_3K_b in endothelial cells.[110] Interestingly, AKT phosphorylation of $S1P_1$ is necessary for S1P-induced Rac activation and cell migration, indicating that S1P and AKT signaling are mutually activating systems. In mouse embryonic fibroblasts (MEFs) and other cell types, S1P activation of $S1P_3$ led to phosphorylation of the PDGFR and phosphorylation of AKT on Ser473. Both $S1P_3$ and PDGFR were required for AKT activation by S1P, indicating crosstalk between S1P and PDGF receptors.[111] In some ovarian cell lines, S1P mediates its effects on PI_3K-dependent AKT activation through the stress-activated kinases MEK and p38.[112] Finally, there is evidence suggesting that S1P activation of $S1P_2$ inhibits cell migration and enhances vascular permeability through Rho-dependent activation of PTEN, leading to inhibition of AKT signaling through reduction of PIP_3.[113]

Sphingolipids and Autophagy

In addition to regulating apoptosis and protein translation, the PI_3K/AKT signaling pathway plays a central role in regulating autophagy, another important mechanism of cell death upon which sphingolipids exert their influence. Autophagy is the evolutionarily conserved lysosomal degradation of a cell's own macromolecules and organelles. Autophagy is important in energy homeostasis, providing a means by which cells and organisms recycle their own parts when nutritional intake is low and energy stores are depleted. However, autophagy also plays a role in innate immunity, stress responses, neurodegeneration, aging and cancer. The latter has been shown by the finding that some tumor suppressors induce autophagy.[114]

During autophagy, a doubled-membraned vesicle called an autophagosome sequesters the target molecules or organelles followed by the fusion of the outer membrane with lysosomes and culminating in the degradation of the sequestered material via lysosomal enzymes.[115] This process is mediated by the actions of a protein complex involving beclin-1, which serves as a platform for recruitment of autophagy activators and inhibitors. For example, the anti-apoptosis protein Bcl-2 inhibits autophagy by interacting with beclin-1. In response to growth deprivation, Jnk1 phosphorylates Bcl-2 on T69, S70 and S87, disrupting its interaction with beclin in favor of Bcl-2/Bad interactions (Patrice Codogno, personal communication).

Under favorable growth conditions the AKT pathway inhibits the onset of autophagy via activation of mTOR kinase.[116] Once a cell senses the lack of nutrients or growth factors, signaling through the AKT pathway ceases and the block on autophagy is lifted. Autophagy can be induced in malignant glioma cells by inhibition of PI_3K (by LY294002), AKT (by UCN-01) or mTOR (by rapamycin).[117] In contrast, overexpression of constitutively active AKT affords the cells more resistance to autophagic cell death induced by rapamycin treatment.

C2-ceramide was shown to induce autophagy by upregulating beclin-1 in MCF-7 cells.[118] In the same study it was shown that tamoxifen induced autophagy and concurrent beclin-1 upregulation. Both of these endpoints were inhibited by the SPT inhibitor myriocin, indicating a role for the de novo ceramide synthesis in mediating tamoxifen-induced autophagy. Ceramide has also been implicated in mediating autophagy in human umbilical vein endothelial cells exposed to glycated collagen I.[119] Daido and colleagues showed that C2-ceramide induced autophagic cell death in U373-MG and T98G glioblastoma cells.[120] The cell death was linked to an increase in the transcription of BNIP3 gene, a Bcl-2 family member that induces nonapoptotic cell death.

Whereas ceramide and S1P usually exert opposing effects on cell fate, both lipids function as positive regulators of autophagy. It has been show that overexpression of SK1 induces autophagy and that SK1 is activated when cells are starved, leading to increased autophagy.[121] However, the autophagy induced by S1P seems to be independent of its effects on AKT. Rather, it is exerted by inhibition of mTOR and, in contrast to the effect of ceramide, the increase in beclin-1 is nominal. Furthermore, the autophagy induced by SK1 upregulation during starvation seems to be cytoprotective, as siRNA against SK1 inhibited autophagy and increased cell death during starvation.[122] It remains unclear to what extent sphingolipid metabolism affects autophagy in physiological circumstances and in cancer, where autophagy may initially promote cell survival in the harsh environment of a tumor but may eventually be a mechanism for targeted cancer cell death and induction of apoptotic cell death by chemotherapy and radiation.

In summary, sphingolipid metabolism wields a double-edged sword with respect to cell fate. In response to stressful conditions or cytotoxic therapy, ceramide generated from either de novo biosynthetic or recycling metabolic pathways triggers activation of signals affecting intrinsic and extrinsic apoptotic pathways and cell cycle control, while simultaneously inhibiting signaling events that promote cell growth and survival mediated by the AKT pathway. In contrast, S1P generally reverses the effects of ceramide on these pathways through the stimulation of receptor-mediated signaling events and activation of the AKT pathway. The two effectors meet, however, in the autophagy response, where both lipids serve as activators of the pathway.

Other Signaling Pathways Influenced by Sphingolipids

In addition to effects on the AKT pathway, ceramide been shown to induce members of the stress-activated protein kinases such as JNK,[123] kinase suppressor of Ras (KSR)[124] and the atypical protein kinase C (PKC) isoform, PKC ζ.[92] Each of these proteins influences cell survival and apoptosis. S1P opposes the actions of ceramide, suppressing JNK activation mediated by ceramide and activating ERK. Both these effects of S1P can contribute to cell survival,[125] whereas in other cases, S1P blocks ceramide-induced apoptosis through effects that appear to be independent of ERK.[126] S1P can also stimulate production and secretion of EGF, PDGF and VEGF.[127] The increase in levels of these pro-growth and -angiogenesis factors in turn results in transactivation of their cognate receptors, leading to downstream signals that regulate cell proliferation, migration and vascular remodeling. Hsieh and colleagues showed that S1P induced both transcription and translation of EGFR via activation of p42/p44 MAPK, AP-1, PI_3K/AKT and NF-κB-signaling pathways in rat primary cultured vascular smooth muscle cells.[128] High levels of the EGFR have been detected in various human cancers, including glioblastomas and ovarian cervical and kidney tumors.[129-131] S1P treatment enhanced proliferation of these cells in a process that could be inhibited either by treatment with EGFR kinase inhibitor AG1478 or EGFR shRNA.

COX-2 is a critical enzyme required for synthesis of prostaglandins and the target of the anti-inflammatory and chemopreventive agent celecoxib. In HT29 colon cancer cells, downregulation of SK1 expression by siRNA significantly inhibited TNFα- or IL-1β-induced COX-2 expression and PGE_2 production. These findings suggest that S1P signaling may promote carcinogenic inflammatory effects through upregulation of COX-2 and generation of prostaglandin metabolites.

Ceramide Regulates Cell Cycle Progression

In some cells, ceramide has been shown to prevent cell proliferation through inhibition of cell cycle progression. Treatment of MOLT4 leukemia cells with ceramide or its metabolite sphingosine results in dephosphorylation of the retinoblastoma gene product, leading to cell cycle arrest.[132] Ceramide has also been shown to regulate cyclin dependent kinase 2 (CDK2) activity.[22] Ceramide can cause G2 cell cycle arrest via induction of p21 expression, even in the absence of DNA damage.[133] Interestingly, the negative regulator of p53, MDM2 was found to abrogate ceramide-induced growth arrest and apoptosis.[133] In RD rhabdomyosarcoma cells, overexpression of MDM2 prevented ceramide-induced G2 arrest, p21 induction and apoptosis, whereas siRNA against MDM2 sensitized the cells to ceramide-induced G2 arrest. Accumulation of p21 as well as reduction in cyclin D1 and CDK7 were responsible for ceramide-induced cell cycle arrest in Be7402 hepatocarcinoma cells.[134]

Ceramide and Telomerases

Telomerase is a ribonucleoprotein composed of a reverse transcriptase (hTERT) and an RNA partner (hTR). The complex is responsible for addition of TTAGGG tandem repeat sequences to the ends of chromosomes.[135-138] Telomerase is overexpressed in approximately 90% of human cancers. Its expression, usually absent in somatic cells, allows malignant cells to maintain telomere length, thereby insuring infinite replicative capacity. Increased telomerase activity has been suggested to offer protection against apoptosis. Both exogenous short-chain ceramide and stimulation of endogenous ceramide production were found to inhibit telomerase expression.[139] Ceramide affected the promoter activity of hTERT, decreasing the half-life of its trans-activator, cMyc, via increased ubiquitination and rapid proteolysis. More recent findings suggest that ceramide modulates the promoter activity of hTERT via a pathway that involves epigenetic effects on the hTERT promoter, culminating in transcriptional repression.[140]

Ceramide and S1P in Cancer Stem Cells

Cancer cells are usually a heterogeneous population, of which only a small fraction referred to as "cancer stem cells" is capable of continuous self-renewal, as demonstrated by colony formation in soft agar or development into tumors following transplantation into a host organism. A recent study found mammary cancer stem cells to be more resistant than their nonstem cell counterparts to serum deprivation and ceramide-induced apoptosis.[141] This characteristic of the cancer stem cells was attributed to the unique overexpression of sphingomyelin synthase 1 and Bcl-2 in the stem cells, two properties that would serve to lower ceramide levels.

Effects of S1P on Migration and Metastasis

S1P activation of its receptors influences endothelial cell migration by promoting changes in cytoskeletal organization that are needed for the formation of lamellipodia and attachment to the basement membrane.[142] These effects may be important targets for the prevention of pathological angiogenesis, including tumor angiogenesis, which is an important contributor to the progression of cancer.

S1P has also been shown to influence epithelial cancer cell migration, which could contribute to metastatic potential. S1P was found to stimulate chemotaxis and invasion of ovarian cancer cell line (OVCAR3) in a receptor-dependent fashion that involved activation of ERK, AKT and p38.[143] A link between S1P signaling, cell migration and tumor metastasis was uncovered when it was observed that overexpression of the *KAI1* gene downregulates SK1. The *KAI1* gene was originally isolated as a prostate-specific tumor metastasis suppressor gene that inhibited metastases without affecting primary tumor formation. When KAI1 was overexpressed using an adenovirus vector in pancreatic carcinoma cells (PANC I) both SK-1 activity and migration of PANC I cells was significantly reduced.[144]

Cancer Cells Exhibit Molecular and Genetic Changes in Sphingolipid Metabolism

Considering the central role of ceramide in mediating physiological as well as pharmacologically stimulated apoptosis, ceramide could be considered a tumor-suppressor lipid. If ceramide-mediated apoptosis functions as a form of cancer surveillance, one would predict that the tumor cells might evolve mechanisms to avoid ceramide accumulation either by inhibiting the production of ceramide or by enhancing its degradation. Recent research has provided evidence that both these types of molecular changes occur in cancer cells.

Ceramide levels in fresh human specimens of primary and metastatic colon cancer were found to be half that of normal colon mucosa from the same patient.[145] The difference seemed to stem from higher ceramidase activity in the tumor tissue. Importantly, ceramidase inhibition increased the ceramide content of tumor cells and caused activation of the apoptotic cascade, whereas this maneuver had no effect on normal cells.

Consistent with this notion, acid ceramidase was found to be over-expressed in 70% of head and neck squamous cell tumors compared with normal tissues.[146] Forced expression of acid ceramidase in tumor-derived cells increased resistance to Fas-induced apoptosis, whereas its down-regulation using specific siRNA sensitized the cells to Fas. Inhibition of acid ceramidase using the small molecule LCL204 sensitized head and neck squamous cancer cell lines to Fas-induced apoptosis both in vitro and in a xenograft model. The increased expression of acid ceramidase has also been observed in prostate cancer cells.[146] Treatment of DU145 prostate cancer cells with LCL204 resulted in increased ceramide and decreased sphingosine levels along with rapid destabilization of lysosomes and the release of lysosomal proteases into the cytosol, culminating in the activation of the intrinsic apoptotic pathway.[147]

Another mechanism that cancer cells utlize to increase drug resistance is the enhancement of ceramide transport via the actions of a protein called CERT (**cer**amide **t**ransporter). CERT is involved in ATP-dependent ER-to-Golgi trafficking of ceramide molecules. CERT expression was recently shown to be elevated in paclitaxel-resistant cancer cell lines and in ovarian cancer cells that survived paclitaxel treatment. Reduction of CERT expression using siRNA increased cellular sensitivity to paclitaxel. CERT may attenuate the effects of ceramide via its removal from ER, where it has been postulated to potentiate the drug-induced ER stress response.[148]

Given the role of S1P as a mitogen and inhibitor of apoptotic pathways, genetic changes resulting in S1P accumulation in cancer cells might also be selected for during carcinogenesis and cancer progression. In fact, overexpression of SK1 has been observed in a variety of solid tumors including lung, colon and other cancers.[149-152] SK1 upregulation was also observed in a transgenic mouse model of erythroleukemia caused by overexpression of the transcription factor Spi-1.[153] In this model, transcriptional upregulation of the SK1 gene was a recurrent event in the development of malignant erythroblasts, where SK1 expression and activity conferred a proliferative advantage, increased clonogenicity, resistance to apoptosis and in vivo tumorigenic potential. Conversely, Kohno et al[154] reported that an $SK1^{-/-}$ genotype reduces adenoma size and attenuates epithelial cell proliferation in the $Apc^{Min/+}$ mouse model of colon cancer, implicating SK1 as a mediator of adenoma progression. Interestingly, this effect was independent of S1P receptor expression and therefore extracellular S1P signaling, since polyp incidence or size was unaltered in $Apc^{Min/+}$ in which S1P receptor genes were disrupted.

Observations of SK1 upregulation in human cancer specimens and the tumor suppressive effects of mutations preventing S1P accumulation in mouse models of intestinal tumorigenesis suggest that conversion of sphingosine to S1P may promote tumor progression, whereas S1P catabolism might inhibit it.[149,154] Consistent with this hypothesis is the recent finding that SPL and S1P phosphatase, two enzymes responsible for degradation of S1P, are downregulated in colon cancer and in mouse models of intestinal tumorigenesis.[155] Downregulation of SPL and reduced SPL enzyme activity were consistent hallmarks of early adenomas in the $Apc^{Min/+}$ mouse model and were associated with increased S1P levels in polyp compared to surrounding tissue. Sphingosine derived from dietary sphingolipids has been shown to have protective effects against colon cancer in rodent models.[156-158] Thus, while loss of SPL or phosphatase has not yet been shown to directly contribute to tumorigenesis, it seems plausible that neoplastic intestinal tissue may be under selection pressure to facilitate conversion of dietary sphingosine to S1P and to prevent S1P degradation, thereby converting a growth-inhibitory pathway into a growth activating and angiogenic stimulus.

The inhibitory effects of S1P on apoptotic pathways may be responsible for mechanisms of cancer cell drug resistance. In HL60 leukemia cells, drug sensitivity correlated with low SK activity and the ability to generate ceramide.[159] In contrast, drug-resistant derivatives of the parent cell line exhibited greater SK activity, failure to accumulate ceramide and inability to activate the mitochondrial apoptotic pathway in response to drug treatment. Inhibition of SK in these cells restored drug sensitivity. The opposing actions of ceramide and S1P are illustrated by a study in which apoptosis-resistant human melanoma cell lines M221- and Mel-2a were found to have higher SK activity, lower ceramide levels and higher S1P-to-ceramide ratios than the corresponding drug-sensitive cell lines A-375 and M186.[160] Overexpression of functional (but not mutant) SK1 increased resistance of A-375 melanoma cells to short-chain ceramide, Fas antibody and doxorubicin-induced apoptosis. In contrast, siRNA-mediated downregulation of SK1 expression rendered ceramide-resistant Mel-2a cells sensitive to ceramide- and Fas-induced cell death. Similar effects of S1P were observed in response to imatinib, a small molecule inhibitor of Bcr-Abl tyrosine kinase.[161] Forced SK1 downregulation and reduction of S1P levels resulted in imatinib-resistant cells becoming markedly more sensitive to the drug, whereas overexpression of SK1 prevented the apoptotic response to imatinib.[162]

Targeting Sphingolipids for Cancer Therapy

The crucial role that sphingolipids play in regulating cell fate in combination with the genetic alteration of sphingolipid metabolic pathways observed in cancer specimens and in association with cellular drug resistance have prompted efforts to harness sphingolipid signaling for antitumor therapy. Approaches have included the development of: (1) synthetic ceramide analogs, (2) small molecule inhibitors of enzymes that catalyze ceramide catabolism or its conversion to other molecular species, (3) inhibitors of SK and (4) S1P receptor antagonists. Additional approaches that have been proposed include reactivation of genes such as SPL, S1P phosphatase and SMase that have been silenced in cancer tissue, targeting dihydroceramide desaturase (see below) and activation of SMase enzymes using small molecules. Pharmacological modulation of sphingolipid targets singly, as adjuvant therapy in combination with conventional cytotoxic approaches and in combinations that serve to enhance ceramide generation while preventing auto-regulatory effects caused by conversion of ceramide to S1P are all being considered.

Numerous ceramide analogs with unique physicochemical and biological properties have been developed, including cell-permeable, short chain ceramides, analogs with additional double bonds in the long chain base, analogs containing disulfide linkages, conjugates of ceramides and dihydroceramides with pyridinium salts and a variety of alkyl- acyl and urea-analogs.[163] Some of these molecules have the advantage of improved solubility, efficient cellular uptake and the ability to be targeted to different cellular compartments such as mitochondria and lysosomes for enhanced efficacy. Some have been shown to promote mitochondrial apoptosis, deplete cellular glutathione and cause cytotoxicity in malignant cells. However, it is important to keep in mind that ceramide analogs have the potential to interact with numerous cellular and biochemical targets in the sphingolipid metabolic pathway. Thus, the toxicity profiles and efficacy of these novel drugs will need to be tested in vivo before conclusions about their utility in the clinic can be drawn.

As discussed above, analysis of the gene expression patterns associated with the acquisition of cytotoxic drug resistance phenotypes suggest that preventing ceramide accumulation through enhanced degradation or glycosylation is an important mechanism of drug resistance. In contrast to the use of ceramide analogs, which may exert effects that are not specific to cancer cells, targeting genes that are overexpressed in cancer may afford some selectivity. Interestingly, the pharmacological inhibition of glucosylceramide synthase by miglustat and similar agents has been used for many years in "substrate depletion" approaches to treat sphingolipidosis storage diseases. These agents act by preventing formation of the first intermediate in the synthesis of many higher order sphingolipids, glucosylceramide.[164] While these drugs have not been tested as anticancer agents in humans, targeting glucosylceramide synthase has shown some efficacy in cancer models. For example, the imino sugar OGT2378, an inhibitor of glucosylceramide synthase, inhibits melanoma tumor growth in vitro and in a syngeneic orthotopic mouse model.[165] In the case of melanoma, the positive effects of the drug may be related to prevention of ganglioside formation, rather than ceramide accumulation, as the former contributes to tumorigenicity in melanoma.[166] Dbaibo and colleagues reported that arsenic trioxide, an agent used in the treatment of promyelocytic leukemia, induces apoptosis via a combination of glucosylceramide synthase inhibition and induction of de novo ceramide synthesis.[167] As these and other specific small molecule inhibitors of ceramide-metabolizing enzymes are tested in a range of tumor types, the utility of enhancing endogenous ceramide accumulation to reactivate apoptotic pathways in drug-resistant cancer cells should become clear.

Inhibition of S1P synthesis by blocking SK activity has proven useful in enhancing cancer cell responses to cytotoxic therapy in vitro and in animal models.[152,159,168-170] French and colleagues developed a panel of nonlipid inhibitors of SK1 that prevented proliferation and induced apoptosis in various tumor cell lines, including some that displayed multidrug resistant phenotypes.[151] A recent study[171] provided evidence suggesting that apoptosis induced by the SK inhibitor dimethylsphingosine occurred in a Bcl-2-independent fashion in human prostatic adenocarcinoma cell lines. Since Bcl-2 upregulation is a common event in cancer, Bcl-2 independence could be advantageous in treating drug-resistant cancers. A preclinical study demonstrating the efficacy of inhibiting S1P signaling using blocking antibodies to prevent tumorigenesis/tumor progression was also recently reported.[78] In this hallmark study, a specific monoclonal antibody recognizing S1P was administered to mice harboring human cancer xenografts. The intervention reduced and in some cases completely eliminated, tumor formation and accompanying tumor angiogenesis. These studies suggest that blocking tumor angiogenesis through antibody-mediated inhibition of S1P signaling, similar to or in combination with VEGF-directed antibodies, may be developed as a strategy for treating advanced malignant disease. In a different approach to blocking S1P signaling, the immunomodulatory drug FTY720 promoted apoptosis in drug resistant multiple myeloma cells by inducing mitochondrial membrane potential changes, Bax cleavage and caspase activation.[172] FTY720 acts by downregulating S1P receptors and interfering with S1P signaling. This study suggests that additional small molecule inhibitors of S1P signaling may be useful targets to enhance cancer cell apoptosis and inhibit tumor angiogenesis. Due to the immunomodulatory effects of FTY720, development of S1P receptor-specific agonists and antagonists is an area of active research. As these agents become available, testing for anticancer efficacy could be included in biological assays.

Another promising chemotherapeutic agent is retinimide or fenretinide (4-HPR). The drug is a synthetic retinoid N-(4-hydroxyphenyl) that is cytotoxic to a variety of cancers including neuroblastoma. Interestingly, the drug does not appear to be dependent on retinoid receptors, for which it has a very low binding affinity.[173,174] Recently, fenretinide was found to increase ceramide generation via the de novo pathway, through activation of both SPT and ceramide synthase.[175] Importantly, fenretinide does not increase ceramide levels in nonmalignant cells. Furthermore, it induces apoptosis in a p53- and caspase-independent manner, making the drug useful even in cancer cells harboring mutations in these molecular pathways.[176] Fenretinide was also found to be cytotoxic to endothelial cells, in which it induced ceramide accumulation and caspase-dependent apoptosis. The ceramide generation could be suppressed by ceramide synthesis inhibitors fumonisin B1, myriocin and L-cycloserine but not by the A-SMase inhibitor desipramine. These studies suggest that fenretinide's antitumor mechanism of action may involve an anti-angiogenic component.[177]

Interestingly, one report has provided evidence that fenretinide functions as an inhibitor of ceramide desaturase, suggesting that dihydroceramide or dihydrosphingosine rather than ceramide may be the cytotoxic intermediate.[178] This possibility is consistent with the finding that γ-tocopherol induces apoptosis in prostate cancer cells by causing accumulation of dihydrosphingosine and dihydroceramide.[179] If these studies hold true, dihydroceramide desaturase may become another important anticancer drug target. Additional studies have explored the role of the dihydrosphingosine analog safingol, which is also toxic to human prostate cancer cell lines and other malignant cell lines.[176,180] Safingol acts through inhibition of PKC and has been shown to promote a caspase-independent form of apoptosis. However, a wide range of toxicities are associated with safingol, its N-methyl metabolites and potentially its conversion to dihydroceramide-like molecules on murine liver, kidney and blood.

Sphingomyelinases are activated by cytotoxic agents, resulting in the generation of ceramide through the recycling pathway and the promotion of mitochondrial apoptosis. Overexpression of A-SMase and other forms of sphingomyelinase have been shown to sensitize glioma cells and other cancer cell types to chemotherapy.[181] Activation of sphingomyelinases by small molecules, thus, represents another potential avenue by which to enhance cancer cell death. Aside from conventional chemotherapeutic agents and radiation, sphingomyelinase activation by reactive oxygen species, ursolic acid and the histone deacetylase inhibitor LAQ824 have been reported.[182-184] Interestingly, however, treatment of normal and cancer cells with a thiourea derivative of sphingomyelin resulted in accumulation of ceramide and induction of apoptotic cell death, despite inhibition of acid and neutral sphingomyelinase activity by the agent.[185] These results appear to be due to the drug's ability to inhibit formation of sphingomyelin from ceramide, illustrating the fact that sphingolipid analogs may have multiple targets and may induce unexpected effects on cell fate.

S1P Signaling to Protect Normal Tissues from Therapy-Related Cytotoxicity

Many cancer interventions exhibit a high therapeutic index, causing unacceptable toxicity to normal tissues. Identification of ways to protect normal cells from iatrogenic apoptosis in response to radiation and chemotherapy would improve overall outcomes for cancer patients. Interestingly, S1P administration to mice has been shown to suppress male germ cell apoptosis and ovarian failure/infertility in response to radiation and other cancer therapies.[60,186] Similar observations suggest that S1P pathways are involved in regulating cardiac cell fate in response to oxidative stress.[187] While still untested, these findings suggest the possibility that S1P may be able to protect against radiation and anthracycline-induced cardiotoxicity as well. These promising findings suggest that targeted protection of normal tissues may be yet another method for improving the therapeutic index of commonly employed chemotherapeutic drugs and radiation.

Conclusion

Sphingolipids are an important class of membrane lipids that give rise to signaling metabolites such as ceramide and S1P, which exert control over cell fate and senescence. Through effects on mitochondrial membrane potential, Bcl-2 family proteins, cell survival and stress-response signaling pathways, telomerase and other critical cellular machinery, these two molecules modulate cell fate in response to environmental stress, chemotherapy, radiation and genotoxic stress associated with continual DNA replication. Not surprisingly, genes involved in metabolism and signaling of S1P and ceramide have been implicated in tumorigenesis, tumor progression and patterns of drug resistance. The expression of these genes and their protein products may become useful indicators of tumor progression and modulation of these pathways through pharmacological means holds promise as novel approaches for the treatment of cancer.

Acknowledgements

This work was supported by National Insitutes of Health Grants CA77528 and CA129438 (JDS).

We apologize to the many colleagues whose work could not be cited directly in this review due to space limitations. We thank Tom McElderry for expert administrative assistance.

References

1. Morales A, Lee H, Goñi F et al. Sphingolipids and cell death. Apoptosis 2007; 12:923-39.
2. Futerman A, Riezman H. The ins and outs of sphingolipid synthesis. Trends Cell Biol 2005; 15:312-8.
3. Kihara A, Mitsutake S, Mizutani Y et al. Metabolism and biological functions of two phosphorylated sphingolipids, sphingosine 1-phosphate and ceramide 1-phosphate. Prog Lipid Res 2007; 46:126-44.
4. Prinetti A, Chigorno V, Mauri L et al. Modulation of cell functions by glycosphingolipid metabolic remodeling in the plasma membrane. J Neurochem 2007; 103(Suppl 1):113-25.
5. Zeidan Y, Hannun Y. Translational aspects of sphingolipid metabolism. Trends Mol Med 2007; 13:327-36.
6. Kitatani K, Idkowiak-Baldys J, Hannun YA. The sphingolipid salvage pathway in ceramide metabolism and signaling. Cell Signal 2007.
7. Murph M, GB M. Targeting the lipids LPA and S1P and their signalling pathways to inhibit tumour progression. Expert Rev Mol Med 2007; 9:1-18.

8. Kolesnick RN, Goni FM, Alonso A. Compartmentalization of ceramide signaling: physical foundations and biological effects. J Cell Physiol 2000; 184(3):285-300.
9. Itoh M, Kitano T, Watanabe M et al. Possible role of ceramide as an indicator of chemoresistance: decrease of the ceramide content via activation of glucosylceramide synthase and sphingomyelin synthase in chemoresistant leukemia. Clin Cancer Res 2003; 9(1):415-23.
10. Okazaki T, Bell R, Hannun Y. Sphingomyelin turnover induced by vitamin D3 in HL-60 cells. J Biol Chem 1989; 264(32):19076-80.
11. Liu B, Obeid L, Hannun Y. Sphingomyelinases in cell regulation. Semin Cell Dev Biol 1998; 8(3):311-22.
12. Valaperta R, Chigorno V, Basso L et al. Plasma membrane production of ceramide from ganglioside GM3 in human fibroblasts. FASEB J 2006; 20:1227-9.
13. Nikolova-Karakashian M, Merril AH Jr. Ceramidases. Methods Enzymol 2000; 311(Part A):194-207.
14. Olivera A, Spiegel S. Sphingosine kinase: a mediator of vital cellular functions. Prostaglandins Other Lipid Mediat 2001; 64(1-4):123-34.
15. Le Stunff H, Peterson C, Liu H et al. Sphingosine-1-phosphate and lipid phosphohydrolases. Biochim Biophys Acta 2002; 1582(1-3):8-17.
16. Mandala SM, Thornton R, Tu Z et al. Sphingoid base 1-phosphate phosphatase: a key regulator of sphingolipid metabolism and stress response. Proc Natl Acad Sci USA 1998; 95(1):150-5.
17. Bandhuvula P, Saba J. Sphingosine-1-phosphate lyase in immunity and cancer: silencing the siren. Trends Mol Med 2007; 13(5):210-7.
18. Watanabe M, Kitano T, Kondo T et al. Increase of nuclear ceramide through caspase-3-dependent regulation of the "sphingomyelin cycle" in Fas-induced apoptosis. Cancer Res 2004; 64(3):1000-7.
19. Clarke C, Snook C, Tani M et al. The extended family of neutral sphingomyelinases. Biochemistry 2006; 45:11247-56.
20. Jayadev S, Liu B, Bielawska A et al. Role for ceramide in cell cycle arrest. J Biol Chem 1995; 270(5):2047-52.
21. Yoshimura S, Banno Y, Nakashima S et al. Ceramide formation leads to caspase-3 activation during hypoxic PC12 cell death. Inhibitory effects of Bcl-2 on ceramide formation and caspase-3 activation. J Biol Chem 1998; 273(12):6921-7.
22. Liu Y, Wada R, Yamashita T et al. Edg-1, the G protein-coupled receptor for sphingosine-1-phosphate, is essential for vascular maturation. J Clin Invest 2000; 106(8):951-61.
23. Takeda Y, Tashima M, Takahashi A et al. Ceramide generation in nitric oxide-induced apoptosis. Activation of magnesium-dependent neutral sphingomyelinase via caspase-3. J Biol Chem 1999; 274(15):10654-60.
24. Mansat V, Laurent G, Levade T et al. The protein kinase C activators phorbol esters and phosphatidylserine inhibit neutral sphingomyelinase activation, ceramide generation and apoptosis triggered by daunorubicin. Cancer Res 1997; 57(23):5300-4.
25. Sawada M, Nakashima S, Banno Y et al. Ordering of ceramide formation, caspase activation and Bax/Bcl-2 expression during etoposide-induced apoptosis in C6 glioma cells. Cell Death Differ 2000; 7(9):761-72.
26. Hara S, Nakashima S, Kiyono T et al. p53-Independent ceramide formation in human glioma cells during gamma-radiation-induced apoptosis. Cell Death Differ 2004; 11:853-61.
27. Jayadev S, Linardic C, Hannun Y. Identification of arachidonic acid as a mediator of sphingomyelin hydrolysis in response to tumor necrosis factor a. J Biol Chem 1994; 269(8):5757-63.
28. Schütze S, Machleidt T, Krönke M. The role of diacylglycerol and ceramide in tumor necrosis factor and interleukin-1 signal transduction. J Leukoc Biol 1994; 56:533-41.
29. Lei X, Zhang S, Bohrer A et al. The group VIA calcium-independent phospholipase A2 participates in ER stress-induced INS-1 insulinoma cell apoptosis by promoting ceramide generation via hydrolysis of sphingomyelins by neutral sphingomyelinase. Biochemistry 2007; 46:10170-85.
30. Bruno A, Laurent G, Averbeck D et al. Lack of ceramide generation in TF-1 human myeloid leukemic cells resistant to ionizing radiation. Cell Death Differ 1998; 5:172-82.
31. Maddens S, Charruyer A, Plo I et al. Kit signaling inhibits the sphingomyelin-ceramide pathway through PLC gamma 1: implication in stem cell factor radioprotective effect. Blood 2002; 100:1294-301.
32. Liu B, Hannun Y. Inhibition of the neutral magnesium-dependent sphingomyelinase by glutathione. J Biol Chem 1997; 272:16281-7.
33. Marchesini N, Hannun Y. Acid and neutral sphingomyelinases: roles and mechanisms of regulation. Biochem Cell Biol 2004; 82:27-44.
34. Schneider PB, Kennedy EP. Sphingomyelinase in normal human spleens and in spleens from subjects with Niemann-Pick disease. J Lipid Res 1967; 8(3):202-9.
35. Santana P, Pena L, Haimovitz-Friedman A et al. Acid sphingomyelinase-deficient human lymphoblasts and mice are defective in radiation-induced apoptosis. Cell 1996; 86:189-99.

36. Zundel W, Giaccia A. Inhibition of the anti-apoptotic PI(3)K/Akt/Bad pathway by stress. Genes Dev 1998; 12(13):1941-6.
37. Schutze S, Potthoff K, Machleidt T et al. TNF activates NF-kB by phosphatidylcholine-specific phospholipase C-induced "acidic" sphingomyelin breakdown. Cell 1992; 71:765-76.
38. Kashkar H, Wiegmann K, Yazdanpanah B et al. Acid sphingomyelinase is indispensable for UV light-induced Bax conformational change at the mitochondrial membrane. J Biol Chem 2005; 280(21):20804-13.
39. Yu ZF, Nikolova-Karakashian M, Zhou D et al. Pivotal role for acidic sphingomyelinase in cerebral ischemia-induced ceramide and cytokine production and neuronal apoptosis. J Mol Neurosci 2000; 15(2):85-97.
40. Schissel S, Keesler G, Schuchman E et al. The cellular trafficking and zinc dependence of secretory and lysosomal sphingomyelinase, two products of the acid sphingomyelinase gene. J Biol Chem 1998; 273:18250-9.
41. Sathishkumar S, Boyanovsky B, Karakashian AA et al. Elevated sphingomyelinase activity and ceramide concentration in serum of patients undergoing high dose spatially fractionated radiation treatment: implications for endothelial apoptosis. Cancer Biol Ther 2005; 4(9):979-86.
42. Wright K, Messing E, Reeder J. Increased expression of the acid sphingomyelinase-like protein ASML3a in bladder tumors. J Urol 2002; 168:2645-9.
43. Nilsson A. The presence of spingomyelin- and ceramide-cleaving enzymes in the small intestinal tract. Biochim Biophys Acta 1969; 176(2):339-47.
44. Duan RD. Alkaline sphingomyelinase: an old enzyme with novel implications. Biochim Biophys Acta 2006; 1761(3):281-91.
45. Duan RD. Anticancer compounds and sphingolipid metabolism in the colon. In Vivo 2005; 19(1):293-300.
46. Bose R, Verheij M, Haimovitz-Friedman A et al. Ceramide synthase mediates daunorubicin-induced apoptosis: an alternative mechanism for generating death signals. Cell 1995; 82(3):405-14.
47. Son JH, Yoo HH, Kim DH. Activation of de novo synthetic pathway of ceramides is responsible for the initiation of hydrogen peroxide-induced apoptosis in HL-60 cells. J Toxicol Environ Health A 2007; 70(15-16):1310-8.
48. Petrache I, Natarajan V, Zhen L et al. Ceramide upregulation causes pulmonary cell apoptosis and emphysema-like disease in mice. Nat Med 2005; 11:491-8.
49. Turpin S, Lancaster G, Darby I et al. Apoptosis in skeletal muscle myotubes is induced by ceramides and is positively related to insulin resistance. Am J Physiol Endocrinol Metab 2006; 291:E1341-350.
50. Itoh G, Tamura J, M S et al. DNA fragmentation of human infarcted myocardial cells demonstrated by the nick end labeling method and DNA agarose gel electrophoresis. Am J Pathol 1995; 146:1325-31.
51. Seumois G, Fillet M, Gillet L et al. De novo C16- and C24-ceramide generation contributes to spontaneous neutrophil apoptosis. J Leukoc Biol 2007; 81:1477-86.
52. Itoh Y, Yano T, Sendo T et al. Involvement of de novo ceramide synthesis in radiocontrast-induced renal tubular cell injury. Kidney Int 2006; 69:288-97.
53. Moussavi M, Assi K, Gomez-Munoz A et al. Curcumin mediates ceramide generation via the de novo pathway in colon cancer cells. Carcinogenesis 2006; 27(8):1636-44.
54. Signorelli P, Ghidoni R. Resveratrol as an anticancer nutrient: molecular basis, open questions and promises. J Nutr Biochem 2005; 16:449-66.
55. Scarlatti F, Sala G, Somenzi G et al. Resveratrol induces growth inhibition and apoptosis in metastatic breast cancer cells via de novo ceramide signaling. FASEB J 2003; 17(15):2339-41.
56. Chan T, Morin P, Vogelstein B et al. Mechanisms underlying nonsteroidal antiinflammatory drug-mediated apoptosis. Proc Natl Acad Sci USA 1998; 95:681-6.
57. Ruvolo PP, Clark W, Mumby M et al. A functional role for the B56 alpha-subunit of protein phosphatase 2A in ceramide-mediated regulation of Bcl2 phosphorylation status and function. J Biol Chem 2002; 277(25):22847-52.
58. Stoica BA, Movsesyan VA, Lea PMt et al. Ceramide-induced neuronal apoptosis is associated with dephosphorylation of Akt, BAD, FKHR, GSK-3beta and induction of the mitochondrial-dependent intrinsic caspase pathway. Mol Cell Neurosci 2003; 22(3):365-82.
59. von Haefen C, Wieder T, Gillissen B et al. Ceramide induces mitochondrial activation and apoptosis via a Bax-dependent pathway in human carcinoma cells. Oncogene 2002; 21(25):4009-19.
60. Kolesnick R, Fuks Z. Radiation and ceramide-induced apoptosis. Oncogene 2003; 22(37):5897-906.
61. Scorrano L, Oakes SA, Opferman JT et al. BAX and BAK regulation of endoplasmic reticulum Ca2+: a control point for apoptosis. Science 2003; 300(5616):135-9.
62. Kim HJ, Mun JY, Chun YJ et al. Bax-dependent apoptosis induced by ceramide in HL-60 cells. FEBS Lett 2001; 505(2):264-8.

63. Boise L, González-García M, Postema C et al. Bcl-x, a bcl-2-related gene that functions as a dominant regulator of apoptotic cell death. Cell 1993; 74:579-608.
64. Chalfant C, Rathman K, Pinkerman R et al. De novo ceramide regulates the alternative splicing of caspase 9 and Bcl-x in A549 lung adenocarcinoma cells. Dependence on protein phosphatase-1. J Biol Chem 2002; 277:12587-95.
65. Heinrich M, Wickel M, Schneider-Brachert W et al. Cathepsin D targeted by acid sphingomyelinase-derived ceramide. EMBO J 1999; 18(19):5252-63.
66. Heinrich M, Neumeyer J, Jakob M et al. Cathepsin D links TNF-induced acid sphingomyelinase to Bid-mediated caspase-9 and -3 activation. Cell Death Differ 2004; 11(5):550-63.
67. Grassme H, Jekle A, Riehle A et al. CD95 signaling via ceramide-rich membrane rafts. J Biol Chem 2001; 276(23):20589-96.
68. Kilkus J, Goswami R, Testai FD et al. Ceramide in rafts (detergent-insoluble fraction) mediates cell death in neurotumor cell lines. J Neurosci Res 2003; 72(1):65-75.
69. Spiegel S, AH Merrill J. Sphingolipid metabolism and cell growth regulation. FASEB J 1996; 10:1388-97.
70. Yatomi Y, Ohmori T, Rile G et al. Sphingosine 1-phosphate as a major bioactive lysophospholipid that is released from platelets and interacts with endothelial cells. Blood 2000; 96(10):3431-8.
71. Radeff-Huang J, Seasholtz TM, Matteo RG et al. G protein mediated signaling pathways in lysophospholipid induced cell proliferation and survival. J Cell Biochem 2004; 92(5):949-66.
72. Hobson JP, Rosenfeldt HM, Barak LS et al. Role of the sphingosine-1-phosphate receptor EDG-1 in PDGF-induced cell motility. Science 2001; 291(5509):1800-3.
73. Brinkmann V, Cyster J, Hla T. FTY720: sphingosine 1-phosphate receptor-1 in the control of lymphocyte egress and endothelial barrier function. Am J Transplant 2004; 4:1019-25.
74. Spiegel S, Milstien S. Functions of a new family of sphingosine-1-phosphate receptors. Biochim Biophys Acta 2000; 1484(2-3):107-16.
75. Hla T, Lee MJ, Ancellin N et al. Lysophospholipids-receptor revelations. Science 2001; 294(5548):1875-8.
76. Payne SG, Milstien S, Spiegel S. Sphingosine-1-phosphate: dual messenger functions. FEBS Lett 2002; 531(1):54-7.
77. Xia P, Gamble JR, Wang L et al. An oncogenic role of sphingosine kinase. Curr Biol 2000; 10(23):1527-30.
78. Visentin B, Vekich J, Sibbald B et al. Validation of an anti-sphingosine-1-phosphate antibody as a potential therapeutic in reducing growth, invasion and angiogenesis in multiple tumor lineages. Cancer Cell 2006; 9:225-38.
79. Cuvillier O, Levade T. Sphingosine 1-phosphate antagonizes apoptosis of human leukemia cells by inhibiting release of cytochrome c and Smac/DIABLO from mitochondria. Blood 2001; 98:2828-36.
80. Goetzl E, Kong Y, Mei B. Lysophosphatidic acid and sphingosine 1-phosphate protection of T-cells from apoptosis in association with suppression of Bax. J Immunol 1999; 162(4):2049-56.
81. Cuvillier O, Rosenthal DS, Smulson ME et al. Sphingosine 1-phosphate inhibits activation of caspases that cleave poly(ADP-ribose) polymerase and lamins during Fas- and ceramide- mediated apoptosis in Jurkat T-lymphocytes. J Biol Chem 1998; 273(5):2910-6.
82. Betito S, Cuvillier O. Regulation by sphingosine 1-phosphate of Bax and Bad activities during apoptosis in a MEK-dependent manner. Biochem Biophys Res Commun 2006; 340(4):1273-7.
83. Taha TA, Kitatani K, Bielawski J et al. Tumor necrosis factor induces the loss of sphingosine kinase-1 by a cathepsin B-dependent mechanism. J Biol Chem 2005; 280(17):17196-202.
84. Liu H, Toman RE, Goparaju S et al. Sphingosine kinase type 2 is a putative BH3-Only protein that induces apoptosis. J Biol Chem 2003; 278:40330-6.
85. Le Stunff H, Giussani P, Maceyka M et al. Recycling of sphingosine is regulated by the concerted actions of sphingosine-1-phosphate phosphohydrolase 1 and sphingosine kinase 2. J Biol Chem 2007; 282:34372-80.
86. Spiegel S, Milstien S. Functions of the multifaceted family of sphingosine kinases and some close relatives. J Biol Chem 2006.
87. Woodgett JR. Recent advances in the protein kinase B signaling pathway. Curr Opin Cell Biol 2005; 17(2):150-7.
88. Ruggero D, Sonenberg N. The Akt of translational control. Oncogene 2005; 24(50):7426-34.
89. Tohma Y, Gratas C, Biernat W et al. PTEN (MMAC1) mutations are frequent in primary glioblastomas (de novo) but not in secondary glioblastomas. J Neuropathol Exp Neurol 1998; 57(7):684-9.
90. Carpten JD, Faber AL, Horn C et al. A transforming mutation in the pleckstrin homology domain of AKT1 in cancer. Nature 2007; 448(7152):439-44.
91. Vivanco I, Sawyers CL. The phosphatidylinositol 3-Kinase AKT pathway in human cancer. Nat Rev Cancer 2002; 2(7):489-501.

92. Fox TE, Houck KL, O'Neill SM et al. Ceramide recruits and activates protein kinase C zeta (PKC zeta) within structured membrane microdomains. J Biol Chem 2007; 282(17):12450-7.
93. Powell DJ, Hajduch E, Kular G et al. Ceramide disables 3-phosphoinositide binding to the pleckstrin homology domain of protein kinase B (PKB)/Akt by a PKCzeta-dependent mechanism. Mol Cell Biol 2003; 23(21):7794-808.
94. Salinas M, Lopez-Valdaliso R, Martin D et al. Inhibition of PKB/Akt1 by C2-ceramide involves activation of ceramide-activated protein phosphatase in PC12 cells. Mol Cell Neurosci 2000; 15(2):156-69.
95. Stratford S, Hoehn K, Liu F et al. Regulation of insulin action by ceramide: dual mechanisms linking ceramide accumulation to the inhibition of Akt/protein kinase B. J Biol Chem 2004; 279:36608-15.
96. Law B, Rossie S. The dimeric and catalytic subunit forms of protein phosphatase 2A from rat brain are stimulated by C2-ceramide. J Biol Chem 1995; 270(21):12808-13.
97. Dobrowsky R, Kamibayashi C, Mumby M et al. Ceramide activates a heterotrimeric protein phosphatase 2A. J Biol Chem 1993; 268(21):15523-30.
98. Dey R, Majumder N, Bhattacharjee S et al. Leishmania donovani-induced ceramide as the key mediator of Akt dephosphorylation in murine macrophages: role of protein kinase Czeta and phosphatase. Infect Immun 2007; 75:2136-42.
99. Schmitz-Peiffer C, Craig D, Biden T. Ceramide generation is sufficient to account for the inhibition of the insulin-stimulated PKB pathway in C2C12 skeletal muscle cells pretreated with palmitate. J Biol Chem 1999; 274:24202-10.
100. Guan L, Song K, Pysz M et al. Protein kinase C-mediated down-regulation of cyclin D1 involves activation of the translational repressor 4E-BP1 via a phosphoinositide 3-kinase/Akt-independent, protein phosphatase 2A-dependent mechanism in intestinal epithelial cells. J Biol Chem 2007; 282:14213-25.
101. Goswami R, Singh D, Phillips G et al. Ceramide regulation of the tumor suppressor phosphatase PTEN in rafts isolated from neurotumor cell lines. J Neurosci Res 2005; 81:541-50.
102. Gómez-Muñoz A, Kong J, Parhar K et al. Ceramide-1-phosphate promotes cell survival through activation of the phosphatidylinositol 3-kinase/protein kinase B pathway. FEBS Lett 2005; 579:3744-50.
103. Means C, Xiao C, Li Z et al. Sphingosine 1-phosphate S1P2 and S1P3 receptor-mediated Akt activation protects against in vivo myocardial ischemia-reperfusion injury. Am J Physiol Heart Circ Physiol 2007; 292:H2944-H2951.
104. Igarashi J, Bernier SG, Michel T. Sphingosine 1-phosphate and activation of endothelial nitric-oxide synthase. differential regulation of Akt and MAP kinase pathways by EDG and bradykinin receptors in vascular endothelial cells. J Biol Chem 2001; 276(15):12420-6.
105. Levine Y, Li G, Michel T. Agonist-modulated regulation of AMP-activated protein kinase (AMPK) in endothelial cells. Evidence for an AMPK -> Rac1 -> Akt -> endothelial nitric-oxide synthase pathway. J Biol Chem 2007; 282:20351-64.
106. Fieber C, Eldgridge J, Taha T et al. Modulation of total Akt kinase by increased expression of a single isoform: requirement of the sphingosine-1-phosphate receptor Edg3/S1P3, for the VEGF-dependent expression of Akt3 in primary endothelial cells. Exp Cell Res 2006; 312:1164-73.
107. Tanimoto T, Jin ZG, Berk BC. Transactivation of vascular endothelial growth factor (VEGF) receptor Flk-1/KDR is involved in sphingosine 1-phosphate-stimulated phosphorylation of Akt and endothelial nitric-oxide synthase (eNOS). J Biol Chem 2002; 277(45):42997-3001.
108. Limaye V, Li X, Hahn C et al. Sphingosine kinase-1 enhances endothelial cell survival through a PECAM-1-dependent activation of PI-3K/Akt and regulation of Bcl-2 family members. Blood 2005; 105:3169-77.
109. Banno Y, Takuwa Y, Akao Y et al. Involvement of phospholipase D in sphingosine 1-phosphate-induced activation of phosphatidylinositol 3-kinase and Akt in Chinese hamster ovary cells overexpressing EDG3. J Biol Chem 2001; 276:35622-8.
110. Igarashi J, Michel T. Sphingosine 1-phosphate and isoform-specific activation of phosphoinositide 3-kinase beta. Evidence for divergence and convergence of receptor-regulated endothelial nitric-oxide synthase signaling pathways. J Biol Chem 2001; 276:36281-8.
111. Baudhuin LM, Jiang Y, Zaslavsky A et al. S1P3-mediated Akt activation and cross-talk with platelet-derived growth factor receptor (PDGFR). Faseb J 2004; 18(2):341-3.
112. Baudhuin L, Cristina K, Lu J et al. Akt activation induced by lysophosphatidic acid and sphingosine-1-phosphate requires both mitogen-activated protein kinase kinase and p38 mitogen-activated protein kinase and is cell-line specific. Mol Pharmacol 2002; 62:660-71.
113. Sanchez T, Skoura A, Wu M et al. Induction of vascular permeability by the sphingosine-1-phosphate receptor-2 (S1P2R) and its downstream effectors ROCK and PTEN. Arterioscler Thromb Vasc Biol 2007; 27:1312-8.
114. Liang C, Feng P, Ku B et al. Autophagic and tumour suppressor activity of a novel Beclin1-binding protein UVRAG. Nat Cell Biol 2006; 8:688-99.
115. Mizushima N. Autophagy: process and function. Genes Dev 2007; 21(22):2861-73.

116. Wullschleger S, Loewith R, Hall MN. TOR signaling in growth and metabolism. Cell 2006; 124(3):471-84.
117. Takeuchi H, Kondo Y, Fujiwara K et al. Synergistic augmentation of rapamycin-induced autophagy in malignant glioma cells by phosphatidylinositol 3-kinase/protein kinase B inhibitors. Cancer Res 2005; 65(8):3336-46.
118. Scarlatti F, Bauvy C, Ventruti A et al. Ceramide-mediated macroautophagy involves inhibition of protein kinase B and up-regulation of beclin 1. J Biol Chem 2004; 279:18384-91.
119. Patschan S, Chen J, Polotskaia A et al. Lipid mediators of autophagy in stress-induced premature senescence of endothelial cells. Am J Physiol Heart Circ Physiol 2008; 294:H1119-H29.
120. Daido S, Kanzawa T, Yamamoto A et al. Pivotal role of the cell death factor BNIP3 in ceramide-induced autophagic cell death in malignant glioma cells. Cancer Res 2004; 64(12):4286-93.
121. Lavieu G, Scarlatti F, Sala G et al. Regulation of autophagy by sphingosine kinase 1 and its role in cell survival during nutrient starvation. J Biol Chem 2006; 281:8518-27.
122. Boya P, Gonzalez-Polo RA, Casares N et al. Inhibition of macroautophagy triggers apoptosis. Mol Cell Biol 2005; 25(3):1025-40.
123. Verheij M, Bose R, Lin X et al. Requirement for ceramide-initiated SAPK/JNK signalling in stress-induced apoptosis. Nature 1996; 380:75-9.
124. Basu S, Bayoumy S, Zhang Y et al. BAD enables ceramide to signal apoptosis via Ras and Raf-1. J Biol Chem 1998; 273(46):30419-26.
125. Cuvillier O, Pirianov G, Kleuser B et al. Suppression of ceramide-mediated programmed cell death by sphingosine-1-phosphate. Nature 1996; 381(6585):800-3.
126. Edsall LC, Cuvillier O, Twitty S et al. Sphingosine kinase expression regulates apoptosis and caspase activation in PC12 cells. J Neurochem 2001; 76(5):1573-84.
127. Usui S, Sugimoto N, Takuwa N et al. Blood lipid mediator sphingosine 1-phosphate potently stimulates platelet-derived growth factor-A and -B chain expression through S1P1-Gi-Ras-MAPK-dependent induction of Kruppel-like factor 5. J Biol Chem 2004; 279(13):12300-11.
128. Hsieh HL, Sun CC, Wu CB et al. Sphingosine 1-phosphate induces EGFR expression via Akt/NF-kappaB and ERK/AP-1 pathways in rat vascular smooth muscle cells. J Cell Biochem 2007.
129. Libermann TA, Nusbaum HR, Razon N et al. Amplification, enhanced expression and possible rearrangement of EGF receptor gene in primary human brain tumours of glial origin. Nature 1985; 313(5998):144-7.
130. Kohler M, Janz I, Wintzer HO et al. The expression of EGF receptors, EGF-like factors and c-myc in ovarian and cervical carcinomas and their potential clinical significance. Anticancer Res 1989; 9(6):1537-47.
131. Yoshida K, Tosaka A, Takeuchi S et al. Epidermal growth factor receptor content in human renal cell carcinomas. Cancer 1994; 73(7):1913-8.
132. Dbaibo G, Pushkareva M, Jayadev S et al. Retinoblastoma gene product as a downstream target for a ceramide-dependent pathway of growth arrest. Proc Natl Acad Sci USA 1995; 92:1347-51.
133. Phillips DC, Hunt JT, Moneypenny CG et al. Ceramide-induced G2 arrest in rhabdomyosarcoma (RMS) cells requires p21Cip1/Waf1 induction and is prevented by MDM2 overexpression. Cell Death Differ 2007; 14(10):1780-91.
134. Wang J, Lv XW, Shi JP et al. Mechanisms involved in ceramide-induced cell cycle arrest in human hepatocarcinoma cells. World J Gastroenterol 2007; 13(7):1129-34.
135. Blackburn EH, Greider CW, Szostak JW. Telomeres and telomerase: the path from maize, Tetrahymena and yeast to human cancer and aging. Nat Med 2006; 12(10):1133-8.
136. Greider CW, Blackburn EH. Telomeres, telomerase and cancer. Sci Am 1996; 274(2):92-7.
137. Blackburn EH, Greider CW, Henderson E et al. Recognition and elongation of telomeres by telomerase. Genome 1989; 31(2):553-60.
138. Weinrich SL, Pruzan R, Ma L et al. Reconstitution of human telomerase with the template RNA component hTR and the catalytic protein subunit hTRT. Nat Genet 1997; 17(4):498-502.
139. Ogretmen B, Kraveka JM, Schady D et al. Molecular mechanisms of ceramide-mediated telomerase inhibition in the A549 human lung adenocarcinoma cell line. J Biol Chem 2001; 276(35):32506-14.
140. Wooten-Blanks LG, Song P, Senkal CE et al. Mechanisms of ceramide-mediated repression of the human telomerase reverse transcriptase promoter via deacetylation of Sp3 by histone deacetylase 1. FASEB J 2007; 21(12):3386-97.
141. Xu JX, Morii E, Liu Y et al. High tolerance to apoptotic stimuli induced by serum depletion and ceramide in side-population cells: high expression of CD55 as a novel character for side-population. Exp Cell Res 2007; 313(9):1877-85.
142. Ozaki H, Hla T, Lee MJ. Sphingosine-1-phosphate signaling in endothelial activation. J Atheroscler Thromb 2003; 10(3):125-31.

143. Park KS, Kim MK, Lee HY et al. S1P stimulates chemotactic migration and invasion in OVCAR3 ovarian cancer cells. Biochem Biophys Res Commun 2007; 356(1):239-44.
144. Guo XZ, Zhang WW, Wang LS et al. [Adenovirus-mediated overexpression of KAI1 suppresses sphingosine kinase activation and metastasis of pancreatic carcinoma cells]. Zhonghua Nei Ke Za Zhi 2006; 45(9):752-4.
145. Selzner M, Bielawska A, Morse MA et al. Induction of apoptotic cell death and prevention of tumor growth by ceramide analogues in metastatic human colon cancer. Cancer Res 2001; 61(3):1233-40.
146. Elojeimy S, Liu X, McKillop JC et al. Role of acid ceramidase in resistance to FasL: Therapeutic approaches based on acid ceramidase inhibitors and FasL gene therapy. Mol Ther 2007; 15(7):1259-63.
147. Holman DH, Turner LS, El-Zawahry A et al. Lysosomotropic acid ceramidase inhibitor induces apoptosis in prostate cancer cells. Cancer Chemother Pharmacol 2008; 61(2):231-42.
148. Swanton C, Marani M, Pardo O et al. Regulators of mitotic arrest and ceramide metabolism are determinants of sensitivity to paclitaxel and other chemotherapeutic drugs. Cancer Cell 2007; 11(6):498-512.
149. Kawamori T, Osta W, Johnson KR et al. Sphingosine kinase 1 is up-regulated in colon carcinogenesis. FASEB J 2006; 20(2):386-8.
150. Johnson KR, Johnson KY, Crellin HG et al. Immunohistochemical distribution of sphingosine kinase 1 in normal and tumor lung tissue. J Histochem Cytochem 2005; 53(9):1159-66.
151. French K, Schrecengost R, Lee B et al. Discovery and evaluation of inhibitors of human sphingosine kinase. Cancer Res 2003; 63:5962-9.
152. French K, Upson J, Keller S et al. Antitumor activity of sphingosine kinase inhibitors. J Pharmacol Exp Ther 2006; 318:596-603.
153. Le Scolan E, Pchejetski D, Banno Y et al. Overexpression of sphingosine kinase 1 is an oncogenic event in erythroleukemic progression. Blood 2005; 106(5):1808-16.
154. Kohno M, Momoi M, Oo M et al. Intracellular role for sphingosine kinase 1 in intestinal adenoma cell proliferation. Mol Cell Biol 2006; 26:7211-23.
155. Oskouian B, Sooriyakumaran P, Borowsky A et al. Sphingosine-1-phosphate lyase potentiates apoptosis via p53- and p38-dependent pathways and is downregulated in colon cancer. Proc Natl Acad Sci USA 2006; 103:17384-9.
156. Symolon H, Schmelz E, Dillehay D et al. Dietary soy sphingolipids suppress tumorigenesis and gene expression in 1,2-dimethylhydrazine-treated CF1 mice and ApcMin/+ mice. J Nutr 2004; 134:1157-61.
157. Schmelz EM, Dillehay DL, Webb SK et al. Sphingomyelin consumption suppresses aberrant colonic crypt foci and increases the proportion of adenomas versus adenocarcinomas in CF1 mice treated with 1,2-dimethylhydrazine: implications for dietary sphingolipids and colon carcinogenesis. Cancer Res 1996; 56(21):4936-41.
158. Schmelz EM, Roberts PC, Kustin EM et al. Modulation of intracellular beta-catenin localization and intestinal tumorigenesis in vivo and in vitro by sphingolipids. Cancer Res 2001; 61(18):6723-9.
159. Bonhoure E, Pchejetski D, Aouali N et al. Overcoming MDR-associated chemoresistance in HL-60 acute myeloid leukemia cells by targeting sphingosine kinase-1. Leukemia 2006; 20(1):95-102.
160. Bektas M, Jolly PS, Muller C et al. Sphingosine kinase activity counteracts ceramide-mediated cell death in human melanoma cells: role of Bcl-2 expression. Oncogene 2005; 24(1):178-87.
161. Schindler T, Bornmann W, Pellicena P et al. Structural mechanism for STI-571 inhibition of abelson tyrosine kinase. Science 2000; 289(5486):1938-42.
162. Baran Y, Salas A, Senkal CE et al. Alterations of ceramide/sphingosine 1-phosphate rheostat involved in the regulation of resistance to imatinib-induced apoptosis in K562 human chronic myeloid leukemia cells. J Biol Chem 2007; 282(15):10922-34.
163. Bittman R. Synthetic sphingolipids as bioactive molecules: roles in regulation of cell function. In: Wiley, editor. Wiley Encyclopedia 2008.
164. McCormack P, Goa K. Miglustat. Drugs 2003; 63:2427-36.
165. Weiss M, Hettmer S, Smith P et al. Inhibition of melanoma tumor growth by a novel inhibitor of glucosylceramide synthase. Cancer Res 2003; 63(13):3654-8.
166. Guerrera M, Ladisch S. N-butyldeoxynojirimycin inhibits murine melanoma cell ganglioside metabolism and delays tumor onset. Cancer Lett 2003; 201(1):31-40.
167. Dbaibo GS, Kfoury Y, Darwiche N et al. Arsenic trioxide induces accumulation of cytotoxic levels of ceramide in acute promyelocytic leukemia and adult T-cell leukemia/lymphoma cells through de novo ceramide synthesis and inhibition of glucosylceramide synthase activity. Haematologica 2007; 92(6):753-62.
168. Billich A, Bornancin F, Mechtcheriakova D et al. Basal and induced sphingosine kinase 1 activity in A549 carcinoma cells: function in cell survival and IL-1beta and TNF-alpha induced production of inflammatory mediators. Cell Signal 2005; 17:1203-17.

169. Gamble J, Xia P, Hahn C et al. Phenoxodiol, an experimental anticancer drug, shows potent antiangiogenic properties in addition to its antitumour effects. Int J Cancer 2006; 118:2412-20.
170. Leroux M, Auzenne E, Evans R et al. Sphingolipids and the sphingosine kinase inhibitor, SKI II, induce BCL-2-independent apoptosis in human prostatic adenocarcinoma cells. Prostate 2007; 67:1699-717.
171. Leroux ME, Auzenne E, Evans R et al. Sphingolipids and the sphingosine kinase inhibitor, SKI II, induce BCL-2-independent apoptosis in human prostatic adenocarcinoma cells. Prostate 2007; 67(15):1699-717.
172. Yasui H, Hideshima T, Raje N et al. FTY720 induces apoptosis in multiple myeloma cells and overcomes drug resistance. Cancer Res 2005; 65(16):7478-84.
173. Sani BP, Shealy YF, Hill DL. N-(4-hydroxyphenyl)retinamide: interactions with retinoid-binding proteins/receptors. Carcinogenesis 1995; 16(10):2531-4.
174. Sheikh MS, Shao ZM, Li XS et al. N-(4-hydroxyphenyl)retinamide (4-HPR)-mediated biological actions involve retinoid receptor-independent pathways in human breast carcinoma. Carcinogenesis 1995; 16(10):2477-86.
175. Wang H, Maurer BJ, Reynolds CP et al. N-(4-hydroxyphenyl)retinamide elevates ceramide in neuroblastoma cell lines by coordinate activation of serine palmitoyltransferase and ceramide synthase. Cancer Res 2001; 61(13):5102-5.
176. Maurer BJ, Melton L, Billups C et al. Synergistic cytotoxicity in solid tumor cell lines between N-(4-hydroxyphenyl)retinamide and modulators of ceramide metabolism. J Natl Cancer Inst 2000; 92(23):1897-909.
177. Erdreich-Epstein A, Tran LB, Bowman NN et al. Ceramide signaling in fenretinide-induced endothelial cell apoptosis. J Biol Chem 2002; 277(51):49531-7.
178. Kraveka J, Li L, Szulc Z et al. Involvement of dihydroceramide desaturase in cell cycle progression in human neuroblastoma cells. J Biol Chem 2007; 282:16718-28.
179. Jiang Q, Wong J, Fyrst H et al. gamma-Tocopherol or combinations of vitamin E forms induce cell death in human prostate cancer cells by interrupting sphingolipid synthesis. Proc Natl Acad Sci USA 2004; 101(51):17825-30.
180. Morales PR, Dillehay DL, Moody SJ et al. Safingol toxicology after oral administration to TRAMP mice: demonstration of safingol uptake and metabolism by N-acylation and N-methylation. Drug Chem Toxicol 2007; 30(3):197-216.
181. Grammatikos G, Teichgraber V, Carpinteiro A et al. Overexpression of acid sphingomyelinase sensitizes glioma cells to chemotherapy. Antioxid Redox Signal 2007; 9(9):1449-56.
182. Won JS, Singh I. Sphingolipid signaling and redox regulation. Free Radic Biol Med 2006; 40(11):1875-88.
183. Andersson D, Cheng Y, Duan RD. Ursolic acid inhibits the formation of aberrant crypt foci and affects colonic sphingomyelin hydrolyzing enzymes in azoxymethane-treated rats. J Cancer Res Clin Oncol 2008; 134(1):101-7.
184. Rosato RR, Maggio SC, Almenara JA et al. The histone deacetylase inhibitor LAQ824 induces human leukemia cell death through a process involving XIAP down-regulation, oxidative injury and the acid sphingomyelinase-dependent generation of ceramide. Mol Pharmacol 2006; 69(1):216-25.
185. Darroch PI, Dagan A, Granot T et al. A lipid analogue that inhibits sphingomyelin hydrolysis and synthesis, increases ceramide and leads to cell death. J Lipid Res 2005; 46(11):2315-24.
186. Suomalainen L, Hakala JK, Pentikainen V et al. Sphingosine-1-phosphate in inhibition of male germ cell apoptosis in the human testis. J Clin Endocrinol Metab 2003; 88(11):5572-9.
187. Pchejetski D, Kunduzova O, Dayon A et al. Oxidative stress-dependent sphingosine kinase-1 inhibition mediates monoamine oxidase A-associated cardiac cell apoptosis. Circ Res 2007; 100(1):41-9.

Chapter 14

Therapeutic Strategies for Diabetes and Complications: A Role for Sphingolipids?

Todd E. Fox and Mark Kester*

Abstract

Diabetes is a debilitating chronic disease that has no cure and can only be managed by pharmaceutical or nutritional interventions. Worldwide, the incidence of diabetes and diabetic complications is dramatically increasing. This may reflect the incomplete knowledge base underlying the role of inflammatory or nutritional stresses to exacerbate diabetic complications. Despite the knowledge that hyperlipidemia is a cardinal feature of both Type 1 and 2 diabetes, the actual lipid species that contribute to complications such as diabetic nephropathy, retinopathy, neuropathy and cardiovascular disease have not been well defined, or have not elucidated new treatment strategies. Sphingolipids comprise only a fraction of total lipids but a body of evidence has now identified dysfunctional sphingolipid metabolism and/or generation of specific sphingolipid metabolites as contributors to diabetic complications. This review suggests that pharmacological therapies that target dysfunctional sphingolipid metabolism and/or signaling may prove beneficial in decreasing the chronic pathology of hyperglycemia and hyperlipidemia. Moreover, the review suggests that these treatment options may also prove beneficial to ameliorate or delay pancreatic beta cell failure.

Diabetes and Insulin Resistance

Diabetes affects over 246 million people worldwide and is increasing at an alarming rate in both developed and developing countries.[1] Worldwide, diabetes is one of the top five leading causes of disease-related death.[1] Despite insulin treatment, good nutritional support as well as new treatment modalities to increase the effectiveness of insulin; diabetic complications continue to plague diabetic patients. Moreover, a worldwide obesity epidemic contributes to the increase in metabolic syndrome and diabetic patients.

Simply stated, diabetes is the failure of the body to correctly utilize glucose due to diminished insulin signaling. Insulin is an essential anabolic hormone and growth factor that promotes cell function and survival. Diminished insulin levels and/or insulin signaling lead to the catabolic features of diabetes; hyperglycemia, hyperlipidemia and accelerated protein degradation. This failure in insulin signaling can reflect either impaired insulin production (Type 1 diabetes) or impaired insulin responsiveness (Type 2 diabetes). Both of these pathologies lead to hyperglycemia, which contributes to serious pathological complications, including nephropathy, neuropathy, retinopathy or cardiovascular disease. The underlying etiology of both diseases are still unknown, but may reflect an autoimmune disease targeting pancreatic beta islet cells (Type 1) and an inflammatory

*Corresponding Author: Mark Kester—Penn State College of Medicine, Department of Pharmacology, Hershey, Pennsylvania 17033, USA. Email: mkester@psu.edu

Sphingolipids as Signaling and Regulatory Molecules, edited by Charles Chalfant and Maurizio Del Poeta.

condition exacerbated by nutritional stress or obesity (Type 2) that impairs insulin signaling in insulin responsive cell types.[2] Even though the initial insult may be distinctly different for Type 1 and 2 patients, the resultant insulin resistant phenotype contributes to diabetic complications in both diseases.

Type 1 patients develop secondary insulin resistance in insulin responsive tissues like Type 2 patients.[3] Type 2 patients develop secondary pancreatic symptoms, similar to Type 1 patients, which further worsens clinical outcomes. Several clinical studies are now reporting that global insulin resistance and resultant diabetic complications is more than just an increase in glucose levels and may reflect nutritional stresses. In fact, the EURODIAB Prospective Complications Study showed that insulin resistance is a major predictive factor for the development of complications in Type 1 diabetes.[4] These findings have been confirmed in other studies showing dramatic increased risk of complications with body mass index >25 kg/m^2.[5] In addition, persons with Type 1 diabetes and the metabolic syndrome have a 3.7 fold increased risk of severe retinopathy, a 22.8 fold increased risk of stroke and a 7.3 fold increased risk of peripheral vascular disease.[6] A recent study shows that the estimated glucose disposal rate, but not metabolic syndrome or insulin dose, is highly related with the development of retinopathy, nephropathy and macrovascular disease in Diabetes Control and Complications Trial (DCCT) subjects.[7] This finding is in line with the observation that insulin doses parallel plasma triglyceride concentrations in children with Type 1 diabetes.[8] Together, these findings indicate that complications of diabetes depend on factors other than hyperglycemia and include insulin resistance and overall nutritional status.

The contribution of hyperlipidemia to diabetes is being defined. While sphingolipids typically constitute a relatively minor component of the lipid environment compared to glycerolipids and sterols, altered sphingolipid metabolism has been implicated in diabetes pathogenesis. In vitro studies are beginning to define the biochemical and biophysical signaling mechanisms by which sphingolipid metabolites contribute to diabetic complications. Particular areas in which sphingolipids are being investigated include the detrimental effects of the generation of reactive oxygen species,[9] induction of the hexosamine biosynthetic pathway,[10,11] inflammation[12] and increases in advanced glycated end products,[13-15] all of which may contribute to diabetic complications. As these mechanisms, including signaling, have been reviewed elsewhere,[16-19] we have focused the present review on the in vivo animal and human studies that have implicated dysfunctional sphingolipid metabolism and/or accumulation of sphingolipid metabolites in insulin resistance and/or diabetic complications.

Insulin Resistance and Altered Sphingolipid Metabolism

Classically, insulin resistance is typically associated with the liver, skeletal muscle and adipose tissue. Consequences of insulin resistance in these tissues include glucose intolerance and hyperlipidemia. Thus, it is not surprising that altered sphingolipid metabolism has been observed in multiple obese rodent models of Type 2 diabetes. What is intriguing is the restoration of insulin sensitivity by pharmacological strategies that target dysfunctional sphingolipid metabolism.

Recent studies have begun to define the role of hyperlipidemia to alter sphingolipid metabolites and regulate insulin resistance.[20] Ceramide accumulates in skeletal muscle and liver after infusion of lard oil emulsions into the bloodstream of Sprague-Dawley rats. Treating these animals with inhibitors of serine palmitoyltransferase (SPT) prevented the lard oil-induced increases in ceramide accumulation and maintained insulin-stimulated glucose utilization. Using an ex vivo approach, administration of saturated fats, including palmitate, a precursor necessary for serine palmitoyltransferase-induced generation of de novo sphingolipids, to isolated rodent muscles induced ceramide-dependent insulin resistance and impaired 2-deoxyglucose uptake. Knock-out animal models have also been quite useful to define a role for sphingolipid metabolism in insulin resistance. Palmitate-induced insulin resistance was abrogated in isolated soleus muscles from mice lacking one allele of dihydroceramide desaturase 1.[20] Deficiency of the acid sphingomyelinase gene product protected against saturated fat diet-induced hyperglycemia, insulin resistance and impaired hepatic triacylglyceride accumulation.[21] Similarly, hyperlipidemia and altered sphingolipid metabolites was noted in the liver of *ob/ob* mice where $C_{16:0}$ and $C_{18:0}$ ceramides positively correlated with liver triglycerides.[22] Yet, despite evidence that dysfunctional sphingolipid metabolism contributes to insulin resistance, the role of specific glucosylated or phosphorylated sphingolipid metabolites to contribute to insulin resistance are only recently being identified.

It should not be surprising that glycosphingolipids accumulate in obese animal models of diabetes. As examples, Zucker diabetic fatty *fa/fa* rats (ZDF) and *ob/ob* mice display increased levels of glucosylceramide in liver,[23] consistent with an increase in GM3 synthase expression in adipose tissue.[24] Zucker diabetic fatty rats also have elevated skeletal muscle GM3 ganglioside levels.[25] However, the latter finding is in contrast to the first mentioned group,[23] which found that neither glucosylceramide nor GM3 gangliosides were elevated in muscle or liver of ZDF rats.[23] These apparent discrepancies are usually dismissed as consequences of different tissues or animal models. Yet, these contradictions may also reflect different analytical strategies to characterize and quantify sphingolipid metabolites in addition to age and diets of the animals. Alternatively, they may reflect a "global" increase in sphingolipid metabolism, manifested by transcriptional, translational and posttranslational modifications of sphingolipid and glycosphingolipid metabolic enzymes. As an example, TNFα-induced insulin resistance can be reversed by inhibiting ceramide production[26] or via inhibition of glucosylceramide synthase.[24] Alluding to the contribution of glycosphingolipids to diabetes complications, the ganglioside, GM3, can impair insulin signaling[24,27] and GM3 knockout mice have increased insulin sensitivity.[28] Similarly, GD3 has been implicated in TNFα and CD95 (Fas)-induced apoptosis[29] and GT1b has been demonstrated to inhibit pro-survival Akt activation in diabetic organs.[30,31] Of clinical interest, two recent publications have demonstrated increased glycemic control and insulin sensitivity upon pharmacological inhibition of glucosylceramide synthase.[23,25] In addition to Type 2 obese models, elevated concentrations of GM3 have also been observed in the Type 1 streptozotocin-induced diabetic rat model.[31,32] Finally, accumulating sphingolipid metabolites are not limited to glycosylated species, as elevated levels of sphingosine, at the expense of sphingomyelin or ceramide, has been observed in adipose tissue of diabetic *ob/ob* mice.[33]

Studies of human populations have not been as conclusive as the above animal experiments. Obese insulin-resistant human subjects displayed higher concentrations of ceramide, but not other sphingolipids, in skeletal muscle compared with lean subjects.[34] Similarly, a relationship was observed between decreased insulin sensitivity and increased ceramides in human skeletal muscle.[34,35] Consistent with these human studies, are additional studies demonstrating increased ceramide content in the skeletal muscle of insulin-resistant rats.[36,37] In contrast, a recent study observed no significant alterations in ceramide levels in skeletal muscle of Type 2 patients.[38] These disparate results for ceramide mass in diabetic populations may, in part, reflect overall dysfunctional sphingolipid metabolism and altered levels of phosphorylated or glycosylated sphingolipid metabolites. For example, it has also been demonstrated that plasma sphinganine and sphingosine were elevated in Type 2 diabetics compared with healthy control subjects, which may be consistent with elevated ceramide levels.[39] The presence of sulfated lactosylceramide in blood samples of Type 2 patients and low blood levels of sulfatides may also be linked to insulin resistance.[40] Thus far, these studies have involved analysis of relatively small numbers of patients or volunteers and have not revealed whether ceramide accumulation and/or altered sphingolipid metabolism predicts insulin resistance in lean or obese individuals. Again, biopsied tissues subjected to lipidomic analyses will be essential in defining the "sphingolipidsome" in controlled and poorly controlled diabetic patients, and establishing the contributions of specific sphingolipid metabolites to diabetic complications. Analysis of multiple sphingolipid species has the potential to define diabetic complications as a consequence of global dysfunctional sphingolipid metabolism. Moreover, the integrated use of lipidomics and metabolomics offers the potential to identify novel surrogate biomarkers of diabetes that can be monitored during pharmacological or nutritional treatment regimens.

Diabetic Pancreatic Dysfunction and Sphingolipids

The primary defect in Type 1 patients is autoimmune-mediated destruction of pancreatic beta cells. Deterioration of beta cell function in Type 2 diabetes, commonly referred to as beta cell exhaustion, occurs as a result of a compensatory mechanism by which the pancreas responds to global insulin resistance. Similar to insulin responsive tissues, altered sphingolipid metabolism or metabolites have been observed in pancreas tissue from diabetic animal models. Again, more importantly, pancreatic beta cell destruction can be reduced by pharmacological strategies that reduce sphingolipid metabolism. Serine palmitoyltransferase is upregulated in pancreatic islet cells of ZDF rats and inhibition of serine palmitoyltransferase reduces beta cell death.[41,42]

Consistent with elevated sphingolipid metabolism, ceramide is elevated in prediabetic and diabetic islets of ZDF rats. Mechanistically, palmitate (a de novo ceramide synthesis precursor) blocks insulin gene expression in primary rat islets[43] and induces beta-cell death,[44] possibly via de novo ceramide production. An alternative mechanism implicates reactive oxygen species in ceramide-induced pancreatic dysfunction, Treatment of ZDF rats with the antioxidant, NAC (N-acetylcysteine) between 6 and 12 weeks of age prevented beta-cell failure and ameliorated the hyperglycemic state.[45] Sphingolipid metabolites contribute to the regulation of cellular redox homeostasis, and reactive oxygen species scavengers have been demonstrated to prevent ceramide-induce cell death.[9] It is of note that while ceramide may be destructive to the pancreas that sphingosine-1-phosphate (S1P) and dihydroS1P may actually promote beta-cell survival and insulin secretion.[46,47] Yet, these S1P data must be interpreted with caution, as discussed later S1P may also exacerbate diabetic complications.

Altered levels of specific glycosylated metabolites of ceramide may ultimately be shown to modulate pancreatic islet cell destruction. In the pancreas of STZ-diabetic rats, c-series gangliosides (GT3, GT2, GQ1c, GP1c) are drastically reduced.[32] Similarly, in *ob/ob* and *db/db* mice, models of Type 2 diabetes, $C_{16:0}$ sulfatides are drastically diminished in the pancreas.[48] More importantly, treatment of ZDF rats with $C_{16:0}$ sulfatides increased the amount of insulin in the blood and improved the first-phase insulin response.[49] Such an effect may be mediated through activation of potassium channels to increase insulin secretion[50] and/or serving as a molecular chaperone for insulin.[51] In addition to decreases or increases in specific sphingolipid metabolites, pancreatic dysfunction may also reflect an immunological component that targets these altered glycosphingolipid metabolites. Autoantibodies to sulfatides, GT3, GD3 and GM2-1 have been identified in patients with Type 1 diabetes and may target immune cells to the pancreas and contribute to their destruction.[52] Similarly, ganglioside-specific autoantibodies to GM3 have also been found in patients with Type 2 diabetes.[53] Autoimmunity may not be restricted to the pancreas. GM1 autoantibodies have been found in both Type 1 and Type 2 patients with peripheral neuropathy.[54] In fact, increased titers of autoantibodies to GM1, GD1b, GD1a and sulfatides correlated with more pronounced neuropathic changes.[55]

Diabetic Cardiovascular Dysfunction and Sphingolipids

Cardiovascular disease is the major cause of death of patients with diabetes. Both Type 1 and 2 patients have an increased incidence and severity of atherosclerosis, myocardial infarctions and strokes that is exacerbated by poor glycemic control or obesity. Importantly, a recent clinical trial has shown that the sphingolipid metabolite, S1P, is more predictive of obstructive coronary artery disease than other well-established risk factors, including age, sex, family history, diabetes, lipid profile and hypertension.[56] The heart and aorta from the Type 1 STZ-induced diabetic rat demonstrated increased sphingosine kinase activity, which was diminished in animals on an insulin pump.[57] Mechanistically, hyperglycemia induces sphingosine kinase activity, corresponding S1P levels and high glucose induced-leukocyte adhesion to endothelial cells was blocked by overexpression of a kinase-dead sphingosine kinase 1 mutant.[57] In contrast to these reports, in the Type 1 nonobese diabetic (NOD) mouse model, S1P prevented monocyte/endothelial interactions through S1P1 receptor activation.[58] Also, in Type 2 *KK/Ay* diabetic mice, sphingosine kinase 1 gene delivery by adenoviruses via intravenous injection, reduced blood glucose and improved plasma lipid profiles (reduced cholesterol, triglycerides, LDL, nonesterfied fatty acids and increased HDL), and prevented cardiac injury.[59] Moreover, subcutaneous injection of sphingosine-1-phosphate also has been shown to improve wound healing through improved vascularization of the wounded tissue.[60] The protective or detrimental actions of S1P have not as yet been completely defined in diabetes and will need to be interpreted in light of future studies that will utilize well defined selective S1P receptor modulators in clinical models.

The role of ceramide to contribute to cardiovascular disease in diabetic patients is also not well documented. Acid sphingomyelinase activity levels are elevated in the blood of Type 2 diabetic patients.[61] While the link to elevated circulating sphingomyelinase in diabetes is unknown, sphingomyelinase (and presumably ceramide) has been linked to the development of atherosclerosis.[62] In addition, sphingomyelinase inhibition as well as diminishing ceramide levels through preconditioning can diminish apoptosis and infarct size after cardiac ischemia.[63] Ceramide and other sphingolipid metabolites has also been shown to be a major component of lipoprotein vesicles.[64] Yet, intriguing data have suggested that ceramide nanofilms coated on balloon embolectomy catheters actually prevent coronary and carotid neointimal hyperplasia after vascular trauma in porcine and rabbit models.[65,66] These disparate findings might suggest that ceramide itself decreases vascular smooth muscle cell growth, while ceramide metabolites, including ceramide-1-phosphate or sphingosine-1-phosphate exacerbate restenotic or atherosclerotic injury. In fact, ceramide coated catheters promote wound healing responses, possibly as a consequence of metabolism of ceramide to other sphingoid metabolites within the endothelium.[66] Consistent with these interpretations, recent studies have suggested a novel anti-inflammatory antileukocytic effect for nanoliposomal ceramide, but not ceramide-1-phosphate or sphingosine-1-phosphate) in models of corneal keritites.[67]

Diabetic Nephropathy and Sphingolipids

Diabetic nephropathy is one of the most frequent causes of renal dysfunction. Diabetic nephropathy results from angiopathy of the capillaries in the glomeruli. Diabetes also causes an initial mesangial cell proliferation followed by growth arrest and hypertrophy. This contributes to an overproduction of extracellular matrix proteins, resulting sequentially in impaired blood filtration, increased proteinuria and renal failure.

Again, evidences implicate altered sphingolipids contributing to diabetic dysfunction of peripheral organs. The paradigm continues that pharmacological strategies to restore sphingolipid metabolism to basal levels will dissipate diabetic complications in the kidney. Accumulation of glucosylceramides at the expense of ceramide in streptozotocin (STZ)-induced diabetes was shown to contribute to renal hypertrophy.[68] Elevation of the ganglioside GM3 was also observed. More importantly, an inhibitor of glucosylceramide syhthase, PPMP, limited diabetes-induced mesangial cell hypertrophy and decreased glomeruli volume.[68] Such increases in glycosphingolipids have also been implicated in renal hypertrophy/diabetic nephropathy, possibly mediated by advanced glycation end products (AGEs).[15,68] Yet, controversy again exists in the literature, as ganglioside content, particularly GM3 and sialic acid content, is decreased from glomeruli of STZ-induced diabetic rats,[69] in contrast to *Zador* et al.[68] The answers to these disparities may again need to include a more comprehensive, integrated lipidomic approach, reflecting simultaneous measurements of multiple sphingolipid metabolites. Similar to glucosylceramide metabolites, phosphorylated sphingosine metabolites have also been implicated in glomerular mesangial cell proliferation, concomitant with increased neutral ceramidase and sphingosine kinase activities.[70]

Diabetic Retinopathy and Sphingolipids

Diabetic retinopathy is the leading cause of blindness among working age adults. A misconception about diabetic retinopathy is that it is solely a microvascular disease of the retina based upon visual clinical manifestations such as hemorrhages, microaneurysms, exudates, edema and neovascularization. A large body of work has demonstrated that diabetic retinopathy is a complex complication that, in addition to the microvasculature (endothelial cells and pericytes), also affects macroglial cells (Müller and astrocytes), microglia cells and neurons.[71] The vascular changes actually occur later on in the pathogenesis of this disease and thus understanding mechanisms that contribute to earlier pathology may reveal new therapeutic targets.

As already discussed, altered sphingolipids may be involved in insulin resistance in peripheral organs. In fact, the EURODIAB Prospective Complications Study showed that insulin resistance is a major predictive factor for the development of retinopathy in Type 1 diabetes.[4] It should be noted that the retina is an insulin responsive tissue and exhibits high basal insulin receptor activity, reminiscent of postprandial livers.[72] Furthermore, retinal insulin receptor and kinases that can be downstream (PI_3K, Akt and p70 S6 kinase) are impaired in models of Type 1 diabetes[73,74] and can only be partially resored with insulin therapy.[75] In addition, retina-specific insulin receptor deletion impairs the ability of the retina to withstand light induced stress.[76] The roles of sphingolipids in this insulin-signaling impairment are unknown, but in vitro evidence in a retinal neuronal cell line demonstrates that glucosylceramide synthase inhibitors can augment insulin signaling.[10] In neuronal tissues, glycosphingolipids have been implicated in increasing sensitivity to neurotoxic agents such as the excitatory amino acid neurotransmitter, glutamate,[77] a potential contributor to diabetic retinopathy.[75,78] Thus, it is not surprising that lipidomic analysis has recently revealed dysregulated sphingolipid metabolism in retinal tissue from Type 1 diabetic models. Specifically, in the streptozotocin-induced model of diabetes[10] as well as the $Ins2^{Akita}$ mouse model (unpublished observations), a decrease in ceramides with a corresponding increase in glucosylceramides[10] has been observed. The in vivo consequence of this is still unknown; however, as discussed above, mounting in vitro evidence has implicated elevated glucosylceramide metabolites in the inflammatory or immunomodulatory regulation of retinal and pancreatic cell death, cardiovascular abnormalities and dysfunctional insulin signaling.

Similar to glycosphingolipid metabolites, phosphorylated metabolites (S1P) may also contribute to diabetic retinopathy. For example, a sphingosine kinase inhibitor has been demonstrated to inhibit retinal vascular permeability in streptozotocin-induced diabetic rats.[79] Knockout of the S1P2 receptor, but not S1P1 or S1P3 receptors, suppressed some inflammatory mediators (COX-2 and eNOS) and shifted the phenotype of vascular changes from pathogenic to normal in a model of retinopathy of prematurity.[80] Furthermore, a monoclonal antibody that binds to S1P (sonepcizumab), suppressed laser-induced choroidal revascularization in mice and did not demonstrate any overt toxicology to the retina in nonhuman primates.[81] Often, these models of induced vascularization serve as surrogate models to study the mechanisms by which dysfunctional vasculature contribute to diabetic retinopathy. Our studies have not found significantly differences in retinal S1P, but we (unpublished observations from the streptozocin-induced diabetic rats and $Ins2^{Akita}$ diabetic mice) have observed elevated S1P from the plasma of these Type 1 diabetes models. Thus, it may be that retinal vascular alterations may be influenced by S1P in the blood. In fact, elevated sphingosine kinase activity within the retina may not be sufficient to induce retinopathy, as intraocular administration of adenoviruses carrying the sphingosine kinase gene was unable to induce neovascularization.[82]

Therapeutics That Target Sphingolipid Metabolism or Sphingolipid Signaling in Diabetes

Given the large body of in vitro evidence that glycosphingolipids exacerbate diabetic complications, it is not surprising that treatment strategies that inhibit glucosylceramide synthase (GCS) have proven effective in diabetic animal models. Treatment of *ob/ob* mice with the selective GCS inhibitor, AMP-DNM, lowered blood glucose levels, improved oral glucose tolerance, reduced hemoglobin A1C and improved insulin sensitivity in muscle and liver.[23] Similar beneficial metabolic effects were observed in high fat—fed mice and ZDF rats.[23] Furthermore, AMP-DMN in normal mice lowered triglyceride and cholesterol levels in plasma.[83] Analogous results were also observed with a structurally distinct glucosylceramide synthase inhibitor Genz-123346 in ZDF rats and diet-induced obese mice, including diminished blood glucose, improved glucose tolerance and reduced A1C levels. Furthermore, this inhibitor limited the loss of pancreatic beta-cell function.[25] Yet, interpretation of these provacative in vivo studies must be cautioned, as administration of specific exogenous glycosphingolipids could have beneficial effects. For example, GM1 administration to NOD mice reduced the incidence of diabetes onset and the degree of pancreatic islet injury.[84] Treatment with other glycosphingolipids, including sulfatide and galactosylceramide, were also able to decrease the incidence of diabetes in NOD mice.[85] Likewise, synthetic homologs of nonphysiological alpha-galactosylceramides, such as OCH and KRN7000, have also proven beneficial in NOD mice, possibly through several mechanisms which include increasing T_H2 responses and recruitment of tolerogenic dendritic cells.[86-90]

Likewise, treatment regimens to regulate S1P mass, sphingosine kinase activity or S1P receptor activity must be coached in light of the conflicting evidence suggesting pancreatic islet survival despite some evidence supporting exacerbation of cardiovascular and microvascular complications by S1P. FTY720 (Fingolimod) is a myriocin analog with structural similarity to sphingosine, that when phosphorylated may serve as a S1P receptor modulator.[91] FTY720 has shown significant immunomodulatory actions, limiting lymphocyte egress from lymphoid tissues.[91] FTY720 has been shown to limit autoimmune diabetes in rats[92] and mice.[93-95] Furthermore, survival of both islet allografts and xenografts in rodents and nonhuman primates were augmented when treated with FTY720.[96-99] Further studies will prove if the immunological or antiinflammatory actions of FTY720 in diabetic models are manifested by alterations in S1P receptor signaling or though inhibition of cytosolic phospholipase A_2[100] or ceramide synthase[101] and/or possibly other mechanisms.

Several studies are now correlating the actions of patient approved therapies with altered sphingolipid metabolism. In addition to tight insulin control and proper diet, Type 2 patients are further managed by strategies to increase pancreatic insulin secretion (sulfonylureas or meglintinides), decrease hepatic gluconeogenesis (biguanides(metformin)), improve insulin sensitivity in peripheral tissues (thiazolidinediones (Peroxisome proliferator-activated receptor-gamma (PPAR-gamma)) and to prevent breakdown of complex carbohydrates into simple sugars (alpha-glucosidase inhibitors). It is not surprising that thiazolidinediones, which regulate a transcriptional complex for insulin-responsive genes that control glucose and lipid metabolism, could affect sphingolipid metabolism. Yet, the literature reveal inconsistent findings. Various thiazolinediones, (pioglitizone, troglitazone and rosiglitazone) can reduce ceramide levels in rat and mouse skeletal muscle, likely through de novo inhibition.[102-104] Similarly, troglitazone reduced ceramide level in the heart of ZDF rats.[105] In contrast, pioglitizone has been shown to increase de novo ceramide synthesis in rat hearts.[106] Also, transcriptome analyses of rosiglitazone-treated cardiac tissue of *db/db* mice revealed increases in ASAH2 (ceramidase), ST3GAL5 (GM3 biosynthesis) and B4GALNT1 (GM2 and GD2 biosynthesis).[107] Thus, it is possible that specific thiazolidinediones could increase sphingolipid metabolism and possibly explain recent observations that thiazolidinediones have been linked to increased cardiovascular complications in diabetic patients.[108] In a similar manner, Metformin also alters sphingolipid metabolism, reducing ceramide content in skeletal muscle of high fat-fed rats, which was further augmented with exercise.[109] Again, a comprehensive "omic" approach, integrating LC/MS and NMR-based lipidomics and metabolomics, has the power to define sphingolipid and other metabolites as novel surrogate biomarkers that can be monitored during therapy.

Conclusion

The review identifies dysfunctional sphingolipid metabolism as contributors to diabetes and diabetic complications. Novel therapeutic strategies that target dysfunctional sphingolipid metabolism are identified that could decrease diabetic complications. Additional targets and/or biomarkers can be identified in animal models by integrating lipidomic approaches with functional genomics, proteomics and metabolomics, coupled with detailed organ-specific natural histories. In additional, integrated "omic" approaches coupled with therapeutic interventions can be used to firmly determine mechanisms underlying diabetic complications.

References

1. Diabetes Atlas, Third Edition: International Diabetes Federation; 2007.
2. Muoio DM, Newgard CB. Mechanisms of disease: molecular and metabolic mechanisms of insulin resistance and beta-cell failure in type 2 diabetes. Nat Rev Mol Cell Biol 2008; 9(3):193-205.
3. Greenbaum CJ. Insulin resistance in type 1 diabetes. Diabetes Metab Res Rev 2002; 18(3):192-200.
4. Chaturvedi N, Sjoelie AK, Porta M et al. Markers of insulin resistance are strong risk factors for retinopathy incidence in type 1 diabetes. Diabetes Care 2001; 24(2):284-289.
5. De Block CE, De Leeuw IH, Van Gaal LF. Impact of overweight on chronic microvascular complications in type 1 diabetic patients. Diabetes Care 2005; 28(7):1649-1655.
6. McGill M, Molyneaux L, Twigg SM et al. The metabolic syndrome in type 1 diabetes: does it exist and does it matter? J Diabetes Complications 2008; 22(1):18-23.
7. Kilpatrick ES, Rigby AS, Atkin SL. Insulin resistance, the metabolic syndrome and complication risk in type 1 diabetes: "double diabetes" in the Diabetes Control and Complications Trial. Diabetes Care 2007; 30(3):707-712.
8. Balkau B, Tichet J, Caces E et al. Insulin dose and cardiovascular risk factors in type 1 diabetic children and adolescents. Diabetes Metab 1998; 24(2):143-150.
9. Won JS, Singh I. Sphingolipid signaling and redox regulation. Free Radic Biol Med 2006; 40(11):1875-1888.
10. Fox TE, Han X, Kelly S et al. Diabetes alters sphingolipid metabolism in the retina: a potential mechanism of cell death in diabetic retinopathy. Diabetes 2006; 55(12):3573-3580.
11. Kralik SF, Liu P, Leffler BJ et al. Ceramide and glucosamine antagonism of alternate signaling pathways regulating insulin- and osmotic shock-induced glucose transporter 4 translocation. Endocrinology 2002; 143(1):37-46.
12. El Alwani M, Wu BX, Obeid LM et al. Bioactive sphingolipids in the modulation of the inflammatory response. Pharmacol Ther 2006; 112(1):171-183.
13. Natalizio A, Ruggiero D, Lecomte M et al. Glycosphingolipid changes induced by advanced glycation end-products. Biochem Biophys Res Commun 2001; 281(1):78-83.
14. Geoffroy K, Wiernsperger N, Lagarde M et al. Bimodal effect of advanced glycation end products on mesangial cell proliferation is mediated by neutral ceramidase regulation and endogenous sphingolipids. J Biol Chem 2004; 279(33):34343-34352.
15. Masson E, Troncy L, Ruggiero D et al. a-Series gangliosides mediate the effects of advanced glycation end products on pericyte and mesangial cell proliferation: a common mediator for retinal and renal microangiopathy? Diabetes 2005; 54(1):220-227.
16. Nilsson J, Bengtsson E, Fredrikson GN et al. Inflammation and immunity in diabetic vascular complications. Curr Opin Lipidol 2008; 19(5):519-524.
17. Schenk S, Saberi M, Olefsky JM. Insulin sensitivity: modulation by nutrients and inflammation. J Clin Invest 2008; 118(9):2992-3002.
18. Basta G, Schmidt AM, De Caterina R. Advanced glycation end products and vascular inflammation: implications for accelerated atherosclerosis in diabetes. Cardiovasc Res 2004; 63(4):582-592.
19. Holland WL, Summers SA. Sphingolipids, insulin resistance and metabolic disease: new insights from in vivo manipulation of sphingolipid metabolism. Endocr Rev 2008; 29(4):381-402.
20. Holland WL, Brozinick JT, Wang LP et al. Inhibition of ceramide synthesis ameliorates glucocorticoid-, saturated-fat- and obesity-induced insulin resistance. Cell Metab 2007; 5(3):167-179.
21. Deevska GM, Rozenova KA, Giltiay NV et al. Acid sphingomyelinase deficiency prevents diet-induced hepatic triacylglycerol accumulation and hyperglycemia in mice. J Biol Chem 2008.
22. Yetukuri L, Katajamaa M, Medina-Gomez G et al. Bioinformatics strategies for lipidomics analysis: characterization of obesity related hepatic steatosis. BMC Syst Biol 2007; 1:12.
23. Aerts JM, Ottenhoff R, Powlson AS et al. Pharmacological inhibition of glucosylceramide synthase enhances insulin sensitivity. Diabetes 2007; 56(5):1341-1349.
24. Tagami S, Inokuchi Ji J, Kabayama K et al. Ganglioside GM3 participates in the pathological conditions of insulin resistance. J Biol Chem 2002; 277(5):3085-3092.
25. Zhao H, Przybylska M, Wu IH et al. Inhibiting glycosphingolipid synthesis improves glycemic control and insulin sensitivity in animal models of type 2 diabetes. Diabetes 2007; 56(5):1210-1218.
26. Grigsby RJ, Dobrowsky RT. Inhibition of ceramide production reverses TNF-induced insulin resistance. Biochem Biophys Res Commun 2001; 287(5):1121-1124.
27. Kabayama K, Sato T, Kitamura F et al. TNFalpha-induced insulin resistance in adipocytes as a membrane microdomain disorder: involvement of ganglioside GM3. Glycobiology 2005; 15(1):21-29.
28. Yamashita T, Hashiramoto A, Haluzik M et al. Enhanced insulin sensitivity in mice lacking ganglioside GM3. Proc Natl Acad Sci USA 2003; 100(6):3445-3449.
29. De Maria R, Lenti L, Malisan F et al. Requirement for GD3 ganglioside in CD95- and ceramide-induced apoptosis. Science 1997; 277(5332):1652-1655.

30. Abregu AV, Genta SB, Sanchez Riera AN et al. Immunohistochemical detection of hepatic GM1 and GM2 gangliosides in streptozotocin-induced diabetic rats. Hepatol Res 2002; 24(3):256.
31. Sanchez SS, Abregu AV, Aybar MJ et al. Changes in liver gangliosides in streptozotocin-induced diabetic rats. Cell Biol Int 2000; 24(12):897-904.
32. Saito M, Ito M, Sugiyama K. A specific loss of C-series gangliosides in pancreas of streptozotocin-induced diabetic rats. Life Sci 1999; 64(20):1803-1810.
33. Samad F, Hester KD, Yang G et al. Altered adipose and plasma sphingolipid metabolism in obesity: a potential mechanism for cardiovascular and metabolic risk. Diabetes 2006; 55(9):2579-2587.
34. Adams JM, 2nd, Pratipanawatr T, Berria R et al. Ceramide content is increased in skeletal muscle from obese insulin-resistant humans. Diabetes 2004; 53(1):25-31.
35. Straczkowski M, Kowalska I, Nikolajuk A et al. Relationship between insulin sensitivity and sphingomyelin signaling pathway in human skeletal muscle. Diabetes 2004; 53(5):1215-1221.
36. Turinsky J, O'Sullivan DM, Bayly BP. 1,2-Diacylglycerol and ceramide levels in insulin-resistant tissues of the rat in vivo. J Biol Chem 1990; 265(28):16880-16885.
37. Gorska M, Dobrzyn A, Zendzian-Piotrowska M et al. Effect of streptozotocin-diabetes on the functioning of the sphingomyelin-signalling pathway in skeletal muscles of the rat. Horm Metab Res 2004; 36(1):14-21.
38. Skovbro M, Baranowski M, Skov-Jensen C et al. Human skeletal muscle ceramide content is not a major factor in muscle insulin sensitivity. Diabetologia 2008; 51(7):1253-1260.
39. Gorska M, Dobrzyn A, Baranowski M. Concentrations of sphingosine and sphinganine in plasma of patients with type 2 diabetes. Med Sci Monit 2005; 11(1):CR35-38.
40. Buschard K, Fredman P, Bog-Hansen E et al. Low serum concentration of sulfatide and presence of sulfated lactosylceramid are associated with Type 2 diabetes. The Skaraborg Project. Diabet Med 2005; 22(9):1190-1198.
41. Shimabukuro M, Higa M, Zhou YT et al. Lipoapoptosis in beta-cells of obese prediabetic fa/fa rats. Role of serine palmitoyltransferase overexpression. J Biol Chem 1998; 273(49):32487-32490.
42. Shimabukuro M, Zhou YT, Levi M et al. Fatty acid-induced beta cell apoptosis: a link between obesity and diabetes. Proc Natl Acad Sci USA 1998; 95(5):2498-2502.
43. Kelpe CL, Moore PC, Parazzoli SD et al. Palmitate inhibition of insulin gene expression is mediated at the transcriptional level via ceramide synthesis. J Biol Chem 2003; 278(32):30015-30021.
44. Maedler K, Oberholzer J, Bucher P et al. Monounsaturated fatty acids prevent the deleterious effects of palmitate and high glucose on human pancreatic beta-cell turnover and function. Diabetes 2003; 52(3):726-733.
45. Tanaka Y, Gleason CE, Tran PO et al. Prevention of glucose toxicity in HIT-T15 cells and Zucker diabetic fatty rats by antioxidants. Proc Natl Acad Sci USA 1999; 96(19):10857-10862.
46. Shimizu H, Okajima F, Kimura T et al. Sphingosine 1-phosphate stimulates insulin secretion in HIT-T 15 cells and mouse islets. Endocr J 2000; 47(3):261-269.
47. Laychock SG, Sessanna SM, Lin MH et al. Sphingosine 1-phosphate affects cytokine-induced apoptosis in rat pancreatic islet beta-cells. Endocrinology 2006; 147(10):4705-4712.
48. Blomqvist M, Osterbye T, Mansson JE et al. Selective lack of the C16:0 fatty acid isoform of sulfatide in pancreas of type II diabetic animal models. APMIS 2003; 111(9):867-877.
49. Blomqvist M, Carrier M, Andrews T et al. In vivo administration of the C16:0 fatty acid isoform of sulfatide increases pancreatic sulfatide and enhances glucose-stimulated insulin secretion in Zucker fatty (fa/fa) rats. Diabetes Metab Res Rev 2005; 21(2):158-166.
50. Buschard K, Hoy M, Bokvist K et al. Sulfatide controls insulin secretion by modulation of ATP-sensitive K(+)-channel activity and Ca(2+)-dependent exocytosis in rat pancreatic beta-cells. Diabetes 2002; 51(8):2514-2521.
51. Osterbye T, Jorgensen KH, Fredman P et al. Sulfatide promotes the folding of proinsulin, preserves insulin crystals and mediates its monomerization. Glycobiology 2001; 11(6):473-479.
52. Misasi R, Dionisi S, Farilla L et al. Gangliosides and autoimmune diabetes. Diabetes Metab Rev 1997; 13(3):163-179.
53. Morano S, Tiberti C, Cristina G et al. Autoimmune markers and neurological complications in non-insulin-dependent diabetes mellitus. Hum Immunol 1999; 60(9):848-854.
54. Milicevic Z, Newlon PG, Pittenger GL et al. Anti-ganglioside GM1 antibody and distal symmetric "diabetic polyneuropathy" with dominant motor features. Diabetologia 1997; 40(11):1364-1365.
55. Mata S, Betti E, Masotti G et al. Motor nerve damage is associated with anti-ganglioside antibodies in diabetes. J Peripher Nerv Syst 2004; 9(3):138-143.
56. Deutschman DH, Carstens JS, Klepper RL et al. Predicting obstructive coronary artery disease with serum sphingosine-1-phosphate. Am Heart J 2003; 146(1):62-68.
57. Wang L, Xing XP, Holmes A et al. Activation of the sphingosine kinase-signaling pathway by high glucose mediates the proinflammatory phenotype of endothelial cells. Circ Res 2005; 97(9):891-899.

58. Whetzel AM, Bolick DT, Srinivasan S et al. Sphingosine-1 phosphate prevents monocyte/endothelial interactions in type 1 diabetic NOD mice through activation of the S1P1 receptor. Circ Res 2006; 99(7):731-739.
59. Ma MM, Chen JL, Wang GG et al. Sphingosine kinase 1 participates in insulin signalling and regulates glucose metabolism and homeostasis in KK/Ay diabetic mice. Diabetologia 2007; 50(4):891-900.
60. Kawanabe T, Kawakami T, Yatomi Y et al. Sphingosine 1-phosphate accelerates wound healing in diabetic mice. J Dermatol Sci 2007; 48(1):53-60.
61. Gorska M, Baranczuk E, Dobrzyn A. Secretory Zn2+-dependent sphingomyelinase activity in the serum of patients with type 2 diabetes is elevated. Horm Metab Res 2003; 35(8):506-507.
62. Marathe S, Kuriakose G, Williams KJ et al. Sphingomyelinase, an enzyme implicated in atherogenesis, is present in atherosclerotic lesions and binds to specific components of the subendothelial extracellular matrix. Arterioscler Thromb Vasc Biol 1999; 19(11):2648-2658.
63. Argaud L, Prigent AF, Chalabreysse L et al. Ceramide in the antiapoptotic effect of ischemic preconditioning. Am J Physiol Heart Circ Physiol 2004; 286(1):H246-251.
64. Merrill AH, Jr., Lingrell S, Wang E et al. Sphingolipid biosynthesis de novo by rat hepatocytes in culture. Ceramide and sphingomyelin are associated with, but not required for, very low density lipoprotein secretion. J Biol Chem 1995; 270(23):13834-13841.
65. Charles R, Sandirasegarane L, Yun J et al. Ceramide-coated balloon catheters limit neointimal hyperplasia after stretch injury in carotid arteries. Circ Res 2000; 87(4):282-288.
66. O'Neill SM, Olympia DK, Fox TE et al. C6-ceramide-coated catheters promote re-endothelialization of stretch-injured arteries. Vasc Dis Prev 2008; 5(3):200-210.
67. Sun Y, Fox T, Adhikary G et al. Inhibition of corneal inflammation by liposomal delivery of short-chain, C-6 ceramide. J Leukoc Biol 2008; 83(6):1512-1521.
68. Zador IZ, Deshmukh GD, Kunkel R et al. A role for glycosphingolipid accumulation in the renal hypertrophy of streptozotocin-induced diabetes mellitus. J Clin Invest 1993; 91(3):797-803.
69. Kwak DH, Rho YI, Kwon OD et al. Decreases of ganglioside GM3 in streptozotocin-induced diabetic glomeruli of rats. Life Sci 2003; 72(17):1997-2006.
70. Geoffroy K, Troncy L, Wiernsperger N et al. Glomerular proliferation during early stages of diabetic nephropathy is associated with local increase of sphingosine-1-phosphate levels. FEBS Lett 2005; 579(5):1249-1254.
71. Gardner TW, Antonetti DA, Barber AJ et al. Diabetic retinopathy: more than meets the eye. Surv Ophthalmol 2002; 47 Suppl 2:S253-262.
72. Reiter CE, Sandirasegarane L, Wolpert EB et al. Characterization of insulin signaling in rat retina in vivo and ex vivo. Am J Physiol Endocrinol Metab 2003; 285(4):E763-774.
73. Reiter CE, Wu X, Sandirasegarane L et al. Diabetes reduces basal retinal insulin receptor signaling: reversal with systemic and local insulin. Diabetes 2006; 55(4):1148-1156.
74. Barber AJ, Antonetti DA, Kern TS et al. The Ins2Akita mouse as a model of early retinal complications in diabetes. Invest Ophthalmol Vis Sci 2005; 46(6):2210-2218.
75. Barber AJ, Lieth E, Khin SA et al. Neural apoptosis in the retina during experimental and human diabetes. Early onset and effect of insulin. J Clin Invest 1998; 102(4):783-791.
76. Rajala A, Tanito M, Le YZ et al. Loss of neuroprotective survival signal in mice lacking insulin receptor gene in rod photoreceptor cells. J Biol Chem 2008; 283(28):19781-19792.
77. Pelled D, Shogomori H, Futerman AH. The increased sensitivity of neurons with elevated glucocerebroside to neurotoxic agents can be reversed by imiglucerase. J Inherit Metab Dis 2000; 23(2):175-184.
78. Ambati J, Chalam KV, Chawla DK et al. Elevated gamma-aminobutyric acid, glutamate and vascular endothelial growth factor levels in the vitreous of patients with proliferative diabetic retinopathy. Arch Ophthalmol 1997; 115(9):1161-1166.
79. Maines LW, French KJ, Wolpert EB et al. Pharmacologic manipulation of sphingosine kinase in retinal endothelial cells: implications for angiogenic ocular diseases. Invest Ophthalmol Vis Sci 2006; 47(11):5022-5031.
80. Skoura A, Sanchez T, Claffey K et al. Essential role of sphingosine 1-phosphate receptor 2 in pathological angiogenesis of the mouse retina. J Clin Invest 2007; 117(9):2506-2516.
81. Xie B, Shen J, Dong A et al. Blockade of sphingosine-1-phosphate reduces macrophage influx and retinal and choroidal neovascularization. J Cell Physiol 2009; 218(1):192-198.
82. Oshima Y, Takahashi K, Oshima S et al. Intraocular gutless adenoviral-vectored VEGF stimulates anterior segment but not retinal neovascularization. J Cell Physiol 2004; 199(3):399-411.
83. Bijl N, van Roomen CP, Triantis V et al. Reduction of glycosphingolipid biosynthesis stimulates biliary lipid secretion in mice. Hepatology 2008.
84. Vieira KP, de Almeida e Silva Lima Zollner AR, Malaguti C et al. Ganglioside GM1 effects on the expression of nerve growth factor (NGF), Trk-A receptor, proinflammatory cytokines and on autoimmune diabetes onset in non-obese diabetic (NOD) mice. Cytokine 2008; 42(1):92-104.

85. Buschard K, Hanspers K, Fredman P et al. Treatment with sulfatide or its precursor, galactosylceramide, prevents diabetes in NOD mice. Autoimmunity 2001; 34(1):9-17.
86. Mizuno M, Masumura M, Tomi C et al. Synthetic glycolipid OCH prevents insulitis and diabetes in NOD mice. J Autoimmun 2004; 23(4):293-300.
87. Wang B, Geng YB, Wang CR. CD1-restricted NK T-cells protect nonobese diabetic mice from developing diabetes. J Exp Med 2001; 194(3):313-320.
88. Naumov YN, Bahjat KS, Gausling R et al. Activation of CD1d-restricted T-cells protects NOD mice from developing diabetes by regulating dendritic cell subsets. Proc Natl Acad Sci USA 2001; 98(24):13838-13843.
89. Sharif S, Arreaza GA, Zucker P et al. Activation of natural killer T-cells by alpha-galactosylceramide treatment prevents the onset and recurrence of autoimmune Type 1 diabetes. Nat Med 2001; 7(9):1057-1062.
90. Hong S, Wilson MT, Serizawa I et al. The natural killer T-cell ligand alpha-galactosylceramide prevents autoimmune diabetes in non-obese diabetic mice. Nat Med 2001; 7(9):1052-1056.
91. Chiba K. FTY720, a new class of immunomodulator, inhibits lymphocyte egress from secondary lymphoid tissues and thymus by agonistic activity at sphingosine 1-phosphate receptors. Pharmacol Ther 2005; 108(3):308-319.
92. Popovic J, Kover KL, Moore WV. The effect of immunomodulators on prevention of autoimmune diabetes is stage dependent: FTY720 prevents diabetes at three different stages in the diabetes-resistant biobreeding rat. Pediatr Diabetes 2004; 5(1):3-9.
93. Maki T, Gottschalk R, Monaco AP. Prevention of autoimmune diabetes by FTY720 in nonobese diabetic mice. Transplantation 2002; 74(12):1684-1686.
94. Maki T, Gottschalk R, Ogawa N et al. Prevention and cure of autoimmune diabetes in nonobese diabetic mice by continuous administration of FTY720. Transplantation 2005; 79(9):1051-1055.
95. Yang Z, Chen M, Fialkow LB et al. The immune modulator FYT720 prevents autoimmune diabetes in nonobese diabetic mice small star, filled. Clin Immunol 2003; 107(1):30-35.
96. Hering BJ, Wijkstrom M, Graham ML et al. Prolonged diabetes reversal after intraportal xenotransplantation of wild-type porcine islets in immunosuppressed nonhuman primates. Nat Med 2006; 12(3):301-303.
97. Liu L, Wang C, He X et al. Long-term effect of FTY720 on lymphocyte count and islet allograft survival in mice. Microsurgery 2007; 27(4):300-304.
98. Truong W, Emamaullee JA, Merani S et al. Human islet function is not impaired by the sphingosine-1-phosphate receptor modulator FTY720. Am J Transplant 2007; 7(8):2031-2038.
99. Fu F, Hu S, Deleo J et al. Long-term islet graft survival in streptozotocin- and autoimmune-induced diabetes models by immunosuppressive and potential insulinotropic agent FTY720. Transplantation 2002; 73(9):1425-1430.
100. Payne SG, Oskeritzian CA, Griffiths R et al. The immunosuppressant drug FTY720 inhibits cytosolic phospholipase A2 independently of sphingosine-1-phosphate receptors. Blood 2007; 109(3):1077-1085.
101. Berdyshev EV, Gorshkova I, Skobeleva A et al. FTY720 inhibits ceramide synthases and upregulates dihydrosphingosine-1-phosphate formation in human lung endothelial cells. J Biol Chem 2009.
102. Planavila A, Alegret M, Sanchez RM et al. Increased Akt protein expression is associated with decreased ceramide content in skeletal muscle of troglitazone-treated mice. Biochem Pharmacol 2005; 69(8):1195-1204.
103. Todd MK, Watt MJ, Le J et al. Thiazolidinediones enhance skeletal muscle triacylglycerol synthesis while protecting against fatty acid-induced inflammation and insulin resistance. Am J Physiol Endocrinol Metab 2007; 292(2):E485-493.
104. Zendzian-Piotrowska M, Baranowski M, Zabielski P et al. Effects of pioglitazone and high-fat diet on ceramide metabolism in rat skeletal muscles. J Physiol Pharmacol 2006; 57 Suppl 10:101-114.
105. Zhou YT, Grayburn P, Karim A et al. Lipotoxic heart disease in obese rats: implications for human obesity. Proc Natl Acad Sci USA 2000; 97(4):1784-1789.
106. Baranowski M, Blachnio A, Zabielski P et al. Pioglitazone induces de novo ceramide synthesis in the rat heart. Prostaglandins Other Lipid Mediat 2007; 83(1-2):99-111.
107. Wilson KD, Li Z, Wagner R et al. Transcriptome alteration in the diabetic heart by rosiglitazone: implications for cardiovascular mortality. PLoS ONE 2008; 3(7):e2609.
108. Nissen SE, Wolski K. Effect of rosiglitazone on the risk of myocardial infarction and death from cardiovascular causes. N Engl J Med 2007; 356(24):2457-2471.
109. Smith AC, Mullen KL, Junkin KA et al. Metformin and exercise reduce muscle FAT/CD36 and lipid accumulation and blunt the progression of high-fat diet-induced hyperglycemia. Am J Physiol Endocrinol Metab 2007; 293(1):E172-181.

Chapter 15

Roles for Sphingolipids in *Saccharomyces cerevisiae*

Robert C. Dickson*

Abstract

Studies using *Saccharomyces cerevisiae*, the common baker's or brewer's yeast, have progressed over the past twenty years from knowing which sphingolipids are present in cells and a basic outline of how they are made to a complete or nearly complete directory of the genes that catalyze their anabolism and catabolism. In addition, cellular processes that depend upon sphingolipids have been identified including protein trafficking/exocytosis, endocytosis and actin cytoskeleton dynamics, membrane microdomains, calcium signaling, regulation of transcription and translation, cell cycle control, stress resistance, nutrient uptake and aging. These will be summarized here along with new data not previously reviewed. Advances in our knowledge of sphingolipids and their roles in yeast are impressive but molecular mechanisms remain elusive and are a primary challenge for further progress in understanding the specific functions of sphingolipids.

Introduction

Besides providing many satiating bakery and brewery products for human pleasures, *Saccharomyces cerevisiae*, has been a remarkably informative host for discovering how sphingolipids are made and degraded and the cellular functions they perform. Carter's laboratory identified the long-chain bases (LCBs) dihydrosphingosine (DHS, sphinganine) and phytosphingosine (PHS) in the 1950s and early 1960s.[1] In the 1970s and 1980s the Lester laboratory identified the three classes of inositol phosphate-containing complex sphingolipids in yeast and their mode of synthesis and in doing so devised excellent methods for their extraction and analysis.[2,3] These advances and others enabled yeast genes to be identified and manipulated in every imaginable way and have permitted the elucidation of most, perhaps all, genes for making and degrading sphingolipids in *S. cerevisiae*, the first organism for which this has been achieved (reviewed in ref. 4-6). Many of these genes made possible the identification of homologs in organisms ranging from bacteria to plants to man. In this chapter I try to summarize our knowledge of sphingolipids in *S. cerevisiae* and suggest where future research is needed.

Sphingolipid Metabolism in *S. cerevisiae*

Sphingolipids are abundant in *S. cerevisiae* representing about 7% of the mass of the plasma membrane or 30% of phospholipids.[7] *S. cerevisiae* sphingolipid metabolism including metabolites, enzymes and their cognate genes are diagramed in Figure 1. Synthesis begins with condensation of serine and a fattyacyl-CoA, typically palmitoyl-CoA, in every organism that has been examined and generates the short-lived intermediate 3-ketodihydrosphingosine,

*Robert C. Dickson—Department of Molecular and Cellular Biochemistry and the Lucille P. Markey Cancer Center, University of Kentucky College of Medicine, 741 S. Limestone St., BBSRB, B173, Lexington, Kentucky 40536-0509, USA. Email: bobd@uky.edu

Sphingolipids as Signaling and Regulatory Molecules, edited by Charles Chalfant and Maurizio Del Poeta.

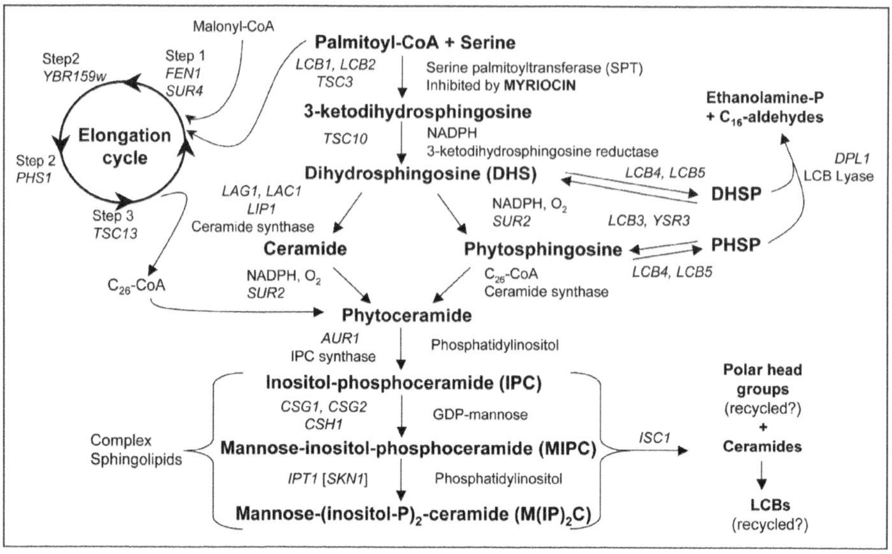

Figure 1. Pathways of sphingolipid metabolism in *Saccharomyces cerevisiae*. Metabolic intermediates and complex sphingolipids are shown in bold, genes are indicated by italics and enzyme names are in regular lettering. When grown aerobically the fatty acid in complex sphingolipids is often hydroxylated at C_2 and sometimes at C_3 (not shown), a reaction that requires Scs7 (not shown).[107] Ceramides can be hydrolyzed by two ceramidases, Ydc1 and Ypc1 (not shown), to yield a fatty acid and an LCB.[35,36] Structures of the indicated compounds are presented in previous publications.[4,8] Adapted from reference 6.

which is reduced to yield DHS, the first LCB in the pathway. DHS can either be N-acylated with a fatty acid to give ceramide or hydroxylated on C4 to give PHS which is N-acylated to yield phytoceramide. The C1 hydroxyl of phytoceramide is decorated with polar head groups by the three sequential reactions diagramed in Figure 1 which yield three species of complex sphingolipids including inositol phosphoceramide (IPC), mannose inositol phosphoceramide (MIPC) and mannose-(inositol-P)$_2$-ceramide M(IP)$_2$C. The genes and enzymes for the steps in sphingolipid metabolism have been discussed thoroughly in previous reviews.[2-5,8-10]

Sphingolipid synthesis begins in the endoplasmic reticulum and generates ceramides which are transported by both vesicular and nonvesicular transport to the Golgi apparatus for addition of the polar head groups.[11] Most enzymes involved in sphingolipid synthesis have been localized to these two compartments although there are exceptions.[12] Movement of sphingolipids between these compartments has been examined in detail although much remains to be elucidated (reviewed in refs. 8,9,13,14). Most complex sphingolipids are then transported to the plasma membrane but small amounts are found in other membrane compartments[15,16] including mitochondria.[17] The function of sphingolipids in these other cellular compartments is not understood. About three-fourths of the mass of the sphingolipids in *S. cerevisiae* cells is M(IP)$_2$C with the rest being equal parts of IPC and MIPC.[18] It is not known how these ratios are determined, but they are probably important in ways that remain to be identified. The shear mass of sphingolipids and their negative charge are likely to affect processes dependent upon the plasma membrane. Many types of sphingolipids in mammals are localized to one or the other leaflet of the plasma membrane, but this has not been determined for yeast sphingolipids.

A unique and distinguishing feature of *S. cerevisiae* sphingolipids is a C_{26} fatty acid, although a small percentage of C22 and C24 fatty are also present.[16,19] These long-chain fatty acids are synthesized by an elongation system[19-21] whose components have now been identified and this has led to a model for how the length of the fatty acid is determined.[22] The functions of such a long-chain fatty acid in sphingolipids are not well defined, although one function may have something to do with the fact that they span both leaflets of a membrane bilayer. They have been suggested to play roles in nuclear membrane pores.[23] Others have argued that sphingolipids do not need them but that they are necessary to perform some function that remains to be identified.[24] While it is true that cells making sphingolipids with shorter acyl chains can survive in laboratory situations, it is highly unlikely that such strains would survive in the wild, thus arguing that the C_{26} fatty acid is an essential component of yeast sphingolipids under natural conditions.

While *S. cerevisiae* only makes inositolphosphoceramides and no other type of sphingolipids such as glycosylceramides, many other fungi make both classes of sphingolipids[3] as do plants.[25,26] Phosphoinositol-containing sphingolipids have not been identified in mammals. It probably is no accident that *S. cerevisiae* and other fungi have sphingolipids that are more like plants than mammals given that the natural habitat of *S. cerevisiae* appears to be forests, particularly oaks and other broadleaf trees.[27]

Like most other membrane components, sphingolipids are broken down as a normal part of membrane remodeling. The need for turnover is apparent in mammals where defects in enzymes that catalyze turnover lead to debilitating diseases termed sphingolipidoses.[28] Such a need has been much less apparent in yeast but this notion is changing. One type of sphingolipidoses is Niemann Pick type C, a fatal neurodegenerative disorder caused by a defect in the human *NPC1* gene. The yeast homolog of this gene, *NCR1*, has been suggested to play a role in recycling of sphingolipids.[29] The yeast sequence similarity between a bacterial neutral sphingomyelinase and the Isc1 protein of *S. cerevisiae* lead to the demonstration that Isc1 has phospholipase C-type activity and cleaves the polar head group from yeast sphingolipids, much like mammalian sphingomyelinases cleave the polar head group from sphingomyelin.[30,31]

Unraveling the physiological importance of sphingolipid turnover in *S. cerevisiae* has been challenging and much remains to be learned. At least in some strains backgrounds, Isc1 enzyme activity is necessary for growth on nonfermentable carbon sources, implying a role in mitochondrial respiration[32] and several types of experiments support this mitochondrial connection as discussed in detail in a previous review.[6] Another role for Isc1 is to regulate the concentration of IPC which is toxic above its physiological concentration.[33,34] In response to heat stress and perhaps other stress that perturb the plasma membrane, Isc1 appears to be activated to breakdown complex sphingolipids (we don't know their location) by a pathway involving synthesis of phosphatidylinositol 4,5-bisphosphate (PIP_2) as described below along.

A good deal of effort has gone into trying to determine the physiological functions of the other enzymes that degrade sphingolipids including the ceramidases Ypc1[35] and Ydc1,[36] the LCB kinases Lcb4 and Lcb5, the LCB-phosphate phosphatases Lcb2 and Ysr3 and the LCB lyase Dpl1. Some may disagree, but I think it is fair to say that other than their basic catalytic function, we do not know the real physiological roles of these enzymes and their precursors and products in *S. cerevisiae* (reviewed in refs. 2-5,8,9). They must be important because most are conserved in fungi and other organisms.

Both natural and synthetic inhibitors of several enzymes in the sphingolipid biosynthesis pathway have been identified and have been very useful in a range of experiments (reviewed in refs.[3,37]). These include myriocin, sphingofungins, lipoxamycin and viridiofungins that inhibit serine palmitoyltransferase (SPT), the first enzyme in sphingolipid biosynthesis (Fig. 1), australifungin and the fumonisins that inhibit ceramide synthase and aureobasidin and khafrefungin that inhibit IPC synthase.

Membrane-Associated Functions and Processes

A screen for mutants defective in endocytosis gave the first indication that sphingolipids are necessary for this process.[38] The screen identified *end8-1*, later shown to be allelic with *LCB1* and renamed *lcb1-100*.[39] The *lcb1-100* mutation causes serine palmitoyltransferase (Fig. 1) activity to be labile even in cells grown at a permissive temperature[40] and sphingolipid synthesis becomes limiting for growth at the restrictive temperature of 37°C,[41] most likely because LCB levels drop quickly and limit sphingolipid synthesis.[42] The *lcb1-100* mutation has been extremely useful for studying sphingolipid metabolism and functions. Sphingolipids affect multiple aspects of endocytosis including the actin cytoskeleton and these will be described below in the section on signal transduction. Other experiments implicating sphingolipids in endocytosis used strains defective in *SNC1* and/or *SNC2* that encode v-SNARE proteins.[43] Sphingolipids have been shown to play several roles in endocytosis of the uracil transporter Fur4 (reviewed in ref. 9) as have the Slm proteins[44] (see below).

Sphingolipids along with sterols (cholesterol in mammals and ergosterol in fungi and plants) are critical for formation of microdomains within membranes that have been referred to as lipid rafts and are typically isolated by treating a yeast extract with detergent at low temperature to form what are referred to as detergent-insoluble complexes or detergent-resistant membranes.[45,46] In *S. cerevisiae* rafts are vital for sorting and delivering membrane-bound proteins to their proper cellular address and they are also necessary for fusion of cells during mating.[45,47,48] Such proteins include Pma1, which exports protons across the plasma membrane to maintain intracellular pH and also to generate a proton gradient necessary for cells to take up nutrients from their surroundings.[49-54] Other membrane proteins requiring rafts/sphingolipids include Gas1, a Beta-1,3-glucanosyltransferase and Nce2,[49] Fus2, Fig1, Sho1, Ste1 and Prm1,[55] Fur4[42,56] and Can1, an arginine transporter.[57] The general amino acid permease Gap1 depends upon sphingolipids for its transport to the plasma membrane in an active conformation capable of amino acid transport and that can resist degradation.[58]

Roles for sphingolipids in exocytosis based upon suppressor mutant analyses were reviewed previously.[4,9] In wild-type cells the C_{26}-fatty acid in sphingolipids is required for raft association and stable surface transport of newly made Pma1 to the plasma membrane.[59] Recently, a large-scale visual screen for genes that play roles in the sorting of proteins in the trans-Golgi network for delivery by the exocytosis pathway to the cell surface identified genes in sphingolipid metabolism (*SUR2, SUR4* and *YPC1*).[48] It is not clear why the Sur2, Sur3 and Ypc1 proteins are required for protein sorting. Ypc1 is particularly interesting because it degrades phytoceramides.[35]

In mammals, sphingosine-1-phosphate is involved in intracellular calcium signaling through an unknown mechanism.[60] Likewise, sphingolipids may regulate calcium fluxes and signaling pathways in yeast, but the mechanisms are unknown (reviewed in ref. 2) and see also ref. [61] Cells defective in *CSG1* or *CSG2* (Fig. 1) are sensitive to 100 mM calcium and this phenotype has played a key role in identifying genes in sphingolipid metabolism.[33,34,62,63] Why mutants defective in *CSG1* or *CSG2* are sensitive to calcium is unclear, but it indicates a connection of some type between sphingolipids and calcium metabolism or signal transduction and these potential connections are presented below in the section on Signal Transduction Pathways.

Nutrient transport is affected by LCBs as first suggested by the extreme sensitivity of auxotrophic strains to LCBs in the culture medium.[64] PHS, but not other LCBs, was found to block uptake of tryptophan, leucine, histidine and uracil.[65] It is not entirely clear how LCBs are regulating nutrient transport, except in the case of uracil where PHS has been shown to be important for heat-induced, ubiquitin-mediated breakdown of Fur4, the uracil transporter.[40] Other data also support the idea that heat-induced LCBs promote ubiquitination of proteins[66] and that the immunosuppressant drug, FYT720, a synthetic sphingolipid-like molecule, acts in ways very similar to PHS to inhibit yeast growth.[29] As described below in Signal Transduction Pathways, LCBs maybe regulating nutrient transporters via activation of the Pkh1/2 kinases (Fig. 2) or their downstream kinases such as Ypk1 or Sch9, which play roles in nutrient sensing.[67-69]

Interactions of some sort or cross-talk between sphingolipids and both ergosterol and glycerophospholipids has been observed. For example, defects in ergosterol synthesis are suppressed by mutations in *SUR4*[70] and a decrease in ergosterol content is compensated by an increase in sphingolipids.[71] Other data show that defects in ergosterol synthesis affect hydroxylation of yeast sphingolipids.[72] Yeast cells also have mechanisms for maintaining asymmetry in the distribution of sphingolipids and glycerophospholipids in the two leaflets of the plasma membrane so that a change in one class of lipids is compensated by a change in another class.[73]

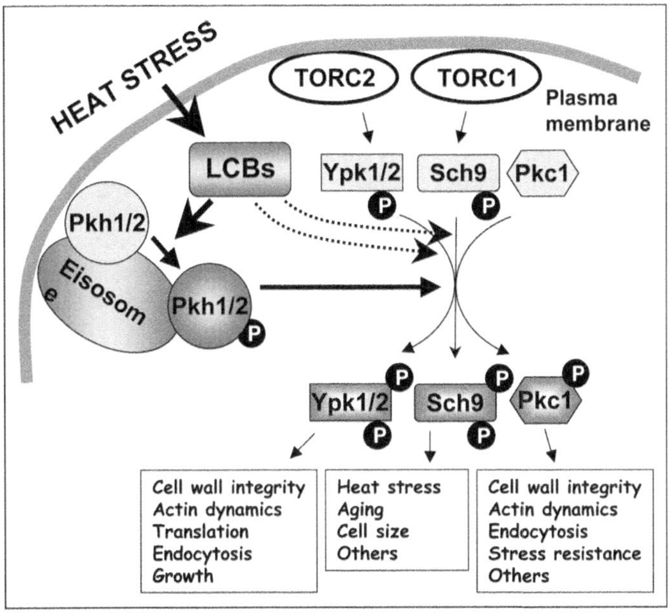

Figure 2. *S. cerevisiae* signal transduction pathways regulated by LCBs. LCBs transiently increase during a heat stress and are hypothesized to activate Pkh1 and Pkh2. As discussed in the text, Pkh2 is probably more dependent upon LCBs than is Pkh1 although this has not been defined for each cellular response. Kinase assays with purified proteins suggest that LCBs can directly trigger a small increase in Ypk1, Ypk2 and Sch9 activity (indicated by a dotted line).[97] Pkh1/2 phosphorylate Ypk1, Ypk2, Sch9 and Pkc1 in their activation loop (PDK1 site) but the proteins are not enzymatically active. To become active they also need to be phosphorylated in a hydrophobic region (PDK2 site) and in a turn motif in their C-terminus. Phosphorylation of these sites in Ypk2 is mediated by TORC2[124] and for Sch9 phosphorylation is mediated by TORC1.[94] Ypk1/2 and Pkc1 are shown working in parallel pathways to control cell wall integrity, but data also support an alternative pathway in which Ypk1/2 work upstream of Pkc1.[90,91] Adapted from reference 6.

While the complex yeast sphingolipids, IPCs, MIPCs and $M(IP)_2Cs$, are abundant, progress in identifying unique functions for them has been slow. The antifungal action of the plant defensin DmAMP1, a peptide defensin produced by *Dahlia merckii*, requires $M(IP)_2C$, which serves as a high affinity receptor.[74] DmAMP1 is postulated to bind or interact with $M(IP)_2C$-containing lipid rafts making the plasma membrane more permeable thereby disrupting essential cellular processes.[75] Syringomycin E is an antifungal cyclic lipodepsinonapeptide that interacts with the plasma membrane and inhibits growth of *S. cerevisiae* cells by forming ion channels. Yeast mutants defective in *ipt1*, *fen1* or *sur4*, *scs7* and *sur2* are drug resistant, showing that $M(IP)_2C$ with a C_{26}-fatty acid and PHS but not DHS is essential for the antifungal action of syringomycin E.[76-78] While interesting, these results do not represent normal functions of the complex sphingolipids beyond their role as structural components of the plasma membrane and essential elements of lipid rafts. One would imagine that they interact in physiologically essential ways with plasma membrane proteins, but evidence is lacking and will require advances in three dimensional structure of membrane proteins bound to natural lipids.

Sphingolipids play roles in exocytosis of glycosylphosphatidylinositol-anchored proteins (reviewed in ref. 79). For example, transport of Gas1p from the endoplasmic reticulum to the Golgi apparatus requires ceramide or a related sphingolipid for transport.[39,80,81] Furthermore, the diacylglycerol moiety in glycosylphosphatidylinositol anchors is often replaced by ceramide.[79]

Sphingolipids and the Slm protein (see below) are required for exocytosis of arginine transporter Can1[82] and references therein). Sphingolipids have also been shown to play roles in generating a functional V1 component of the vacuolar ATPase.[83]

Signal Transduction Pathways That Require Sphingolipids

One of the most exciting and extensively explored functions of sphingolipids is the regulation of signal transduction pathways and ceramides and sphingosine-1-phosphate in mammals are the most well documented signaling species.[84-86] There is no firm evidence that ceramides and long-chain base phosphates play similar roles in yeast. Instead, LCBs appear to regulate signaling pathways. The one piece of evidence that would firmly establish LCBs as regulators of signaling pathways is to show that they bind to specific proteins such as the protein kinases discussed below that are implicated to be regulated by LCBs (Fig. 2).

The first clue that LCBs might be intracellular signaling molecules or second messengers was the observation that they rapidly but transiently increase following heat stress typically executed in the laboratory by shifting cells from 25°C to 37°C or 39°C[87,88] and reviewed in detail in ref.[4,9,10]

Insight into LCB signaling pathways was first found during a screen to identify genes whose overexpression bypassed growth inhibition by myriocin,[89] an inhibitor of serine palmitoyltransferase (Fig. 1). The *YPK1* gene bypassed the myriocin block. The Ypk1 protein kinase has a role in cell wall maintenance and actin cytoskeleton dynamics,[90,91] endocytosis[92] and translation during nitrogen starvation and nutrient sensing.[67] Ypk1 and its paralog Ypk2 are structural and functional homologs of mammalian serum and glucocorticoid-inducible kinase (SGK).[93] Ypk1 is phosphorylated and activated by Pkh1[93] and multiple copies of *PKH1* were found to also bypass the myriocin block.[89] One explanation for these results is that a sphingolipid acts upstream of and activates the Pkh1/Ypk1-Ypk2 the signaling pathway. Myriocin-treated cells lacked a phosphorylated and presumably active form of Ypk1 but this form reappeared in vivo following PHS-treatment of cells. The original and still accepted view of these experiments is that PHS activates Pkh1 or its homolog Pkh2, which then phosphorylate and activate Ypk1 and probably Ypk2. Many laboratories (reviewed in ref. 9,10) have contributed data supporting this hypothesis and expanded it to include activation of other kinases including Sch9[94] and Pkc1 which are activated by the LCB-Pkh1/2 pathway (Fig.2). To be active, Sch9 and Ypk1/2 must be phosphorylated at other residues by the Target of Rapamycin (TOR) pathways as indicated in Figure 2.

The diagram shown in Figure 2 suggests that Pkh1/2 and the kinases and cellular processes downstream of them are equally regulated by LCBs, but current data suggest that this is probably an oversimplification. First, initial analysis in vitro showed that purified Pkh2 was more strongly activated by LCBs than was Pkh1.[95] Second, as discussed below, the Slm2 protein is more responsive to sphingolipids (likely to be LCBs) than is Slm1, consistent with the finding that *slm1*Δ cells are more sensitive to growth inhibition by myriocin than are *slm2*Δ cells.[44] Likewise, homozygous diploid *pkh1*Δ cells were 1000-fold more sensitive to myriocin than *pkh2*Δ cells, consistent with LCBs having a stronger affect on Pkh2 than on Pkh1. Additionally, this screen found both heterozygous and homozygous diploid *ypk1*Δ cells to be extremely sensitive to myriocin while *ypk2*Δ cells were not sensitive, implying that LCBs regulate Ypk2 but not Ypk1 activity. These myriocin data only apply to a process or processes that are necessary for growth: nonessential processes may depend upon LCBs to regulate Pkh1 and Ypk1, as suggested by the finding that in vitro the activity of Pkh1 and both Ypk1 and Ypk2 is stimulated by LCBs.[97] As mentioned above, demonstrating which proteins bind LCBs would help to clarify their physiological roles.

Pkh1/2 are primarily found on eisosomes,[98] newly described very large structures, estimated to contain about 2000 copies each of the Pil1 and Lsp1 proteins, which bind to the cytoplasmic face of the plasma membrane.[99] Pil1 and Lsp1 are highly conserved but seem to be found only in fungi. Eisosomes play roles in endocytosis of lipids and some proteins and localize at cites of endocytosis where they may physically interact with the actin cytoskeleton based on confocal fluorescent microscopy and genetic interaction studies.[99] Pil[98,100] and Lsp1 (Dickson, et al unpublished data) are highly phosphorylated by Pkh1/2 and phosphorylation plays roles in assembly and disassembly of eisosomes. Eisosomes are also thought to be involved in sensing changes in the plasma membrane caused by stresses such as heat and then signaling the cell to adjust the lipid and protein composition of the membrane by endocytosis and exocytosis in order for cells to be more stress tolerant.[98,100] Earlier data[95] had indicated that Pil1 and Lsp1 indeed play roles in resisting heat stress and they may do so in several ways including serving as binding sites for Pkh1/2 to enable them to properly control downstream kinases with known roles in heat and other stress responses (Fig. 2). It is not known if some fraction of Pkh1 or Pkh2 also resides in other cellular locations or whether they cycle on and off of eisosomes. Clearly there is much to learn about eisosomes and the role they play in regulating cellular processes.

Previous results from genetic suppression experiments[34,101] suggested that sphingolipids interacted in some manner with phosphoinositides, probably phosphatidylinositol-5,4-bisphosphate (PIP_2) and with the TOR pathways. These novel interactions along with the calcineurin signaling pathway have now been shown by genetic and biochemical assays to control phosphorylation and dephosphorylation of the Slm1 and Slm2 proteins during heat stress thereby modulating actin polarization, endocytosis and sphingolipid metabolism.[44,82,102]

The Slm proteins have overlapping functions and at least one is required for viability.[103] They also have a PH domain enabling binding to PIP_2, known to arise transiently during stresses such as during heat shock on the inner leaflet of the plasma membrane.[103,104] Slm binding to the membrane is further strengthened by interactions with the Avo2 and Bit61 subunits of the Target Of Rapamycin Complex 2 (TORC2)[103,105] and these protein-protein interactions promote phosphorylation of Slm1 and Slm2 by TORC2.[103] The phosphorylated Slm proteins then mediate downstream effects of PIP_2 and TORC2 that control roles of the actin cytoskeleton essential for growth, cell wall integrity and receptor-mediated endocytosis (Fig. 3). A recent study suggests that TORC2 is located on eisosomes, if true, then eisosomes may play a role in Slm protein function.[106]

Data supporting a role for sphingolipids in regulating the Slm proteins and how they in turn regulate sphingolipid synthesis and turnover were recently reviewed in ref.[6] and are summarized diagrammatically in Figure 3. The model depicts events during unstressed growth and during heat stress: whether similar events are active during other stresses is unknown.[44,82,102] Central to this model are the Slm proteins functioning downstream of the PIP_2, TORC2 and LCB-Pkh1/2 signaling pathways. Heat stress causes a transient increase in PIP_2 thereby recruiting the Slm proteins to the plasma membrane where they are phosphorylated by TORC2 and also by Pkh1/2, which are activated also by the transient burst of LCBs induced by heat. Activated Slm proteins control movement of the actin cytoskeleton via the Rho1/Pkc1 pathway. Unexpectedly, the Slm proteins also control sphingolipid metabolism by down-regulating Isc1 activity, thus slowing breakdown of complex sphingolipids, especially the very abundant IPC-C (IPC with one hydroxyl group on the C_{26} fatty acid.[107] The Slm proteins also down-regulate calcineurin phosphatase activity, which interacts with Csg2, in an unknown manner, to regulate conversion of IPC-C to MIPC. Hence, during heat stress the Slm proteins are thought to interface between phosphoinositides and sphingolipids and orchestrate changes in membrane lipid composition to promote survival.

Once cells adjust to a heat stress they down-regulate activated Slm proteins by calcineurin-mediated dephosphorylation in what appears to be a negative feedback loop (Fig. 3). It is also likely that IPC, possibly a specific pool of IPC-C, regulates actin organization and viability.[102,108] Identifying this pool of IPC-C could provide important clues for understanding how such regulation occurs.

Recognition that the LCB-Pkh1/2, PIP_2 and TORC2 pathways use the Slm proteins and calcineurin to regulate sphingolipid metabolism provides a framework for deciphering the molecular basis for maintaining sphingolipid levels that are optimal for growth in the absence of stress and for surviving a heat stress. Many questions remain unanswered. Do Pkh1/2 directly phosphorylate Slm1 and Slm2 or does one of the protein kinases regulated by Pkh1/2 (Fig. 2)? Available evidence suggests that Pkh1/2 do not phosphorylate the Slm proteins nor do Ypk1/2, but the data do not exclude these possibilities and more work is needed.[82] How do the Slm proteins

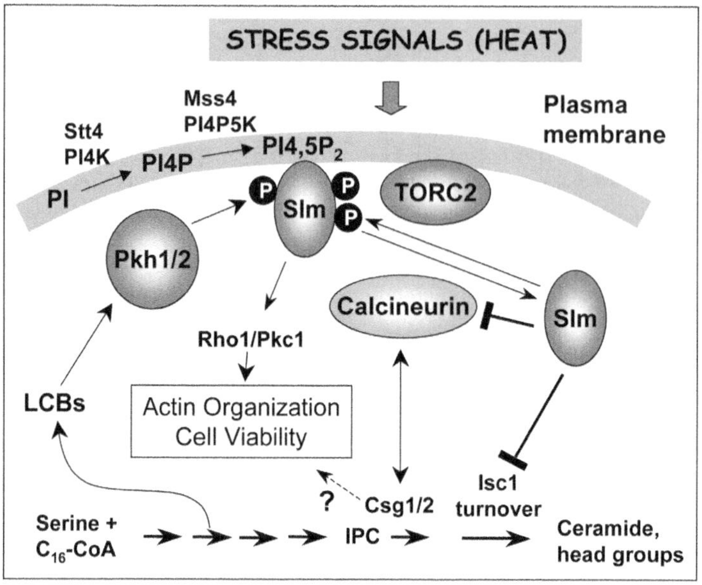

Figure 3. Synthesis and turnover of complex sphingolipids along with actin dynamics are regulated by $PI4,5P_2$ and the Slm1 and Slm2 proteins. $PI4,5P_2$, synthesized by Stt4 and Mss4 on the plasma membrane, in cooperation with the TORC2 and the Pkh1/2 protein kinases are proposed to activate Slm1 and Slm2. The Slm proteins then impair turnover of complex sphingolipids, particularly IPC, by inhibiting Isc1 and they also regulate the calcium/calmodulin-regulated protein phosphatase calcineurin which dephosphorylates and inactivates the Slm proteins and also interacts with Csg1/2 in an unknown manner to regulate conversion of IPC to MIPC by the Csg1/2 enzymes. Regulation of the Rho1/Pkc1 pathway by Slm1/2 works independently from regulation of sphingolipid metabolism. Genetic interactions suggest that IPC plays a role in actin organization, but the mechanism is unknown. Pkh1/2 are probably attached to eisosomes[98] (not shown) as depicted in Figure 2. Adapted from reference 6.

regulate Isc1 activity? Does Isc1 play a direct role in regulating the actin cytoskeleton and how do Csg1\2 interact with calcineurin?

Slm2 seems more dependent upon sphingolipids for function than does Slm1. First, haploid *slm1*Δ cells are more sensitive to growth inhibition by myriocin than *slm2*Δ cells. Second, heat-induced and sphingolipid-dependent endocytosis of the uracil transporter requires Slm2 but not Slm1.[44] Lastly, a recent large-scale screen of deletion mutants found that *slm1*Δ heterozygous and homozygous diploid cells were extremely sensitive to myriocin compared to all other gene deletion mutants including *slm2*Δ cells.[96]

Longevity and Cellular Aging

Long before it was realized that *LAG1* (Longevity-Assurance Gene[109]) encodes a ceramide synthase,[110,111] it had been found to affect replicative lifespan, measured by determining how many times a cell can bud. Deletion of *LAG1* increases replicative lifespan by 50 percent,[109] but the mechanism remains unknown.

Sch9 (Fig. 2) plays roles in both replicative and chronological lifespan, measured by how long cells survive in stationary phase when division has ceased. Deletion of *SCH9* increases chronological lifespan by 300%[112] and increases both the mean and maximal replicative lifespan.[113] No one has yet determined if LCBs play a role in lifespan by regulating Pkh1/2 which in turn phosphorylate Sch9 on threonine 570 in its activation domain.[94,97] To be active, Sch9 must also be phosphorylated by TORC1 at several serine and threonine residues located in its C-terminus.[94]

Isc1, which cleaves polar head groups from complex sphingolipids to generate ceramides, has been shown to play a role in oxidative stress resistance (hydrogen peroxide) and chronological lifespan.[114] For lifespan measurements, cells were grown to log phase or early stationary phase in YP medium containing glycerol as the carbon source and then transferred to water, a severe form of calorie restriction. Both log and stationary phase *isc1*Δ cells died extremely rapidly compared to *ISC1* cells, indicating that viability under these experimental conditions depends on Isc1 function. The oxidative stress theory of aging argues that cells die because of oxidative stress and several but not all measures of oxidative stress were elevated in *isc1*Δ cells. Apoptosis is induced by oxidative stress and during aging[115-118] and *isc1*Δ cells showed classic signs of apoptosis including DNA fragmentation and activation of Yca1 metacaspase activity. The rapid loss of viability in calorie-restricted stationary phase *isc1*Δ cells was completely prevented by deletion of *YCA1*, arguing that Isc1 is involved in a pathway that regulates Yca1 and apoptosis.

To begin to elucidate the role of Isc1 in oxidative stress resistance and chronological lifespan, a global analysis of mRNAs was performed using microarray technology.[114] Expression of 72 genes was found to increase and sorting these into biological processes showed that six were involved in iron uptake. The authors reasoned that induction of these genes during iron abundance could cause iron overloading and enhance oxidative stress. They found iron increased in both log and stationary phase *isc1*Δ cells and data from several types of experiments supported the notion that such accumulation contributed to but was not entirely responsible for increased oxidative stress and cell death. At this time it is not clear how Isc1 protects cells against oxidative stress and apoptosis. A basic question is whether the ceramide generated by Isc1 activity is necessary for stress protection or whether accumulation of one or more complex sphingolipids such as IPC1-C in *isc1*Δ cells causes cell death. Others have argued that the increased level of M(IP_2)C found in *isc1*Δ cells[30,114] is the cause of cell death. It would be interesting and novel if ceramide production by Isc1 protected against oxidative stress, since ceramide plays an opposite role in mammals during stresses and promotes cell death.[84,119,120] Other roles for Isc1 were recently reviewed and readers are encouraged to examine previous work on this interesting enzyme.[6,9]

Finally, recent studies showed that overexpression of the Ydc1 ceramidase (Fig. 1) shortened chronological lifespan, but this shortening probably results from fragmentation of mitochondria and vacuoles and increased apoptosis and it is not clear how these studies relate to normal physiological functions of Ydc1 and sphingolipids.[121] Analysis of ceramides in Ydc1 overproducing cells showed that both dihydroceramides and phytoceramides were reduced, suggesting that the enzyme does not specifically hydrolyze dihydroceramides in vivo. These studies are supportative of a role for ceramide as an inducer of apoptosis in yeast similar to what transpires in mammals.

Regulation of Sphingolipid Biosynthesis

I think it is safe to say that we do not really understand how de novo sphingolipid synthesis is regulated in any organism. Readers should consult previous reviews describing what we do know about the regulation of sphingolipid biosynthesis including regulation of gene transcription as only the latest studies will be discussed herein.[4-6,8-10]

Recently it was shown that TORC2 controls ceramide synthase activity.[122] The TOR protein kinases sense nutrients and stresses and coordinate metabolism both temporarily and spatially to regulate cell growth.[123] All eukaryotes that have been examined contain two TOR protein complexes, TORC1, which is inhibited by rapamycin, a bacterial macrocyclic lactone and TORC2, which is rapamycin-insensitive. A genetic screen had implied a link between sphingolipids, the TOR proteins and calcium homeostasis in yeast. This screen used the sensitivity of a *csg2* mutant to 100 mM Ca^{++} to identify temperature-sensitive mutations that bypassed the calcium-sensitivity.[62] Bypass mutations occurred in several genes including *TOR2* and *AVO3/TSC11* that encode components of TORC2.[34] TORC2 specifically controls the organization of the actin cytoskeleton in yeast and mammals.[123]

A critical advance in understanding the connection between TOR signaling and sphingolipids relied upon the isolation of a temperature-sensitive allele of *AVO3* (*avo3-30*) that diminished growth even at 30°C. Three hours after shifting *avo3-30* cells from 25°C to 30°C the major yeast ceramide species containing PHS and a C_{26}-fatty acid was 5-fold lower in concentration compared to wild-type cells and minor ceramide species having shorter fatty acid chains were reduced about 10-fold.[122] Even when grown at 25°C, the concentration of the major ceramides was reduced 2-fold in *avo3-30* cells. These results implied a reduction in ceramide synthase activity in *avo3-30* cells and this prediction was confirmed by measuring enzyme activity in cells grown at 30°C.

How might TORC2 influence ceramide synthase activity? The Ypk2 protein kinase was known to be activated by TORC2[124] and mutant *ypk2* and *avo3-30* cells showed similar defects in cell wall integrity and actin polarization, suggesting that Ypk2 working downstream of TORC2 and Pkh1/2 (Fig. 2) might regulate ceramide synthase activity. This possibility was supported by data showing that a constitutively active allele of *YPK2* reversed all *avo3-30* phenotypes including the ceramide deficiency.[122] These results argue that Ypk2 acts downstream of TORC2 to activate ceramide synthase activity. The molecular details of how Ypk2 governs ceramide synthase activity remains to be determined. Further analysis implicated calcineurin as a regulator of ceramide levels. Deletion of the *CNB1* gene, which encodes the calcineurin regulatory subunit B, restored ceramide levels in *avo3-30* cells and promoted growth at 30°C, suggesting that calcineurin down-regulates ceramide synthase activity. Perhaps calcineurin dephosphorylates ceramide synthase and one wonders if the Slm proteins (Fig. 3) are somehow involved is such regulation. Surprisingly, deletion of *CNB1* did not restore actin polarization to *avo3-30* cells, implying that it is controlled by TORC2 in a distinctly different manner from ceramide synthesis activity.

A reduction in ceramide synthase activity predicts an increase in DHS and PHS based upon previous work (reviewed in ref. 6) and *avo3-30* cells did indeed have elevated DHS and PHS levels.[122] These elevated LCBs could act in a feed-forward manner to activate ceramide synthase activity and promote synthesis of complex sphingolipids. This may occur because DHS and PHS activate Pkh1 and Pkh2 which then phosphorylate Ypk2 in its activation domain while TORC2 phosphorylate residues in the C-terminus (Fig. 2). To be active, Ypk2 must be phosphorylated in both domains. The possibility of similar types of regulation of ceramide synthase in mammals has been discussed.[125]

These studies form a basis for understanding how yeast cells coordinate ceramide and sphingolipid synthesis with nutrient availability and cell growth and how synthesis is reduced during stress when cells need to shift from growth to survival mode. But there is likely to be a lot more to this story than is apparent now and there are likely to be many more layers of regulation including regulation of Isc1 as depicted in Figure 3.

Conclusion and Future Developments

Since sphingolipids are abundant in the plasma membrane and also present in smaller amounts in other cellular membranes and compartments it is not surprising that they have been found to play roles in many processes occurring in or on membranes (Figs. 2 and 3 and see also Fig. 3 in ref. 9). Most of what we know about sphingolipid involvement in these processes has come from directed, small-scale experiments involving one or a few proteins or cellular processes as readouts and such studies will continue to be required to elucidate molecular mechanisms. Larger-scale, genome or proteome-wide experiments have started to identify new roles for sphingolipids and proteins that require sphingolipids for function.[12,48,96,126-130] More large-scale experiments that interrogate all yeast proteins need to be devised to determine if there are general rules for why proteins depend upon sphingolipids for function or whether such functional dependence is unique for each protein. The ability to analyze and quantify ("profile") of nearly all species of sphingolipids in a single sample by mass spectrometry is a great advance in methodology that will likely become the standard way to measure sphingolipids in yeast[16,24,131,132] much as it has become the standard way to analyze sphingolipids in mammals.[131,133] One of the most challenging problems will be to advance our understanding of the roles that sphingolipids play in cells from phenomenology and associations to molecular mechanisms. This will require techniques to measure sphingolipid binding to proteins either in vivo or in vitro and to demonstrate that binding promotes or inhibits a cellular process. For example, do LCBs directly interact with Pkh1/2 to stimulate their activity or do they act indirectly such as by binding to eisosomes which then activate Pkh1/2?

Acknowledgements

Work in the author's laboratory was supported by research grant from the National Institutes of Health (AG024377) and by core facilities supported by Grant P20-RR020171 from the National Center for Research Resources, a component of the National Institutes of Health.

References

1. Carter HE, Hendrickson HS. Biochemistry of the sphingolipids. XV. Structure of phytosphingosine and dehydrophytospingosine. Biochemistry 1963; 2:389-93.
2. Dickson RC, Lester RL. Metabolism and selected functions of sphingolipids in the yeast Saccharomyces cerevisiae. Biochim Biophys Acta 1999; 1438(3):305-21.
3. Dickson RC, Lester RL. Yeast sphingolipids. Biochim Biophys Acta 1999; 1426(2):347-357.
4. Dickson RC, Lester RL. Sphingolipid functions in Saccharomyces cerevisiae. Biochim Biophys Acta 2002; 1583(1):13-25.
5. Sims KJ, Spassieva SD, Voit EO et al. Yeast sphingolipid metabolism: clues and connections. Biochem Cell Biol 2004; 82(1):45-61.
6. Dickson RC. Thematic review series: sphingolipids. New insights into sphingolipid metabolism and function in budding yeast. J Lipid Res 2008; 49(5):909-21.
7. Patton JL, Lester RL. The phosphoinositol sphingolipids of Saccharomyces cerevisiae are highly localized in the plasma membrane. J Bacteriol 1991; 173(10):3101-3108.
8. Funato K, Vallee B, Riezman H. Biosynthesis and trafficking of sphingolipids in the yeast Saccharomyces cerevisiae. Biochemistry 2002; 41(51):15105-15114.
9. Dickson RC, Sumanasekera C, Lester RL. Functions and metabolism of sphingolipids in Saccharomyces cerevisiae. Prog Lipid Res 2006; 45(6):447-65.
10. Cowart LA, Obeid LM. Yeast sphingolipids: recent developments in understanding biosynthesis, regulation and function. Biochim Biophys Acta 2007; 1771(3):421-31.
11. Funato K, Riezman H. Vesicular and nonvesicular transport of ceramide from ER to the Golgi apparatus in yeast. J Cell Biol 2001; 155(6):949-59.
12. Natter K, Leitner P, Faschinger A et al. The spatial organization of lipid synthesis in the yeast Saccharomyces cerevisiae derived from large scale green fluorescent protein tagging and high resolution microscopy. Mol Cell Proteomics 2005; 4(5):662-672.
13. Hechtberger P, Daum G. Intracellular transport of inositol-containing sphingolipids in the yeast, Saccharomyces cerevisiae. FEBS Lett 1995; 367(2):201-204.
14. Futerman AH, Riezman H. The ins and outs of sphingolipid synthesis. Trends Cell Biol 2005; 15(6):312-8.
15. Hechtberger P, Zinser E, Saf R et al. Characterization, quantification and subcellular localization of inositol-containing sphingolipids of the yeast, Saccharomyces cerevisiae. Eur J Biochem 1994; 225(2):641-649.
16. Schneiter R, Brugger B, Sandhoff R et al. Electrospray ionization tandem mass spectrometry (ESI-MS/MS) analysis of the lipid molecular species composition of yeast subcellular membranes reveals acyl chain-based sorting/remodeling of distinct molecular species en route to the plasma membrane. J Cell Biol 1999; 146(4):741-54.
17. Kitagaki H, Cowart LA, Matmati N et al. Isc1 regulates sphingolipid metabolism in yeast mitochondria. Biochim Biophys Acta 2007; 1768(11):2849-61.
18. Smith SW, Lester RL. Inositol phosphorylceramide, a novel substance and a chief member of a major group of yeast sphingolipids containing single inositol phosphate. Journal of Biological Chemistry 1974; 249:3395-3405.
19. Oh CS, Toke DA, Mandala S et al. ELO2 and ELO3, homologues of the Saccharomyces cerevisiae ELO1 gene, function in fatty acid elongation and are required for sphingolipid formation. J Biol Chem 1997; 272(28):17376-84.
20. Kohlwein SD, Eder S, Oh CS et al. Tsc13p is required for fatty acid elongation and localizes to a novel structure at the nuclear-vacuolar interface in Saccharomyces cerevisiae. Mol Cell Biol 2001; 21(1):109-25.
21. Rossler H, Rieck C, Delong T et al. Functional differentiation and selective inactivation of multiple Saccharomyces cerevisiae genes involved in very-long-chain fatty acid synthesis. Mol Genet Genomics 2003; 269(2):290-8.
22. Denic V, Weissman JS. A molecular caliper mechanism for determining very long-chain fatty acid length. Cell 2007; 130(4):663-77.
23. Schneiter R, Hitomi M, Ivessa AS et al. A yeast acetyl coenzyme a carboxylase mutant links very-long-chain fatty acid synthesis to the structure and function of the nuclear membrane-pore complex. Mol Cell Biol 1996; 16(12):7161-7172.
24. Cerantola V, Vionnet C, Aebischer OF et al. Yeast sphingolipids do not need to contain very long chain fatty acids. Biochem J 2007; 401(1):205-16.

25. Lynch V, Dunn T. An introduction to plant sphingolipids and a review of recent advances in understanding their metabolism and function. New Phytologist 2004; 16(3):677-702.
26. Markham JE, Li J, Cahoon EB et al. Separation and identification of major plant sphingolipid classes from leaves. J Biol Chem 2006; 281(32):22684-94.
27. Replansky T, Koufopanou V, Greig D et al. Saccharomyces sensu stricto as a model system for evolution and ecology. Trends Ecol Evol 2008; 23(9):494-501.
28. Raas-Rothschild A, Pankova-Kholmyansky I, Kacher Y et al. Glycosphingolipidoses: beyond the enzymatic defect. Glycoconj J 2004; 21(6):295-304.
29. Welsch CA, Roth LW, Goetschy JF et al. Genetic, biochemical and transcriptional responses of Saccharomyces cerevisiae to the novel immunomodulator FTY720 largely mimic those of the natural sphingolipid phytosphingosine. J Biol Chem 2004; 279(35):36720-31.
30. Sawai H, Okamoto Y, Luberto C et al. Identification of ISC1 (YER019w) as inositol phosphosphingolipid phospholipase C in Saccharomyces cerevisiae. J Biol Chem 2000; 275(50):39793-8.
31. Wells GB, Dickson RC, Lester RL. Isolation and composition of inositolphosphorylceramide-type sphingolipids of hyphal forms of Candida albicans. Journal of Bacteriology 1996; 178(21):6223-6226.
32. Vaena de Avalos S, Su X, Zhang M et al. The phosphatidylglycerol/cardiolipin biosynthetic pathway is required for the activation of inositol phosphosphingolipid phospholipase C, Isc1p, during growth of Saccharomyces cerevisiae. J Biol Chem 2005; 280(8):7170-7.
33. Zhao C, Beeler T, Dunn T. Suppressors of the Ca(2+)-sensitive yeast mutant (csg2) identify genes involved in sphingolipid biosynthesis. Cloning and characterization of SCS1, a gene required for serine palmitoyltransferase activity. J Biol Chem 1994; 269(34):21480-8.
34. Beeler T, Bacikova D, Gable K et al. The Saccharomyces cerevisiae TSC10/YBR265w gene encoding 3-ketosphinganine reductase is identified in a screen for temperature-sensitive suppressors of the Ca2+-sensitive csg2 Delta mutant. J Biol Chem 1998; 273(46):30688-94.
35. Mao C, Xu R, Bielawska A et al. Cloning of an alkaline ceramidase from Saccharomyces cerevisiae. An enzyme with reverse (CoA-independent) ceramide synthase activity. J Biol Chem 2000; 275(10):6876-84.
36. Mao C, Xu R, Bielawska A et al. Cloning and characterization of a Saccharomyces cerevisiae alkaline ceramidase with specificity for dihydroceramide. J Biol Chem 2000; 275(40):31369-78.
37. Pruett ST, Bushnev A, Hagedorn K et al. Biodiversity of sphingoid bases ("sphingosines") and related amino alcohols. J Lipid Res 2008; 49(8):1621-39.
38. Munn AL, Riezman H. Endocytosis is required for the growth of vacuolar H(+)-ATPase- defective yeast: identification of six new END genes. J Cell Biol 1994; 127(2):373-86.
39. Sutterlin C, Doering TL, Schimmoller F et al. Specific requirements for the ER to Golgi transport of GPI-anchored proteins in yeast. Journal of Cell Science 1997; 110(Pt 21):2703-2714.
40. Chung N, Jenkins G, Hannun YA et al. Sphingolipids signal heat stress-induced ubiquitin-dependent proteolysis. J Biol Chem 2000; 275(23):17229-32.
41. Zanolari B, Friant S, Funato K et al. Sphingoid base synthesis requirement for endocytosis in Saccharomyces cerevisiae. EMBO J 2000; 19(12):2824-2833.
42. Hearn JD, Lester RL, Dickson RC. The uracil transporter Fur4p associates with lipid rafts. J Biol Chem 2003; 278(6):3679-3686.
43. Grote E, Vlacich G, Pypaert M et al. A snc1 endocytosis mutant: phenotypic analysis and suppression by overproduction of dihydrosphingosine phosphate lyase. Mol Biol Cell 2000; 11(12):4051-65.
44. Bultynck G, Heath VL, Majeed AP et al. Slm1 and slm2 are novel substrates of the calcineurin phosphatase required for heat stress-induced endocytosis of the yeast uracil permease. Mol Cell Biol 2006; 26(12):4729-45.
45. Bagnat M, Simons K. Lipid rafts in protein sorting and cell polarity in budding yeast Saccharomyces cerevisiae. Biol Chem 2002; 383(10):1475-80.
46. Simons K, Vaz WL. Model systems, lipid rafts and cell membranes. Annu Rev Biophys Biomol Struct 2004; 33:269-95.
47. Lisman Q, Pomorski T, Vogelzangs C et al. Protein sorting in the late Golgi of Saccharomyces cerevisiae does not require mannosylated sphingolipids. J Biol Chem 2004; 279(2):1020-9.
48. Proszynski TJ, Klemm RW, Gravert M et al. A genome-wide visual screen reveals a role for sphingolipids and ergosterol in cell surface delivery in yeast. Proc Natl Acad Sci USA 2005; 102(50):17981-6.
49. Bagnat M, Keranen S, Shevchenko A et al. Lipid rafts function in biosynthetic delivery of proteins to the cell surface in yeast. Proc Natl Acad Sci USA 2000; 97(7):3254-9.
50. Wang Q, Chang A. Sphingoid base synthesis is required for oligomerization and cell surface stability of the yeast plasma membrane ATPase, Pma1. Proc Natl Acad Sci USA 2002; 99(20):12853-8.
51. Lee MC, Hamamoto S, Schekman R. Ceramide biosynthesis is required for the formation of the oligomeric H+-ATPase Pma1p in the yeast endoplasmic reticulum. J Biol Chem 2002; 277(25):22395-401.
52. Eisenkolb M, Zenzmaier C, Leitner E et al. A specific structural requirement for ergosterol in long-chain fatty acid synthesis mutants important for maintaining raft domains in yeast. Mol Biol Cell 2002; 13(12):4414-28.

53. Gaigg B, Timischl B, Corbino L et al. Synthesis of sphingolipids with very long chain fatty acids but not ergosterol is required for routing of newly synthesized plasma membrane ATPase to the cell surface of yeast. J Biol Chem 2005; 280(23):22515-22.
54. Toulmay A, Schneiter R. Lipid-dependent surface transport of the proton pumping ATPase: a model to study plasma membrane biogenesis in yeast. Biochimie 2007; 89(2):249-54.
55. Bagnat M, Simons K. Cell surface polarization during yeast mating. Proc Natl Acad Sci USA 2002; 99(22):14183-8.
56. Dupre S, Haguenauer-Tsapis R. Raft partitioning of the yeast uracil permease during trafficking along the endocytic pathway. Traffic 2003; 4(2):83-96.
57. Opekarova M, Malinska K, Novakova L et al. Differential effect of phosphatidylethanolamine depletion on raft proteins: further evidence for diversity of rafts in Saccharomyces cerevisiae. Biochim Biophys Acta 2005; 1711(1):87-95.
58. Lauwers E, Grossmann G, Andre B. Evidence for coupled biogenesis of yeast Gap1 permease and sphingolipids: essential role in transport activity and normal control by ubiquitination. Mol Biol Cell 2007; 18(8):3068-80.
59. Gaigg B, Toulmay A, Schneiter R. Very long-chain fatty acid-containing lipids rather than sphingolipids per se are required for raft association and stable surface transport of newly synthesized plasma membrane ATPase in yeast. J Biol Chem 2006; 281(45):34135-45.
60. Olivera A, Zhang H, Carlson RO et al. Stereospecificity of sphingosine-induced intracellular calcium mobilization and cellular proliferation. Journal of Biological Chemistry 1994; 269(27):17924-17930.
61. Birchwood CJ, Saba JD, Dickson RC et al. Calcium influx and signaling in yeast stimulated by intracellular sphingosine 1-phosphate accumulation. J Biol Chem 2001; 276(15):11712-8.
62. Beeler T, Gable K, Zhao C et al. A novel protein, CSG2p, is required for Ca2+ regulation in Saccharomyces cerevisiae. J Biol Chem 1994; 269(10):7279-84.
63. Lisman Q, Urli-Stam D, Holthuis JC. HOR7, a multicopy suppressor of the Ca2+-induced growth defect in sphingolipid mannosyltransferase-deficient yeast. J Biol Chem 2004; 279(35):36390-6.
64. Skrzypek MS, Nagiec MM, Lester RL et al. Inhibition of amino acid transport by sphingoid long chain bases in Saccharomyces cerevisiae. J Biol Chem 1998; 273(5):2829-34.
65. Chung N, Mao C, Heitman J et al. Phytosphingosine as a specific inhibitor of growth and nutrient import in Saccharomyces cerevisiae. J Biol Chem 2001; 276(38):35614-35621.
66. Friant S, Meier KD, Riezman H. Increased ubiquitin-dependent degradation can replace the essential requirement for heat shock protein induction. EMBO J 2003; 22(15):3783-91.
67. Gelperin D, Horton L, DeChant A et al. Loss of ypk1 function causes rapamycin sensitivity, inhibition of translation initiation and synthetic lethality in 14-3-3-deficient yeast. Genetics 2002; 161(4):1453-1464.
68. Kraakman L, Lemaire K, Ma P et al. A Saccharomyces cerevisiae G-protein coupled receptor, Gpr1, is specifically required for glucose activation of the cAMP pathway during the transition to growth on glucose. Mol Microbiol 1999; 32(5):1002-12.
69. Roosen J, Engelen K, Marchal K et al. PKA and Sch9 control a molecular switch important for the proper adaptation to nutrient availability. Molecular Microbiology 2005; 55(3):862-880.
70. Valachovic M, Wilcox LI, Sturley SL et al. A mutation in sphingolipid synthesis suppresses defects in yeast ergosterol metabolism. Lipids 2004; 39(8):747-52.
71. Valachovic M, Bareither BM, Shah Alam Bhuiyan M et al. Cumulative mutations affecting sterol biosynthesis in the yeast Saccharomyces cerevisiae result in synthetic lethality that is suppressed by alterations in sphingolipid profiles. Genetics 2006; 173(4):1893-908.
72. Swain E, Baudry K, Stukey J et al. Sterol-dependent regulation of sphingolipid metabolism in Saccharomyces cerevisiae. J Biol Chem 2002; 277(29):26177-84.
73. Kihara A, Igarashi Y. Cross talk between sphingolipids and glycerophospholipids in the establishment of plasma membrane asymmetry. Mol Biol Cell 2004; 15(11):4949-59.
74. Thevissen K, Francois IE, Takemoto JY et al. DmAMP1, an antifungal plant defensin from dahlia (Dahlia merckii), interacts with sphingolipids from Saccharomyces cerevisiae. FEMS Microbiol Lett 2003; 226(1):169-73.
75. Thevissen K, Cammue BP, Lemaire K et al. A gene encoding a sphingolipid biosynthesis enzyme determines the sensitivity of Saccharomyces cerevisiae to an antifungal plant defensin from dahlia (Dahlia merckii). Proc Natl Acad Sci USA 2000; 97(17):9531-6.
76. Stock SD, Hama H, Radding JA et al. Syringomycin E inhibition of Saccharomyces cerevisiae: requirement for biosynthesis of sphingolipids with very-long-chain fatty acids and mannose- and phosphoinositol-containing head groups. Antimicrob Agents Chemother 2000; 44(5):1174-1180.
77. Hama H, Youngb DA, Raddingb JA et al. Requirement of sphingolipid alpha-hydroxylation for fungicidal action of syringomycin E. FEBS Lett 2000; 478(1-2):26-8.
78. Grilley MM, Stock SD, Dickson RC et al. Syringomycin action gene SYR2 is essential for sphingolipid 4-hydroxylation in Saccharomyces cerevisiae. J Biol Chem 1998; 273(18):11062-11068.

79. Bosson R, Conzelmann A. Multiple functions of inositolphosphorylceramides in the formation and intracellular transport of glycosylphosphatidylinositol-anchored proteins in yeast. Biochem Soc Symp 2007(74):199-209.
80. Horvath A, Sutterlin C, Manning-Krieg U et al. Ceramide synthesis enhances transport of GPI-anchored proteins to the Golgi apparatus in yeast. Embo J 1994; 13(16):3687-95.
81. Skrzypek M, Lester RL, Dickson RC. Suppressor gene analysis reveals an essential role for sphingolipids in transport of glycosylphosphatidylinositol-anchored proteins in Saccharomyces cerevisiae. J Bacteriol 1997; 179(5):1513-20.
82. Daquinag A, Fadri M, Jung SY et al. The yeast PH domain proteins Slm1 and Slm2 are targets of sphingolipid signaling during the response to heat stress. Mol Cell Biol 2007; 27(2):633-50.
83. Chung JH, Lester RL, Dickson RC. Sphingolipid requirement for generation of a functional v1 component of the vacuolar ATPase. J Biol Chem 2003; 278(31):28872-81.
84. Carpinteiro A, Dumitru C, Schenck M et al. Ceramide-induced cell death in malignant cells. Cancer Lett 2008; 264(1):1-10.
85. Hannun YA, Obeid LM. Principles of bioactive lipid signalling: lessons from sphingolipids. Nat Rev Mol Cell Biol 2008; 9(2):139-50.
86. Hait NC, Oskeritzian CA, Paugh SW et al. Sphingosine kinases, sphingosine 1-phosphate, apoptosis and diseases. Biochim Biophys Acta 2006; 1758(12):2016-26.
87. Dickson RC, Nagiec EE, Skrzypek M et al. Sphingolipids are potential heat stress signals in Saccharomyces. Journal of Biological Chemistry 1997; 272:30196-30200.
88. Jenkins GM, Richards A, Wahl T et al. Involvement of yeast sphingolipids in the heat stress response of Saccharomyces cerevisiae. J Biol Chem 1997; 272(51):32566-32572.
89. Sun Y, Taniguchi R, Tanoue D et al. Sli2 (Ypk1), a homologue of mammalian protein kinase SGK, is a downstream kinase in the sphingolipid-mediated signaling pathway of yeast. Mol Cell Biol 2000; 20(12):4411-4419.
90. Schmelzle T, Helliwell SB, Hall MN. Yeast protein kinases and the RHO1 exchange factor TUS1 are novel components of the cell integrity pathway in yeast. Mol Cell Biol 2002; 22(5):1329-1339.
91. Roelants FM, Torrance PD, Bezman N et al. Pkh1 and Pkh2 differentially phosphorylate and activate Ypk1 and Ykr2 and define protein kinase modules required for maintenance of cell wall integrity. Mol Biol Cell 2002; 13(9):3005-3028.
92. deHart AK, Schnell JD, Allen DA et al. The conserved Pkh-Ypk kinase cascade is required for endocytosis in yeast. J Cell Biol 2002; 156(2):241-248.
93. Casamayor A, Torrance PD, Kobayashi T et al. Functional counterparts of mammalian protein kinases PDK1 and SGK in budding yeast. Current Biology 1999; 9(4):186-197.
94. Urban J, Soulard A, Huber A et al. Sch9 is a major target of TORC1 in Saccharomyces cerevisiae. Mol Cell 2007; 26(5):663-74.
95. Zhang X, Lester RL, Dickson RC. Pil1p and Lsp1p negatively regulate the 3-phosphoinositide-dependent protein kinase-like kinase Pkh1p and downstream signaling pathways Pkc1p and Ypk1p. J Biol Chem 2004; 279(21):22030-22038.
96. Hillenmeyer ME, Fung E, Wildenhain J et al. The chemical genomic portrait of yeast: uncovering a phenotype for all genes. Science 2008; 320(5874):362-5.
97. Liu K, Zhang X, Lester RL et al. The sphingoid long chain base phytosphingosine activates AGC-type protein kinases in Saccharomyces cerevisiae including Ypk1, Ypk2 and Sch9. J Biol Chem 2005; 280(24):22679-22987.
98. Walther TC, Aguilar PS, Frohlich F et al. Pkh-kinases control eisosome assembly and organization. Embo J 2007; 26(24):4946-55.
99. Walther TC, Brickner JH, Aguilar PS et al. Eisosomes mark static sites of endocytosis. Nature 2006; 439(7079):998-1003.
100. Luo G, Gruhler A, Liu Y et al. The sphingolipid long-chain base-Pkh1/2-Ypk1/2 signaling pathway regulates eisosome assembly and turnover. J Biol Chem 2008; 283(16):10433-44.
101. Kobayashi T, Takematsu H, Yamaji T et al. Disturbance of sphingolipid biosynthesis abrogates the signaling of Mss4, phosphatidylinositol-4-phosphate 5-kinase, in yeast. J Biol Chem 2005; 280(18):18087-94.
102. Tabuchi M, Audhya A, Parsons AB et al. The phosphatidylinositol 4,5-biphosphate and TORC2 binding proteins Slm1 and Slm2 function in sphingolipid regulation. Mol Cell Biol 2006; 26(15):5861-75.
103. Audhya A, Loewith R, Parsons AB et al. Genome-wide lethality screen identifies new PI4,5P2 effectors that regulate the actin cytoskeleton. Embo J 2004; 23(19):3747-57.
104. Yu JW, Mendrola JM, Audhya A et al. Genome-wide analysis of membrane targeting by S. cerevisiae pleckstrin homology domains. Mol Cell 2004; 13(5):677-88.
105. Fadri M, Daquinag A, Wang S et al. The pleckstrin homology domain proteins Slm1 and Slm2 are required for actin cytoskeleton organization in yeast and bind phosphatidylinositol-4,5-bisphosphate and TORC2. Mol Biol Cell 2005; 16(4):1883-900.

106. Sturgill TW, Cohen A, Diefenbacher M et al. TOR1 and TOR2 have distinct locations in live cells. Eukaryot Cell 2008; 7(10):1819-30.
107. Dunn TM, Haak D, Monaghan E et al. Synthesis of monohydroxylated inositolphosphorylceramide (IPC-C) in Saccharomyces cerevisiae requires Scs7p, a protein with both a cytochrome b5-like domain and a hydroxylase/desaturase domain. Yeast 1998; 14(4):311-21.
108. Brace JL, Lester RL, Dickson RC et al. SVF1 regulates cell survival by affecting sphingolipid metabolism in Saccharomyces cerevisiae. Genetics 2007; 175(1):65-76.
109. D'Mello NP, Childress AM, Franklin DS et al. Cloning and characterization of LAG1, a longevity-assurance gene in yeast. J Biol Chem 1994; 269(22):15451-9.
110. Guillas I, Kirchman PA, Chuard R et al. C26-CoA-dependent ceramide synthesis of Saccharomyces cerevisiae is operated by Lag1p and Lac1p. Embo J 2001; 20(11):2655-65.
111. Schorling S, Vallee B, Barz WP et al. Lag1p and Lac1p are essential for the Acyl-CoA-dependent Ceramide synthase reaction in Saccharomyces cerevisae. Mol Biol Cell 2001; 12(11):3417-27.
112. Fabrizio P, Pozza F, Pletcher SD et al. Regulation of longevity and stress resistance by Sch9 in yeast. Science 2001; 292(5515):288-290.
113. Kaeberlein M, Powers RW 3rd, Steffen KK et al. Regulation of yeast replicative life span by TOR and Sch9 in response to nutrients. Science 2005; 310(5751):1193-1196.
114. Almeida T, Marques M, Mojzita D et al. Isc1p plays a key role in hydrogen peroxide resistance and chronological lifespan through modulation of iron levels and apoptosis. Mol Biol Cell 2008; 19(3):865-876.
115. Madeo F, Frohlich E, Ligr M et al. Oxygen stress: a regulator of apoptosis in yeast. J Cell Biol 1999; 145(4):757-67.
116. Madeo F, Herker E, Maldener C et al. A caspase-related protease regulates apoptosis in yeast. Mol Cell 2002; 9(4):911-7.
117. Fabrizio P, Battistella L, Vardavas R et al. Superoxide is a mediator of an altruistic aging program in Saccharomyces cerevisiae. J Cell Biol 2004; 166(7):1055-67.
118. Herker E, Jungwirth H, Lehmann KA et al. Chronological aging leads to apoptosis in yeast. J Cell Biol 2004; 164(4):501-7.
119. Ogretmen B, Hannun YA. Biologically active sphingolipids in cancer pathogenesis and treatment. Nat Rev Cancer 2004; 4(8):604-16.
120. Taha TA, Mullen TD, Obeid LM. A house divided: ceramide, sphingosine and sphingosine-1-phosphate in programmed cell death. Biochim Biophys Acta 2006; 1758(12):2027-36.
121. Aerts AM, Zabrocki P, Francois IE et al. Ydc1p ceramidase triggers organelle fragmentation, apoptosis and accelerated ageing in yeast. Cell Mol Life Sci 2008; 65(12):1933-42.
122. Aronova S, Wedaman K, Aronov PA et al. Regulation of ceramide biosynthesis by TOR complex 2. Cell Metab 2008; 7(2):148-58.
123. Wullschleger S, Loewith R, Hall MN. TOR signaling in growth and metabolism. Cell 2006; 124(3):471-84.
124. Kamada Y, Fujioka Y, Suzuki NN et al. Tor2 directly phosphorylates the AGC kinase Ypk2 to regulate actin polarization. Mol Cell Biol 2005; 25(16):7239-48.
125. Dickson RC. More chores for TOR: de novo ceramide synthesis. Cell Metab 2008; 7(2):99-100.
126. Bammert GF, Fostel JM. Genome-wide expression patterns in Saccharomyces cerevisiae: Comparison of drug treatments and genetic alterations affecting biosynthesis of ergosterol. Antimicrob Agents Chemother 2000; 44(5):1255-1265.
127. Kwast KE, Lai LC, Menda N et al. Genomic analyses of anaerobically induced genes in Saccharomyces cerevisiae: functional roles of Rox1 and other factors in mediating the anoxic response. J Bacteriol 2002; 184(1):250-65.
128. Cowart LA, Okamoto Y, Pinto FR et al. Roles for sphingolipid biosynthesis in mediation of specific programs of the heat stress response determined through gene expression profiling. J Biol Chem 2003; 278(32):30328-38.
129. Baetz K, McHardy L, Gable K et al. Yeast genome-wide drug-induced haploinsufficiency screen to determine drug mode of action. Proc Natl Acad Sci USA 2004; 101(13):4525-30.
130. Laun P, Ramachandran L, Jarolim S et al. A comparison of the aging and apoptotic transcriptome of Saccharomyces cerevisiae. FEMS Yeast Res 2005; 5(12):1261-72.
131. Bielawski J, Szulc ZM, Hannun YA et al. Simultaneous quantitative analysis of bioactive sphingolipids by high-performance liquid chromatography-tandem mass spectrometry. Methods 2006; 39(2):82-91.
132. Guan XL, Wenk MR. Mass spectrometry-based profiling of phospholipids and sphingolipids in extracts from Saccharomyces cerevisiae. Yeast 2006; 23(6):465-77.
133. Sullards MC, Allegood JC, Kelly S et al. Structure-specific, quantitative methods for analysis of sphingolipids by liquid chromatography-tandem mass spectrometry: "inside-out" sphingolipidomics. Methods Enzymol 2007; 432:83-115.

Chapter 16

Sphingolipid Signaling in Fungal Pathogens

Ryan Rhome and Maurizio Del Poeta*

Abstract

Sphingolipid involvement in infectious disease is a new and exciting branch of research. Various microbial pathogens have been shown to synthesize their own sphingolipids and some have evolved methods to "hijack" host sphingolipids for their own use. For instance, *Sphingomonas* species are bacterial pathogens that lack the lipopolysaccharide component typical but instead contain glycosphingolipids (Kawahara 1991, 2006). In terms of sphingolipid signaling and function, perhaps the best-studied group of microbes is the pathogenic fungi.

Pathogenic fungi still represent significant problems in human disease, despite treatments that have been used for decades. Because fungi are eukaryotic, drug targets in fungi can have many similarities to mammalian processes. This often leads to significant side effects of antifungal drugs that can be dose limiting in many patient populations. The search for fungal-specific drugs and the need for better understanding of cellular processes of pathogenic fungi has led to a large body of research on fungal signaling. One particularly interesting and rapidly growing field in this research is the involvement of fungal sphingolipid pathways in signaling and virulence. In this chapter, the research relating to sphingolipid signaling pathogenic fungi will be reviewed and summarized, in addition to highlighting pathways that show promise for future research.

Sphingolipid Synthesis

Sphingolipid synthesis in pathogenic fungi is largely conserved among species. Early steps in the process, such as the condensation of palmitoyl-CoA with serine to form 3-ketodihydrosphingosine, are the same in *Saccharomyces cerevisiae*. *S. cerevisiae* is often the gold standard for model systems, but in this context, that system is limited by the fact that *S. cerevisiae* has very different sphingolipid metabolism from most of the significant fungal pathogens. Both *S. cerevisiae* and fungal pathogens like *Candida albicans* make phytoceramide for instance (a modified ceramide with a carbon 4 hydroxylation on its backbone instead of the 3,4 desaturation). Phytoceramide is often conjugated with very long chain fatty acids and can be used to make more complex sphingolipids such as inositol-phosphoryl ceramide (IPC). In addition to this pathway, most pathogenic fungi also synthesize a different ceramide that contains the 3,4 desaturation and cannot be referred to as phytoceramide. This ceramide is typically conjugated to a 16-18 carbon acyl chain and undergoes additional modifications. These modifications include an additional sphingoid backbone desaturation between the 8th and 9th carbons and a methylation of that backbone on the 9th carbon. Typically, the acyl chain is hydroxylated at the α-carbon position. This modified ceramide is the substrate for an enzyme called glucosylceramide synthase

*Corresponding Author: Maurizio Del Poeta—Departments of Biochemistry and Molecular Biology, Microbiology and Immunology and Division of Infectious Diseases, Medical University of South Carolina, Charleston, South Carolina, 29425. Email: delpoeta@musc.edu

Sphingolipids as Signaling and Regulatory Molecules, edited by Charles Chalfant and Maurizio Del Poeta.

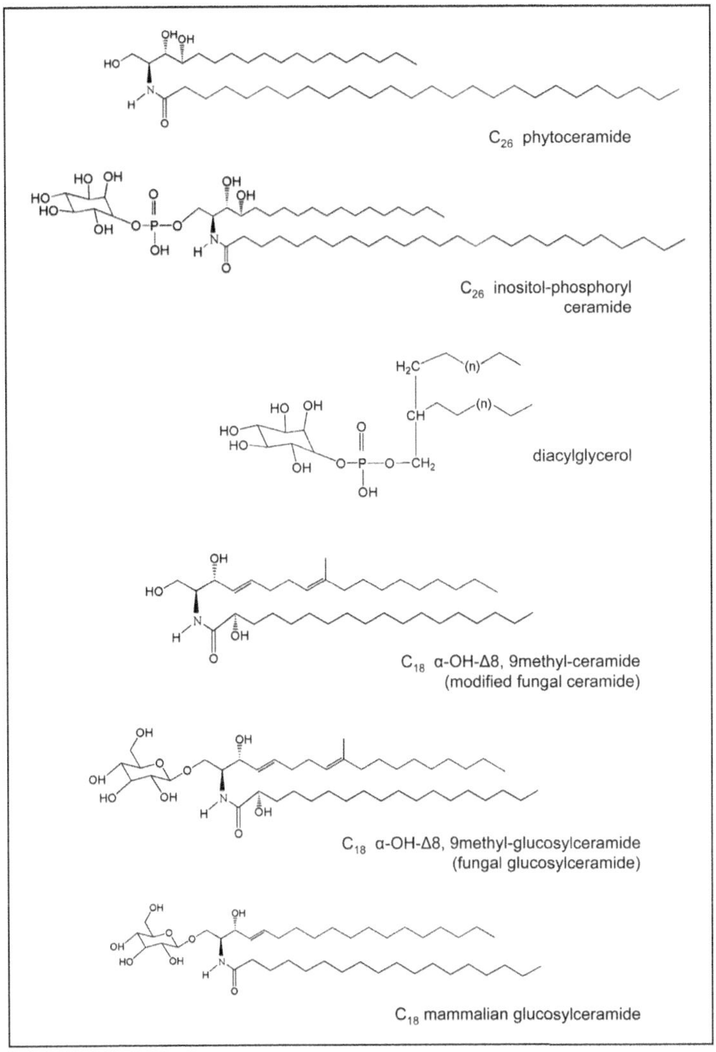

Figure 1. Chemical structures of C26 phytoceramide, C26 inositol phosphoryl ceramide, diacylglycerol, C18 α-hydroxy-Δ8, 9methyl-ceramide, C18 α-hydroxy-Δ8, 9methyl-glucosylceramide (fungal glucosylceramide) and C18 glucosylceramide (mammalial glucosylceramide).

(Gcs1), which glycosylates this molecule on the hydroxyl group of the 1st carbon, creating glucosylceramide (GlcCer). Though there is no direct evidence of cross talk between the glucosylceramide synthase pathway and the pathway leading to the synthesis of inositol-phosphoryl ceramide-containing lipids, this possibility cannot be ruled out. The structures of major fungal sphingolipids are found in Figure 1.

Cryptococcus Neofomans: Model of Sphingolipid Signaling in Fungi

Cryptococcus neoformans is an encapsulated fungal pathogen that primarily affects immunocompromised patients (ex. HIV/AIDS patients, transplant candidates and patients on long-term steroid treatments). This environmental yeast is inhaled to the lungs, where it can live extracellularly or intracellularly (inside the phagolysosomes of alveolar macrophages). Infection with this organism is known as cryptococcosis. In some patients, the fungus disseminates to the bloodstream, seeding many organ systems, but eventually proliferating in the central nervous system. This scenario represents a significant medical emergency, as *C. neoformans* is the leading cause of fungal meningoencephalitis in the world and the disease is lethal if left untreated. *Cryptococcus* has several recognized virulence factors, such as the production of melanin and the polysaccharide capsule.

Sphingolipid studies in *C. neoformans* have revealed some interesting implications for signaling pathways involving these molecules. Beyond understanding fungal biology on a cellular and biochemical level, the study of the fungal sphingolipid pathway is advantageous due to the fact that many of the enzymes and products are distinct in structure and function from their mammalian counterparts. This distinction makes them great candidates for drug targets.

The best-studied example of this paradigm is inositol-phosphoryl ceramide synthase (Ipc1). This enzyme uses phytoceramide and phosphatidylinositol (PI) as substrates, transferring the phosphorylinositol moiety to phytoceramide. In addition to the generation of inositol-phosphoryl ceramide (IPC), diacylglycerol (DAG) is also released as a product of this reaction. Early studies on Ipc1 in *C. neoformans* implicated the enzyme in virulence pathways. In strains where Ipc1 is downregulated, melanin production is impaired and the strain has growth deficits when inside alveolar macrophages. When tested in mouse models of cryptococcosis, the strain lacking Ipc1 was less virulent in comparison to the wildtype *C. neoformans*. Studies on IPC metabolism have also given clues to the role of this reaction in virulence. Inositol phosphosphingolipid-phospholipase C (Isc1) is the enzyme that catalyzes the reverse reaction of Ipc1, which is to remove the phosphotidylinositol component from IPC. A strain of *C. neoformans* in which this enzyme is deleted (Δ*isc1*) shows reduced virulence in immunocompromised mouse models. However, when macrophages are depleted in this model, Δ*isc1* will disseminate and cause meningoencephalitis. The Ipc1/Isc1 balance seems to play a role in the interaction between *C. neoformans* and the alveolar macrophages.

Further studies into the mechanism underlying the connection between Ipc1 and virulence of *C. neoformans* have shown that the production of DAG is a common step of at least two separate determinants of virulence in this fungus. As mentioned, early studies downregulating Ipc1 showed impairments in melanin production. This interaction was found to be mediated by cryptococcal protein kinase C (Pkc1). DAG, the byproduct of Ipc1 activity, was found to bind to the C1 domain of Pkc1. This binding led to an increase in Pkc1 activity and that activation was abolished by the selective deletion of the C1 domain. It is known that Pkc1 and several pathway components are required for proper cell wall integrity, including the function of some cell wall-associated enzymes. One such enzyme is laccase, which is responsible for the synthesis of melanin. The defect in melanin synthesis observed in Ipc1-downregulated strains was caused by improper localization of laccase to the cell wall, due to reduction in DAG-dependent Pkc1 activity.

Another way in which DAG production has been linked to virulence involves the fungal-macrophage interaction. *C. neoformans* has methods to avoid phagocytosis by alveolar macrophages when necessary, including the polysaccharide capsule. One such method is the production of antiphagocytic protein 1 (App1). When App1 is deleted, the resulting strain (Δ*app1*) shows reduced virulence in immunocompromised mice. The production of App1 is driven, transcriptionally, by the presence of DAG. DAG binds and activates the transcription factor Atf2. Atf2 activation promotes the transcription of App1 and thus evasion of phagocytosis leading to increased virulence. App1, in addition to regulating phagocytosis, has been shown to bind host complement receptors CR2 and CR3, suggesting even more complex fungal-host interactions affected by production of DAG in *C. neoformans*. The downstream effects of Ipc1 activity are summarized in Figure 2.

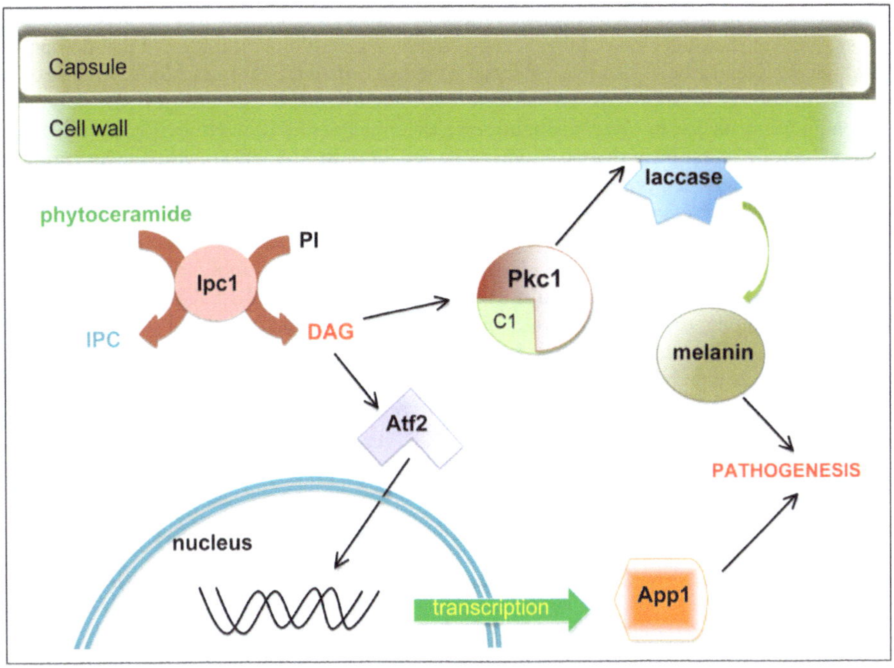

Figure 2. Sphingolipid signaling in *Cryptococcus neoformans*. Inositol phosphoryl ceramide synthase 1 (Ipc1) in *C. neoformans* produces diacylglycerol (DAG) in addition to inositol phosphoryl ceramide IPC). DAG binds to the C1 domain of protein kinase C1 (Pkc1), which is important for cell wall integrity. This integrity is crucial for localization of laccase, the enzyme responsible for melanin synthesis. In addition, DAG also activates the transcription factor Atf2, which leads to transcription of the antiphagocytic protein 1 (App1). Both App1 and melanin regulate pathogenicity of *C. neoformans*.

Sphingolipid Signaling in Other Pathogenic Fungi

While many groups have discovered and characterized sphingolipid components of other pathogenic fungi, few have delved into the role of these lipids in signaling or cellular processes. One such fungus is *Candida albicans*. *C. albicans* is a dimorphic fungus that normally lives as a commensal organism in the human gut and urogenital tract. In immunocompromised patients, *C. albicans* can cause significant disease, including systemic dissemination. Recent evidence in *Candida albicans* has shown possible sphingolipid involvement in endocytosis and plasma membrane functions. Sur7 is a membrane bound enzyme known to be involved in sphingolipid membrane makeup in *S. cerevisiae*. When the homolog of Sur7 was deleted in *C. albicans*, the resulting strain showed defects in hyphal morphogenesis, endocytosis and cell wall formation. Confirming the role of sphingolipids in this process, blocking sphingolipid synthesis resulted in disruption of Sur7 patches in the plasma membrane. Though this has yet to show the definitive role of sphingolipids in this process, their involvement is clear. Also, the function and localization of some multidrug resistance proteins in *C. albicans* have been shown to be dependent on membrane sphingolipid composition.

Conclusion

The signaling pathway involving Ipc1 and the production of DAG is clearly related to virulence in more ways than one. What other sphingolipid metabolic pathways could be involved in virulence of *C. neoformans* and other pathogenic fungi? Recall that in addition to IPC-based sphingolipids, most pathogenic fungi synthesize glucosylceramide that is not based on a phytoceramide backbone. The synthesis of these GlcCers requires enzymes that introduce the desaturation of the sphingoid backbone between carbons 8 and 9 as well as the methylation of the ninth carbon. Examinations into the function of these enzymes as well as glucosylceramide synthase have suggested major roles in biology and virulence. In *Candida,* for instance, the sphingolipid Δ8-desaturase, responsible for the backbone desaturation at that position, is required for proper hyphal growth. The function of the methyltransferase responsible for the C9 methylation seen in most fungi was studied in the plant pathogen *Fusarium graminerum.* Disruption of the enzyme encoding this enzyme resulted in a strain that showed defects in virulence, growth and differentiation. When the gene for glucosylceramide synthase is deleted in *C. neoformans,* the resulting strain shows condition-dependent growth defects as well as a lack of virulence in inhalation mouse models. Taken together, these studies are beginning to uncover the roles of the "glucosylceramide branch" of sphingolipids in pathogenic fungi. Like enzymes involved in IPC production, many of these enzymes are unique to fungi and thus represent attractive possibilities as therapeutic targets. Though the signaling mechanisms involved are unclear, these observations may represent the beginning of new understandings into the role of sphingolipid signaling in pathogenic fungi.

References

1. Hanada K. Sphingolipids in infectious diseases. Jpn J Infect Dis 2005; 58(3):131-148.
2. Heung LJ, Luberto C, Del Poeta M. Role of sphingolipids in microbial pathogenesis. Infect Immun 2006; 74(1):28-39.
3. Kawahara K, Sato N, Tsuge K, Seto Y. Confirmation of the anomeric structure of galacturonic acid in the galacturonosyl-ceramide of Sphingomonas yanoikuyae. Microbiol Immunol 2006; 50(1):67-71.
4. Kawahara K, Moll H, Knirel YA et al. Structural analysis of two glycosphingoliids from the lipopolysaccharide-lacking bacterium Sphingomonas capsulata. Eur J Biochem 2000; 267(6):1837-1846.
5. Heitman J, G FS, Edwards JEJ, Mitchell AP. Molecular Principles of Fungal Pathogenesis. Washington American Society of Microbiology; 2006.
6. Marr KA. New approaches to invasive fungal infections. Curr Opin Hematol 2003; 10(6):445-450.
7. Rapp RP. Changing strategies for the management of invasive fungal infections. Pharmacotherapy 2004; 24(2 Pt 2):4S-28S; quiz 29S-32S.
8. McQuiston TJ, Haller C, Del Poeta M. Sphingolipids as targets for microbial infections. Mini Rev Med Chem 2006; 6(6):671-680.
9. Rhome R, McQuiston T, Kechichian T et al. Biosynthesis and immunogenicity of glucosylceramide in Cryptococcus neoformans and other human pathogens. Eukaryot Cell 2007; 6(10):1715-1726.
10. Matmati N, Hannun YA. Thematic review series: sphingolipids. ISC1 (inositol phosphosphingolipid-phospholipase C), the yeast homologue of neutral sphingomyelinases. J Lipid Res 2008; 49(5):922-928.
11. Obeid LM, Okamoto Y, Mao C. Yeast sphingolipids: metabolism and biology. Biochim Biophys Acta 2002; 1585(2-3):163-171.
12. Sims KJ, Spassieva SD, Voit EO, Obeid LM. Yeast sphingolipid metabolism: clues and connections. Biochem Cell Biol 2004; 82(1):45-61.
13. Garcia J, Shea J, Alvarez-Vasquez F et al. Mathematical modeling of pathogenicity of Cryptococcus neoformans. Mol Sys Biol 2008; 4:183-195.
14. Rittershaus PC, Kechichian TB, Allegood J et al. Glucosylceramide is an essential regulator of pathogenicity of Cryptococcus neoformans. J Clin Invest 2006; 116(6):1651-1659.
15. Casadevall A, Perfect JR. Cryptococcus neoformans. Washington, DC, 381-405: ASM Press; 1998.
16. Luberto C, Toffaletti DL, Wills EA et al. Roles for inositol-phosphoryl ceramide synthase 1 (IPC1) in pathogenesis of C. neoformans. Genes Dev 2001; 15(2):201-212.
17. Shea J, Kechichian TB, Luberto C, Del Poeta M. The cryptococcal enzyme inositol phosphosphingolipid-phospholipase C (Isc1) confers resistance to the antifungal effects of macrophages and promotes fungal dissemination to the central nervous system. Infect Immun 2006; 74(10):5977-5988.
18. Heung LJ, Luberto C, Plowden A et al. The sphingolipid pathway regulates protein kinase C 1 (Pkc1) through the formation of diacylglycerol (DAG) in Cryptococcus neoformans. J Biol Chem 2004; 279(20):21144-21153.
19. Heung LJ, Kaiser AE, Luberto C, Del Poeta M. The role and mechanism of diacylglycerol-protein kinase C1 signaling in melanogenesis by Cryptococcus neoformans. J Biol Chem 2005; 280(31):28547-28555.
20. Gerik KJ, Donlin MJ, Soto CE et al. Cell wall integrity is dependent on the PKC1 signal transduction pathway in Cryptococcus neoformans. Mol Microbiol 2005; 58(2):393-408.

21. Luberto C, Martinez-Marino B, Taraskiewicz D et al. Identification of App1 as a regulator of phagocytosis and virulence of Cryptococcus neoformans. J Clin Invest 2003; 112(7):1080-1094.
22. Mare L, Iatta R, Montagna MT et al. APP1 transcription is regulated by IPC1-DAG pathway and is controlled by ATF2 transcription factor in Cryptococcus neoformans. J Biol Chem 2005; 280(43):36055-36064.
23. Tommasino N, Villani M, Qureshi A et al. Atf2 transcription factor binds to the APP1 promoter in Cryptococcus neoformans: stimulatory effect of diacylglycerol. Eukaryot Cell 2008;7(2):294-301.
24. Stano P, Williams V, Villani M et al. App1: an antiphagocytic protein that binds to complement receptors 3 and 2. J Immunol 2009; 182(1):84-91.
25. Alvarez FJ, Douglas LM, Rosebrock A et al. The Sur7 protein regulates plasma membrane organization and prevents intracellular cell wall growth in Candida albicans. Mol Biol Cell 2008; 19(12):5214-5225.
26. Pasrija R, Panwar SL, Prasad R. Multidrug transporters CaCdr1p and CaMdr1p of Candida albicans display different lipid specificities: both ergosterol and sphingolipids are essential for targeting of CaCdr1p to membrane rafts. Antimicrob Agents Chemother 2008; 52(2):694-704.
27. Oura T, Kajiwara S. Disruption of the sphingolipid Delta8-desaturase gene causes a delay in morphological changes in Candida albicans. Microbiology 2008; 154(Pt 12):3795-3803.
28. Ramamoorthy V, Cahoon EB, Thokala M et al. Sphingolipid C-9 methyltransferases are important for growth and virulence but not for sensitivity to antifungal plant defensins in Fusarium graminearum. Eukaryot Cell 2009; 8(2):217-229.
29. Kechichian TB, Shea J, Del Poeta M. Depletion of alveolar macrophages decreases the dissemination of a glucosylceramide-deficient mutant of Cryptococcus neoformans in immunodeficient mice. Infect Immun 2007; 75(10):4792-4798.

CHAPTER 17

Sphingolipids in Parasitic Protozoa

Kai Zhang, James D. Bangs and Stephen M. Beverley*

Abstract

The surface of most protozoan parasites relies heavily upon lipid-anchored molecules, to form protective barriers and play critical functions required for infectivity. Sphingolipids (SLs) play important roles through their abundance and involvement in membrane microdomain formation, as well as serving as the lipid anchor for many of these molecules and in some but possibly not all species, as important signaling molecules. Interactions of parasite sphingolipid metabolism with that of the host may potentially contribute to parasite survival and/or host defense. In this chapter we summarize current knowledge of SL structure, synthesis and function in several of the major parasitic protozoan groups.

Introduction

Protozoan pathogens have tremendous negative impact on human health and prosperity. Malaria, a mosquito-borne disease caused by *Plasmodium* parasites, is responsible for 350-500 million cases and more than one million deaths each year. Trypanosomatid parasites including *Trypanosoma cruzi, Trypanosoma brucei* and *Leishmania spp* infect 20-30 million people worldwide, causing a spectrum of devastating diseases from disfiguring skin lesions to lethal visceral, cardiac and cerebral infections.[1-3] Other parasitic protozoans such as *Toxoplasma gondii, Giardia spp.* and *Entameoba histolytica* are widely distributed pathogens capable of causing severe diseases in humans. As yet, there are no safe vaccines for any of these parasites, leaving drug treatments as the major strategy for control. Available drugs are compromised by low efficacy, high toxicity and wide spread resistance.[2,4] It is important, therefore, to identify parasite-specific virulence pathways and develop novel inhibitors that target them.

Sphingolipids (SLs) are ubiquitous membrane components in pathogenic protozoans (Table 1). In comparison to mammals and fungi, knowledge about metabolism, structure and function of SLs in parasitic protozoa is limited and an area of active research. Trypanosomatids such as *Trypanosoma* and *Leishmania spp.* synthesize large amounts of unglycosylated inositol phosphorylceramide (IPC), a lipid found widely among fungi and plants. Trypanosomes also synthesize sphingomyelin (SM), a lipid commonly found in mammals. Apicomplexan parasites including *Plasmodium* spp mostly synthesize SM and glycosyl-SLs, although a recent report suggests inositol-SLs occur in *Toxoplasma gondii*.[5]

In this chapter, we review recent findings about the roles of SL metabolism in pathogenic protozoa.

Leishmania

Leishmania represents one of the three major lineages of trypanosomatid protozoans, which may have separated as long as 250-500 million years ago.[24] Parasites within the genus *Leishmania* are transmitted by Phlebotomine sand flies and cause a spectrum of diseases, varying from mild

*Corresponding Author: Stephen M. Beverley—Department of Molecular Microbiology, Campus Box 8230, Washington University School of Medicine, St. Louis, Missouri 63110 USA. Email: beverley@borcim.wustl.edu

Sphingolipids as Signaling and Regulatory Molecules, edited by Charles Chalfant and Maurizio Del Poeta.

Table 1. Composition of higher order SLs in parasitic protozoa

Species	Sphingolipids Identified	References
Leishmania major	IPC	6, 7
Trypanosoma brucei	IPC (procyclics only)	8, 9
	EPC (blood stream forms only)	10
	SM	10, 11
	Glycosyl-cer	12
Trypanosoma cruzi	IPC*	13, 14
	SM	15
Plasmodium falciparum	Glycosyl-cer, SM	16, 17
Toxoplasma gondii	Glycosyl-cer, SM, IPC?	5, 18
Trichomonas vaginalis	IPC*, SM	19-21
Giardia lamblia	SM	22, 23

*: Glycosylated and used as anchors for lipopeptidophosphoglycan (LPPG) and some surface proteins.

self resolving cutaneous lesions, to fatal visceral infections and disfiguring mucosal manifestations. There is a strong albeit imperfect tendency for these phenotypes to be associated with individual species, for example *L. major* and *L. mexicana* with cutaneous disease, *L. donovani* with visceral disease and *L. braziliensis* with mucocutaneous disease. The biochemical basis underlying these different disease manifestations is largely unknown. *Leishmania* live extracellularly within the sand fly midgut, but within an acidified, fusogenic phagolysosome of the macrophage within the mammalian host.

Sphingolipids were reported first in *L. donovani*[25,26] and thereafter in all species studied. Notably the predominant SL was IPC, more typical to those found in fungi than in the mammalian hosts. Methods for mass spectrometric analysis and quantification of IPCs were developed and have shown that the structure within both the promastigote and amastigote stage of *Leishmania major* are IPC-d16:1 (phosphoryl inositol N-stearoylhexadecesphing-4-enine, d16:1/18:0-PI-Cer) and IPC-d18:1 (phosphoryl inositol N-stearoylsphingosine, d18:1/18:0-PI-Cer), respectively, with IPC-d16:1 being more abundant. IPCs are present in an abundance of about 2×10^8 molecules per cell, nearly 10% of the total membrane phospholipids.[6,7,27] Additionally, several studies have reported the presence of mammalian type sphingolipids on *Leishmania* amastigote residing in mammalian macrophages. Current data from diverse species suggest that these arise in some manner by transfer from the mammalian host rather than direct synthesis by the parasite.[28-30]

The occurrence of the parasite specific IPC suggested that *Leishmania* organisms encode a complete, functional SL sphingolipid pathway[31] (Fig. 1). In the de novo pathway, activities and/or genes have been found for serine palmitoyl transferase (*SPT1/SPT2*), ceramide synthase (*LAG1/CERS*) and IPC synthase *IPCS1*.[6,27,32,33] On the degradative side, a sphingosine 1-P lyase *SPL*[33] has been functionally identified; parasites also encode a homolog of the IPC phospholipase C *ISC1* (Lmj08.0200) and sphingosine kinase (LmjF26.0710). There are potential homologs encoding dihydroceramide desaturase, 3-ketosphingosine reductase, ceramidase and sphingosine 1-P phosphatases as well, although the evolutionary distances make firm conclusions premature. Future studies will likely fill in missing steps and/or confirm the assignments above. Orthologs of all of these genes are found in the other trypanosomatid genomes.

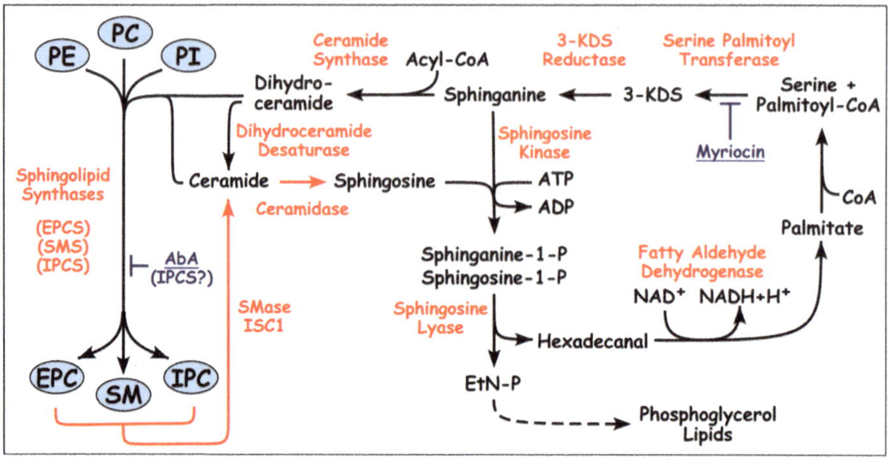

Figure 1. Kinetoplastid sphingolipid pathways. Synthetic pathways are indicated by black arrows, degradative pathways by red arrows. Names of key enzymes are indicated in red. Sites of inhibition by myriocin and aureobasidin A (AbA) are also indicated. Abbreviations: 3-KDS, 3-ketodihydrosphinganine; PE, phosphatidylethanolamine; PC, phosphatidylcholine; PI, phosphatidylinositol; EPC, ethanolaminephosphoryl ceramide; SM, sphingomyelin; IPC, inositolphosphoryl ceramide; EtN, ethanolamine; EPCS/SMS/IPCS, sphingolipid synthases as indicated; SMase, sphingomyelinase; ISC1, inositol phosphosphingolipid phospholipase-C. The attachment and role of SLs in the attachment of membrane proteins in glycoconjugates is not shown in this figure.

SL Pathway Genetics

Genetic studies of SL metabolism have yielded some surprises in *Leishmania major*. Two studies reported the generation of null mutants of one of the subunits (*SPT2*) required for serine palmitoyl transferase activity in *L. major* promastigotes, the stage normally carried by the sand fly vector.[6,27] Mass spectrometry confirmed these parasites completely lacked SLs including IPC, yet these organisms grew normally in logarithmic phase and continued to make 'lipid rafts' as judged by several criteria. This was remarkable as SLs are considered essential membrane components in all eukaryotes. The ability of *Leishmania* to survive in their absence may reflect their rather unique membrane composition, with ergosterol substituting for cholesterol and high levels of plasmalogen ether phospholipids, both of which strongly promote raft formation.

While normal in logarithmic phase, upon entry into stationary phase *spt2*⁻ parasites showed severe defects including loss of viability and abnormalities including accumulation of small vesicles reminiscent of multivesicular bodies, lipid inclusions and defects in the formation of acidocalcisomes.[34] While originally attributed to a defect in SL synthesis, subsequent studies of a null mutant (*spl*⁻) lacking the SL degradative enzyme sphingosine-1-P lyase (SPL) revealed a very similar set of phenotypes to those of *spt2*⁻. Genetically this suggests that both defects lay upstream of a critical metabolite, which was predicted from metabolic pathway analysis and then shown to be ethanolamine (EtN).[33] Provision of EtN completely reversed most phenotypes of the *spt2*⁻ and *spl*⁻ mutants, except a modest difference in log phase cell shape. Thus, in *L. major* promastigotes the primary role of SL synthesis appears to be focused on provision of EtN, rather than the more typical roles in other eukaryotes critical to cell signaling and raft formation.[33]

SL Salvage by Amastigotes

The requirement for SLs in the promastigote stage when propagated in its native sand fly vector has not been reported thus far. However, despite the complete absence of SLs, *spt2*⁻ mutant parasites remain able to induce infections in the mammalian host more or less normally (when grown in EtN), consistent with the down regulation of SPT2 seen in parasite development.[6,27,34] Remarkably, when isolated from mammalian hosts, *spt2⁻* amastigotes showed normal IPC levels.[34] This mandates some form of SL salvage to occur within the amastigote-bearing phagolysosomes, either as sphingoid bases, ceramide, or complex SLs followed by some form of remodeling into parasite IPCs. The transfer of mammalian SLs to the amastigote surface mentioned earlier is one likely route for this to occur, although this has not been proven.

Given the role of SLs in mammalian cell biology and signaling, several studies have found alterations in mammalian SL metabolism in *Leishmania*-infected macrophages in vitro, including alterations in ceramide levels that could contribute to parasite survival.[35,36] These studies suggest the possibility of a fascinating interplay between host and parasite SL metabolism, extending beyond nutritional factors into signaling pathways relevant to parasite survival and/or host defense.

Other *Leishmania* Species

Genetic studies of SL metabolism have concentrated thus far on *L. major*. One might reasonably predict from similarities in SL pathway gene content and organization amongst *Leishmania* species that these species might behave similarly. However the same might have been said for comparisons with trypanosomes, yet unlike *L. major* de novo SL synthesis is clearly essential in these species (discussed below). There are precedents for major differences the dependency upon glycoconjugate pathways in *Leishmania* despite seemingly identical biochemical functions, one example being the Golgi GDP-mannose transporter which is required for amastigote virulence in *L. major* but not *L. mexicana*.[37,38]

Inhibition of SL Synthetic Pathways

The lack of SL dependency for promastigote growth and the reliance upon salvage rather than de novo synthesis in the amastigote stage may argue that inhibition of de novo SL synthesis, or at least that of sphingoid bases, is unlikely to be a good target for inhibition in these species. In contrast, *L. major* amastigotes maintain IPC levels and in combination with the potential for interactions with host SL pathways, suggesting that inhibition of IPC synthesis may offer a valid chemotherapeutic target, as proposed in fungi.

The best studied IPC synthase (IPCS) inhibitor is aureobasidin A (AbA), which inhibits fungal growth at in the nanomolar range (~50 nM) and fungal IPC synthase in the subnanomolar range.[39,40] *Leishmania* species are much less sensitive, requiring 1-10 μM.[32,34,41] While AbA inhibits *L. major* promastigote growth with an EC_{50} of 0.6 μM, MS analysis revealed no effect on IPC synthesis at this concentration and IPC synthesis inhibition was not seen until 5.0 μM AbA.[6] Consistent with these findings, studies of heterologous expressed IPCS suggests an enzymatic IC_{50} on the order of 100 μM.[32] Lastly, AbA was similarly toxic to the *spt2⁻* parasites, which lack IPC.[27,34] These findings suggest that the major inhibitory activity of AbA against *Leishmania major* in vivo may involve a target other than IPCS, while not yet excluding IPCS as a potential amastigote target for more effective inhibitors in the future.

Trypanosoma brucei (*ssp*) and *Trypanosoma cruzi*

These related trypanosome species cause two distinct human diseases: African Trypanosomiasis, or Sleeping Sickness, (*T. brucei ssp*) and South American Trypanosomiasis, or Chagas disease (*T. cruzi*). Because of the predominance of IPC in the related kinetoplastid genus *Leishmania*, both groups of trypanosomes have also been generally thought to contain mostly IPC, despite early compositional studies indicating the presence of SM in each (Table 1). However, as is the case with *Leishmania*, there are marked differences within the genus *Trypanosoma* in regard to geographical distribution, life cycle, biochemistry and disease pathology. Consequently, it is not surprising that there are significant differences in sphingolipid composition and metabolism between the trypanosomal species.

Trypanosoma brucei

African trypanosomes have a life cycle that alters between the bloodstream of the mammalian host and the insect vector, the tsetse fly (genus *Glossina*). There are two replicative forms that are amenable to in vitro study, the pathogenic bloodstream form and procyclic insect form, which is found in the tsetse midgut. Early compositional studies reported the presence of SM in both stages[11,42] and the presence of IPC was suggested,[11] although this was not confirmed with known standards or enzymatic treatments. More recently high-resolution mass spectroscopy has confirmed the presence of both SM and IPC, as well as the unusual phosphosphingolipid, ethanolamine phosphorylceramide (EPC).[8,10,43] In addition, neutral glycosphingolipids have also been indentified in bloodstream stage parasites.[12] Interestingly, synthesis of phosphosphingolipids is developmentally regulated in *T. brucei*.[10] Both compositional analyses and biosynthetic assays measuring the incorporation of fluorescent ceramide analogues indicate that SM is actively synthesized in both bloodstream and procyclic stages, but IPC is only synthesized in procyclic trypanosomes. Furthermore EPC is only found in bloodstream stage parasites. However, this finding may be a matter of sensitivity since EPC levels are very low even in the bloodstream stage. Another aspect of developmental regulation is the ceramide platform used in sphingolipid synthesis. A fairly standard mixture of dihydroceramide (d18:0/16:0, *N*-palmitoylsphinganine) and ceramide (d18:1/16:0, *N*-palmitoylsphingosine) are found in both stages, but in both the pool of ceramide precursors and mature phosphosphingolipids the ratio favors dihydroceramide in procyclic parasites and ceramide in bloodstream forms. This difference likely derives from differential expression or regulation of dihydroceramide desaturase, the enzyme responsible for conversion of dihydroceramide to ceramide. Sphingolipid synthesis is essential in *T. brucei*. Pharmacological[44] and genetic[43] interdiction of the first enzyme of sphingoid base synthesis, serine palmitoyl transferase, is lethal and can be rescued with the immediate downstream product, 3-ketodihydrosphingosine. Exogenous EtN cannot rescue lethality however, indicating that *T. brucei* relies on sphingoid base synthesis for something other than production of essential EtN, unlike *L. major* (Fig. 1). Presumably this dependency is synthesis of complex sphingolipids.

Trypanosoma cruzi

South American trypanosomes have a more complicated life cycle with multiple morphological forms, all of which can be maintained in vitro. Log phase cultured epimastigotes mimic the replicative form found in the insect vector, reduvid bugs. Upon entry into stationary phase, parasites differentiate to a trypomastigote form that closely mimics the infectious metacyclic trypomastigotes found in the reduvid hindgut. These can invade cultured mammalian cells where they differentiate into replicative amastigotes. Nonreplicative trypomastigotes are subsequently released and go on to infect new host cells. These culture-derived trypomastigotes correspond to the circulating trypomastigotes that are responsible for maintaining the natural mammalian infection and for transmission back to reduvids during a blood meal. Several compositional studies with epimastigotes reported the presence of SM, but failed to note the presence of IPC.[45-47] However, radiolabeling with (^{3}H)palmitate and mass chemical analyses revealed the presence of free IPC in the cellular pools of phospholipids in both epimastigotes, tissue culture derived trypomastigotes and in vitro differentiated amastigotes.[14,48,49] The level of IPC synthesis increases during differentiation of trypomastigotes to amastigotes, both in infected myoblasts and during in vitro differentiation.[50] These studies failed to detect SM, but they did reveal the ceramide content of IPC to be a mixture of dihydroceramide (d18:0/16:0, *N*-palmitoylsphinganine) and ceramide (d18:1/16:0, *N*-palmitoylsphingosine), with lesser amounts of each containing stearate as the *N*-acyl group (d18:0/18:0 and d18:1/18:0). No ceramides containing longer chain fatty acids, e.g., lignoceric acid (C24:0) were detected, but it must be noted that mass chemical analyses were only performed with epimastigotes[48] and conclusions based solely on radiolabeling can be misleading. Neutral glycosphingolipids have also been identified in epimastigotes by mass chemical approaches.[51] These include gluco-, galacto- and lactosylceramides composed of sphingosine with *N*-acyl mixtures of saturated, unsaturated and hydroxylated fatty acids of predominantly 16 carbon (palmitate) and 24 carbon (lignocerate) lengths. Similar structures are found in neutral glycosphinolipids of heart muscle and it has been proposed that this mimicry may contribute to pathogenesis associated with chronic Chagas disease.[52]

As in *Leishmania*, the role of IPC in *T. cruzi* has been investigated using the fungal IPCS inhibitor AbA. High concentrations of AbA (~20 μM) inhibited in vitro differentiation of amastigotes[50] and the generation of culture-derived trypanosomatids.[53] However, it is has no effect on growth of epimastigotes, even at excessive concentration (~50 μM) and likewise is completely ineffective against IPC synthase in a direct enzymatic assay with epimastigote membranes.[53] These findings strongly suggest that any inhibition of trypomastigotes/amastigote differentiation is likely due to secondary off-target effects and furthermore that there are considerable differences between fungal and kinetoplastid IPC synthases.

Unlike *Leishmania* and the African trypanosomes, IPC is also used as a membrane anchor for two abundant cell surface glycoconjugates: lipopeptidophosphoglycan (LPPG) and the mucin-like glycoproteins. First characterized in stationary phase epimastigotes, LPPG is a free glycosylphosphatidylinositol (GPI)-like molecule unusual in being constructed on IPC rather than phosphatidylinositol, as is typical for GPI structures.[54-57] Strikingly, mucins from metacyclic trypomastigotes, derived from stationary phase epimastigote cultures, are anchored by an almost identical structure, differing primarily in the absence of two β-linked galactofuranose moieties attached to the tetramannose core of LPPG and the presence of the ethanolamine phosphate bridge to the mucin polypeptide chain.[58] In each case the ceramide glycoinositol lipid platform is a mixture of ceramide and dihydroceramide (d18:1 and d18:0), *N*-acylated with either palmitate (C16:0) or lignoceric acid (C24:0). Furthermore, both structures are developmentally regulated. LPPG and mucin glycoinositol lipid anchors from log phase epimastigotes are constructed on a standard phosphatidylinositol platform, but in all other regards resemble the equivalent IPC-based structures from stationary cultures.[58-60] This developmental shift in glycoinositol lipid anchor structure is not universal since 1G7-antigen, a metacyclic-specific surface marker, has a glycerolipid anchor, even as the mucins are being overwhelmingly converted to a ceramide-based anchor.[61] Ceramide-based glycoinositol lipid anchors are also found on the amastigote-specific Ssp4 antigen[49] and the trans-sialidase of culture derived trypomastigotes. It seems likely that these ceramide-based structures are made by remodeling of a de novo synthesized glycerolipid GPI structure, as is known to happen in yeast,[62] since no evidence could be found for de novo synthesis of GPI-like structures on an IPC platform.[63] A possible remodeling pathway could be transfer of the entire glycoinositol head group from the initial glycerolipid anchor to a ceramide accepter in a reaction analogous to the synthesis of phosphosphingolipids by head group exchange with phosphoglycerol lipids. Subsequent remodeling of the *N*-acyl group could introduce lignoceric acid, which is not found in the general pools of ceramide and IPC, into a subset of the ceramide anchors. Several of the enzymatic activities required for ceramide fatty acid remodeling have been reported in membrane extracts of *T. cruzi*.[50]

Trypanosomatid Sphingolipid Synthases

The trypanosomatid protozoa have an orthologous gene family of documented sphingolipid synthases.[10,32] *L. major* has a single gene (*LmjIPCS*), *T. brucei* has four tandemly arrayed genes (*TbSLS1-4*) and *T. cruzi* has a single gene (*TcSLS*). The evolutionary relatedness of these genes is confirmed by their syntenic location on the chromosomes of the three species. All the gene products, have clear homology to mammalian sphingomyelin synthases[64] by reciprocal Blast searching, but no obvious sequence relationship to yeast IPC synthase.[39] Nevertheless, LmjIPCS has been biochemically characterized as a true IPC synthase.[32] One of the *T. brucei* gene products (TbSLS4) is a bifunctional SM/EPC synthase[10] and while it is possible that one or more of the other genes (TbSLS1-3) may be an IPC synthase, this seems unlikely given the high degree of sequence identity for the predicted proteins. There is currently no evidence concerning the biosynthetic specificity of the *T. cruzi* enzyme. The *T. brucei* genes are constitutively expressed and simultaneous knockdown of the entire locus by pan-specific RNAi is rapidly lethal for bloodstream parasites, likely due to toxic effects of a 4-5 fold increase in cellular levels of ceramide. No data are currently available for the essentiality of the *T. cruzi* ortholog. One might predict from the viability of the *L. major* de novo SL synthetic mutant *spt2-* that *Leishmania* IPCS would not be essential; this is under investigation currently.

All members of the gene family contain the catalytic residues and four signature sequence motifs (D1-4) defined for mammalian SM synthases.[65] Two of these motifs (D1-2) are absent from yeast IPCS, further emphasizing the lack of direct relatedness with the kinetoplastid enzymes. All of the parasite sphingolipid synthases conform to a six transmembrane topology model predicted for mammalian SM synthase 2.[10] Collectively these data suggest that the kinetoplastid sphingolipid synthase family evolved from a common ancestral SM synthase gene and that subsequently the Leishmania gene diverged to encode an IPC synthase activity. If true this implies that IPC synthetic capability has evolved at least twice in eukaryotes, once in the kinetoplastids and once in a fungal lineage.

Plasmodium falciparum

Malaria parasites of the genus *Plasmodium* are obligate intracellular pathogens residing in mammalian red blood cells, hepatocytes, or mosquito midgut epithelial cells. Biosynthesis of sphingomyelin (SM) and glycosphingolipids by *P. falciparum* parasites (intraerythrocytic stages) has been demonstrated using radiolabeled or fluorescent ceramides.[16,66] A recent report revealed the presence of an active glucosylceramide synthase, which glycosylates only dihydroceramide in vitro.[67] Species of glycosphingolipids in *P. falciparum* include monohexosylceramides, globotriaosylceramides, globotetraosyl-ceramides and sulfated glycosphingolipids.[17,67] No inositol SLs have been described in *P. falciparum* parasites.

Biosynthesis of SLs can be inhibited with low concentrations of *threo*-phenyl-2-palmitoylamino-3-morpholino-1-propanol (PPMP) or its derivative *threo*-phenyl-2-decanoylamino-3-morpholino-1-propanol (PDMP), potent inhibitors of glucosylceramide synthase and/or sphingomyelin synthase.[67] Interestingly, PPMP disrupts tubovesicular membrane (TVM), an interconnected network that extends from the parasite's vacuolar membrane to the periphery of the red cell. TVM is essential for the uptake of nutrients and the treatment of PPMP prevents the parasite from importing vital nutrients such as amino acids. Consequently, the intraerythrocytic development of *P. falciparum* is severely inhibited.[68,69] The link between SL biosynthesis and TVM formation is not well understood, although the traffic of nutrients appears to be dependent upon detergent resistant membrane (DRM) domains in TVM, which are SL- and cholesterol-rich.[70]

Another interesting finding is that de novo ceramide synthesis is not required for TVM homeostasis; instead, ceramide derived from the turnover of host SLs are likely to be the substrate for SL biosynthesis in *P. falciparum*.[16] A recently cloned *Pf*SMase, the enzyme responsible for SM hydrolysis to ceramide and phosphocholine, was found to be essential for the parasite and its inhibition impaired the maturation of *P. falciparum* trophozoites into schizonts.[71] Together, these findings suggest SL metabolism could provide targets for new antimalarial drugs. Indeed, ceramide analogs including PPMP and PDMP have shown potent anti- *P. falciparum* activity in vitro, although their mechanism of action is still under investigation.[72,73]

Toxoplasma gondii

Toxoplasma gondii is the most versatile and widespread member of the Apicomplexan family. It is capable of infecting a wide array of host cells, causing a disease (toxoplasmosis) of major medical and veterinary importance worldwide. *De novo* synthesis of ceramide, glycosylated ceramide and SM in *T. gondii* has been demonstrated by tritiated serine and galactose labeling, followed by thin layer chromatography analysis.[18] Two well-documented SL synthesis inhibitors, L-cycloserine (which blocks the activity of SPT) and *threo*-PPMP, strongly inhibit the synthesis of these SLs. AbA treatment inhibits *T. gondii* replication irreversibly and induces morphological changes in the parasite cell shape and integrity, with increased cell vacuolization.[5] Importantly, analyses of parasite lipids did reveal an inositol-containing species that was resistant to alkaline treatment and whose synthesis was inhibited by AbA treatment—all properties of IPC. It is not yet clear whether IPC synthesis is required for parasite replication and more definitive characterizations of *T. gondii* SLs and enzymes are underway.

Trichomonas vaginalis and *Giardia lamblia*

Trichomonas vaginalis is an extracellular aerotolerant flagellated protozoan responsible for trichomoniasis, a common sexually transmitted disease in industrialized countries. Characterized SLs in *T. vaginalis* include SM and IPC.[19,21] Importantly, IPC anchors a group of lipophosphoglycan (LPG)-like glycoconjugates to the cell surface of *T. vaginalis*.[74] The detailed polysaccharide structure of *T. vaginalis* LPG is still unknown, but it contains 50 to 54 monosaccharide residues with galactose and glucosamine being the most common species. To study the role of LPG in *T. vaginalis*, two LPG mutants were recently generated by chemical mutagenesis.[75] Both mutants had truncated form of LPG and failed to bind *Ricinnus comunis* agglutinin I (RCA120) and wheat germ agglutinin, indicating alterations in surface galactose and glucosamine residues. Although mutants grew normally in culture, they were less adherent and less toxic to human vaginal ectocervical cells than the parental strain. Together these results suggest that *T. vaginalis* adherence to host cells is mediated by LPG.

To further explore the contribution of IPC-anchored LPG to *T. vaginalis* pathogenesis, the effect of purified LPG on human reproductive tract l epithelial cells was tested.[76] Briefly, *T. vaginalis* LPG triggered an increased expression of interleukin 8 (IL-8) by human endocervical, ectocervical and vaginal epithelial cells. IL-8 promotes the transmigration of neutrophils across the endothelium and the production of macrophage inflammatory protein 3α, a chemoattractant for immune cells. Such effects were dose dependent and sustained in the absence of cytotoxicity, suggesting that early involvement of chemokine production by epithelial cells induced by nontoxic doses of LPG precedes cytotoxic effects.

Similar to *T. vaginalis*, *Giardia lamblia* is a flagellated protozoan parasite. This aerotolerant anaerobe colonizes and reproduces in human small intestine, causing giardiasis, a common cause of severe diarrhea infecting approximately 200 million people worldwide. It is believed that *G. lamblia* has only limited ability to synthesize phospholipids, long-chain fatty acids and sterols de novo; therefore, this protozoan has developed a special process to acquire lipids from the lipid-rich environment of the human small intestine, i.e., via deacylation/reacylation reactions (the Lands cycle).[77,78] It is known that *G. lamblia* synthesizes SM,[22,79] although more detailed structure characterization has yet to be performed.

A recent report indicates that PPMP blocks *Giardia* replication in vitro and causes a cytokinesis arrest.[80] In addition, PPMP induced a significant (~90%) reduction in *G. lamblia* differentiation into cysts, the parasite stage responsible for the transmission of the disease. In another study, the giardial ceramide glucosyltransferase 1 gene (*gglct-1*) was transcribed only in encysting cells and the treatment of PPMP altered the expression of cyst wall protein transcripts in encysting cells.[23] These combined data suggest that SL synthesis genes are involved in key events in giardial pathobiology. However, effect of PPMP and other SL-synthesis inhibitors on the lipid metabolism of *G. lamblia* has yet to be elucidated.

Conclusion

- Our understanding of SL structure and function in pathogenic protozoans is still in its infancy.
- Current data suggest SL metabolites play important roles in multiple aspects of cellular life including differentiation, replication, trafficking and the synthesis of virulence factors.
- SLs can be synthesized de novo and via salvage pathways.
- Future directions include: (1) to determine the composition of SL species in pathogenic protozoans; (2) to identify and characterize key enzymes involved in SL metabolism; (3) to define the molecular basis by which SLs contribute to infections; and (4) to develop specific inhibitors of SL metabolism and evaluate their potential as novel drugs against parasites.

Acknowledgements

Supported by United States Public Health Service Grants R01AI031078 (S.M.B.), R01-AI35739 (J.D.B.), R15AI076909 (K.Z.) and the Texas Tech University Office of Vice President for Research (K.Z.). We thank Drs. Fong-Fu Hsu and John Turk of the Dept of

Medicine at Washington University School of Medicine for their invaluable assistance in the mass spectrometric analysis of trypanosomatid sphingolipids, which were supported by USPHS grants P41-RR00954, P60-DK20579 and P30-DK56341 to their laboratories.

References

1. Schofield CJ, Jannin J, Salvatella R. The future of Chagas disease control. Trends Parasitol 2006; 22:583-588.
2. Simarro PP, Jannin J, Cattand P. Eliminating human African trypanosomiasis: where do we stand and what comes next? PLoS Med 2008; 5:e55.
3. Cunningham AC. Parasitic adaptive mechanisms in infection by Leishmania. Exp Mol Pathol 2002; 72:132-141.
4. Croft SL, Coombs GH. Leishmaniasis—current chemotherapy and recent advances in the search for novel drugs. Trends Parasitol 2003; 19:502-508.
5. Sonda S, Sala G, Ghidoni R et al. Inhibitory effect of aureobasidin A on Toxoplasma gondii. Antimicrob Agents Chemother 2005; 49:1794-1801.
6. Zhang K, Showalter M, Revollo J et al. Sphingolipids are essential for differentiation but not growth in Leishmania. EMBO J 2003; 22:6016-6026.
7. Hsu FF, Turk J, Zhang K et al. Characterization of inositol phosphorylceramides from Leishmania major by tandem mass spectrometry with electrospray ionization. J Am Soc Mass Spectrom 2007; 18:1591-1604.
8. Guther ML, Lee S, Tetley L et al. GPI-anchored proteins and free GPI glycolipids of procyclic form Trypanosoma brucei are nonessential for growth, are required for colonization of the tsetse fly and are not the only components of the surface coat. Mol Biol Cell 2006; 17:5265-5274.
9. Fridberg A, Olson CL, Nakayasu ES et al. Sphingolipid synthesis is necessary for kinetoplast segregation and cytokinesis in Trypanosoma brucei. J Cell Sci 2008; 121:522-535.
10. Sutterwala SS, Hsu FF, Sevova ES et al. Developmentally regulated sphingolipid synthesis in African trypanosomes. Mol Microbiol 2008; 70:281-296.
11. Patnaik PK, Field MC, Menon AK et al. Molecular species analysis of phospholipids from Trypanosoma brucei bloodstream and procyclic forms. Mol Biochem Parasitol 1993; 58:97-105.
12. Uemura A, Watarai S, Kushi Y et al. Analysis of neutral glycosphingolipids from Trypanosoma brucei. Vet Parasitol 2006; 140:264-272.
13. Bertello LE, Goncalvez MF, Colli W et al. Structural analysis of inositol phospholipids from Trypanosoma cruzi epimastigote forms. Biochem J 1995; 310(Pt 1):255-261.
14. Uhrig ML, Couto AS, Colli W et al. Characterization of inositolphospholipids in Trypanosoma cruzi trypomastigote forms. Biochim Biophys Acta 1996; 1300:233-239.
15. Quinones W, Urbina JA, Dubourdieu M et al. The glycosome membrane of Trypanosoma cruzi epimastigotes: protein and lipid composition. Exp Parasitol 2004; 106:135-149.
16. Gerold P, Schwarz RT. Biosynthesis of glycosphingolipids de-novo by the human malaria parasite Plasmodium falciparum. Mol Biochem Parasitol 2001; 112:29-37.
17. Landoni M, Duschak VG, Peres VJ et al. Plasmodium falciparum biosynthesizes sulfoglycosphingolipids. Mol Biochem Parasitol 2007; 154:22-29.
18. Azzouz N, Rauscher B, Gerold P et al. Evidence for de novo sphingolipid biosynthesis in Toxoplasma gondii. Int J Parasitol 2002; 32:677-684.
19. Costello CE, Glushka J, van Halbeek H et al. Structural characterization of novel inositol phosphosphingolipids of Tritrichomonas foetus and Trichomonas vaginalis. Glycobiology 1993; 3:261-269.
20. Singh BN, Costello CE, Beach DH. Structures of glycophosphosphingolipids of Tritrichomonas foetus: a novel glycophosphosphingolipid. Arch Biochem Biophys 1991; 286:409-418.
21. Beach DH, Holz GG Jr, Singh BN et al. Phospholipid metabolism of cultured Trichomonas vaginalis and Tritrichomonas foetus. Mol Biochem Parasitol 1991; 44:97-108.
22. Kaneda Y, Goutsu T. Lipid analysis of Giardia lamblia and its culture medium. Ann Trop Med Parasitol 1988; 82:83-90.
23. Hernandez Y, Shpak M, Duarte TT et al. Novel role of sphingolipid synthesis genes in regulating giardial encystation. Infect Immun 2008; 76:2939-2949.
24. Fernandes AP, Nelson K, Beverley SM. Evolution of nuclear ribosomal RNAs in kinetoplastid protozoa: perspectives on the age and origins of parasitism. Proc Natl Acad Sci USA 1993; 90:11608-11612.
25. Wassef MK, Fioretti TB, Dwyer DM. Lipid analyses of isolated surface membranes of Leishmania donovani promastigotes. Lipids 1985; 20:108-115.
26. Kaneshiro ES, Jayasimhulu K, Lester RL. Characterization of inositol lipids from Leishmania donovani promastigotes: identification of an inositol sphingophospholipid. J Lipid Res 1986; 27:1294-1303.
27. Denny PW, Goulding D, Ferguson MA et al. Sphingolipid-free Leishmania are defective in membrane trafficking, differentiation and infectivity. Mol Microbiol 2004; 52:313-327.

28. McConville MJ, Blackwell JM. Developmental changes in the glycosylated phosphatidylinositols of Leishmania donovani. Characterization of the promastigote and amastigote glycolipids. J Biol Chem 1991; 266:15170-15179.
29. Winter G, Fuchs M, McConville MJ et al. Surface antigens of Leishmania mexicana amastigotes: characterization of glycoinositol phospholipids and a macrophage-derived glycosphingolipid. J Cell Sci 1994; 107 (Pt 9):2471-2482.
30. Schneider P, Rosat JP, Ransijn A et al. Characterization of glycoinositol phospholipids in the amastigote stage of the protozoan parasite Leishmania major. Biochem J 1993; 295:555-564.
31. Ivens AC, Peacock CS, Worthey EA et al. The genome of the kinetoplastid parasite, Leishmania major. Science 2005; 309:436-442.
32. Denny PW, Shams-Eldin H, Price HP et al. The protozoan inositol phosphorylceramide synthase: a novel drug target that defines a new class of sphingolipid synthase. J Biol Chem 2006; 281:28200-28209.
33. Zhang K, Pompey JM, Hsu FF et al. Redirection of sphingolipid metabolism toward de novo synthesis of ethanolamine in Leishmania. EMBO J 2007; 26:1094-1104.
34. Zhang K, Hsu FF, Scott DA et al. Leishmania salvage and remodelling of host sphingolipids in amastigote survival and acidocalcisome biogenesis. Mol Microbiol 2005; 55:1566-1578.
35. Ghosh S, Bhattacharyya S, Das S et al. Generation of ceramide in murine macrophages infected with Leishmania donovani alters macrophage signaling events and aids intracellular parasitic survival. Mol Cell Biochem 2001; 223:47-60.
36. Ghosh S, Bhattacharyya S, Sirkar M et al. Leishmania donovani suppresses activated protein 1 and NF-kappaB activation in host macrophages via ceramide generation: involvement of extracellular signal-regulated kinase. Infect Immun 2002; 70:6828-6838.
37. Spath GF, Lye LF, Segawa H et al. Persistence without pathology in phosphoglycan-deficient Leishmania major. Science 2003; 301:1241-1243.
38. Ilg T, Demar M, Harbecke D. Phosphoglycan repeat-deficient Leishmania mexicana parasites remain infectious to macrophages and mice. J Biol Chem 2001; 276:4988-4997.
39. Nagiec MM, Nagiec EE, Baltisberger JA et al. Sphingolipid synthesis as a target for antifungal drugs. Complementation of the inositol phosphorylceramide synthase defect in a mutant strain of Saccharomyces cerevisiae by the AUR1 gene. J Biol Chem 1997; 272:9809-9817.
40. Takesako K, Ikai K, Haruna F et al. Aureobasidins, new antifungal antibiotics: taxonomy, fermentation, isolation and properties. J Antibiot 1991; 44:919-924.
41. Tanaka AK, Valero VB, Takahashi HK et al. Inhibition of Leishmania (Leishmania) amazonensis growth and infectivity by aureobasidin A. J Antimicrob Chemother 2007; 59:487-492.
42. Godfrey DG. Phospholipids of Trypanosoma lewisi, T. vivax, T. congolense and T. brucei. Exp Parasitol 1967; 20:106-118.
43. Fridberg A, Olsen CL, Nakayasu ES et al. Sphingolipid synthesis is necessary for kinetoplast segregation and cytokinesis in Trypanosoma brucei. J Cell Sci 2008; 121:522-535.
44. Sutterwala SS, Creswell CH, Sanyal S et al. De novo sphingolipid synthesis is essential for viability, but not transport of glycosylphosphatidylinositol-anchored proteins in African trypanosomes. Eukaryotic Cell 2007; 6:454-464.
45. Da Silveira JF, Colli W. Chemical composition of the plasma membrane from epimastigote forms of Trypanosoma cruzi. Biochimica et Biophysica Acta 1981; 644:341-350.
46. Oliveira MM, Timm SL, Costa SCG. Lipid composition of Trypanosoma cruzi. Comp Biochem Physiol 1977; 58B:195-199.
47. Quiñones W, Urbina JA, Dubourdieu M et al. The glycosome membrane of Trypansoma cruzi epimastigotes: protein and lipid composition. Exp Parasitol 2004; 106:135-149.
48. Bertello L, Goncalvez MF, Colli W et al. Structural analysis of inositol phospholipids from Trypanosoma cruzi epimastigote forms. Biochem J 1995; 310:255-261.
49. Bertello LE, Andrews NW, Lederkremer RM. Developmentally regulated expression of ceramide in Trypanosoma cruzi. Mol Biochem Parasitol 1996:143-151.
50. Salto ML, Bertello LE, Vieira M et al. Formation and remodeling of inositolphosphoceramide during differentiation of Trypanosoma cruzi from trypomastigote to amastigote. Eukaryot Cell 2003; 2:756-768.
51. Barreto-Bergter E, Vermelho AB, Hartmann R et al. Structural characterization of neutral glycosphingolipids from Trypanosoma cruzi. Mol Biochem Parasitol 1992; 51:263-270.
52. Vermelho AB, Meirelles MNL, Pereira MC et al. Heart muscle cells share common neutral glycosphingolipids with Trypanosoma cruzi. Acta Tropica 1997; 64:131-143.
53. Figueiredo JM, Dias WB, Mendonca-Previato L et al. Characterization of the inositol phosphorylceramide synthase activity from Trypanosoma cruzi. Biochem J 2005; 387:519-529.
54. Lederkremer RM, Casal OL, Tanaka CT et al. Ceramide and inositol content of the lipopeptidophosphoglycan from Trypanosoma cruzi. Biochem Biophys Res Commun 1978; 85:1268-1274.

55. Lederkremer RM, Lima C, Ramirez MI et al. Structural features of the lipopeptidophosphoglycan from Trypanosoma cruzi common with the glycophosphatidylinositol anchors. Eur J Biochem 1990; 192:337-345.
56. Lederkremer RM, Lima C, Ramirez MI et al. Complete structure of the glycan of lipopeptidophosphoglycan from Trypanosoma cruzi epimastigotes. J Biol Chem 1991; 266:23670-23675.
57. Previato JO, Gorin PAJ, Mazurek M et al. Primary structure of the oligosaccharide chain of lipopeptidophosphoglycan of epimastigote forms of Trypanosoma cruzi. J Biol Chem 1990; 265:2518-2526.
58. Acosta-Serrano A, Schenkman SNY, Mehlert A et al. The lipid structure of the glycosylphosphatidylinositol-anchored mucin-like sialic acid acceptors of Trypanosoma cruzi changes during parasite differentiation from epimastigotes to infective metacyclic trypomastigote forms. J Biol Chem 1995; 270:27244-27253.
59. Lederkremer RM, Lima CE, Ramirez MI et al. Hexadecylpalmitoylglycerol or ceramide is linked to similar glycophosphoinositol anchor-like structures in Trypanosoma cruzi. Eur J Biochem 1993; 218:929-936.
60. Previato JO, Jones C, Xavier MT et al. Structural characterization of the major glycosylphosphatidylinositol membrane-anchored glycoprotein form epimastigote forms of Trypanosoma cruzi Y-strain. J Biol Chem 1995; 270:7241-7250.
61. Heise N, Cardoso de Almeida ML, Ferguson MAJ. Characterization of the lipid moiety of the glycosylphosphatidyl inositol anchor of Trypanosoma cruzi 1G7-antigen. Mol Biochem Parasitol 1995; 70:71-84.
62. Pittet M, Conzelman A. Biosynthesis and function of GPI proteins in the yeast Saccharomyces cerevisiae. Biochimica et Biophysica Acta 2007; 1771:405-420.
63. Bertello LE, Alves MJ, Colli W et al. Inositolphosphoceramide is not a substrate for the first steps in the biosynthesis of glycoinositolphospholipids in Trypanosoma cruzi. Mol Biochem Parasitol 2004; 133:71-80.
64. Tafesse FG, Ternes P, Holthuis JC. The multigenic sphingomyelin synthase family. J Biol Chem 2006; 281:29421-29425.
65. Huitema K, van den Dikkenberg J, Brouwers JF et al. Identification of a family of animal sphingomyelin synthases. Eur Mol Biol Organ J 2004; 23:33-44.
66. Haldar K, Uyetake L, Ghori N et al. The accumulation and metabolism of a fluorescent ceramide derivative in Plasmodium falciparum-infected erythrocytes. Mol Biochem Parasitol 1991; 49:143-156.
67. Couto AS, Caffaro C, Uhrig ML et al. Glycosphingolipids in Plasmodium falciparum. Presence of an active glucosylceramide synthase. Eur J Biochem 2004; 271:2204-2214.
68. Lauer SA, Ghori N, Haldar K. Sphingolipid synthesis as a target for chemotherapy against malaria parasites. Proc Natl Acad Sci USA 1995; 92:9181-9185.
69. Lauer SA, Rathod PK, Ghori N et al. A membrane network for nutrient import in red cells infected with the malaria parasite. Science 1997; 276:1122-1125.
70. Lauer S, VanWye J, Harrison T et al. Vacuolar uptake of host components and a role for cholesterol and sphingomyelin in malarial infection. EMBO J 2000; 19:3556-3564.
71. Hanada K, Palacpac NM, Magistrado PA et al. Plasmodium falciparum phospholipase C hydrolyzing sphingomyelin and lysocholinephospholipids is a possible target for malaria chemotherapy. J Exp Med 2002; 195:23-34.
72. Pankova-Kholmyansky I, Flescher E. Potential new antimalarial chemotherapeutics based on sphingolipid metabolism. Chemotherapy 2006; 52:205-209.
73. Labaied M, Dagan A, Dellinger M et al. Anti-Plasmodium activity of ceramide analogs. Malar J 2004; 3:49.
74. Singh BN, Beach DH, Lindmark DG et al. Identification of the lipid moiety and further characterization of the novel lipophosphoglycan-like glycoconjugates of Trichomonas vaginalis and Trichomonas foetus. Arch Biochem Biophys 1994; 309:273-280.
75. Bastida-Corcuera FD, Okumura CY, Colocoussi A et al. Trichomonas vaginalis lipophosphoglycan mutants have reduced adherence and cytotoxicity to human ectocervical cells. Eukaryot Cell 2005; 4:1951-1958.
76. Fichorova RN, Trifonova RT, Gilbert RO et al. Trichomonas vaginalis lipophosphoglycan triggers a selective upregulation of cytokines by human female reproductive tract epithelial cells. Infect Immun 2006; 74:5773-5779.
77. Gibson GR, Ramirez D, Maier J et al. Giardia lamblia: incorporation of free and conjugated fatty acids into glycerol-based phospholipids. Exp Parasitol 1999; 92:1-11.
78. Das S, Castillo C, Stevens T. Phospholipid remodeling/generation in Giardia: the role of the Lands cycle. Trends Parasitol 2001; 17:316-319.
79. Mohareb EW, Rogers EJ, Weiner EJ et al. Giardia lamblia: phospholipid analysis of human isolates. Ann Trop Med Parasitol 1991; 85:591-597.
80. Sonda S, Stefanic S, Hehl AB. A sphingolipid inhibitor induces a cytokinesis arrest and blocks stage differentiation in Giardia lamblia. Antimicrob Agents Chemother 2008; 52:563-569.

CHAPTER 18

Biosynthesis of Sphingolipids in Plants (and Some of Their Functions)

Simone Zäuner, Philipp Ternes and Dirk Warnecke*

Abstract

Our knowledge of plant sphingolipid metabolism and function has significantly increased over the past years. This applies mainly to the identification and the functional characterization of genes and enzymes involved in sphingolipid biosynthesis. In addition a number of plant mutants have provided new insights into sphingolipid functions. Very little is still known about intracellular transport, spatial distribution, degradation and signaling functions of sphingolipids. However, combination of *Arabidopsis* genetics with lipidomics and cell biology will soon bring our understanding of these issues to a new level.

Introduction

In 2003 and 2004 plant sphingolipids have been extensively reviewed.[1-4] Strikingly, each of these four reviews covered almost all aspects of plant sphingolipids: They described the structural features of nearly all known sphingolipids present in plants, the enzymes and genes involved in their biosynthesis in *Arabidopsis thaliana* (using the genes known from baker's yeast as a guideline) and some of their biological functions. In a nutshell: These reviews, despite their encyclopedic nature, disclosed our embarrassingly limited knowledge of plant sphingolipids at that time.

Plant science could neither keep up with the large scientific community making progress using mammalian systems nor with the small community using the power of yeast genetics to dissect nearly all aspects of sphingolipids in these organisms (see all other chapters of this book).

Five years after the above mentioned reviews, it is, fortunately, not possible any more to compile all data on plant sphingolipids into a single review. There is still restricted manpower for this topic—here we will cite the papers of about 25 groups only a few of which dedicate their work exclusively to plant sphingolipids. However, considerable progress has been made which is worth being presented to the students, the postdocs and other readers of this book. The main achievements have been obtained by using a systematic approach to unravel the pathways of sphingolipid biosynthesis and to find first functions of the sphingolipid compounds: (i) Sphingolipid genes have been frequently identified by sequence comparisons, screening procedures or bioinformatic techniques. (ii) The gene products have been characterized by functional expression in wild-type cells or genetically engineered lines of the yeasts *Saccharomyces cerevisiae* and *Pichia pastoris*. The enzymatic functions of the plant proteins were uncovered by rescuing a physiological phenotype of the yeast mutant, by the biosynthesis of novel lipids or by novel enzyme activities determined in vitro with extracts of the cells. (iii) T-DNA insertion mutants and RNAi suppression lines mainly from *Arabidopsis* have been generated to elucidate the functions of genes involved in sphingolipid metabolism *in planta*. These mutants have been subjected to analyses by biochemical, cell biological and sphingolipidomics techniques.

*Corresponding Author: Dirk Warnecke—University of Hamburg, Biocenter Klein Flottbek, Ohnhorststrasse 18, 22609 Hamburg, Germany. Email: warnecke@botanik.uni-hamburg.de

Sphingolipids as Signaling and Regulatory Molecules, edited by Charles Chalfant and Maurizio Del Poeta.

By this approach and other techniques, some functions have been ascribed to plant sphingolipids in different biological contexts: programmed cell death (PCD), freezing stress and acclimation, lipid and protein transport, plant development, formation of detergent resistant membrane domains and stomata closure. However, many of the papers published during the past five years represent first or second descriptions of novel *Arabidopsis* mutants. Thus, for most of these physiological subjects there are less than ten papers available, making it impossible to present a comprehensive picture of sphingolipid functions in any of these topics.

Therefore, this review will report on the progress of plant sphingolipids in an old-fashioned style: We will give an overview of plant sphingolipid pathways and describe gene by gene novel identifications, functional characterizations and plant mutants. If known, biological functions will be given for each gene. We are convinced that during the next years our knowledge will significantly increase and that future reviews will focus on specific biological functions of plant sphingolipid metabolites.

In order to keep this text as compact as possible, we will not review the following:

- The role of glucosylceramides (GlcCers) in freezing stress and acclimation. Their roles should be described excusively in the background of additional lipid and protein alterations at the plasma membrane and metabolite changes in the cytoplasm (which is out of the scope of this review). Many contributions to investigate the role of plasma membrane lipids in freezing stress have been made by Peter Steponkus, Daniel Lynch, Matsuo Uemura and their coworkers.
- The involvement of plant sphingolipids in PCD. This link will be mentioned in the appropriate sections, but sphingolipid aspects of PCD have briefly been reviewed elsewhere.[5-6]
- The role of sphingolipids in the formation and function of lipid microdomains. This topic is too complex to be covered in appropriate details here. A short list of papers includes references 7-12.
- The biosynthesis of very-long-chain fatty acids (VLCFA) which are prominent in ceramides, glycosphingolipids and waxes of plants and other eukaryotic organisms.
- Sphingolipid transport (proteins).[13-17]

Pathway of Plant Sphingolipid Biosynthesis

The pathways of sphingolipid biosynthesis in mammals, plants and fungi are very similar indicating that they are a very ancient eukaryotic property. Several hundred million years of separate evolution have resulted in surprisingly few differences. Although mammals display a diversity of different sphingoid long-chain bases (LCBs), (*E*)-sphing-4-enine (commonly known as sphingosine) is regarded as the prevalent LCB in simple and complex sphingolipids.[18] In contrast, in many plants this LCB is quantitatively only a minor compound of the nine different C_{18}-LCBs which are present in different proportions (Fig. 1). Besides sphinganine and 4-hydroxysphinganine (phytosphingosine) which are also found in mammals, plants possess Δ8-unsaturated LCBs. They are synthesized by the activity of sphingolipid-Δ8(*E/Z*)-desaturases which introduce the double bond between C8 and C9 both in (*E*) (trans) and (*Z*) (cis) configuration.[19] The physiological consequences of this LCB heterogeneity are not known and will be discussed below. All these LCBs have been found in simple and complex sphingolipids in different proportions depending on the plant species, the tissue and the physiological conditions.

The major problems in determining the biological functions of sphingolipids are that each sphingolipid class (e.g., LCP-1-P, LCB, Cer, GlcCer and GIPC) plays its role either on its own or in homoeostasis with other (sphingo)lipid classes, that sphingolipid metabolites are being continuously converted into each other via biosynthetic and degradative pathways and that their relative abundance comprises several orders of magnitude.

The latter seems to be a minor problem at first sight, but still proves to be challenging even for advanced analytical techniques. It would have been helpful to visualize the relative abundance of various sphingolipid classes, but it was not possible for us to depict such large differences between

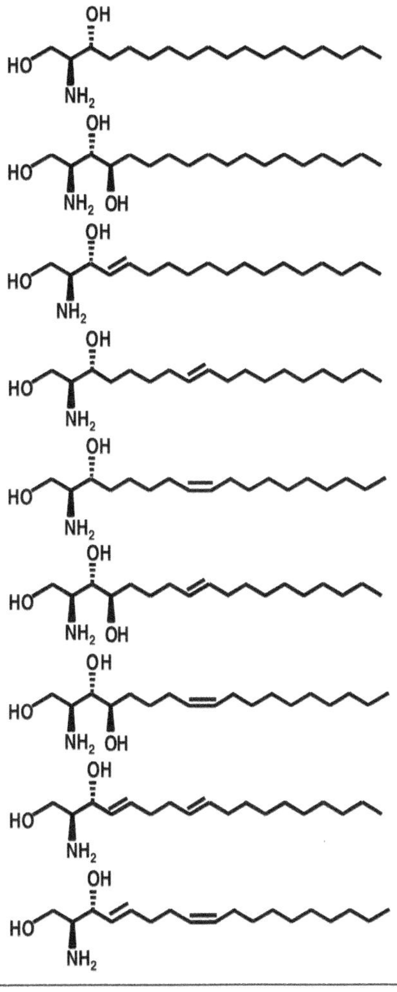

sphinganine, **d18:0**, dihydrosphingosine, (*2S,3R*)-D-*erythro*-2-aminooctadecane-1,3-diol

4-hydroxysphinganine, **t18:0**, phytosphingosine, (*2S,3S,4R*)-D-*ribo*-2-aminooctadecane-1,3,4-triol

(*E*)-sphing-4-enine, **d18:1**^{4E}, sphingosine, (*2S,3R,4E*)-D-*erythro*-2-aminooctadec-4-ene-1,3-diol

(*E*)-sphing-8-enine, **d18:1**^{8E}, (*2S,3R,8E*)-D-*erythro*-2-aminooctadec-8-ene-1,3-diol

(*Z*)-sphing-8-enine, **d18:1**^{8Z}, (*2S,3R,8Z*)-D-*erythro*-2-aminooctadec-8-ene-1,3-diol

(*E*)-4-hydroxysphing-8-enine, **t18:1**^{8E}, (*2S,3S,4R,8E*)-D-*ribo*-2-aminooctadec-8-ene-1,3,4-triol

(*Z*)-4-hydroxysphing-8-enine, **t18:1**^{8Z}, (*2S,3S,4R,8Z*)-D-*ribo*-2-aminooctadec-8-ene-1,3,4-triol

(*4E,8E*)-sphinga-4,8-dienine, **d18:2** ^{4E,8E}, (*2S,3R,4E,8E*)-D-*erythro*-2-aminooctadeca-4,8-diene-1,3-diol

(*4E,8Z*)-sphinga-4,8-dienine, **d18:2** ^{4E,8Z}, (*2S,3R,4E,8Z*)-D-*erythro*-2-aminooctadeca-4,8-diene-1,3-diol

Figure 1. Common LCBs from plants. Trivial names, shorthand designations and systematic names are given on the right.

the lipids in a column diagram. Therefore you will find calculations of the quantities of sphingolipid classes performed by Ernst Heinz[2] and by Yusuf Hannun and Lina Obeid[20] in Table 1. There is a clear trend: The abundance of complex sphingolipids (GIPC, GlcCer and SM) is about one order of magnitude higher than that of ceramide, while the amounts of simple sphingolipids, LCB and LCB-1-P, are more than tenfold lower. However, there remain some discrepancies, e.g., the diverging amounts of LCB-1-P, which might result from collecting the values from different samples that have been analyzed by different methods. In contrast, the studies of Markham and Jaworski[21] and of Michaelson et al[22] used one object, *Arabidopsis* and the complete analysis was performed using one technique: HPLC coupled to electrospray ionization-tandem mass spectrometry. Interestingly, the amounts of GIPC, GlcCer and Cer have been found to be in the same order of magnitude. However there was a huge variation in the amounts of Cer (25-300 nmoles per g dry weight, Table 1). In addition, the relative abundance of LCB-1-P compared to ceramide was much higher than in the calculations of Heinz, Hannun and Obeid.

Table 1. Amounts and relative abundance of different types of sphingolipids in plants

	Sperling et al[2]		Markham and Jaworski[21] Michaelson et al[22]		Hannun and Obeid[20]
	Different Plant sources		*A. thaliana*		Mammals
	nmoles per g Dry Weight	Relative Amounts*	nmoles per g Dry Weight	Relative Amounts*	Relative Amounts*
GIPC	1000	5	240-500	1	-
GlcCer	2000	10	160-500	1	-
SM	-	-	-	-	10
Cer	200	1	25-300	1	1
LCB	10	5×10^{-2}	1.5-15	10^{-2}-10^{-1}	3×10^{-2}
LCB-1-P	0.0003	10^{-6}	0.1-4	10^{-3}-10^{-2}	3×10^{-4}

*The amount of Cer was set as 1 in each column (we set 300 nmol per g dry weight as 1 from the study of Michaelson et al).[22] Abbreviations: GIPC: glycosylinositolphosphoryl ceramide; GlcCer: glycosylceramide; SM: sphingomyeline; Cer: ceramide; LCB: long-chain base; LCB-1-P: long-chain base-1-phosphate.

To make the situation even more confusing, Sperling et al[23] showed that in plasma membranes from *A. thaliana* and four other dicotyledonous plants GlcCer represents only a minor sphingolipid compared to another complex sphingolipid class, presumably GIPC. The LCB composition of GlcCer from purified plasma membranes was analyzed and found to be characteristic for this lipid: It contains high amounts of LCBs with (8*Z*)-double bonds. Total LCB analysis of the same membrane preparation revealed a completely different LCB composition. The proportion of LCBs with (8*E*)-double bonds exceeded those of LCBs typical for GlcCer by several factors. These data suggest that one or more hypothetical complex sphingolipids, e.g., free GIPC and GIPC protein anchors, represent the quantitatively most important sphingolipids in the plasma membranes of many dicotyledonous plants.

These discrepancies are still a challenge for sphingolipid analyses. We think that further improvement of the mass spectrometric techniques will solve at least some of the inconsistencies and will provide the most reliable data in the near future.

Figure 2 shows a hypothetical pathway of sphingolipid metabolism. Although many genes and enzymes of this pathway have been identified and characterized, the value of such a Figure is rather limited. It reflects exclusively formal enzymatic reactions, but not metabolic pathways present in vivo. A famous example is ceramide which is regarded as a hub in sphingolipid metabolism. Although a lot of different enzymes use ceramide as a substrate or form ceramide, they do not necessarily work on the same lipid pool. Ceramide synthases and sphingomyelinases generate ceramide at different subcellular localizations, i.e., the ER and the plasma membrane, respectively.[20] Thus, generic ceramide represents an unknown number of different intracellular pools and this is probably also true for other sphingolipid classes such as LCBs, LCB-1-Ps and GlcCers.

In addition, the generic ceramide represents a multitude of more than one hundred molecular species composed of nine different LCBs and roughly the same number of C_{16}-C_{26} hydroxylated or nonhydroxylated fatty acids. Even the simple sphingolipids LCB and LCB-1-P could be nine different molecular species of C_{18}-LCB. There are hints that the various modifications of the sphingolipid backbone play a role in the spatial separation of different sphingolipid pools and that these molecules fulfill distinct biological functions (see below).

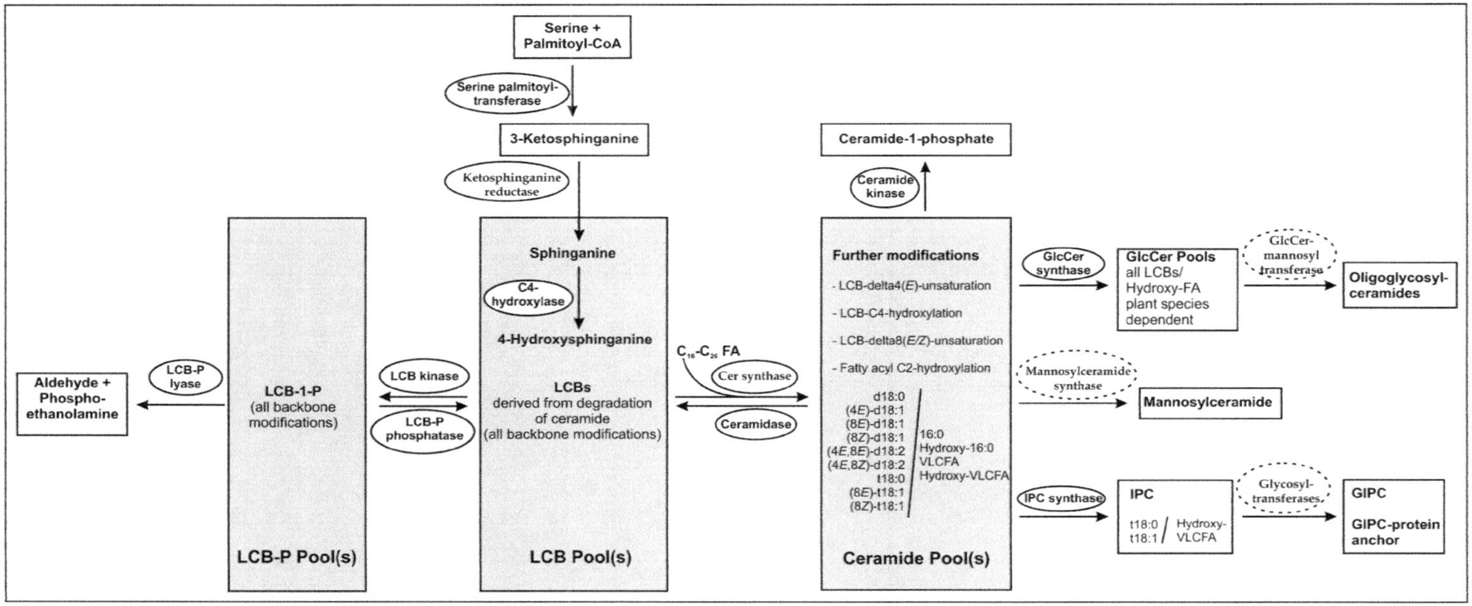

Figure 2. Hypothetical pathway of sphingolipid biosynthesis in plants. Functionally characterized enzymes are framed by a continuous line. Enzymes identified by sequence similarity alone are indicated by a different typeface (experimental data are missing). Hypothetical enzymes whose existence is predicted based on the presence of sphingolipid metabolites are framed by dotted lines. Some sphingolipid classes have been indicated as pool(s), since it is not known whether they represent a biochemical entity or different spatially separated pools. In the case of ceramide the existence of several different possible molecular species are indicated by LCB/FA combinations. Almost nothing is known about degradation of complex sphingolipids. No homolog of the yeast IPCase Isc1p can be found in the *Arabidopsis* genome. GIPC glycosylinositolphosphoryl ceramide, GlcCer glycosylceramide, SM sphingomyeline, Cer ceramide, LCB long-chain base, LCB-1-P long-chain base-1-phosphate.

Consequently, the exact quantitative determination of the different sphingolipid classes by mass spectrometric methods (see above) is not adequate to describe the sphingolipid metabolic pathways in vivo. Rather the analysis of intracellular pools of sphingolipid classes and the localization of each enzyme involved in sphingolipid biosynthesis, transport and degradation is required.

These are the challenges for the next years, but the next chapter describes gene by gene the progress in cloning and characterization of genes and proteins of sphingolipid metabolism (Table 2).

Table 2. **A. thaliana** ***genes involved in sphingolipid biosynthesis (except for VLCFA-biosynthesis)***

Sphingolipid Biosynthesis Gene	*S. cerevisiae*	*A. thaliana* Locus**	References
Serine palmitoyltransferase	*LCB1*	**At4g36480**	24,26,27
	LCB2	**At5g23670** **At3g48780**	25,28,29 (*Nicotiana*)
3-Ketosphinganine reductase	*TSC10*	At3g06060 At5g19200	
Sphingolipid-C4-hydroxylase	*SUR2*	**At1g14290** **At1g69640**	30,31,32 (localization)
Ceramide synthase	*LAC1/LAG1*	At3g25540 At3g19260 At1g13580	33, (tomato), 34 (tomato), 32 (localization)
Alkaline phytoceramidase	*YPC/YDC*	At4g22330	32 (localization)
Neutral ceramidase	None	At1g07380 At2g38010 At5g58980	35 (*Oryza sativa*)
Fatty acyl α-hydroxylase	*SCS7*	**At2g34770** **At4g20870**	36
Sphingolipid-Δ4(*E*)-desaturase	None	**At4g04930**	22
Sphingolipid-Δ8(*E/Z*)-desaturase	None	**At3g61580** At2g46210	19
Glucosylceramide synthase	None	At2g19880	37 (*Gossypium arboreum*)
LCB kinase	*LCB4/LCB5*	**At5g23450** **At4g21540** At2g46090	38,39,40,32 (localization)
LCB phosphate phosphatase	*LCB3*	**At3g58490**	40,32 (localization)
LCB phosphate lyase	*DPL1*	**At1g27980**	41,42,32 (localization)
Ceramide kinase	None	**At5g51290**	43
Inositol phosphorylceramide synthase (IPCS)	(*AUR1*)	**At2g37940*** At2g29525 At3g54020	44

*The IPCS from *Arabidopsis* is more similar to the sphingomyeline synthases from mammals and the IPCS from protozoans than to Aur1p from *S. cerevisiae*.
** Loci printed in boldface have been functionally characterized.

Functional Characterization of Genes and Enzymes Involved in Plant Sphingolipid Biosynthesis (2004-2008)

Serine Palmitoyltransferases

As the first step in sphingolipid biosynthesis, serine palmitoyltransferase (SPT) generates 3-ketosphinganine by the condensation of the amino acid serine with palmitoyl CoA. The plant SPT, like all known eukaryotic SPTs, is a heterodimer and is composed of LCB1 and LCB2 subunits.[24] One of the two *LCB2* genes, At5g23670, has been identified by its expression in a *S. cerevisiae lcb2* knockout strain leading to restoration of 3-ketosphinganine biosynthesis. However, the plant *LCB2* did not complement the LCB auxotrophy of the mutants pointing to very low activity or instability of the plant enzyme in the yeast.[25] The stabilizing subunit of the SPT, LCB1 (At4g36480), was characterized by coexpression with the *Arabidopsis LCB2* which rescued the LCB auxotrophy of the yeast mutant.[24]

Chen et al[24] showed that the embryos from At*lcb1-1* homozygous seeds (obtained from selfed T-DNA insertion plants, *LCB1/lcb1-1*) did not reach the globular stage. Homozygous plants could not be recovered. Furthermore, RNAi mediated suppression of *LCB1* expression led to reduced growth of the plants and their cells as well as an altered leaf morphogenesis. The sphingolipid content in the plants was not altered indicating that their decreased capacity to produce sphingolipids was compensated by reduced growth.[24]

The group of Zuo investigated two *Arabidopsis* mutant lines which also express altered *LCB1* variants. One mutation, *fbr11-1*, results in resistance towards the ceramide synthase inhibitor fumonisin B_1.[27] The amount of free-long chain bases in the mutant was the same as in wild-type plants. Fumonisin B_1 treatment of wild-type plants led to increased levels of free LCBs, followed by the production of reactive oxygen intermediates (ROI) which cause PCD. In contrast, the level of LCB was only moderately increased in the mutant and PCD was not induced. Feeding experiments with *fbr11-1* showed that the ROI-increase and PCD can be induced by supplementation with sphinganine, (*E*)-sphing-4-enine and 4-hydroxysphinganine. No induction has been observed by feeding with phosphorylated LCBs indicating that maintenance of homeostasis between free and phosphorylated LCBs is critical for cell fate.[27]

The second mutant, *fbr11-2*, showed male sterility.[26] Haploid pollen grains displayed different developmental abnormalities after the second mitosis and PCD was induced. This phenotype was very similar to that of a double mutant with T-DNA insertions in both *LCB2* genes.[26] The pollen lethality observed in *fbr11-2* plants is in contrast to the embryo lethality (with fertile pollen) observed with *lcb1-1*.[24] Teng et al argue that the *lcb1-1* allele very likely is not completely inactive.

The *LCB2* genes, At5g23670 and At3g48780, were also analyzed by Dietrich et al[28] with similar results. Homozygous double mutants could not be recovered due to the inability to transmit the double mutant genotype through the haploid pollen. The unicellular microspore was formed but it could not proceed to the bicellular pollen grain. Ultrastructural analysis showed that the pollen contained aberrant endomembranes and lacked an intine layer.[28] By inducible RNAi suppression of one *LCB2* gene in a knockout line for the other, sphingolipid biosynthesis was shown to be essential also in the sporophyte. *LCB2* silencing at various stage of development resulted in chlorosis, extensive necrosis and the loss of viability.[28] This phenotype could be partially rescued by feeding the plants with a LCB with an odd carbon number (d17:0) as a traceable probe.[28]

In *Nicotiana benthamiana*, overexpression of *LCB2* resulted in hypersensitive response-like cell death.[29] This is consistent with the data from *A. thaliana* that fumonisin B_1 results in PCD due to impaired homeostasis between free and phosphorylated LCBs.[27] It was, however, not possible to determine the amounts of these lipids in the *N. benthamiana* mutants. Infection of wild-type plants with the nonhost pathogen *Pseudomonas cichorii* resulted in an increase in *LCB2* expression. *N. benthamiana* is resistant to this bacterium, but its resistance was compromised by treatment with an SPT inhibitor and by silencing both *LCB2* genes.[29] These data suggest that de novo synthesis of LCBs plays a role in pathogene defense.

Long-Chain Base C4-Hydroxylases

Arabidopsis contains two LCB C4-hydroxylases which were identified by heterologous expression of their cDNAs in a baker's yeast *sur2* knockout strain.[30] Preliminary acceptance of the name of the enzymes bases upon the in vitro determination of direct LCB C4-hydroxylating activity in microsomes isolated from *Zea mays.*[45] In addition, an independent ceramide C4-hydroxylating activity was detected in the corn microsomes.[45] It is unclear, whether this activity can be attributed to one or both C4-hydroxylase(s) or whether it is a side reaction of the plant sphingolipid-$\Delta4(E)$-desaturase (see next paragraph). In addition, it is not know whether 4-hydroxysphinganine is incorporated into ceramides by the one or more of the three plant ceramide synthase candidates (Table 1).

Both *Arabidopsis* C4-hydroxylase genes, At1g69640 and At1g14290, have been investigated by the generation of T-DNA insertion and RNAi suppression lines.[31] To our knowledge, this very nice work is the first time that a lipidomics approach was applied to analyze plant sphingolipid mutants. Sphingolipids were determined by HPLC coupled to electrospray ionization-tandem mass spectrometry (ESI-MS/MS) methods.[21,46] Also glycerolipids were analyzed by ESI-MS/MS and only sterols were not investigated. The T-DNA double mutants did not contain any C4-hydroxylated sphingolipid. Interestingly, the total content of sphingolipids was 2.5-fold higher in the mutants compared to the wild-type. This result suggests that the requirement for C4-hydroxylated sphingolipids in the mutant plants was not satisfied and that the lack of these molecular species upregulated sphingolipid biosynthesis by a feedback mechanism. The double mutant was dwarfed, not able to perform reproduction and showed increased levels of PCD-associated genes. The RNAi suppression lines displayed growth inhibition depending on the content of C4-hydroxylated sphingolipids.[31]

In the wild-type plants, dihydroxy LCBs were preferentially linked to C_{16} fatty acids, while trihydroxy LCBs (e.g., 4-hydroxysphinganine) were combined with either C_{16} or VLC fatty acids. The inability of the mutants to perform C4-hydroxylation led to a strong increase of sphingolipids containing dihydroxy LCBs and C_{16} fatty acids (at the cost of C_{16}-containing galactolipids and phosphatidylglycerol in the chloroplasts). The authors suggest that the ceramide synthases of *A. thaliana* might display different substrate specificities with preference for either dihydroxy LCBs and C_{16} FA or C4-hydroxylated LCBs and VLCFA.[31] This assumption seems to be a coherent interpretation of the data obtained with these plants, but it should be kept in mind that other studies showed a high proportion of GlcCer with C4-hydroxylated LCBs linked to C_{16} FA in *Arabidopsis* plants.[47] In rye (*Secale cereale*) and oats (*Avena sativa*), the glucosylceramides of the plasma membrane contain a significant proportions of dihydroxy diunsaturated LCBs linked to VLCFA.[48,49] Therefore, it will be interesting to determine the substrate specificities of the different ceramide synthase isoenzymes in *A. thaliana* and additional plant species.

Ceramidase

A (neutral) ceramidase from rice has been characterized by functional expression of its coding sequence in a baker's yeast strain impaired in intrinsic ceramidase activity.[35] An in vitro assay showed that the enzyme used C_{12}-NBD-Ceramide containing sphinganine, but not 4-hydroxysphinganine as substrate. However, its heterologous expression in yeast resulted in an increase of ceramides molecular species containing VLCFA and 4-hydroxysphinganine. The enzyme was localized to the ER/Golgi compartment by heterologous expression of DsRed fusion protein in onion epidermal cells. The *Arabidopsis* genome contains three orthologes of the rice gene: At1g07380, At2g38010 and At5g58980.[35] A gene encoding a putative alkaline ceramidase with similarity to Ypc1p and Ydc1p from baker's yeast, At4g22330, is present in *Arabidopsis*.

Fatty Acyl α-Hydroxylase

Nagano et al[36] identified a new mechanism of the regulation of acyl amide α-hydroxylation in plants. They found that the protein AtBI-1 interacts via cytochrome-b_5 (Cytb_5) with the plant fatty acyl α-hydroxylase and suggest the following model of regulation: A PCD-inducing stimulus resulted in an increase of calcium concentration in the cytosol. Subsequently, calcium-activated calmodulin interacts with AtBI-1 which on its part binds Cytb_5. The Cytb_5 binds to the fatty acyl α-hydroxylase providing the electrons for the hydroxylation reaction. This complex may enhance ceramide α-hydroxylation in stress conditions (in nonstress situations, the level of α-hydroxylation in complex sphingolipids is high anyway).[36] The original link between the α-hydroxylation of ceramide and PCD is the finding of reference 50 that nonhydroxy ceramides induce PCD in *Arabidopsis* cells while α-hydroxy-ceramides do not.

Sphingolipid-Δ4(E)-Desaturase

An *Arabidopsis* candidate gene for a plant sphingolipid-Δ4(*E*)-desaturase has been identified in 2002 by Ternes et al[51] using sequence comparisons (At4g04930). Until recently all attempts by him and others to functionally express the orthologs of tomato and *Arabidopsis* in a baker's yeast *sur2* knockout strain failed. Finally, the function of this intractable gene was identified by heterologous expression in a *Pichia pastoris* mutant deficient in intrinsic sphingolipid-Δ4(*E*)-desaturase activity.[22] The gene deletion of the desaturase in *Pichia* had resulted in the expected loss of sphingolipid-Δ4(*E*)-desaturase activity and in the unexpected loss of GlcCers. Expression of the *Arabidopsis* cDNA in *Pichia* restored the biosynthesis of GlcCers which were found to contain the Δ4(*E*)-double bond.[22]

The function of the gene in *Arabidopsis* was confirmed by the generation of homozygous T-DNA insertion mutants. The plants did not show any residual sphingolipid-Δ4(*E*)-desaturase expression. The florescence was subjected to sphingolipidomic analyses, since in wild-type plants transcripts of the gene were detected exclusively in pollen and floral tissue. Although the proportion of LCBs carrying the Δ4(*E*)-double is less than 5% in GlcCers in the florescence of the wild-type, the null mutant showed reduced total amounts of GlcCer (63-77% of the wildtype).[22]

In this context it is interesting that an *Arabidopsis* mutant impaired in fatty acid elongation also contains reduced amounts of GlcCers but not of GIPCs.[52] In conclusion, the fatty acyl chain length, the LCB Δ4(*E*)-double bond and the LCB C4-hydroxy group of ceramides seem to represent structural signals involved in the control of biosynthetic flux of particular ceramide species into GlcCer and GIPC, respectively.

The physiological phenotype of the *A. thaliana* sphingolipid-Δ4(*E*)-desaturase null mutant will be discussed below.

Sphingolipid-Δ8(E/Z)-Desaturases

The first sphingolipid desaturase was cloned and functionally characterized by Sperling et al.[19] Two cDNAs from sunflower and from *Arabidopsis* (At3g61580) were expressed in *S. cerevisiae* both resulting in the biosynthesis of the novel LCB 4-hydroxysphing-8-enine in the yeast. Interestingly, both (*E*)- and (*Z*)-isomeres of the new LCB were found indicating stereounselective unsaturation. This is a very unusual property for a desaturase.[19] *A. thaliana* contains a second sphingolipid-Δ8(*E/Z*)-desaturase (At2g46210) the activity of which has been determined by the same method (Sperling and Heinz, unpublished data). Surprisingly, the heterologous expression of the desaturase in *S. cerevisiae* conferred aluminum tolerance on the transgenic yeast.[53] Ryan et al[54] found that the sphingolipid-Δ8(*E/Z*)-desaturase from the aluminum-tolerant plant *Stylosanthes hamata* in contrast to Δ8(*E/Z*)-desaturases from aluminum-sensitive plants, displayed a preference for the formation of the 8(*Z*)-isomer. Expression of the gene from *S. hamata* in the aluminum-sensitive plant *A. thaliana* shifted the ratio of 8(*Z*):8(*E*)-LCB isomers from 1:4 in the wild-type plants to 1:1 in the transgenic line. In addition, the transgenic lines showed increased aluminum resistance.[54] It will be interesting to know whether the ratio of 8(*Z*)-LCBs to 8(*E*)-LCBs is critical for aluminum resistance and membrane physiology.

Inositolphosphorylceramide Synthase (IPCS)

Recently, a plant inositolphosphorylceramide synthase was identified by a genetic screen and its ability to substitute its functional homolog Aur1p in an *S. cerevisiae aur1* knockout strain.[44] The enzyme transfers inositolphosphate from phosphatidylinositol to ceramide. It showed a broad substrate specificity accepting ceramides containing α-hydroxylated fatty acids or nonhydroxylated fatty acids linked to either sphinganine or 4-hydroxysphinganine. In contrast, the activity with ceramides carrying (*E*)-sphing-4-enine was very low. The corresponding gene in *Arabidopsis* (At2g37940) was inactivated by the generation of a homzygous T-DNA insertion mutant. The mutant plants contain normal amounts of GIPC but elevated levels of ceramide. Considering that the amounts of GIPC exceed those of free ceramide by more than one order of magnitude, a slight reduction in GIPC too small to detect could result in significant relative increase in free ceramide. The mutant plants display about 50% of the inositolphosphorylceramide synthase activity of the wild-type, indicating the presence of additional enzymes. Indeed, two homologs of At2g37940 have been found in the genome of *A. thaliana*: At2g29525 and At3g54020.[44] Interestingly, the deduced polypeptides of all three genes show higher sequence similarity to mammalian sphingomyeline synthases and protozoan IPCS[55] than to fungal IPCS.[56] The plant IPCS was localized to the trans-Golgi network by transient, heterologous expression of YFP and DsRed fusion proteins in onion epidermal cells.

Regulation of the expression level of the IPCS from *Arabidopsis* and rice suggest that they fulfill a double function: Biosynthesis of IPC and control of ceramide levels. It is probably this increase in ceramide that is involved in the induction of PCD as response to pathogen attack.[44]

Long-Chain Base Kinase and Long-Chain Base Phosphate Phosphatase

Sphingoid LCB kinase activity in plants was first characterized in corn shoot microsomes by Crowther and Lynch.[57] Both sphinganine and (*E*)-sphing-4-enine, but not 4-hydroxysphinganine were found to be substrates and the enzymatic activity was localized to the ER and/or Golgi by subcellular fractionation. The first LCB kinase from Arabidopsis, LCBK1 (At5g23450), was cloned by Imai and coworkers[39,58] and characterized by expression of the recombinant protein in *E. coli*. The enzyme was shown to accept a range of naturally occurring LCB with sphinganine being the preferred substrate. LCBK1 transcript levels were increased by low humidity or abscisic acid treatments and were found to be especially high in flowers.

The *Arabidopsis* genome encodes three additional proteins with sequence similarity to LCB kinases. One of them, CERK (At5g51290), is not an LCB kinase but a ceramide kinase that presumably plays a role in the regulation of PCD.[43] Another one, SPHK1 (At4g21540) was able to phosphorylate a wide range of naturally occurring desaturated, saturated, or hydroxylated LCBs in vitro.[40] Therefore, LCBK2 would be a more appropriate name for this enzyme. Expression of SPHK1 was linked to ABA-induced stomatal closure (see below). No function could yet be ascribed to the fourth LCB kinase homolog in the genome, At2g46090.

The most prominent function of plant LCB-1-P is regulation of stomatal closure (Fig. 3). It was first observed by Hetherington and coworkers that microinjecting guard cells of *Commelina communis* with (*E*)-sphing-4-enine-1-phosphate led to stomatal closure that was accompanied by oscillations of cytosolic calcium.[63] These observations were later confirmed and extended in *Arabidopsis*.[60,61] Importantly, it was shown that the effect of (*E*)-sphing-4-enine-1-phosphate depended on the sole G-protein alpha subunit encoded by the *Arabidopsis* genome, GPA1. 4-Hydroxysphinganine-1-phosphate was found to be equally active as sphingosine-1-phosphate when microinjected into guard cells and its effect also depends on GPA1. When the LCB kinase activity in *Arabidopsis* leaves was investigated in vitro, the desaturated LCBs (*E*)-sphing-4-enine and sphinga-4,8-dienine were found to be the preferred substrates. Considerable activity was also found with sphinganine and 4-hydroxysphinganine, but not with 4-hydroxysphing-8-enine.

The functions of LCB kinase in plants were recently revisited by Hetherington and coworkers.[40] Both the induction of stomatal closure and the inhibition of stomatal opening by ABA were enhanced in *Arabidopsis* lines overexpressing SPHK1 and repressed in SPHK1 knockdown lines. Similar effects were seen with the ABA-induced delay of germination in these lines. Because the level of LCB-1-P in a given cell depends on the balance between their synthesis and degradation, the contributions of the LCB-1-P lyase DPL1 (At1g27980, see below) and the putative LCB-1-P phosphatase (At3g58490) were investigated in this context. Although no enzymatic

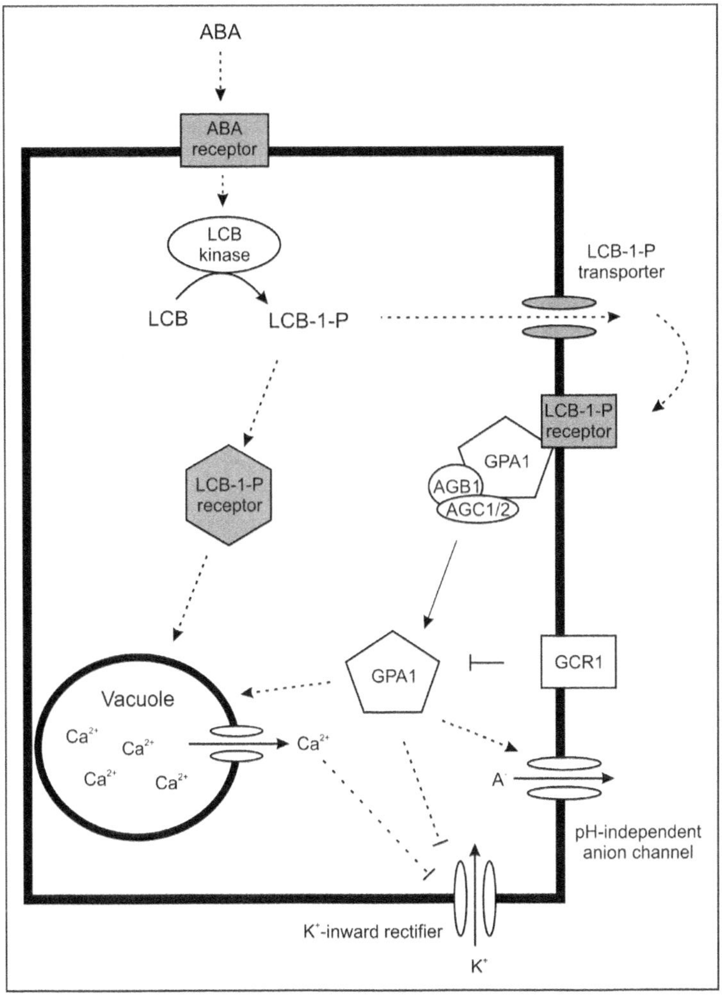

Figure 3. (Partially) hypothetical model of the signaling pathways leading to abscisic acid-mediated stomatal closure/inhibition of stomatal opening. Abscisic acid (ABA) is perceived by a receptor whose identity is still disputed, presumable at the cell surface. Downstream signaling from this receptor leads to increased LCB kinase activity and elevated levels of LCB-1-P. The LCB-1-P is perceived by a yet unknown receptor. Two possibilities can be envisaged in analogy to the situation in mammalian cells: Perception by an unknown intracellular receptor[59] could be directly relayed to the release of calcium from intracellular stores, presumably the vacuole. Alternatively, LCB-1-P could be secreted and perceived by a membrane-bound receptor in an autocrine or paracrine fashion. This model is well established in mammalian cells[20] and support for its validity in plants comes from the finding that the induction of stomatal closure by microinjection of LCB-1-P depends on the sole G-protein alpha-subunit encoded by the *Arabidopsis* genome, GPA1[60,61] Activated GPA1 would then induce downstream responses like calcium release, the opening of pH-dependent anion channels and the inhibition of inward rectifying potassium channels. Inhibition of the latter leads to loss of turgor followed by closing (or inhibition of opening) of the guard cells. GPA1 is negatively regulated by the G protein-coupled receptor protein GCR1.[62] Solid arrows represent known interactions or ion currents, dashed arrows hypothetical or indirect interactions. Grey boxes represent proteins of unknown identity.

characterization of the latter enzyme is available, it appears reasonable to assume that At3g58490 encodes an LCB-1-P phosphatase based on its sequence similarity to LCB-1-P phosphatases from yeast and mammals and the presence of three conserved protein domains.[64]

While neither the phosphatase nor the lyase individual knockdown lines displayed differences in the ABA-mediated inhibition of stomatal opening, a double knockdown line was more sensitive to ABA than wild-type.[40] The ABA-mediated delay in seed germination was enhanced only in the phosphatase but not in the lyase individual knockdown lines. These results point to a context-dependent fine-tuning of LCB-1-P levels by the different enzymes involved in its production and degradation. Worrall et al[40] found that phosphatase transcripts are enhanced in LCB kinase-overexpressing lines and depleted in kinase knockdown lines. This indicates that indeed regulatory mechanisms exist that maintain a proper balance of free and phosphorylated LCB in different situations.

The studies cited above show that LCBK1 and SPHK1, despite showing optimal activity with sphinganine or (*E*)-sphing-4-enine, respectively, both accept a range of different LCB in vitro. The question which are the physiologically relevant LCB-1-Ps in vivo, however, is still open. The initial studies on LCB-1-P signaling in plants[60,63] have focused on the role of (*E*)-sphing-4-enine-1-phosphate, in analogy to the numerous studies in mammalian cells. Coursol et al[61] have later found that 4-hydroxysphinganine-1-phosphate is also active in eliciting stomatal closure. The fact that a certain substance elicits a response when injected into cells, however, does not mean that this substance is the physiologically relevant signal to elicit this response under natural conditions. Michaelson et al[22] have investigated *Arabidopsis* insertional mutants in which the sphingolipid-Δ4(*E*)-desaturase is inactivated (At4g04930). These mutant lines showed no differences compared to wild-type concerning ABA-induced stomatal closure or their performance under drought conditions. Furthermore, the expression of the sphingolipid-Δ4(*E*)-desaturase was restricted to flowers; neither transcripts encoding the desaturase nor Δ4-unsaturated sphingolipids were detectable in *Arabidopsis* leaves. In view of these results, it appears unlikely that (*E*)-sphing-4-enine-1-phosphate, although active when externally applied, is the physiological signal mediating ABA-induced stomatal closure. In our view, a profiling of all naturally occurring LCB-1-Ps and of the transcript levels of all four LCB kinase homologs present in the *Arabidopsis* genome before and after ABA treatment could reveal the true players in ABA-induced LCB-1-P signaling under natural conditions. Such an approach could be complemented by analysis of insertional mutants defective in LCB desaturation or hydroxylation.[22,31]

Long-Chain Base Phosphate Lyase

The genome of *A. thaliana* apparently contains one LCB-1-phosphate lyase gene, At1g27980, which was identified by expression in baker's yeast and by generation of homozygous T-DNA insertion mutants.[41,42] The enzyme was located at the ER and had broad substrate specificity, since it degraded sphinganine-1-P, 4-hydroxysphinganine-1-P, 4-hydroxysphing-8-enine-1-P, sphing-4-enine-1-P and sphinga-4,8-dienine-1-P. In comparison to the wild-type, the sphingolipid composition of the *Arabidopsis* plants devoid of LCB-1-P lyase activity showed no alterations except for an accumulation of 4-hydroxysphing-8-enine-1-P. In contrast to other eukaryotic organisms with impaired LCB-1-P lyase activity, the mutant plants did not display severe developmental abnormalities. However, they showed increased sensitivity towards fumonisin B1 and their leaves showed increased water loss after detachment.[41,42] The role of the LCB-1-P lyase in the context of ABA-mediated stomatal closure has been discussed above.

Conclusion

In the last years, major progress in plant sphingolipid research was made in the identification of genes involved in sphingolipid biosynthesis, the characterization of the respective proteins, the generation and analysis of corresponding *A. thaliana* mutants and the development of sophisticated analytical tools for sphingolipidomics.

The cloning and characterization work needs to be continued to identify those genes involved in the transport and the degradation of sphingolipid as well as the regulation of the metabolic pathway. Researchers in the field will proceed with the systematic generation of T-DNA insertion, RNAi suppression and overexpression mutants until a complete collection of plants will be available required to investigate all aspects of sphingolipid biology.

We have already got a glimpse of the phenotypes of the plant mutants and the functions of sphingolipids, but the interesting plants lines will have to be subjected to much more detailed analysis. Since many sphingolipid metabolites are connected by biosynthetic interconversion and homeostatic control, complete quantitative determination of all sphingolipid metabolites is of particular importance. Interactions with other pathways such as VLCFA biosynthesis, phosphatidylinositol and sterol metabolism could only be investigated by full lipidomic approaches (sphingolipids, glycerolipids, sterols). Applying these analyses to whole leaves of *A. thaliana* will still give insufficient information concerning the physiological functions of these lipids, since generic metabolites such as ceramide or GlcCer do exist in the figures printed in this book, but they do not reflect the situation in vivo. It must be kept in mind that "the function of a lipid depends on its local concentration at a certain time" (Gerrit van Meer). Lipids are present in a number of different pools that are separated in space and/or time. These pools are defined by the function or the metabolic fate of the lipids they contain and they may (but do not have to) show differences in their molecular species composition. To investigate these pools, single tissues or subcellular fractions obtained under a variety of physiological conditions will have to be subjected to lipidomic approaches. Promising techniques to meet this challenge are mass spectrometrical methods which are continuously being improved (see Alfred Merrill's chapter in this book and refs. 21,65 and 66).

These efforts will be accompanied by systematic intracellular localization of all proteins involved in sphingolipid metabolism.[32] Their tissue-specific presence and activity depending on different developmental stages will be determined.

With these techniques and additional biochemical, cell biological and bioinformatical methods, we will be able to describe plant sphingolipid functions in more detail.

Acknowledgements

We thank Sandra Grille for preparation of figures. This work was supported by the Deutsche Forschungsgemeinschaft, SFB470A2 (D.W.), WA2519/1-2 (D.W. and S.Z.) and Te491/3-1 (P.T.).

References

1. Sperling P, Heinz E. Plant sphingolipids: structural diversity, biosynthesis, first genes and functions. Biochim Biophys Acta 2003; 1632:1-15.
2. Sperling P, Warnecke D, Heinz E. Plant sphingolipids. In: Daum G, ed. Lipid metabolism and membrane biogenesis. Berlin, Heidelberg: Springer-Verlag, 2004; 6:337-381.
3. Dunn TM, Lynch DV, Michaelson LV et al. A postgenomic approach to understanding sphingolipid metabolism in Arabidopsis thaliana. Ann Bot (Lond) 2004; 93:483-497.
4. Lynch DV, Dunn TM. An introduction to plant sphingolipids and a review of recent advances in understanding their metabolism and function. New Phytologist 2004; 161:677.
5. Hofius D, Tsitsigiannis DI, Jones JD et al. Inducible cell death in plant immunity. Semin Cancer Biol 2007; 17:166-187.
6. Gadjev I, Stone JM, Gechev TS. Programmed cell death in plants new insights into redox regulation and the role of hydrogen peroxide. Int Rev Cell Mol Biol 2008; 270:87-144.
7. Peskan T, Westermann M, Oelmuller R. Identification of low-density Triton X-100-insoluble plasma membrane microdomains in higher plants. Eur J Biochem 2000; 267:6989-6995.
8. Mongrand S, Morel J, Laroche J et al. Lipid rafts in higher plant cells: purification and characterization of Triton X-100-insoluble microdomains from tobacco plasma membrane. J Biol Chem 2004; 279:36277-36286.
9. Lefebvre B, Furt F, Hartmann MA et al. Characterization of lipid rafts from Medicago truncatula root plasma membranes: a proteomic study reveals the presence of a raft-associated redox system. Plant Physiol 2007; 144:402-418.
10. Roche Y, Gerbeau-Pissot P, Buhot B et al. Depletion of phytosterols from the plant plasma membrane provides evidence for disruption of lipid rafts. FASEB J 2008; 22:3980-3991.
11. Borner GH, Sherrier DJ, Weimar T et al. Analysis of detergent-resistant membranes in Arabidopsis. Evidence for plasma membrane lipid rafts. Plant Physiol 2005; 137:104-116.
12. Laloi M, Perret AM, Chatre L et al. Insights into the role of specific lipids in the formation and delivery of lipid microdomains to the plasma membrane of plant cells. Plant Physiol 2007; 143:461-472.

13. Brodersen P, Petersen M, Pike HM et al. Knockout of Arabidopsis accelerated-cell-death11 encoding a sphingosine transfer protein causes activation of programmed cell death and defense. Genes Dev 2002; 16:490-502.
14. Petersen NH, McKinney LV, Pike H et al. Human GLTP and mutant forms of ACD11 suppress cell death in the Arabidopsis acd11 mutant. FEBS J 2008; 275:4378-4388.
15. Petersen NH, Joensen J, McKinney LV et al. Identification of proteins interacting with Arabidopsis ACD11. J Plant Physiol 2008.
16. West G, Viitanen L, Alm C et al. Identification of a glycosphingolipid transfer protein GLTP1 in Arabidopsis thaliana. FEBS J 2008; 275:3421-3437.
17. Brodersen P, Malinovsky FG, Hematy K et al. The role of salicylic acid in the induction of cell death in Arabidopsis acd11. Plant Physiol 2005; 138:1037-1045.
18. Pruett ST, Bushnev A, Hagedorn K et al. Biodiversity of sphingoid bases ("sphingosines") and related amino alcohols. J Lipid Res 2008; 49:1621-1639.
19. Sperling P, Zähringer U, Heinz E. A sphingolipid desaturase from higher plants. Identification of a new cytochrome b5 fusion protein. J Biol Chem 1998; 273:28590-28596.
20. Hannun YA, Obeid LM. Principles of bioactive lipid signalling: lessons from sphingolipids. Nat Rev Mol Cell Biol 2008; 9:139-150.
21. Markham JE, Jaworski JG. Rapid measurement of sphingolipids from Arabidopsis thaliana by reversed-phase high-performance liquid chromatography coupled to electrospray ionization tandem mass spectrometry. Rapid Commun Mass Spectrom 2007; 21:1304-1314.
22. Michaelson LV, Zäuner S, Markham JE et al. Functional characterization of a higher plant sphingolipid Δ4-desaturase: defining the role of sphingosine and sphingosine-1-phosphate in Arabidopsis. Plant Physiol 2009; 149:487-498.
23. Sperling P, Franke S, Lüthje S et al. Are glucocerebrosides the predominant sphingolipids in plant plasma membranes? Plant Physiol Biochem 2005; 43:1031-1038.
24. Chen M, Han G, Dietrich CR et al. The essential nature of sphingolipids in plants as revealed by the functional identification and characterization of the Arabidopsis LCB1 subunit of serine palmitoyltransferase. Plant Cell 2006; 18:3576-3593.
25. Tamura K, Mitsuhashi N, Hara-Nishimura I et al. Characterization of an Arabidopsis cDNA encoding a subunit of serine palmitoyltransferase, the initial enzyme in sphingolipid biosynthesis. Plant Cell Physiol 2001; 42:1274-1281.
26. Teng C, Dong H, Shi L et al. Serine palmitoyltransferase, a key enzyme for de novo synthesis of sphingolipids, is essential for male gametophyte development in Arabidopsis. Plant Physiol 2008; 146:1322-1332.
27. Shi L, Bielawski J, Mu J et al. Involvement of sphingoid bases in mediating reactive oxygen intermediate production and programmed cell death in Arabidopsis. Cell Res 2007; 17:1030-1040.
28. Dietrich CR, Han G, Chen M et al. Loss-of-function mutations and inducible RNAi suppression of Arabidopsis LCB2 genes reveal the critical role of sphingolipids in gametophytic and sporophytic cell viability. Plant J 2008; 54:284-298.
29. Takahashi Y, Berberich T, Kanzaki H et al. Serine palmitoyltransferase, the first step enzyme in sphingolipid biosynthesis, is involved in nonhost resistance. Mol Plant Microbe Interact 2009; 22:31-38.
30. Sperling P, Ternes P, Moll H et al. Functional characterization of sphingolipid C4-hydroxylase genes from Arabidopsis thaliana. FEBS Lett 2001; 494:90-94.
31. Chen M, Markham JE, Dietrich CR et al. Sphingolipid long-chain base hydroxylation is important for growth and regulation of sphingolipid content and composition in Arabidopsis. Plant Cell 2008; 20:1862-1878.
32. Marion J, Bach L, Bellec Y et al. Systematic analysis of protein subcellular localization and interaction using high-throughput transient transformation of Arabidopsis seedlings. Plant J 2008; 56:169-179.
33. Brandwagt BF, Mesbah LA, Takken FL et al. A longevity assurance gene homolog of tomato mediates resistance to Alternaria alternata f. sp. lycopersici toxins and fumonisin B1. Proc Natl Acad Sci USA 2000; 97:4961-4966.
34. Brandwagt BF, Kneppers TJ, Nijkamp HJ et al. Overexpression of the tomato Asc-1 gene mediates high insensitivity to AAL toxins and fumonisin B1 in tomato hairy roots and confers resistance to Alternaria alternata f. sp. lycopersici in Nicotiana umbratica plants. Mol Plant Microbe Interact 2002; 15:35-42.
35. Pata MO, Wu BX, Bielawski J et al. Molecular cloning and characterization of OsCDase, a ceramidase enzyme from rice. Plant J 2008; 55:1000-1009.
36. Nagano M, Ihara-Ohori Y, Imai H et al. Functional association of cell death suppressor, Arabidopsis Bax inhibitor-1, with fatty acid 2-hydroxylation through cytochrome b. Plant J 2009.
37. Leipelt M, Warnecke D, Zähringer U et al. Glucosylceramide synthases, a gene family responsible for the biosynthesis of glucosphingolipids in animals, plants and fungi. J Biol Chem 2001; 276:33621-33629.
38. Tamura K, Nishiura H, Mori J et al. Cloning and characterization of a cDNA encoding serine palmitoyltransferase in Arabidopsis thaliana. Biochem Soc Trans 2000; 28:745-747.

39. Imai H, Nishiura H. Phosphorylation of sphingoid long-chain bases in Arabidopsis: functional characterization and expression of the first sphingoid long-chain base Kinase gene in plants. Plant Cell Physiol 2005; 46:375-380.
40. Worrall D, Liang YK, Alvarez S et al. Involvement of sphingosine kinase in plant cell signalling. Plant J 2008; 56:64-72.
41. Tsegaye Y, Richardson CG, Bravo JE et al. Arabidopsis mutants lacking long chain base phosphate lyase are fumonisin-sensitive and accumulate trihydroxy-18:1 long chain base phosphate. J Biol Chem 2007; 282:28195-28206.
42. Nishikawa M, Hosokawa K, Ishiguro M et al. Degradation of sphingoid long-chain base 1-phosphates (LCB-1Ps): functional characterization and expression of AtDPL1 encoding LCB-1P lyase involved in the dehydration stress response in Arabidopsis. Plant Cell Physiol 2008; 49:1758-1763.
43. Liang H, Yao N, Song JT et al. Ceramides modulate programmed cell death in plants. Genes Dev 2003; 17:2636-2641.
44. Wang W, Yang X, Tangchaiburana S et al. An inositolphosphorylceramide synthase is involved in regulation of plant programmed cell death associated with defense in Arabidopsis. Plant Cell 2008; 20:3163-3179.
45. Wright BS, Snow JW, O'Brien TC et al. Synthesis of 4-hydroxysphinganine and characterization of sphinganine hydroxylase activity in corn. Arch Biochem Biophys 2003; 415:184-192.
46. Markham JE, Li J, Cahoon EB et al. Separation and identification of major plant sphingolipid classes from leaves. J Biol Chem 2006; 281:22684-22694.
47. Uemura M, Joseph RA, Steponkus PL. Cold acclimation of Arabidopsis thaliana (effect on plasma membrane lipid composition and freeze-induced lesions). Plant Physiol 1995; 109:15-30.
48. Uemura M, Steponkus PL. A contrast of the plasma membrane lipid composition of oat and rye leaves in relation to freezing tolerance. Plant Physiol 1994; 104:479-496.
49. Cahoon EB, Lynch DV. Analysis of glucocerebrosides of rye (Secale cereale L. cv Puma) leaf and plasma membrane. Plant Physiol 1991; 95:58-68.
50. Townley HE, McDonald K, Jenkins GI et al. Ceramides induce programmed cell death in Arabidopsis cells in a calcium-dependent manner. Biol Chem 2005; 386:161-166.
51. Ternes P, Franke S, Zähringer U et al. Identification and characterization of a sphingolipid Δ 4-desaturase family. J Biol Chem 2002; 277:25512-25518.
52. Bach L, Michaelson LV, Haslam R et al. The very-long-chain hydroxy fatty acyl-CoA dehydratase PASTICCINO2 is essential and limiting for plant development. Proc Natl Acad Sci USA 2008; 105:14727-14731.
53. da Silva AL, Sperling P, Horst W et al. A possible role of sphingolipids in the aluminium resistance of yeast and maize. J Plant Physiol 2006; 163:26-38.
54. Ryan PR, Liu Q, Sperling P et al. A higher plant delta8 sphingolipid desaturase with a preference for (Z)-isomer formation confers aluminum tolerance to yeast and plants. Plant Physiol 2007; 144:1968-1977.
55. Denny PW, Shams-Eldin H, Price HP et al. The protozoan inositol phosphorylceramide synthase: a novel drug target that defines a new class of sphingolipid synthase. J Biol Chem 2006; 281:28200-28209.
56. Abbas HK, Tanaka T, Duke SO et al. Fumonisin- and AAL-toxin-induced disruption of sphingolipid metabolism with accumulation of free sphingoid bases. Plant Physiol 1994; 106:1085-1093.
57. Crowther GJ, Lynch DV. Characterization of sphinganine kinase activity in corn shoot microsomes. Arch Biochem Biophys 1997; 337:284-290.
58. Nishiura H, Tamura K, Morimoto Y et al. Characterization of sphingolipid long-chain base kinase in Arabidopsis thaliana. Biochem Soc Trans 2000; 28:747-748.
59. Chalfant CE, Spiegel S. Sphingosine 1-phosphate and ceramide 1-phosphate: expanding roles in cell signaling. J Cell Sci 2005; 118:4605-4612.
60. Coursol S, Fan LM, Le Stunff H et al. Sphingolipid signalling in Arabidopsis guard cells involves heterotrimeric G proteins. Nature 2003; 423:651-654.
61. Coursol S, Le Stunff H, Lynch DV et al. Arabidopsis sphingosine kinase and the effects of phytosphingosine-1-phosphate on stomatal aperture. Plant Physiol 2005; 137:724-737.
62. Pandey S, Assmann SM. The Arabidopsis putative G protein-coupled receptor GCR1 interacts with the G protein alpha subunit GPA1 and regulates abscisic acid signaling. Plant Cell 2004; 16:1616-1632.
63. Ng CK, Carr K, McAinsh MR et al. Drought-induced guard cell signal transduction involves sphingosine-1-phosphate. Nature 2001; 410:596-599.
64. Le Stunff H, Galve-Roperh I, Peterson C et al. Sphingosine-1-phosphate phosphohydrolase in regulation of sphingolipid metabolism and apoptosis. J Cell Biol 2002; 158:1039-1049.
65. Haynes CA, Allegood JC, Park H et al. Sphingolipidomics: Methods for the comprehensive analysis of sphingolipids. J Chromatogr B Analyt Technol Biomed Life Sci 2008.
66. Welti R, Shah J, Li W et al. Plant lipidomics: discerning biological function by profiling plant complex lipids using mass spectrometry. Front Biosci 2007; 12:2494-2506.

CHAPTER 19

Computational Analysis of Sphingolipid Pathway Systems

Eberhard O. Voit,* Fernando Alvarez-Vasquez and Yusuf A. Hannun

Abstract

Sphingolipid metabolism constitutes a complex pathway system that includes biosynthesis of different types of sphingosines and ceramides, the formation and recycling of complex sphingolipids and the supply of materials for remodeling. Many of the metabolites have several roles, for instance, as substrates and as modulators of reactions in other parts of the system. The large number of sphingolipid compounds and the different types of nonlinear interactions among them render it difficult to predict responses of the sphingolipid pathway system to perturbations, unless one utilizes mathematical models. The sphingolipid pathway system not only invites modeling as a useful tool, it is also a very suitable test bed for developing detailed modeling techniques and analyses, due to several features. First, the reaction network is relatively well understood and many of the steps have been characterized, at least in vitro. Second, sphingolipid metabolism constitutes a relatively closed system, such that most reactions occur within the system rather than between the system and other pathways. Third, the basic structure of the pathway is conserved throughout evolution, but some of the details vary among different species. This degree of similarity permits comparative analyses and may one day elucidate the gradual evolution toward superior system designs. We discuss here some reasons that make sphingolipid modeling an appealing companion to experimental research and sketch out applications of sphingolipid models that are different from typical model uses.

Introduction

Most sphingolipid analyses over the past decades have in great detail characterized individual metabolites, genes, or enzymes along with the reactions they catalyze (for example see refs. 1-7). As other chapters in this book attest, these studies have greatly improved our understanding of the components of sphingolipid metabolism and we have by now assembled a fairly good impression of the functionality of biosynthesis, metabolic conversions within the pathway and the ultimate fates of the various sphingolipid compounds. While the detailed characterizations of the genomic, metabolic and regulatory components are of undisputed importance, they do not paint a complete picture of how the integrated metabolic pathway system responds to environmental challenges, such as heat stress.

There are many reasons for why characterizing solely the parts of the sphingolipid pathway system is insufficient for a full understanding. The most obvious is the sheer number of components. In yeast there are roughly twenty to thirty "base" metabolites like sphingosine and phytoceramide.[8] However, many of these may exist in variant forms that differ in the lengths of their fatty acid chains, thereby multiplying the number of possibly relevant "players" several fold. In mammals, many sphingolipids may furthermore bind to various carbohydrates, forming glycosphingolipids

*Corresponding Author: Eberhard O. Voit—Integrative BioSystems Institute and Wallace H. Coulter Department of Biomedical Engineering, Georgia Institute of Technology, Atlanta, Georgia. Email: eberhard.voit@bme.gatech.edu

Sphingolipids as Signaling and Regulatory Molecules, edited by Charles Chalfant and Maurizio Del Poeta.

such as glucosylceramide and lactosylceramide, so that the number of all theoretically possible combinations reaches into the thousands (for example, see ref. 9). It is presently unknown whether every different combination of chain length and carbohydrate group plays a unique role for the functioning of the organism, but it is to be expected that there are at least distinctive differences between sphingolipids with long, as opposed to very long, fatty acid chains. As soon as large numbers of metabolites with subtle differences in their roles and functions are to be considered, the unaided human mind quickly becomes overwhelmed.

A second reason for failure in our intuitive predictive ability of global systems responses is the nonlinear nature of the interactions between the contributing sphingolipids. Specifically, the human mind has problems handling numerical thresholds: a slight increase in a metabolite may lead to a correspondingly slight change in some output measure or to no response at all, whereas a somewhat stronger increase has a distinctly different effect, such as the triggering of apoptosis. If several thresholds are in play, we simply can no longer make reliable predictions on responses. As a pertinent example, it is known that ceramide is a signaling compound associated with apoptosis, while sphingosine-1-phosphate (S1P) is associated with anti-apoptosis.[4,5,10] However, ceramide can easily be converted into S1P and vice versa. Thus, will slight changes in enzyme activities somewhere in the system ultimately lead to apoptosis or anti-apoptosis?

The sphingolipid pathway is usually shown as a more or less linear material flow system, or as a forward pathway with a few branches, beginning with the condensation of palmityl-CoA and serine and leading toward complex sphingolipids (CS), such as MIPC (for example see refs. 2,4,5,11,12). While these representations indeed capture the main flow, closer inspection reveals that there are numerous metabolic or regulatory feedback loops. For instance, dihydrosphingosine (DHS) and phytosphingosine (PHS) are key compounds at the center of the pathway. At the same time, they inhibit the production of phoshphatidylserine (PS), in a reaction that competes with serine palmitoyltransferase (SPT) for the same key substrate serine and of diacylglycerol (DAG), which is involved with the kinetics of complex sphingolipids. Palmitoyl-CoA is not only one of the inputs to biosynthesis, it is also produced in the lyase reaction with DHS-P and PHS-P. Thus, the uni-directionality of the sphingolipid pathway no longer holds as soon as these "loops" are taken into account. It may also happen that metabolites exert competing effects on some other part of the system, for instance, by serving simultaneously as input substrate and as a modulator of one of the catalytic steps in the system (e.g., see pathway diagrams in refs. 13, 14). These complexities render mathematical modeling a valuable tool with unique facilities that are difficult—if not impossible—to match with wet experiments.

A noteworthy feature of complex systems precluding reliable predictions is the fact that a pathway may exhibit new responses when it is put into its regulatory context. As an example, let us look at the simplest of pathways, a linear chain of reactions, which is here considered with one input and three intermediates (Fig. 1A). It does not require much imagination to predict what will happen if the input is raised to a higher level: the intermediates X, Y and Z will correspondingly change in concentration. Similarly, if the input is decreased, X, Y and Z will follow. We may not know how fast the adjustments will happen and what the final values of X, Y and Z might be, but we have a firm grasp of the qualitative response. Indeed, it is very easy to set up a mathematical model that captures the situation. While there are many options for such a model, the typical approach is a set of ordinary differential equations (ODEs), whose right-hand sides may be formulated as mass action kinetics, with Michaelis-Menten rate laws, as power-law functions, or with any number of other representations. For the present discussion, the particular mathematical format is essentially immaterial and we perform the illustration with a power-law model that is designed according to the guidelines of Biochemical Systems Theory (BST[15-19]). The result is shown in Eq. (1).

$$
\begin{aligned}
\dot{X} &= Input - k_1\, E\, X^{0.5} \\
\dot{Y} &= k_1\, E\, X^{0.5} - k_2\, Y^{0.5} \\
\dot{Z} &= k_2\, Y^{0.5} - k_3\, Z^{0.5}
\end{aligned}
\tag{1}
$$

In this formulation, X, Y and Z are generic representations of the metabolites. Their powers of 0.5 are typical in BST.[18] The parameters k_1, k_2 and k_3 are rate constants. E represents (the activity of) the enzyme that catalyzes the conversion of X into Y. We include it explicitly in these equations, because it will be the bridge to the extension following below. We could similarly include enzymes for the conversion of Y into Z and for the degradation of Z, but because their activities are assumed to be constant and because we do not explicitly use them in the following, we merge them with the rate constants k_2 and k_3, respectively. The left-hand sides of Eq. (1) are "dotted" variables that represent their change with respect to time; thus, for instance, $\dot{X} = dX/dt$.

To execute computational analyses, we need to specify parameter values. Because we only intend to illustrate the complications in predicting system responses, the parameters are chosen almost arbitrarily, but so that they are typical and that all variables have a nominal steady-state value of 1 if the Input has a magnitude of 1. Specifically, we set the rate constants k_1, k_2 and k_3 equal to 1.

For a baseline simulation, we start the system at the steady state (1, 1, 1). As expected, the system rests at this point and X, Y and Z do not change in value. Beginning at time $t = 10$, we reduce the input to 0.75 and, not surprisingly, the levels of X, Y and Z decrease. With the given settings, they approach the same value of 0.5625 (Fig. 2A). Doubling the input has the opposite effect: It raises the levels of X, Y and Z to 4 (results not shown). If we ignore the input in our discussion and just study X, Y and Z, we correctly conclude that changes in X cause corresponding changes in Y and Z.

Now suppose that Z is a sphingolipid with signaling function. Specifically, let us assume that it activates a transcription factor TF, which is responsible for the up-regulation of gene G, which codes for the enzyme catalyzing the conversion of X into Y (Fig. 1B). The former pathway is now embedded in a logic loop. This loop is easy to grasp in its organization, but its specific effects and the responses of the metabolic pathway are no longer easy to predict. Intuition may suggest that the positive feedback might magnify any changes in input, but it seems difficult to make specific predictions. For instance, how will changes in input affect the regulatory loop and the steady-state values of X, Y and Z? The answer may be surprising: with the given information alone, particular responses are not predictable with any reliability. It is again straightforward to set up BST equations capturing the situation. A possible implementation is shown in Eq. (2), which is an augmentation of the system in Eq. (1). The new parameters are again chosen so that all variables have "normal" values of 1 at the steady state. For simplicity of discussion, we will keep all parameters constant throughout the next set of simulations, except for the activation of TF. In other words, our "control parameter" is the strength of activation of TF by Z; it is coded here as p. The control parameter p is always positive or 0, because a negative value would represent inhibition rather than activation. For the baseline case $p = 0$, Z has no influence on TF. Gene regulation is in effect decoupled from the metabolic pathway and the reaction between X and Y runs with the former baseline enzyme activity (E = 1); the situation is exactly the same as shown before in Figure 2A.

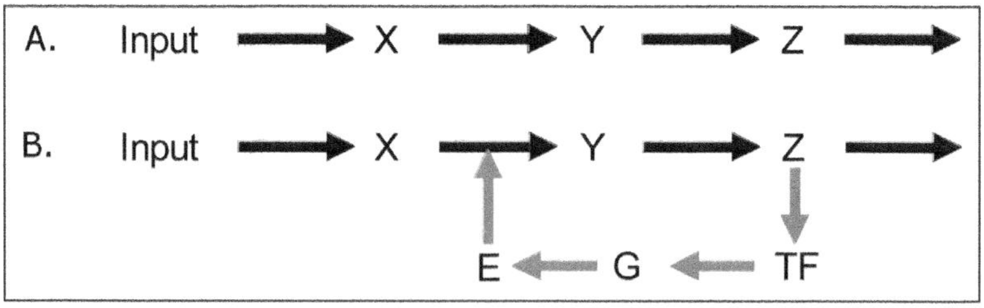

Figure 1. A) Simple linear pathway, in which an Input substrate is sequentially converted into metabolites X, Y and Z. Z is the substrate for another process or transported out of the system. B) Linear pathway from Figure 1A embedded in a "logic loop" consisting of a transcription factor TF, a gene G and an enzyme E. Grey arrows indicate activating effects. See Text for details.

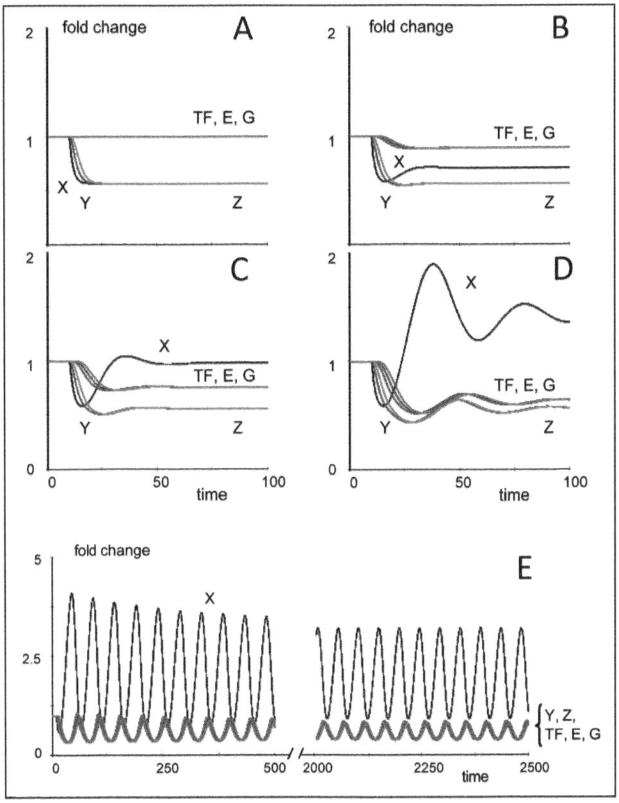

Figure 2. Simulation results for the pathways in Figures 1A,B and Equations (1) and (2). A) Response of the pathway in Figure 1A to a reduction of input to 75%. The situation is equivalent with Figure 1B if the control parameter is set as $p = 0$. B) Response of the pathway if $p = 0.1$. C) Response for $p = 0.24$. D) Response for $p = 0.4$. E) Limit cycle response for $p = 0.51$.

$$\begin{aligned} \dot{TF} &= Z^{p} - TF^{0.5} \\ \dot{G} &= TF^{0.5} - G^{0.5} \\ \dot{E} &= G^{0.5} - E^{0.5} \end{aligned} \tag{2}$$

If p is greater than 0, Z activates TF, which subsequently leads to a change in the expression of gene G and a concordant alteration in enzyme activity E. One should expect the altered enzyme activity to change the balance between X, Y and Z. If so, Z would change in response to changes in X and Y and the change in Z would subsequently affect TF, G, E and, thus, indirectly the balance between X, Y and Z, leading to a cycle of events. Will the system spiral out of control? All the sudden, predictions become doubtful, thus demonstrating the necessity of a mathematical model.

In order to explore the effect of TF activation per simulation, let us set $p = 0.1$. The value is small in magnitude, which implies that the activation is not particularly strong. At the beginning of the experiment, the system is still in the steady state where each variable has a value of 1. As before we reduce the input to 75% at $t = 10$. Metabolite X again decreases initially but, in contrast to the unregulated system, "recovers" to some degree (Fig. 2B). If p is set to a slightly increased value of 0.24, X actually returns close to its original value of 1. Y and Z assume the same values as before (Fig. 2C). Exploration of other values of p show that small magnitudes in p result in values for X that are close to 0.5625, while larger values lead to higher values. Meanwhile, Y and Z always reach the same value of 0.5625, no matter what the value of p; or what the value of X, for that matter. These observations lead to an intriguing conclusion: If we did not know the structure of the pathway, we would surely conclude that Y and Z had nothing to do with X, because they always have the same values, independent of the value of X. Yet, X is their only precursor substrate!

Something quite different happens if p is set higher: The metabolites begin to oscillate, before reaching their new steady state (Fig. 2D). For even larger values, the system "dies." For instance, if $p = 0.8$, the oscillations become so strong that one of the variables vanishes (result not shown). In between the former, reasonable values and these large values lies a small range of values for p ($p \approx 0.51$) where the system exhibits yet another behavior: it oscillates in a stable fashion so that after some while the amplitude and frequency remain constant (Fig. 2E). The so-called limit cycle oscillations in this small range are very interesting mathematically, because they are able to tolerate perturbations, from which they recover. Sustained oscillations have been observed in the expression of actual genomes and in metabolic systems, such as glycolysis.[20-23]

The simulations with the simple pathway demonstrate that it is not necessarily possible to grasp intuitively the full functionality of a system if it is taken out of its context. Considering the complexity of sphingolipid metabolism and its regulation, one must therefore wonder to what degree intuition is sufficient when global responses are to be predicted.

Sphingolipid Models and Their Potential Uses

Over a span of several years, we have been developing a series of increasingly more sophisticated models of sphingolipid metabolism.[13,14,24,25] The models were formulated as systems of nonlinear ordinary differential equations in the format of power-law functions, as suggested in BST.[18] Choosing this framework, it was straightforward to set up symbolic equations that reflect the known or assumed connectivity and regulatory signals of the pathway system. While this part of the model design phase was manageable with reasonable effort, the estimation of suitable parameter values was very challenging. Indeed, our case study confirmed common experience that parameter estimation is the bottleneck of biological modeling. In our case, the estimation was based on literature information, de novo experiments and some default assumptions based on experience with BST.[14] The resulting model was subsequently tested in the typical fashion, namely with stability, sensitivity and robustness analyses, through comparisons with experimental data that had not been used in the estimation phase and through qualitative reality checks based on biological experience. After many iterations and revisions, the model appeared reasonable and was semi-quantitatively validated with additional wet experiments.[13]

It would be counterproductive to use the limited space of this chapter to review the steps of a typical modeling process in general or even within the context of sphingolipid metabolism, because both have been described in recent years and at various levels of sophistication and detail (for example see refs. 10, 14, 18, 24). Instead, it seems more beneficial to ask what we may do with such models, once they are validated. Again, many typical uses of models have been described in the literature and we will simply mention some of them. However, other uses are less typical and will receive more attention in the following.

The first and foremost role of a mathematical model is the integration of diverse data and other pieces of information, such as kinetic characteristics of enzymes, expression profiles of genes, protein abundances and maybe even semi-qualitative clinical observations. This integration often shows very clearly whether we have a good grasp of the functionality of the pathway, because most initial efforts of merging all information into one mathematical construct fail. Typical failures become apparent in lacking stability or robustness of the model that is accompanied by unduly high sensitivities. In the former case, small variations in input or in some variable may cause the model to "crash" in a sense that one or more variables vanish. In the latter case, the system "overreacts" to small changes in parameters. For instance, a 5% increase in some enzyme activity could lead to a 220% increase in some metabolite, which is unreasonable in most cases. Relatively straightforward diagnostic tools weed out such systems (for example, see ref. 18).

If a model appears reasonable, we can further test our intuitive grasp of the system through simulation studies that represent *What-If* scenarios, as we employed them before. For example, the reduction in an enzyme activity is easily implemented in the model and simulation results can possibly be validated with wet experiments. An integrated model of sphingolipid metabolism also allows us to follow the fate of metabolites of interest, many of which are recycled or involved in several reactions. A good model, with slight adaptations, even permits the tracking of labeled substrates, from input to their ultimate fates.[10,14] As a variation on this theme, a reliable model may be used to study the relative contributions of different pathways to a common goal, such as the formation of rafts that become structural elements of membranes.

In a study of a slightly different nature, we asked the question whether a yeast cell mounts a response to stress by up-regulating a small number of enzymes (genes) a lot or whether it changes the activities of many enzymes a little bit. Intuitively one could easily find rationale for either strategy. On one hand, up- or down-regulating only one or two genes or enzymes appears to be the simpler strategy. On the other hand, drastic changes in some part of the pathway could lead to concomitant and undesired side effects. In our test case of the diauxic shift, the modeling analysis suggested that many enzymes are involved in the response.[24]

Sphingolipid metabolism has been analyzed experimentally in different organisms, some of which are phylogenetically close. This similarity permits the cautious extrapolation and use of the model in an untested organism. This type of model transfer was demonstrated by using the original *S. cerevisiae* model with some adaptations to study sphingolipid metabolism in *Cryptococcus neoformans* (*Cn*), an airborne fungal pathogen that may cause life-threatening infections.[25] The main challenge this organism faces is the distinct difference in pH between alkaline or neutral extracellular environments, such as alveolar spaces or the bloodstream and the acidic environment of the intracellular phagolysosome of the host's phagocytic cells, in which the organism lives and grows during a crucial phase of its virulence cycle.[26] Earlier work in Del Poeta's laboratory had suggested the involvement of sphingolipids in growth under acidic conditions, but it had not been possible to characterize the specifics of this process.[27] The model results together with subsequent validation studies led to the very specific proposition that inositol phosphoryl ceramide synthase (Ipc1) and inositol sphingolipid phospholipase C (Isc1) affect the function of the plasma membrane H^+-ATPase pump (Pma1) through modulation of the level of phytoceramides and complex sphingolipids.[25]

The successful use of the yeast model in the investigation of a fungal pathogen suggests that it might even be possible to study the evolution of design principles governing sphingolipid function, based on comparative model analyses of sphingolipid metabolism in closely related and more distant species. Along the same lines, a major future project should convert and test the yeast model for analyses of the mammalian analogues. This "extrapolation" is much more complicated, because mammalian sphingolipids are often bound to carbohydrates, which leads to a multiplication in the number of potentially relevant compounds, as discussed before.[28] This explosion in number may appear overwhelming for a model analysis. However, trying to understand the mammalian systems without computational approaches seems to be incomparably more complicated. It is clear that a mammalian model could be very useful for the exploration of pathways leading from health to diseases such as cancer, where we have strong indication that sphingolipids are involved (for example see ref. 29).

As a first step toward comparative studies, the sphingolipid system of the same organism may be studied at different grades of granularity. For instance, our present yeast model accounts only partially for compartmentalization. As a next step, it might be fruitful to distinguish sphingolipids and precursors with different fatty acid chain lengths. Accounting for this detail will require a substantial increase in model size and require additional biological information about the relevant compounds of different sizes and their respective roles.

As a more detailed example of an atypical model investigation, consider the immediate sphingolipid response in yeast to heat stress. Genome studies have strongly supported the involvement of several genes, such as *MSN2/4* and *YAP1*.[30] However, preliminary concentration

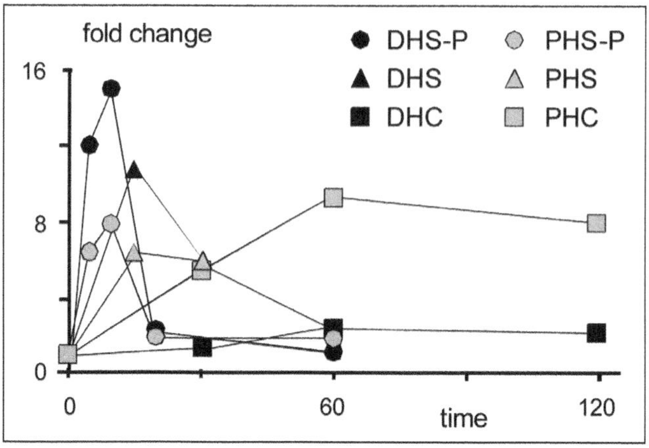

Figure 3. Fold changes in sphingolipids following heat stress at time $t = 0$. Abbreviations: DHS: dihydrosphingosine; PHS: phytoshpingosine; DHC: dihydro-ceramide; PHC: phyroceramide; –P: –phosphate. Data adapted from Jenkins et al[31] and Hannun, Y.A.: *pers. comm.*

measurements of key sphingolipids indicate that the heat stress response is much faster than any gene-regulatory mechanism could accomplish, exhibiting metabolic changes within a few minutes (Fig. 3[31]; and Y.A. Hannun, *pers. comm.*; see also ref. 32). Within about 8 minutes, dihydrosphingosine-phosphate (DHS-P) and phytosphingosine-phosphate (PHS-P) increase to 8- and 15-fold levels, respectively, before resuming almost normal values after about 20 minutes. DSH and PSH respond more slowly, peaking after about 15 minutes at 6- and 11-fold levels. Phytoceramide accumulates gradually over a period of about an hour, peaking at a level that is about 8 times higher than baseline. Dihydroceramide shows the same pattern, but with a peak accumulation of only about two-fold.

This fast change in metabolic profile is intriguing and not explainable with gene regulatory actions. We have seen a similarly quick response to heat stress in the trehalose cycle in yeast, which, according to careful in vivo NMR measurements, begins producing trehalose within two minutes.[30] Again, a gene regulatory response is too slow for such a response. As it turned out in the trehalose case, three key enzymes of the trehalose pathway are heat sensitive. The two enzymes controlling trehalose production are more active at higher temperatures, while trehalase is less active.[33] A preliminary mathematical model analysis suggests that the relatively slight changes in activity are sufficient to mount the fast, observed response.[30]

Given the similarity of the heat response task in the case of trehalose and sphingolipids it is reasonable to ask whether there are heat-sensitive enzymes within the sphingolipid pathway as well. If so, would a single enzyme be sufficient to mount the observed sphingolipid response? Would combinations of two or more enzymes be sufficient?

A mathematical model might help us identify such enzymes. First, it would of course be possible to launch a major simulation study, changing one or a few enzymes at a time and then testing whether metabolic concentration patterns like the one observed can be generated. However, some prudence might help us prescreen some possibilities and favor or disfavor particular hypothetical scenarios. To permit an objective, yet lucid exploration, we reduced the sphingolipid system to a minimum and converted it into a much simplified mathematical model. Thus, consider the reduced core of sphingolipid metabolism that contains only those components that appear most important (Fig. 4).

Looking at once at Figures 3 and 4, we can formulate the following as a framework for preliminary hypotheses. The concentration of phytoceramide at one to two hours exhibits a sustained level at eight times its baseline. This increased level requires that: (1) more material is produced per

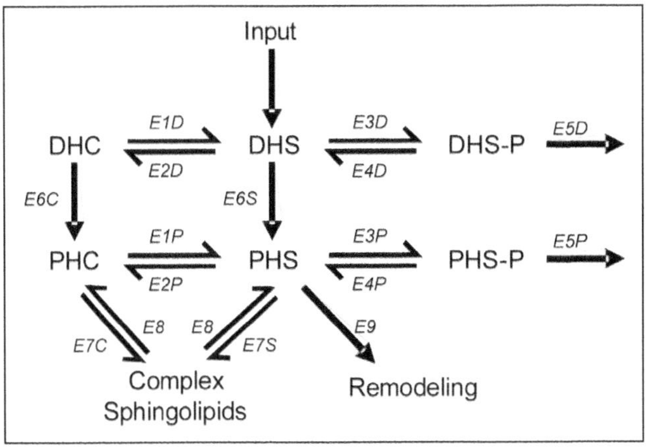

Figure 4. Reduced and simplified diagram of parts of the sphingolipid pathway involved in heat stress response. Abbreviations of metabolites are presented in Figure 3. Names of enzymes are not of relevance, but can be found in.[14] E7C, E7S and E8 are collective representations for the interactions between the simple and complex sphingolipids.

biosynthesis or recycling of complex sphingolipids (*cf.*); (2) material is simply reorganized within the pathway system; (3) the loss of sphingolipids is reduced; or (4) several of the previous options are combined. The second option may be discarded off-hand: A simple reorganization or shift of fluxes in the neighborhood of DSH and PSH would require decreased levels of the substrates of these reactions, but decreases in concentrations are not observed. Although some of the biosynthetic genes are affected by heat, the first option alone also seems questionable, because DSH-P and PSH-P respond most quickly, whereas one would expect a response through increased biosynthesis to cause increases in DHS and PHS first. Besides, if PHC is the target and biosynthesis was the mechanism, why should DSH-P and PSH-P be increased at all? Similar arguments seem to hold for the recycling of complex sphingolipids. One could surmise that a direct change in E7C and/or E8 could be a good strategy toward an increased level of PHC. However, this is apparently not a strategy pursued by the cell. Pursuing the third (or fourth) option, the observed metabolite profile could possibly be achieved through a reduction in the lyase reaction (*E5D, E5P*), which would lead to an accumulation of the phosphorylated forms. This change would have to be followed by increased sphingoid base PPase (*E4D, E4P*) and hydroxylase (*E4D, E4P*) activity, which would gradually shift the increasing amounts of DSH-P and PSH-P toward the dephosphorylated forms and from there to ceramide (*E6C, E6S*). It is interesting to note that the dihydro- and the phyto-forms of sphingolipids essentially form parallel pathways, but that the target profile after heat stress is distinctly different between the two pathways. This observation suggests that the involved enzymes might have different affinities for the dihydro- and the phyto-forms, at least under heat stress conditions.

Complex systems have a way of tricking our intuition and our hand waving arguments could simply turn out to be wrong. Nevertheless, it is possible to test these scenarios with a model. In order to keep our exploration as simple as possible, we created a reduced model reflecting the simplified pathway shown in Figure 4; it is shown in Eq. (3). Because the only purpose of this model is a proof of concept showing whether or not changes in enzyme activities could lead to something like the observed metabolite profile, we set the model up in a minimalistic fashion, again using (almost arbitrary) default parameter values and enzyme activities set to a nominal value of 1, so that the steady state consists of unity values. These intentional simplifications may show us the consequences of introduced changes more clearly than a model that is parameterized from actual data, but of course we cannot expect to obtain numerically valid simulation results.

$$
\begin{aligned}
\dot{DHC} &= 2\ E2D\ DHS^{0.5} - E1D\ DHC^{0.5} - E6C\ DHC^{0.5} \\
\dot{DHS} &= Input + E1D\ DHC^{0.5} - 2\ E2D\ DHS^{0.5} - 2\ E3D\ DHS^{0.5} \\
&\quad + E4D\ DHSP^{0.5} - 5\ E6S\ DHS^{0.5} \\
\dot{DHSP} &= 2\ E3D\ DHS^{0.5} - E4D\ DHSP^{0.5} - E5D\ DHSP^{0.5} \\
\dot{PHC} &= E6C\ DHC^{0.5} + 2\ E2P\ PHS^{0.5} - E1P\ PHC^{0.5} - 3\ E7C\ PHC^{0.5} + E8\ CS^{0.5} \\
\dot{PHS} &= 5\ E6S\ DHS^{0.5} + E1P\ PHC^{0.5} - 2\ E2P\ PHS^{0.5} - 2\ E3P\ PHS^{0.5} \\
&\quad + E4P\ PHSP^{0.5} - 3\ E7S\ PHS^{0.5} + E8\ CS^{0.5} - E9\ PHS^{0.5} \\
\dot{PHSP} &= 2\ E3P\ PHS^{0.5} - E4P\ PHSP^{0.5} - E5P\ PHSP^{0.5} \\
Input &= 7 \\
CS &= 1
\end{aligned}
\tag{3}
$$

As a first simulation, we start the model at the steady state (1, ..., 1). Because all values are 1, all later results automatically represent "fold" changes. At time $t = 2$, we reduce the lyase activity (*E5D*, *E5P*) (Fig. 5A). Even with different magnitudes and different changes for the lyases with respect to DHS-P and PHS-P, the resulting sphingolipid profiles are not even close to the observations in Figure 3; instead, they primarily lead to accumulation of DHS-P and PHS-P (Fig. 5B).

For a second exploration, we double the (biosynthetic) input starting at $t = 2$. The result is increased mass in the system, but the resultant profile is similar to that in Figure 5B (results not shown). As a variation on the same theme, we increase the input more gradually and assume that after a while the substrates for biosynthesis become limiting, thereby reducing the input flux (Fig. 5C). The model now yields a peaking profile, but all metabolites respond in parallel (Fig. 5D).

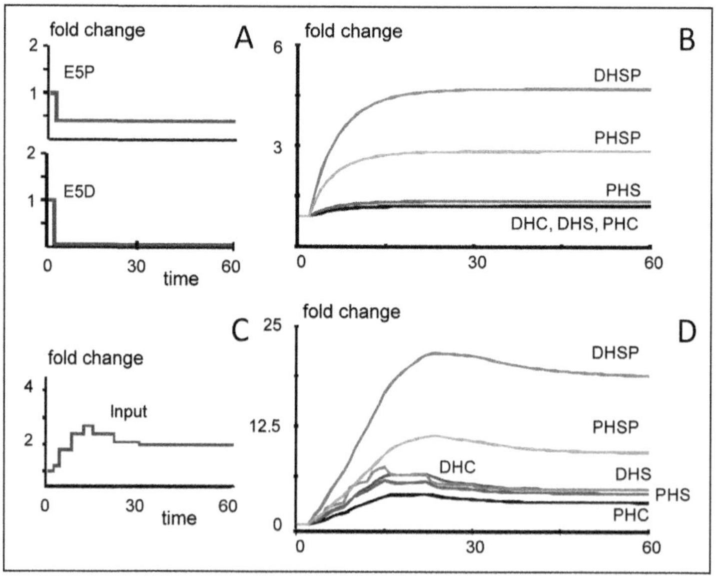

Figure 5. Two simulation scenarios with Equation (3) exploring the feasibility of simple changes in enzyme activities. A) The two lyases are reduced two minutes after initiation of heat stress. As a consequence (B), the phosphorylated sphingolipids accumulate. If in addition the input to the system changes in direct response to heat, the same metabolites increase more. Assuming that increased biosynthesis leads to substrate deprivation, the sphingolipid profiles begin to decrease.

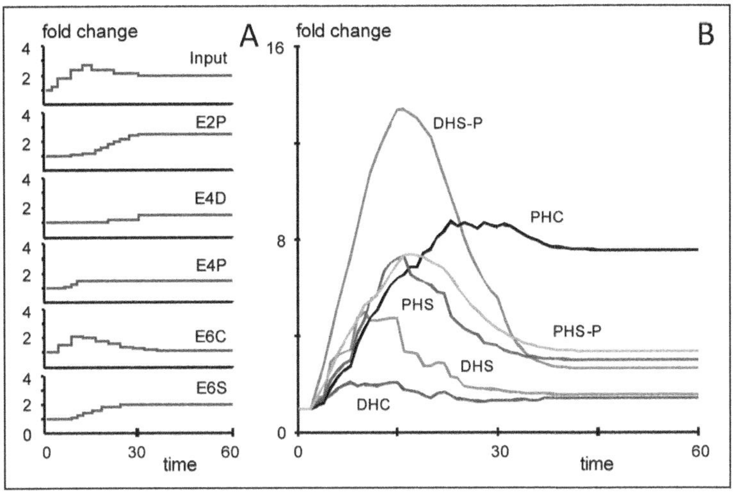

Figure 6. Simulation scenario with Equation (3) in which several enzymes have temperature dependent activities. Profiles of changes in enzymes and input are given in (A); see Figure 3 for abbreviations. The corresponding sphingolipid profiles show a qualitatively similar pattern as it was observed (Fig. 4). The "jagged" appearance of the simulated profiles is due to discrete changes in enzyme activities.

Uncounted simulations of this type may be executed and in the end it would be wise to formulate an optimization program that would guide the progression of simulations. Through manual exploration, we developed a more complex heat stress response pattern in enzyme activities (Fig. 6A) that actually produces a metabolic profile that qualitatively resembles the observations (Fig. 6B). This profile is clearly not unique and not refined or optimized, because it is based on the simplified model with "invented" parameter values, rather than a fully parameterized model. Nonetheless, the simulation constitutes proof of concept that temperature dependent enzymes could be the drivers of the very fast sphingolipid response to heat, as it is the case in the trehalose cycle. Interestingly, "successful" profiles like the ones in Figure 6A seem to require different affinities of the enzymes to the dihydro- and phyto-forms of the metabolites.

In addition to simulation studies, the model format of an S-system within BST could be used to exhaust all possible means of achieving a desired profile at steady state. We have shown in a completely different context how such an analysis could be pursued[10] (see also ref.19). In a nutshell, the mathematical features of the S-system model allow an elegant algebraic analysis of the entire space of steady-state solutions in terms of enzyme activities that are consistent with a desired metabolic concentration profile. Among these consistent solutions, optimal transient solutions could be determined through dynamic analyses or nonlinear optimization studies.

Conclusion

The pathways of sphingolipid biosynthesis, utilization and recycling form an intriguingly complex system whose dynamics is difficult to predict with intuition alone. Mathematical modeling provides an aid in this regard, because it permits the integration of many pieces of information into computational structures that are very easy to diagnose, interrogate and analyze through *What-If* simulations. The bottleneck of setting up such models is the determination of parameter values, which may be based on literature information characterizing genes, enzymes and metabolites in a steady-state situation, or on dynamic time trends, which can be measured with modern methods of mass spectrometry or nuclear magnetic resonance. Parameter estimation is the bridge between wet experimentation and modeling and the need for improved parameter values, which are valid in vivo and under relevant physiological conditions, gives clear indication that mathematical modeling is crucially dependent on solid experimental work. At the same time, once reliable models are established, they become incomparably rich tools for analyses that are often unattainable with wet experimentation. For instance, it is at least theoretically feasible to determine all possible combinations of enzyme activities that lead to an observed metabolite profile at steady state. As we indicated here with an intentionally simplified analysis, it is also possible in principle to determine mathematically which enzymes would have to altered and in what manner, to obtain dynamic metabolite profiles as they are observed in responses to perturbations such as heat stress.

Ultimately, reliable mathematical models will be used as valuable tools for exhaustive prescreening studies for all kinds of scenarios and for creating novel and optimally discerning hypotheses that are then to be tested in the laboratory. If the history of physics is an indication of the future of biology, we might one day execute experiments only once the theory describing the underlying system is sufficiently worked out and understood. Even if this procedure will become the norm in the future, one must expect that it will take many years and dedicated effort, both experimentally and methodologically, to establish models of sufficient scope and reliability. Thus, the rise of mathematical modeling as a biological technique should not be seen as threatening to experimentation, but simply as an additional tool that is able to elucidate different aspects of a system. Modeling has improved tremendously over the past decades and successful collaborations between biological and computational scientists in the recent past have begun to show that their team efforts will be rewarding to both sides and reveal insights that neither side could have obtained without the other.

Acknowledgements

This work was supported in part by a grant from the National Institutes of Health (R01 GM063265; Y.A. Hannun, PI), a Molecular and Cellular Biosciences Grant (MCB-0517135; E.O. Voit, PI) from the National Science Foundation and an endowment from the Georgia Research Alliance. Any opinions, findings and conclusions or recommendations expressed in this material are those of the authors and do not necessarily reflect the views of the sponsoring institutions.

References

1. Carman GM, Henry SA. Phospholipid biosynthesis in the yeast Saccharomyces cerevisiae and interrelationship with other metabolic processes. Prog Lipid Res 1999; 38(5-6):361-99.
2. Dickson RC, Lester RL. Metabolism and selected functions of sphingolipids in the yeast Saccharomyces cerevisiae. Biochim Biophys Acta 1999; 1438:305-321.
3. Hannun YA, Luberto C, Argraves KM. Enzymes of sphingolipid metabolism: from modular to integrative signaling. Biohemistry 2001; 40(16):4893-4903.
4. Merrill AH. De novo sphingolipid biosynthesis: A necessary, but dangerous, pathway. J Biol Chem 2002; 277(29):25843-25846.
5. Hannun YA, Obeid LM. The ceramide-centric universe of lipid-mediated cell regulation: stress encounters on the lipid kind. J Biol Chem 2002; 277(29):25847-25850.
6. Futerman AH, Hannun YA. The complex life of simple sphingolipids. EMBO Rep 2004; 5(8):777-82.
7. Fahy E, Subramaniam S, Brown HA et al. A comprehensive classification system for lipids. J Lipid Res 2005; 46(5):839-61.
8. Guan XL, Wenk MR. Mass spectrometry-based profiling of phospholipids and sphingolipids in extracts from Saccharomyces cerevisiae. Yeast 2006; 23(6):465-77.
9. http://sphingolab.biology.gatech.edu/
10. Alvarez-Vasquez F, Hannun YA, Voit EO. Dynamic of Positional Enrichment: theoretical development and application to carbon labeling in Zymomonas mobilis. Biochem Eng J 2008; 40(1):157-174.
11. Futerman AH, Riezman H. The ins and outs of sphingolipid synthesis. Trends in Cell Biol 2005; 15(2):312-318.
12. Dickson RC. Thematic review series: sphingolipids. New insights into sphingolipid metabolism and function in budding yeast. J Lipid Res 2008; 49(5):909-21.
13. Alvarez-Vasquez F, Sims KJ, Cowart LA et al. Simulation and validation of modelled sphingolipid metabolism in Saccharomyces cerevisiae. Nature 2005; 433(7024):425-30.
14. Alvarez-Vasquez F, Sims KJ, Hannun YA et al. Integration of kinetic information on yeast sphingolipid metabolism in dynamical pathway models. J Theor Biol 2004; 226(3):265-91.

15. Savageau MA. Biochemical systems analysis. I. Some mathematical properties of the rate law for the component enzymatic reactions. J Theor Biol 1969; 25(3):365-9.
16. Savageau MA. Biochemical systems analysis: a study of function and design in molecular biology. Reading, Mass.: Addison-Wesley Pub. Co., Advanced Book Program 1976; xvii:379.
17. Voit EO. ed. Canonical Nonlinear Modeling. S-System Approach to Understanding Complexity. Van Nostrand Reinhold: New York, NY. 1991; xi:365.
18. Voit EO. Computational analysis of biochemical systems: a practical guide for biochemists and molecular biologists. New York: Cambridge University Press 2000; xii:531.
19. Alvarez-Vasquez F, Canovas M, Iborra JL et al. Modeling, optimization and experimental assessment of continuous L-(-)-carnitine production by Escherichia coli cultures. Biotechnol Bioeng 2002; 80(7):794-805.
20. Markus M, Hess B. Input-response relationships in the dynamics of glycolysis. Arch Biol Med Exp (Santiago) 1985; 18(3-4):261-71.
21. Richard P. The rhythm of yeast. FEMS Microbiol Rev 2003; 27(4):547-57.
22. Goldbeter A. Biochemical Oscillations and Cellular Rhythms. 1996, Cambridge: Cambridge University Press.
23. Goldbeter A. Biological rhythms: clocks for all times. Curr Biol 2008; 18(17):R751-R753.
24. Alvarez-Vasquez F, Sims KJ, Voit EO et al. Coordination of the dynamics of yeast sphingolipid metabolism during the diauxic shift. Theor Biol Med Model 2007; 4:42.
25. Garcia J, Shea J, Alvarez-Vasquez F et al. Mathematical modeling of pathogenicity of Cryptococcus neoformans. Mol Syst Biol 2008; 4:183.
26. Feldmesser M, Tucker S, Casadevall A. Intracellular parasitism of macrophages by Cryptococcus neoformans. Trends Microbiol 2001; 9(6):273-8.
27. Shea JM, Kechichian TB, Luberto C et al. The cryptococcal enzyme inositol phosphosphingolipid-phospholipase C confers resistance to the antifungal effects of macrophages and promotes fungal dissemination to the central nervous system. Infect Immun 2006; 74(10):5977-88.
28. Degroote S, Wolthoorn J, van Meer G. The cell biology of glycosphingolipids. Semin Cell Dev Biol 2004; 15(4):375-87.
29. Ogretmen B, Hannun YA. Biologically active sphingolipids in cancer pathogenesis and treatment. Nat Rev Cancer 2004; 4(8):604-16.
30. Fonseca LL, Sanchez C, Wu J et al. Modeling the trehalose cycle in Saccharomyces cerevisiae based on 13C-NMR data of intracellular metabolite pools in response to heat stress. in 10th International Conference on Molecular Systems Biology (ICMSB 2008). 2008. University of Diliman, Philippines 2008. (Poster).
31. Jenkins GM, Richards A. Wahl T et al. Involvement of yeast sphingolipids in the heat stress response of Saccharomyces cerevisiae. J Biol Chem 1997; 272(51):32566-72.
32. Dickson RC, Nagiec EE, Skrzypek M et al. Sphingolipids are potential heat stress signals in Saccharomyces. J Biol Chem 1997; 272(48):30196-200.
33. Neves MJ, Francois J. On the mechanism by which a heat shock induces trehalose accumulation in Saccharomyces cerevisiae. Biochem J 1992; 288 (Pt 3):859-64.

APPENDIX

Introduction to Tools and Techniques for Ceramide-Centered Research

Kazuyuki Kitatani and Chiara Luberto*

Abstract

Sphingolipids are important components of eukaryotic cells, many of which function as bioactive signaling molecules.[1-3] As thoroughly discussed elsewhere in this volume, ceramide, central metabolite of the sphingolipid pathway, plays key roles in a variety of cellular responses. Since the discovery of the bioactive function of ceramide,[4] a growing number of tools and techniques have been and still are being developed in order to better decipher the complexity and implications of ceramide-mediated signaling. With this chapter it is our intention to provide new comers to the sphingolipid arena with a short overview of tools and techniques currently available for the study of sphingolipid metabolism, with the focus on ceramide.

Lipid Extraction

Isolation of ceramide from animal tissues and culture cells is achieved by extraction with organic solvents because of it's nonpolar chemical structure. This is accomplished by using the methods described by Bligh and Dyer[5] or Folch et al.[6] Approximately 100% of the biological ceramide species are extracted into the organic phase under either method. When considering a more comprehensive analysis, it is important to consider the structural heterogeneity of the sphingolipid class members therefore specific lipid extraction methods might be required. For instance, simultaneous isolation of polar (such as sphingoid base phosphates or gangliosides) and nonpolar sphingolipids (such as ceramides) would be better achieved using a monophase extraction.

Identification and Quantification of Steady State Levels of Ceramide

Several methods have been developed to identify and quantify ceramide by mass spectrometry,[7-10] thin layer chromatography (TLC),[11,12] high performance liquid chromatography (HPLC)[13] and enzymatic assay using *Escherichia coli* diacylglycerol kinase (DGK).[14,15] The term ceramide is often used for *N*-acyl-sphingosine, however when investigators use a method that does not distinguish among backbone species (such as analyses by TLC or DGK assay), the term will likely include all N-acyl-sphingoid bases regardless of the back bone.

In the past, the most widely used techniques for the determination of steady state levels of ceramide in cells have been long-term radiolabeling and the diacylglycerol kinase assay.

Radiolabeling of ceramide is based on the use of a radioactive precursor for ceramide synthesis. De novo ceramide synthesis begins with the condensation of L-serine and palmitoyl-CoA through the action of serine palmitoyltransferase to form 3-ketosphinganine, which is then reduced to sphinganine (dihydrosphingosine) (Fig. 1). This sphingoid base is then acylated by

*Corresponding Author: Chiara Luberto—Department of Biochemistry and Molecular Biology, Medical University of South Carolina, Charleston, South Carolina, USA. Email: luberto@musc.edu

Sphingolipids as Signaling and Regulatory Molecules, edited by Charles Chalfant and Maurizio Del Poeta.

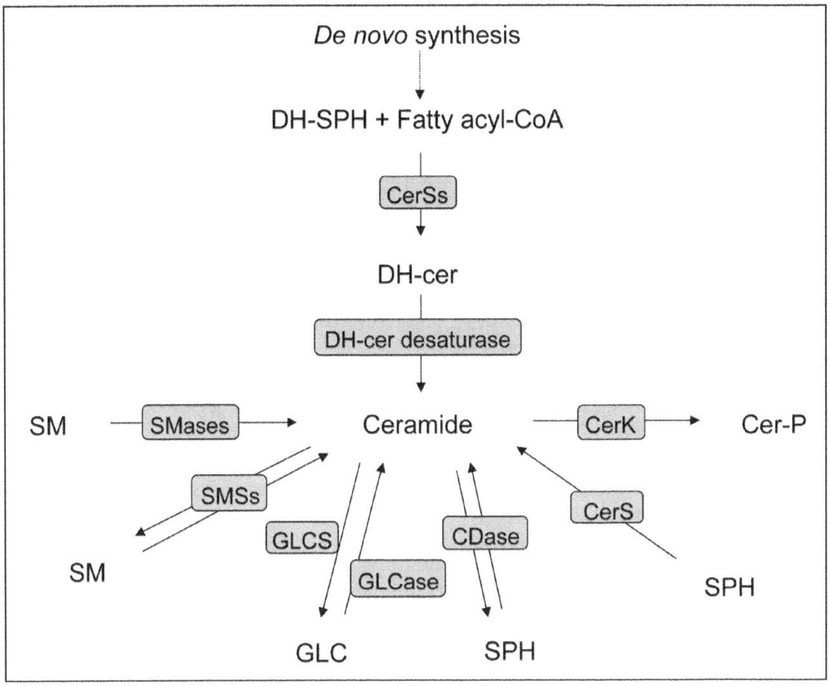

Figure 1. Schematic of ceramide metabolizing enzymes in mammalian cells. Represented are the reactions and the respective enzymes involved in the metabolism of ceramide. Some enzymes can catalyze a reaction both in forward and reverse mode (i.e. SMSs and CDase). DH-cer: dihydroceramide; DH-SPH, dihydrosphingosine; SM: sphingomyelin; GLC: glucosylceramide; SPH: sphingosine; Cer-P: ceramide phosphate; CerSs: ceramide synthases; SMases: sphingomyelinase; SMSs: sphingomyelin synthases; GLCS: glucosylceramide synthase; GLCase: glucosylceramidase; CDase: ceramidase; CerK: ceramide kinase.

(dihydro)ceramide synthases (CerS1 to 6) to form dihydroceramides with distinct fatty acyl moieties which are then desaturated by dihydroceramide desaturase to form the corresponding ceramides. Therefore radioactive palmitate, serine or the more proximal dihydrosphingosine have all been widely used as ceramide precursors. Two critical points must be considered when approaching this technique: the radioactive precursor used should be in trace amounts and the duration of the metabolic labeling should be optimized based on the chosen precursor and the experimental conditions (number of cells, volume of medium, type of medium etc.). The scientific rationale for this technique is based on the fact the radiolabeled precursor should enter the cellular sphingolipid pathway without perturbing it and it is used alongside its unlabeled counterpart present in the cells. Therefore the amount of radiolabeled ceramide present in the cell after steady-state labeling is representative of the total ceramide levels. Since the technique exploits the endogenous metabolic pathway, it is imperative that the amount of radioactive substrate added to the cells does not constitute a significant mass that could alter the flux through the pathway. Cells should be incubated with the radioactive precursors for the amount of time that allows the precursors to be steadily incorporated in the pathway until their conversion into ceramide (equilibrium between ceramide synthesis and ceramide degradation). This implies that a residual pool of not yet metabolized precursor should be still available to the cell and the ratio of radioactive ceramide to total phospholipids or total proteins remains constant with time. The amount of precursor to be added to the cells in order to obtain sufficient sensitivity (enough radioactivity associated with ceramide) greatly depends on the precursor' specific activity. In general, 10-30 uCi of palmitate or dihydrosphingosine added to 2 million cells for 24 hours are enough to achieve all of the above.[16-19] Different solvent systems may be used for separation of ceramide on TLC. We find that a mixture of ethyl acetate/iso-octane/acetic acid (9:5:2, v/v/v) is particularly suitable if ceramide is the lipid of interest. If one, in addition to ceramide, is also interested in more polar sphingolipids, such as sphingomyelin and gangliosides, then a mixture of chloroform/methanol/H_2O (110: 40:6, v/v/v) would be our choice.

The DGK assay exploits the enzymatic promiscuity of the DGK enzyme. The DGK can in fact phosphorylate both DAG and ceramide because of the structural similarities between these two lipids. On the other hand, it needs to be pointed out that the Km for ceramide is almost five times greater than that for DAG and this becomes a critical factor to consider when setting up the proper in vitro conditions for reliable and quantitative ceramide measurements. Thus, after solubilization of total cellular lipid extracts in micelles, addition of recombinant DGK leads to phosphorylation of ceramide (and DAG). The addition of radioactive ATP allows the phosphorylated ceramide to be traced and measured. At completion of the reaction, lipids are extracted, separated by TLC and radioactivity associated with ceramide-P is determined. Conversion of radioactivity into mass is obtained by using a standard curve derived from phosphorylation of known amounts of ceramide. The reliability of this assay depends on the fact that the reaction must be carried out to completion, which means that all the ceramide present in the original extract is converted into ceramide-P and that no factors in the lipid extracts may affect the DGK activity. In order for this to happen, the relative amounts of lipid extracts and the DGK used in the assay have to be carefully chosen. The user should work in the presence of great excess of DGK as compared to the amount of ceramide, so that the effect on the enzyme of potential inhibitors/activators can be dismissed and that the presence of DAG in the lipid extracts is not limiting. A thorough discussion on the use of DGK for ceramide quantification is presented by Dr. Bielawska and colleagues[20] and accurate description of the assay is provided by Dr. Perry and colleagues.[14]

More recently much emphasis is given to the structural diversity of ceramide species. In fact, particular species of ceramide with distinct fatty acyl moieties were shown to selectively increase with cellular responses and generate specific ceramide signals.[21,22] Therefore, the possibility of qualitatively and quantitatively isolate individual ceramide classes has significantly progressed the field. This task is accomplished using mass spectrometry methodology and the specific methods have been extensively described elsewhere.[8,23-25]

Analysis of Ceramide Metabolism

Radiolabeling of cells can also be used to specifically assay ceramide metabolism (either synthesis or break-down). For ceramide synthesis, the goal is to pulse the cells with trace amounts of a radioactive precursor and measure the levels of radioactive ceramide before the radiolabeling has reached steady state equilibrium. Generally few hours of incubation (up to 12 hours) will suffice. Since the pool of ceramide labeled in these conditions is smaller than the one achieved after steady state labeling (see previous section), the specific activity of the precursor is key in order to be able to detect enough radioactive signal associated with newly synthesized ceramide.

Ceramide break-down can be also measured by radiolabeling of cells. After labeling cells to steady state, cells are chased (incubated in absence of radioactive precursor), generally for 1 to 2 hours, to remove most of the still present free precursor. After a final wash, cells can be treated accordingly and subsequently collected and processed for lipid extraction as discussed in the previous section.

A combination of pulse labeling and chase experiments will provide important information on the nature of potential ceramide changes (either modulation of de novo synthesis or turnover).

With the advent of mass spectrometric analysis of sphingolipids, additional specific tools have been developed for detection of de novo ceramide synthesis. One of these is the use of C17-dihydrosphingosine. This particular precursor has a sphingoid backbone of seventeen carbons, which is not found in naturally occurring sphingolipids but it is recognized as substrate for ceramide synthesis. Therefore it can be used as proximal precursor for ceramide synthesis and its conversion into C17-dihydroceramide and more complex C17-sphingolipids can be detected by mass spectrometry.

The Use of Ceramide Analogues

Short-Chain Ceramides

The use of short-chain ceramides provided the first clue of the biologic functions that are affected/regulated by ceramide.[26] Natural ceramides are very hydrophobic, resulting in very poor water-solubility. This physical property prevents their delivery to cells. One of the approaches to overcome this problem has been the development of synthetic water-soluble ceramide analogues that are mostly shortened in the fatty acid moiety (i.e., C_2-ceramide, *D-erythro*-acetylsphingosine; C_6-ceramide, *D-erythro*-hexanoylsphingosine).[27] These short chain ceramides are more soluble and cell membrane-permeable, providing a viable tool to further characterize ceramide biology and identify target molecules using in vitro and in vivo systems.[18,26,28] Exogenously added short-chain ceramides, besides being used for synthesis of more complex sphingolipids, undergo deacylation followed by recycling of the sphingosine backbone by reacylation with a long-chain fatty acid, generating long-chain ceramides.[29] Therefore any target/biological effect identified by the use of these analogues may be in fact due to the accumulation of natural ceramide instead.

Fluorescent Ceramide Analogues

Fluorescent analogues such as Bodipy-ceramide, DMB-ceramide and NBD-ceramide are very useful in studies of ceramide transport and metabolism.[30-32] In addition, short chain NBD-ceramides are also used as substrates for determining in vitro activities of ceramidases[33] or sphingomyelin synthases.[16]

Pharmacological Tools

The discovery of small chemicals capable of inhibiting ceramide metabolizing enzymes in a potent and selective way could offer novel pharmacological tools for studying the biological role of distinct enzymes and new therapeutic reagents. A number of inhibitors for ceramide metabolizing enzymes have been developed or identified (Table 1).

Genetic Tools

A number of gene products contribute to the ceramide metabolism. Recent technologies have allowed selective gene silencing or knockout in mammals, providing powerful new tools for biological research and drug discovery.

RNA Interference

RNA interference (RNAi) represents a natural endogenous mechanism that cells utilize to regulate RNA expression.[34,35] As a research tool, it is evident that RNA interference (RNAi) has revolutionized the biological science by allowing selective silencing of mRNA expression. Gene silencing can be induced by vector-based short hairpin RNA (shRNA) and synthetic small interfering RNA (siRNA) through a sequence-specific cleavage of perfectly complementary mRNA. In the field of sphingolipid research, RNAi has been used as a research tool for understanding metabolic and biological role for the ceramide metabolizing enzymes (Table 1).

Knockout Mice

The mouse is the foremost vertebrate experimental model because its genome can be precisely and variously engineered. Transgenic mice have allowed researchers to observe what happens to an entire organism during the progression of a disease and they have become models for studying human diseases and their treatments. A number of knockout mice of enzymes participating in ceramide metabolism have been generated (Table 1).

Table 1. Ceramide metabolizing enzymes and research tools in mammals

Gene Symbol	Gene Name	Inhibitors	Availabiliy of si/shRNA for Human Genes	KO Mice Availability	Activity Assay
ASAH1	Acid ceramidase	B13[36]	Available[37]	Lethal[38]	39
ASAH2	Neutral/alkaline ceramidase	D-MAPP[40]	Available[41]	Survive[42]	33
ASAH3	Alkaline ceramidase 1		Available[43]		
ASAH3L	Alkaline ceramidase 2		Available[44]		41
PHCA	Alkaline phytoceramidase				45
CERK	Ceramide kinase	NVP-231[46]	Available[47]	Survive[48]	49
DEGS1	Dihydroceramide desaturase	C_8-cyclopropenylceramide[50]	Available[51]		52
GBA1	Glucosylceramidase 1	Conduritol B epoxide[53-55] Iminosugar[54,56]	Available[57,58]	Lethal after birth[59]	55
GBA2	Glucosylceramidase 2	Imino sugar[60]	Available[57,58]	Survive[61]	62
GBA3	Klotho-related protein		Available[63]		63
LASS1/CerS1	Ceramide synthase 1	Fumonisin B1[64]	Available[21]		65,66
LASS2/CerS2	Ceramide synthase 2		Available[67]		65,66
LASS3/CerS3	Ceramide synthase 3				65,66
LASS4/CerS4	Ceramide synthase 4				65,66
LASS5/CerS5	Ceramide synthase 5		Available[21]		65,66
LASS6/CerS6	Ceramide synthase 6		Available[68]		65,66
SGMS1	Sphingomyelin synthase 1	D609[69]	Available[16,70-72]	Survive*	16,72
SGMS2	Sphingomyelin synthase 2		Available[16,70-72]	Survive[73]	16,72
SMPD1	Acid sphingomyelinase	Desipramine[54,74] (in vivo)	Available[37]	Survive[75]	76,77

continued on next page

Table 1. Continued

Gene Symbol	Gene Name	Inhibitors	Availabiliy of si/shRNA for Human Genes	KO Mice Availability	Activity Assay
SMPD2	Neutral sphingomyelinase 1		Available	Survive[78]	79
SMPD3	Neutral sphingomyelinase 2	GW4869[80] Scyphostatin[81]	Available[82]	Survive[78,83]	84
SMPD4	Neutral sphingomyelinase 3		Available[78,85]		
UGCG	Glucosylceramide synthase	PDMP[86] AMP-DNM[87]	Available[88]	Lethal[89]	90,91

*, Unpublished: Dr. Ken Watanabe (National Center for Geriatrics and Gerontology, Aichi, Japan).

Conclusion

Ceramide is at the heart of sphingolipid metabolism. As discussed, ceramide can be generated either de novo, or through the conversion of sphingosine (salvage pathway), or from the break-down of more complex sphingolipids such sphingomyelin (SM), glucosylceramide (Glucer) and ceramide phosphate (Cer-P). Therefore when studying ceramide metabolism, one needs to consider the number of interconnected reactions that contribute to the overall ceramide levels and the number of other metabolites that could themselves contribute to one or more cellular processes. Thus it is important to recognize that the more metabolic information are captured, the more complete and compelling the case will become.

Acknowledgements

This work was supported in part by NIH Grant # P20 RR-017677 from the National Center for Research Resources and Institutional Research Grant from the American Cancer Society IRG 97-219-08 (to C.L.).

References

1. Futerman AH, Riezman H. The ins and outs of sphingolipid synthesis. Trends Cell Biol 2005; 15(6):312-318.
2. Hannun YA, Obeid LM. Principles of bioactive lipid signalling: lessons from sphingolipids. Nat Rev Mol Cell Biol 2008; 9(2):139-150.
3. Merrill AH Jr, Schmelz EM, Dillehay DL et al. Sphingolipids—the enigmatic lipid class: biochemistry, physiology and pathophysiology. Toxicol Appl Pharmacol 1997; 142(1):208-225.
4. Kim MY, Linardic C, Obeid L et al. Identification of sphingomyelin turnover as an effector mechanism for the action of tumor necrosis factor alpha and gamma-interferon. Specific role in cell differentiation. J Biol Chem 1991; 266(1):484-489.
5. Bligh EG, Dyer WJ. A rapid method of total lipid extraction and purification. Can J Biochem Physiol 1959; 37(8):911-917.
6. Folch J, Lees M, Sloane Stanley GH. A simple method for the isolation and purification of total lipides from animal tissues. J Biol Chem 1957; 226(1):497-509.
7. Zheng W, Kollmeyer J, Symolon H et al. Ceramides and other bioactive sphingolipid backbones in health and disease: lipidomic analysis, metabolism and roles in membrane structure, dynamics, signaling and autophagy. Biochim Biophys Acta 2006; 1758(12):1864-1884.
8. Bielawski J, Szulc ZM, Hannun YA et al. Simultaneous quantitative analysis of bioactive sphingolipids by high-performance liquid chromatography-tandem mass spectrometry. Methods 2006; 39(2):82-91.
9. Haynes CA, Allegood JC, Park H et al. Sphingolipidomics: Methods for the comprehensive analysis of sphingolipids. J Chromatogr B Analyt Technol Biomed Life Sci 2008.
10. Ichi I, Nakahara K, Miyashita Y et al. Association of ceramides in human plasma with risk factors of atherosclerosis. Lipids 2006; 41(9):859-863.
11. Uchida Y, Hara M, Nishio H et al. Epidermal sphingomyelins are precursors for selected stratum corneum ceramides. J Lipid Res 2000; 41(12):2071-2082.
12. Perry DK, Carton J, Shah AK et al. Serine palmitoyltransferase regulates de novo ceramide generation during etoposide-induced apoptosis. J Biol Chem 2000; 275(12):9078-9084.
13. Cremesti AE, Fischl AS. Current methods for the identification and quantitation of ceramides: an overview. Lipids 2000; 35(9):937-945.
14. Perry DK, Bielawska A, Hannun YA. Quantitative determination of ceramide using diglyceride kinase. Methods Enzymol 2000; 312:22-31.
15. Perry DK, Hannun YA. The use of diglyceride kinase for quantifying ceramide. Trends Biochem Sci 1999; 24(6):226-227.
16. Villani M, Subathra M, Im YB et al. Sphingomyelin synthases regulate production of diacylglycerol at the Golgi. Biochem J 2008; 414(1):31-41.
17. Triola G, Fabrias G, Dragusin M et al. Specificity of the dihydroceramide desaturase inhibitor N-[(1R,2S)-2-hydroxy-1-hydroxymethyl-2-(2-tridecyl-1-cyclopropenyl)ethyl]o ctanamide (GT11) in primary cultured cerebellar neurons. Mol Pharmacol 2004; 66(6):1671-1678.
18. Kitatani K, Idkowiak-Baldys J, Hannun YA. The sphingolipid salvage pathway in ceramide metabolism and signaling. Cell Signal 2008; 20(6):1010-1018.
19. van Echten-Deckert G, Zschoche A, Bar T et al. cis-4-Methylsphingosine decreases sphingolipid biosynthesis by specifically interfering with serine palmitoyltransferase activity in primary cultured neurons. J Biol Chem 1997; 272(25):15825-15833.

20. Bielawska A, Perry DK, Hannun YA. Determination of ceramides and diglycerides by the diglyceride kinase assay. Anal Biochem 2001; 298(2):141-150.
21. Kitatani K, Idkowiak-Baldys J, Bielawski J et al. Protein kinase C-induced activation of a ceramide/protein phosphatase 1 pathway leading to dephosphorylation of p38 MAPK. J Biol Chem 2006; 281(48):36793-36802.
22. Becker KP, Kitatani K, Idkowiak-Baldys J et al. Selective inhibition of juxtanuclear translocation of protein kinase C betaII by a negative feedback mechanism involving ceramide formed from the salvage pathway. J Biol Chem 2005; 280(4):2606-2612.
23. Shaner RL, Allegood JC, Park H et al. Quantitative analysis of sphingolipids for lipidomics using triple quadrupole and quadrupole linear ion trap mass spectrometers. J Lipid Res 2008.
24. Sullards MC, Allegood JC, Kelly S et al. Structure-specific, quantitative methods for analysis of sphingolipids by liquid chromatography-tandem mass spectrometry: "inside-out" sphingolipidomics. Methods Enzymol 2007; 432:83-115.
25. Merrill AH Jr, Sullards MC, Allegood JC et al. Sphingolipidomics: high-throughput, structure-specific and quantitative analysis of sphingolipids by liquid chromatography tandem mass spectrometry. Methods 2005; 36(2):207-224.
26. Okazaki T, Bielawska A, Bell RM et al. Role of ceramide as a lipid mediator of 1 alpha,25-dihydroxyvitamin D3-induced HL-60 cell differentiation. J Biol Chem 1990; 265(26):15823-15831.
27. Luberto C, Hannun YA. Use of short-chain ceramides. Methods Enzymol 2000; 312:407-420.
28. Ghidoni R, Sala G, Giuliani A. Use of sphingolipid analogs: benefits and risks. Biochim Biophys Acta 1999; 1439(1):17-39.
29. Ogretmen B, Pettus BJ, Rossi MJ et al. Biochemical mechanisms of the generation of endogenous long chain ceramide in response to exogenous short chain ceramide in the A549 human lung adenocarcinoma cell line. Role for endogenous ceramide in mediating the action of exogenous ceramide. J Biol Chem 2002; 277(15):12960-12969.
30. Lipsky NG, Pagano RE. Sphingolipid metabolism in cultured fibroblasts: microscopic and biochemical studies employing a fluorescent ceramide analogue. Proc Natl Acad Sci USA 1983; 80(9):2608-2612.
31. Koval M, Pagano RE. Lipid recycling between the plasma membrane and intracellular compartments: transport and metabolism of fluorescent sphingomyelin analogues in cultured fibroblasts. J Cell Biol 1989; 108(6):2169-2181.
32. Hanada K, Kumagai K, Yasuda S et al. Molecular machinery for nonvesicular trafficking of ceramide. Nature 2003; 426(6968):803-809.
33. Tani M, Okino N, Mitsutake S et al. Specific and sensitive assay for alkaline and neutral ceramidases involving C12-NBD-ceramide. J Biochem 1999; 125(4):746-749.
34. Elbashir SM, Harborth J, Lendeckel W et al. Duplexes of 21-nucleotide RNAs mediate RNA interference in cultured mammalian cells. Nature 2001; 411(6836):494-498.
35. Ghildiyal M, Zamore PD. Small silencing RNAs: an expanding universe. Nat Rev Genet 2009; 10(2):94-108.
36. Samsel L, Zaidel G, Drumgoole HM et al. The ceramide analog, B13, induces apoptosis in prostate cancer cell lines and inhibits tumor growth in prostate cancer xenografts. Prostate 2004; 58(4):382-393.
37. Zeidan YH, Pettus BJ, Elojeimy S et al. Acid ceramidase but not acid sphingomyelinase is required for tumor necrosis factor-{alpha}-induced PGE2 production. J Biol Chem 2006; 281(34):24695-24703.
38. Eliyahu E, Park JH, Shtraizent N et al. Acid ceramidase is a novel factor required for early embryo survival. Faseb J 2007; 21(7):1403-1409.
39. Linke T, Lansmann S, Sandhoff K. Purification of acid ceramidase from human placenta. Methods Enzymol 2000; 311:201-207.
40. Bielawska A, Greenberg MS, Perry D et al. (1S,2R)-D-erythro-2-(N-myristoylamino)-1-phenyl-1-propanol as an inhibitor of ceramidase. J Biol Chem 1996; 271(21):12646-12654.
41. Xu R, Jin J, Hu W et al. Golgi alkaline ceramidase regulates cell proliferation and survival by controlling levels of sphingosine and S1P. Faseb J 2006; 20(11):1813-1825.
42. Kono M, Dreier JL, Ellis JM et al. Neutral ceramidase encoded by the Asah2 gene is essential for the intestinal degradation of sphingolipids. J Biol Chem 2006; 281(11):7324-7331.
43. Sun W, Xu R, Hu W et al. Upregulation of the human alkaline ceramidase 1 and acid ceramidase mediates calcium-induced differentiation of epidermal keratinocytes. J Invest Dermatol 2008; 128(2):389-397.
44. Sun W, Hu W, Xu R et al. Alkaline ceramidase 2 regulates {beta}1 integrin maturation and cell adhesion. FASEB J 2008.
45. Mao C, Xu R, Szulc ZM et al. Cloning and characterization of a novel human alkaline ceramidase. A mammalian enzyme that hydrolyzes phytoceramide. J Biol Chem 2001; 276(28):26577-26588.
46. Graf C, Klumpp M, Habig M et al. Targeting ceramide metabolism with a potent and specific ceramide kinase inhibitor. Mol Pharmacol 2008; 74(4):925-932.

47. Pettus BJ, Bielawska A, Spiegel S et al. Ceramide kinase mediates cytokine- and calcium ionophore-induced arachidonic acid release. J Biol Chem 2003; 278(40):38206-38213.
48. Graf C, Zemann B, Rovina P et al. Neutropenia with impaired immune response to Streptococcus pneumoniae in ceramide kinase-deficient mice. J Immunol 2008; 180(5):3457-3466.
49. Wijesinghe DS, Lamour NF, Gomez-Munoz A et al. Ceramide kinase and ceramide-1-phosphate. Methods Enzymol 2007; 434:265-292.
50. Triola G, Fabrias G, Casas J et al. Synthesis of cyclopropene analogues of ceramide and their effect on dihydroceramide desaturase. J Org Chem 2003; 68(26):9924-9932.
51. Kraveka JM, Li L, Szulc ZM et al. Involvement of dihydroceramide desaturase in cell cycle progression in human neuroblastoma cells. J Biol Chem 2007; 282(23):16718-16728.
52. Schulze H, Michel C, van Echten-Deckert G. Dihydroceramide desaturase. Methods Enzymol 2000; 311:22-30.
53. Premkumar L, Sawkar AR, Boldin-Adamsky S et al. X-ray structure of human acid-beta-glucosidase covalently bound to conduritol-B-epoxide. Implications for Gaucher disease. J Biol Chem 2005; 280(25):23815-23819.
54. Delgado A, Casas J, Llebaria A et al. Chemical tools to investigate sphingolipid metabolism and functions. ChemMedChem 2007; 2(5):580-606.
55. Grabowski GA, Osiecki-Newman K, Dinur T et al. Human acid beta-glucosidase. Use of conduritol B epoxide derivatives to investigate the catalytically active normal and Gaucher disease enzymes. J Biol Chem 1986; 261(18):8263-8269.
56. Yu Z, Sawkar AR, Whalen LJ et al. Isofagomine- and 2,5-anhydro-2,5-imino-D-glucitol-based glucocerebrosidase pharmacological chaperones for Gaucher disease intervention. J Med Chem 2007; 50(1):94-100.
57. Swanton C, Marani M, Pardo O et al. Regulators of mitotic arrest and ceramide metabolism are determinants of sensitivity to paclitaxel and other chemotherapeutic drugs. Cancer Cell 2007; 11(6):498-512.
58. Kitatani K, Sheldon K, Anelli V et al. Acid beta-glucosidase 1 Counteracts p38delta -Dependent Induction of Interleukin-6: Possible Role for Ceramide as an Anti-inflammatory Lipid. J Biol Chem 2009.
59. Tybulewicz VL, Tremblay ML, LaMarca ME et al. Animal model of Gaucher's disease from targeted disruption of the mouse glucocerebrosidase gene. Nature 1992; 357(6377):407-410.
60. Walden CM, Sandhoff R, Chuang CC et al. Accumulation of glucosylceramide in murine testis, caused by inhibition of beta-glucosidase 2: implications for spermatogenesis. J Biol Chem 2007; 282(45):32655-32664.
61. Yildiz Y, Matern H, Thompson B et al. Mutation of beta-glucosidase 2 causes glycolipid storage disease and impaired male fertility. J Clin Invest 2006; 116(11):2985-2994.
62. Boot RG, Verhoek M, Donker-Koopman W et al. Identification of the nonlysosomal glucosylceramidase as beta-glucosidase 2. J Biol Chem 2007; 282(2):1305-1312.
63. Hayashi Y, Okino N, Kakuta Y et al. Klotho-related protein is a novel cytosolic neutral beta-glycosylceramidase. J Biol Chem 2007; 282(42):30889-30900.
64. Wang E, Norred WP, Bacon CW et al. Inhibition of sphingolipid biosynthesis by fumonisins. Implications for diseases associated with Fusarium moniliforme. J Biol Chem 1991; 266(22):14486-14490.
65. Wang E, Merrill AH Jr. Ceramide synthase. Methods Enzymol 2000; 311:15-21.
66. Lahiri S, Lee H, Mesicek J et al. Kinetic characterization of mammalian ceramide synthases: determination of K(m) values towards sphinganine. FEBS Lett 2007; 581(27):5289-5294.
67. Laviad EL, Albee L, Pankova-Kholmyansky I et al. Characterization of ceramide synthase 2: tissue distribution, substrate specificity and inhibition by sphingosine 1-phosphate. J Biol Chem 2008; 283(9):5677-5684.
68. White-Gilbertson S, Mullen T, Senkal C et al. Ceramide synthase 6 modulates TRAIL sensitivity and nuclear translocation of active caspase-3 in colon cancer cells. Oncogene 2009.
69. Meng A, Luberto C, Meier P et al. Sphingomyelin synthase as a potential target for D609-induced apoptosis in U937 human monocytic leukemia cells. Exp Cell Res 2004; 292(2):385-392.
70. Li Z, Hailemariam TK, Zhou H et al. Inhibition of sphingomyelin synthase (SMS) affects intracellular sphingomyelin accumulation and plasma membrane lipid organization. Biochim Biophys Acta 2007; 1771(9):1186-1194.
71. Separovic D, Semaan L, Tarca AL et al. Suppression of sphingomyelin synthase 1 by small interference RNA is associated with enhanced ceramide production and apoptosis after photodamage. Exp Cell Res 2008; 314(8):1860-1868.
72. Tafesse FG, Huitema K, Hermansson M et al. Both sphingomyelin synthases SMS1 and SMS2 are required for sphingomyelin homeostasis and growth in human HeLa cells. J Biol Chem 2007; 282(24):17537-17547.

73. Hailemariam TK, Huan C, Liu J et al. Sphingomyelin synthase 2 deficiency attenuates NFkappaB activation. Arterioscler Thromb Vasc Biol 2008; 28(8):1519-1526.
74. Hurwitz R, Ferlinz K, Sandhoff K. The tricyclic antidepressant desipramine causes proteolytic degradation of lysosomal sphingomyelinase in human fibroblasts. Biol Chem Hoppe Seyler 1994; 375(7):447-450.
75. Horinouchi K, Erlich S, Perl DP et al. Acid sphingomyelinase deficient mice: a model of types A and B Niemann-Pick disease. Nat Genet 1995; 10(3):288-293.
76. Lansmann S, Bartelsen O, Sandhoff K. Purification and characterization of recombinant human acid sphingomyelinase expressed in insect Sf21 cells. Methods Enzymol 2000; 311:149-156.
77. Sawai H, Domae N, Nagan N et al. Function of the cloned putative neutral sphingomyelinase as lyso-platelet activating factor-phospholipase C. J Biol Chem 1999; 274(53):38131-38139.
78. Stoffel W, Jenke B, Block B et al. Neutral sphingomyelinase 2 (smpd3) in the control of postnatal growth and development. Proc Natl Acad Sci USA 2005; 102(12):4554-4559.
79. Zumbansen M, Stoffel W. Neutral sphingomyelinase 1 deficiency in the mouse causes no lipid storage disease. Mol Cell Biol 2002; 22(11):3633-3638.
80. Luberto C, Hassler DF, Signorelli P et al. Inhibition of tumor necrosis factor-induced cell death in MCF7 by a novel inhibitor of neutral sphingomyelinase. J Biol Chem 2002; 277(43):41128-41139.
81. Nara F, Tanaka M, Hosoya T et al. Scyphostatin, a neutral sphingomyelinase inhibitor from a discomycete, Trichopeziza mollissima: taxonomy of the producing organism, fermentation, isolation and physicochemical properties. J Antibiot (Tokyo) 1999; 52(6):525-530.
82. Marchesini N, Osta W, Bielawski J et al. Role for mammalian neutral sphingomyelinase 2 in confluence-induced growth arrest of MCF7 cells. J Biol Chem 2004; 279(24):25101-25111.
83. Stoffel W, Jenke B, Holz B et al. Neutral sphingomyelinase (SMPD3) deficiency causes a novel form of chondrodysplasia and dwarfism that is rescued by Col2A1-driven smpd3 transgene expression. Am J Pathol 2007; 171(1):153-161.
84. Liu B, Hannun YA. Sphingomyelinase assay using radiolabeled substrate. Methods Enzymol 2000; 311:164-167.
85. Clarke CJ, Guthrie JM, Hannun YA. Regulation of neutral sphingomyelinase-2 (nSMase2) by tumor necrosis factor-alpha involves protein kinase C-delta in lung epithelial cells. Mol Pharmacol 2008; 74(4):1022-1032.
86. Inokuchi J, Radin NS. Preparation of the active isomer of 1-phenyl-2-decanoylamino-3-morpholino-1-propanol, inhibitor of murine glucocerebroside synthetase. J Lipid Res 1987; 28(5):565-571.
87. Overkleeft HS, Renkema GH, Neele J et al. Generation of specific deoxynojirimycin-type inhibitors of the nonlysosomal glucosylceramidase. J Biol Chem 1998; 273(41):26522-26527.
88. Diaz-Font A, Chabas A, Grinberg D et al. RNAi-mediated inhibition of the glucosylceramide synthase (GCS) gene: A preliminary study towards a therapeutic strategy for Gaucher disease and other glycosphingolipid storage diseases. Blood Cells Mol Dis 2006; 37(3):197-203.
89. Yamashita T, Wada R, Sasaki T et al. A vital role for glycosphingolipid synthesis during development and differentiation. Proc Natl Acad Sci USA 1999; 96(16):9142-9147.
90. Shayman JA, Abe A. Glucosylceramide synthase: assay and properties. Methods Enzymol 2000; 311:42-49.
91. Marks DL, Paul P, Kamisaka Y et al. Methods for studying glucosylceramide synthase. Methods Enzymol 2000; 311:50-59.

Index

H

I

K

L

M

T

V

W

Y

Batch number: 10762177

Printed by Printforce, the Netherlands